Landfill Leachate Management

Landfill Leachate Management

Edited by
Vinay Kumar Tyagi and C. S. P. Ojha

IWA PUBLISHING

Published by **IWA Publishing**
Unit 104–105, Export Building
1 Clove Crescent
London E14 2BA, UK
Telephone: +44 (0)20 7654 5500
Fax: +44 (0)20 7654 5555
Email: publications@iwap.co.uk
Web: www.iwapublishing.com

First published 2023

British Library Cataloguing in Publication Data
A CIP catalogue record for this book is available from the British Library

ISBN: 9781789063301 (paperback)
ISBN: 9781789063318 (eBook)
ISBN: 9781789063325 (ePub)

This eBook was made Open Access in August 2023.

Doi: 10.2166/9781789063318

Contents

Editors

Dr. Vinay Kumar Tyagi, National Institute of Hydrology, Roorkee, India

Vinay Kumar Tyagi is a scientist D at the Environmental Hydrology Division, National Institute of Hydrology, Roorkee, India. His research interest includes energy-efficient wastewater treatment systems, advanced technologies for sludge pre-treatment, biomass-to-bioenergy recovery, and health-related water microbiology.

Dr. Tyagi has published two-digit refereed articles in SCI journals over the past several years. He has more than 100 publications to his credit (Google Scholar Citation: >4400, h-index: 33), including six full-text books, 30 book chapters and manuals most notably are O&M of Sewerage Works, and Operation and Maintenance Guidelines for UASB-DHS System, jointly published by Government of India and Japan. Dr. Tyagi's research achievement has been receiving many academic honors, including (1) top 2% scientist worldwide (Stanford University list published by Elsevier); (2) outstanding review contribution, Elsevier; (3) research articles are enlisted among the top 1% of highly cited papers of the academic field of engineering worldwide, as well as few of them listed in the most downloaded articles. Dr. Tyagi is fellow and life member of several professional bodies and organizations. He is the management committee member of International Water Association (IWA) Sludge Management Specialist Group and represents the South Asia Division. He is an active reviewer of more than 100 international peer-reviewed journals and editor of IWA publishing (UK), editorial board member of Environmental Science and Eco-technology journal and lead editor of several special issues for international journals.

Dr. Tyagi has extensive international research exposure of working in Taiwan, Spain, Japan, and Singapore in various environmental research domains of wastewater treatment, sludge pre-treatment, and biomass–to-bioenergy recovery.

Professor C.S.P. Ojha, Indian Institute of Technology Roorkee, India

Dr. C.S.P. Ojha (1962–to date), FASCE, FNAE, Ph.D. (Imperial College, London) is a professor in civil engineering, IIT Roorkee since 2004. He has more than 35 years of teaching and research experience in the area of water resources and environmental engineering. He has co-authored more than 250 peer-reviewed journal publications, 50 book chapters and authored/edited nine books. He has supervised 55 Ph.D. students. He was also awarded Alexander von Humboldt Fellowship in 2001, Distinguished Visiting Fellowship of Royal Academy of Engineering in 2008, Visiting Fellowship of ASCE (American Society of Civil Engineers) in 2010 and Senior JSPS Fellowship in 2010–2011. He has received research paper awards from ASCE in 2001, 2009, 2010, 2013, 2018, and 2022, and about half a dozen research paper awards from reputed Indian journals. He was awarded Board-Certified Environmental Engineering Member by American Academy of Environmental Engineering in 2011, Curtis Visiting Professorship by Purdue University in 2012–2013 and State-of-the-Art Award in Civil Engineering by ASCE in 2014.

Contributors

Ahmed Tawfik National Research Centre, Water Pollution Research Department 12622, Dokki, Cairo, Egypt

Alexandre Boscaro França Department of Chemical Engineering, Federal University of Sao João Del Rei, Ouro Branco, Research Group on Waste Treatment and Management Processes, State of Minas Gerais, Brazil

Ali Izzet Cengiz Istanbul Technical University, Civil Engineering Faculty, Environmental Engineering Department, Maslak, 34469 Istanbul, Turkey

Anakha Ambujan Department of Civil Engineering, NITK Surathkal, Mangalore, Karnataka, India

Anjali Bhagwat Hydrological Investigation Division, National Institute of Hydrology, Roorkee, Uttarakhand, India
Department of Civil Engineering, Indian Institute of Technology, Roorkee, Uttarakhand, India

Ankit Singh Environmental Engineering, School of Infrastructure, Indian Institute of Technology Bhubaneswar, Bhubaneswar 752050, Odisha, India

Ankur Rajpal Department of Civil Engineering, Indian Institute of Technology, Roorkee 247667, Uttarakhand, India

Apoorva Upadhyay Centre for Energy and Environment, Malaviya National Institute of Technology, Jaipur 302017, Rajasthan, India

Arun Kumar Thalla Department of Civil Engineering, NITK Surathkal, Mangalore, Karnataka, India

Ashmita Patro CSIR-Institute of Minerals and Materials Technology, Bhubaneswar 751013, Odisha, India
Academy of Scientific and Innovative Research (AcSIR), Ghaziabad, Uttar Pradesh 201002, India

Brahim Arhoun Chemical Engineering Department, Faculty of Sciences, University of Malaga, 29071 Malaga, Spain

Brajesh Kumar Dubey Department of Civil Engineering, Indian Institute of Technology Kharagpur, Kharagpur 721302, West Bengal, India

Busra Cicekalan Istanbul Technical University, Civil Engineering Faculty, Environmental Engineering Department, Maslak, 34469 Istanbul, Turkey

Chandra Shekhar Prasad Ojha Department of Civil Engineering, Indian Institute of Technology, Roorkee, Uttarakhand, India

Clarissa Nascimento de Oliveira Institute of Science and Technology (ICT) Federal University of the Jequitinhonha and Mucuri Valleys (UFVJM) Diamantina - MG, Brazil

Cristiane Medina Finzi Quintão Department of Chemical Engineering, Federal University of Sao João Del Rei, Ouro Branco, Research Group on Waste Treatment and Management Processes, State of Minas Gerais, Brazil

Edilailsa Januário de Melo Faculty of Chemical Engineering (FEQ) UNICAMP Campinas - SP, Brazil

Fabiano Luiz Naves Department of Chemical Engineering, Federal University of Sao João Del Rei, Ouro Branco, Research Group on Waste Treatment and Management Processes, State of Minas Gerais, Brazil

Giovanni Vinti Research Centre for Appropriate Technologies for Environmental Management in Resource-Limited Countries (CeTAmb), University of Brescia, Via Branze 43, 25123 Brescia, Italy

Hale Ozgun Istanbul Technical University, Civil Engineering Faculty, Environmental Engineering Department, Maslak, 34469 Istanbul, Turkey

José Izaquiel Santos da Silva Institute of Science and Technology (ICT) Federal University of the Jequitinhonha and Mucuri Valleys (UFVJM) Diamantina - MG, Brazil

Jose Miguel Rodriguez-Maroto Chemical Engineering Department, Faculty of Sciences, University of Malaga, 29071 Malaga, Spain

Juan Manuel Paz-Garcia Chemical Engineering Department, Faculty of Sciences, University of Malaga, 29071 Malaga, Spain

K.S.D. Premarathna Ecosphere Resilience Research Center, Faculty of Applied Sciences, University of Sri Jayewardenepura, Nugegoda, Sri Lanka

Kaoutar Aboudi Department of Chemical Engineering and Food Technology, Faculty of Sciences, University of Cádiz - International Campus of Excellence (ceiA3), P.O. Box No 40, Cádiz, Puerto Real, 11510, Spain

Madushika Sewwandi Ecosphere Resilience Research Center, Faculty of Applied Sciences, University of Sri Jayewardenepura, Nugegoda, Sri Lanka

Maria Villen-Guzman Chemical Engineering Department, Faculty of Sciences, University of Malaga, 29071 Malaga, Spain

Mateus de Souza Amaral Laboratory of Technological and Environmental Chemistry, Department of Chemistry, Institute of Exact and Biological Sciences, Federal University of Ouro Preto, Ouro Preto, MG, Brazil

Mentore Vaccari Research Centre for Appropriate Technologies for Environmental Management in Resource-Limited Countries (CeTAmb), University of Brescia, Via Branze 43, 25123 Brescia, Italy

Meththika Vithanage Ecosphere Resilience Research Center, Faculty of Applied Sciences, University of Sri Jayewardenepura, Nugegoda, Sri Lanka

Sustainability Cluster, School of Engineering, University of Petroleum and Energy Studies, Dehradun, Uttarakhand 248007, India

The UWA Institute of Agriculture, University of Western Australia, Perth WA 6009, Australia

Mustafa Evren Ersahin Environmental Engineering Department, Civil Engineering Faculty, Istanbul Technical University, Maslak, 34469 Istanbul, Turkey

Neelam Gunjyal Department of Civil Engineering, Indian Institute of Technology, Roorkee, Uttarakhand, India

Nikita Verma Centre for Energy and Environment, Malaviya National Institute of Technology, Jaipur 302017, Rajasthan, India

Prakhar Talwar Centre for Energy and Environment, Malaviya National Institute of Technology, Jaipur 302017, Rajasthan, India

Pratiksha Srivastava Department of Chemical and Environmental Technology, University Rey Juan Carlos, C. Tulipán, s/n, 28933 Móstoles, Madrid, Spain

Pubali Mandal Department of Civil Engineering, Birla Institute of Technology and Science Pilani, 333031, India

Puspendu Bhunia Environmental Engineering, School of Infrastructure, Indian Institute of Technology Bhubaneswar, Bhubaneswar 752 050, Odisha, India

Rahul Mishra CSIR-National Environmental Engineering Research Institute, Nagpur 440 020, Maharashtra, India

Rajesh Roshan Dash Environmental Engineering, School of Infrastructure, Indian Institute of Technology Bhubaneswar, Bhubaneswar 752050, Odisha, India

Rajesh Singh Environmental Hydrology Division, National Institute of Hydrology, Roorkee 247667, Uttarakhand, India

Renata Carolina Zanetti Lofrano Department of Chemical Engineering, Federal University of Sao João Del Rei, Ouro Branco, Research Group on Waste Treatment and Management Processes, State of Minas Gerais, Brazil

Rickwinder Singh Centre for Energy and Environment, Malaviya National Institute of Technology, Jaipur 302017, Rajasthan, India

Sammani Ramanayaka Ecosphere Resilience Research Center, Faculty of Applied Sciences, University of Sri Jayewardenepura, Nugegoda, Sri Lanka

Lancaster Environment Centre, Lancaster University, Lancaster LA1 4YQ, United Kingdom

Sandeep K. Malyan Department of Environmental Studies, Dyal Singh Evening College, University of Delhi, New Delhi 110003, India

Sandeep Singh Environmental Hydrology Division, National Institute of Hydrology, Roorkee 247667, Uttarakhand, India

Sanket Dey Chowdhury Environmental Engineering, School of Infrastructure, Indian Institute of Technology Bhubaneswar, Bhubaneswar 752 050, Odisha, India

Saurabh Dwivedi CSIR-Institute of Minerals and Materials Technology, Bhubaneswar 751013, Odisha, India

Academy of Scientific and Innovative Research (AcSIR), Ghaziabad, Uttar Pradesh 201002, India

Sudeep Kumar Mishra Environmental Engineering, School of Infrastructure, Indian Institute of Technology Bhubaneswar, Bhubaneswar 752 050, Odisha, India

Sujata Kashyap Axa Parenterals Limited, Roorkee 247667, Uttarakhand, India

Sunil Kumar CSIR-National Environmental Engineering Research Institute, Nagpur 440 020, Maharashtra, India

Supriya Gupta CSIR-Institute of Minerals and Materials Technology, Bhubaneswar 751013, Odisha, India

Academy of Scientific and Innovative Research (AcSIR), Ghaziabad, Uttar Pradesh 201002, India

Terry Tudor SusConnect Ltd. Weedon Bec, Northamptonshire NN7 4PS, UK

Thilakshani Atugoda Ecosphere Resilience Research Center, Faculty of Applied Sciences, University of Sri Jayewardenepura, Nugegoda, Sri Lanka

Vijay Kubsad Washington State Department of Ecology, Spokane, Washington, USA

Vinay Kumar Tyagi Environmental Hydrology Division, National Institute of Hydrology, Roorkee 247667, Uttarakhand, India.

Vivekanand Vivekanand Centre for Energy and Environment, Malaviya National Institute of Technology, Jaipur 302017, Rajasthan, India

Xuedong Zhang Jiangsu Key Laboratory of Anaerobic Biotechnology, School of Environmental and Civil Engineering, Jiangnan University, Wuxi 214122, China

Yamini Mittal CSIR-Institute of Minerals and Materials Technology, Bhubaneswar 751013, Odisha, India

Academy of Scientific and Innovative Research (AcSIR), Ghaziabad, Uttar Pradesh 201002, India

doi: 10.2166/9781789063318_xxiii

Foreword

Landfill leachate is a complex mix of organics, inorganics, and heavy metals produced from conventional and engineering landfilling practices. The adverse effects of landfill leachate on human and environmental health have forced the authorities to stipulate stringent disposal requirements, producing the requirement for ground-breaking technological solutions for the effective management of landfill leachate. Researchers and field engineers are still looking for robust options for leachate management. The landfill leachate treatment comprises a sequential arrangement of well-defined unit operations and processes available in the research papers in greater detail, however, non-conforming disposal standards by leachate treatment schemes posed a technical challenge to the engineers, which is not available in the text books. A timely book on landfill leachate management is a valuable addition to this domain. The key features of the book included: Broad range of treatment techniques covered, conventional to advanced technological options discussed, and successful case studies included.

In this book, the latest developments in the characterization techniques of landfill leachate and its treatments are included. The up-to-date methods for leachate characterization and advanced treatment techniques are underlined. The drawbacks and challenges of leachate treatment methods are discussed, together with the future perspectives in the development of treatment approaches for leachate. Thus, the book discusses the conventional and emerging treatments for landfill leachate. Moreover, energy and value-added products recovery from landfill leachate is considered which is a novel addition to the book. The discussion is from the perspective of leachate polishing and recovery of value-added products, future trends and perspective of treatment technologies to achieve stringent disposal standards, and advanced biotechnological concepts for the treatment of landfill leachate for resource recovery. The leachate treatment coupled with various resource recovery processes can arguably help the cyclic economy.

The information thus, gathered can be applied to identify appropriate solutions for landfill leachate treatment and alternatives to existing treatment techniques to advance the efforts for effective landfill leachate management. Moreover, it will help to identify the areas for further research and development opportunities for technological advancements.

Rao Y. Surampalli, Ph.D., P.E., BCEE, HonD.WRE, F.WEF, F.AAAS, Dist.F.IWA, Dist.M.ASCE, NAC

Acknowledgments

Funders
-Department of Biotechnology, DBT (Grant No. BT/RLF/Re-entry/12/2016)

Contributors (Corresponding Authors Alphabetically)
Ahmed Tawfik
Anjali Bhagwat
Arun Kumar Thalla
Brahim Arhoun
Brajesh Kumar Dubey
Fabiano Luiz Naves
Mentore Vaccari
Meththika Vithanage
Mustafa Evren Ersahin
Neelam Gunjyal
Pratiksha Srivastava
Puspendu Bhunia
Rajesh Roshan Dash
Rajesh Singh
Sunil Kumar
Vinay Kumar Tyagi
Vivekanand Vivekanand

doi: 10.2166/9781789063318_0001

Chapter 1

Characteristics of leachate from landfills and dumpsites in Asia, Africa and Latin America: a review update

*Giovanni Vinti[1], Terry Tudor[2] and Mentore Vaccari[1]**

[1]Research Centre for Appropriate Technologies for Environmental Management in Resource-Limited Countries (CeTAmb), University of Brescia, Via Branze 43, 25123 Brescia, Italy
[2]SusConnect Ltd. Weedon Bec, Northamptonshire NN7 4PS, UK
**Corresponding author: mentore.vaccari@ing.unibs.it*

ABSTRACT

Using published data, this chapter updated a previous review to examine differences in pollutant levels in the leachate between landfills and dumpsites in Africa, Asia, and Latin America. A total of 3584 scientific articles published between 2018 and 2022 were initially identified in Scopus. After reviewing the abstracts, only 434 studies were selected, and the full text was examined. Finally, 38 studies were included in the review. However, some studies assessed more than one waste disposal site. Thus, 58 landfills and dumpsites were included. The most significant difference in the leachate between dumpsites and landfills generally occurred in climatic zone A. Indeed, significantly higher values in dumpsites were found for Cr, Ni, and Zn. Comparing the findings with the previous review confirms the differences in the levels of Cd and Pb between landfills and dumpsites. To mitigate these risks, it is vital for there to be investments in improving the waste management infrastructure and systems in Global South countries. In addition, there should be improved governance structures to enhance the enforcement of the existing policies.

Keywords: Leachate, dumpsites, landfills, climatic conditions.

1.1 INTRODUCTION

In recent decades, global solid waste generation rates have increased at a rapid pace (Chen *et al.*, 2020). Due largely to urbanization and rapidly increasing populations, global generation of waste is predicted to rise by 73% between 2020 and 2050 to approximately 3.88 billion tonnes (World Bank, 2022). However, landfilling and open dumping, along with uncontrolled burning of waste, still represent the world's most common solid waste management (SWM) practices (Gómez-Sanabria *et al.*, 2022; Kaza *et al.*, 2018). Indeed, in the Global South, approximately 90% of the waste is either disposed of in dumpsites or openly burnt (World Bank, 2022).

Dumpsites and poorly maintained landfills pose a high risk to human health and the environment (e.g., in terms of greenhouse gas (GHG) emissions, which can also impact on the health of individuals) (Gómez-Sanabria *et al.*, 2022; Pujara *et al.*, 2019; Vinti *et al.*, 2021). The risk posed by dumpsites is higher, as unlike landfills, these structures are not engineered, therefore, leading to higher pollutant flows (Vaccari *et al.*, 2019a).

Despite the existence of global frameworks (e.g., the Basel Convention) and national legislation, countries in the Global South, which are primarily located in equatorial and warmer climates are particularly at risk. For example, waste pickers living on dumpsites are at risk due to direct exposure from hazardous materials, which are often imported into Global South countries (Ferronato & Torretta, 2019).

Thus, understanding the risks posed by different waste disposal sites and the key factors influencing these risks (particularly leachate contaminant levels) is crucial to developing measures to mitigate them (Tesseme *et al.*, 2022). With this in mind, Vaccari *et al.* (2019a) conducted a review of the characterization of leachate from landfills and dumpsites in Asia, Africa, and Latin America. Using these findings as a basis, this study presents an updated analysis.

1.2 MATERIALS AND METHODS

The previous review (Vaccari *et al.*, 2019a) included studies published up to 2017. Thus, this review focused on work from 2018 to 2022. Only peer-reviewed articles were selected. Scopus was used as the search engine. The following search words were adopted: landfill leachate; dumpsite leachate; open dump leachate. As in Vaccari *et al.* (2019a), only the general distinction between landfills and dumpsites was considered, and only sites from Asia, Africa, and Latin America were selected. In particular, dumpsites were defined as open and not regulated areas in the ground with no environmental protection, used for the disposal of waste. Landfills were defined as waste disposal sites characterized by the registration of waste inflow, and typically using daily cover material, surface and groundwater monitoring, infrastructure, and a waterproof liner at the bottom. When it was not possible to establish the nature of the disposal site, the study was discarded. Furthermore, waste disposal sites that had already closed were not included. This approach was utilized because some authors have found a lower leachate contamination level in closed landfills (Wdowczyk & Szymańska-Pulikowska, 2021). Thus, including closed sites could have biased the results.

As in Vaccari *et al.* (2019a, 2019b), the following leachate parameters were selected: chemical oxygen demand (COD), biochemical oxygen demand (BOD_5), NH_3-N, Al, As, Cd, Cr, Cu, Fe, Hg, Mn, Ni, Pb, and Zn. Thus, if the study did not include any of them, it was not included. For each parameter, the mean (μ), standard deviation (σ), median (M_e), and coefficient of variation (CV) were calculated to determine the spread of the factors and the impact of each at the site. As in Vaccari *et al.* (2019a), the significance level (α) was evaluated by using Welch's test. Thus, once t and ν were computed, they were used with the t-distribution to test the null hypothesis that the two population means were equal (p-value).

When more than one measure of a chemical parameter from the same disposal site was reported, the average value was considered. In any paper, when the value assigned to a parameter was given as 'less than…', this value was used.

The site's age was also indicated whenever possible, trying at least to distinguish between more, and less than 10 years old.

Furthermore, as temperature and precipitation can impact the leachate generation and characteristics (Ma *et al.*, 2022), the data were also categorized according to climatic conditions by using the Köppen–Geiger climate classification map. The updated open source version available online (Rubel *et al.*, 2017) was used. Thus, the following climatic zones were included in the classification: A (equatorial), B (arid), C (warm temperate), D (snow) and E (polar).

1.3 RESULTS AND DISCUSSION

1.3.1 Study selection

A total of 3584 studies published between 2018 and 2022 were initially identified in Scopus. After reviewing the abstracts, only 434 studies were selected because they concerned leachate from landfills or dumpsites in Asia, Africa and Latin America; thus, the full text of these studies was examined. Finally, 38 studies were included in this review. This number of sites was based on the fact that most studies did not have enough information, the nature of the site (i.e., dumpsite or landfill) was unclear, the data were inadequate for the purpose of this review, the same study was collected more than once by using different keywords, or the waste disposal site was closed. However, some studies assessed more than one waste disposal site. Thus, 58 landfills and dumpsites were included.

General information about landfills and dumpsites that were selected are available in Tables 1.1 and 1.2. In particular, the reference of each study, the country, the waste disposal site and its age are provided. If the age was not available, the cell was left empty.

1.3.2 Key types of sites and climatic zones

As shown in Figure 1.1, the majority of the sites (46 out of 58) were identified in the most populated continent, that is, Asia. All of the disposal sites identified in Latin America were landfills. However, in Africa, most (55%) came from dumpsites, and in Asia, the same number of dumpsites and landfills was identified.

Furthermore, as shown in Figure 1.2, considering the climatic zone, more than half of the sites (both landfills and dumpsites) were from zone A (equatorial). The remaining sites were from zone B (arid), and zone C (warm temperate).

Table 1.1 General information about the landfills analyzed.

Country	Site	Age of the Site (years)	Source
China	XinFeng landfill, Guangzhou	12	Deng *et al.* (2018)
Malaysia	Rimba Mas Sanitary landfill, Kaki Bukit	n.a.	Kow *et al.* (2022)
Thailand	• Tha Rae landfill, Sakon Nakorn • Sakon Nakhon landfill, Sakon Nakhon • Phang Khon landfill, Sakon Nakhon	n.a.	Ruengruehan *et al.* (2021)
Bangladesh	Matuail sanitary landfill, Dhaka	n.a.	Saeed *et al.* (2022)
China	Sanitary landfill in Xiamen city	≈12	Mao *et al.* (2022)
Pakistan	Lakhodair landfill site, Lahore	12	Ashraf *et al.* (2022)
Oman	• Multaqa landfill, Muscat • Barka landfill, Muscat	n.a.	Siddiqi *et al.* (2022)
Morocco	Sanitary landfill of Fez city	n.a.	El Mrabet *et al.* (2022)
China	Sanitary landfill site in Weifang	n.a.	Deng *et al.* (2021)
Vietnam	Nam Som landfill, Hanoi	>10	Hoai *et al.* (2021)
Malaysia	• Sungai Udang sanitary landfill, Melaka • Ladang Tanah Merah sanitary landfill, Negeri Sembilan	2 2	Hussein *et al.* (2021)
Brazil	Sanitary landfill of Sabará	n.a.	Reis *et al.* (2020)
Turkey	Sanitary landfill of Aksaray	n.a.	Tulun (2020)
India	Turmuri sanitary landfill, Belgaum	n.a.	Desai *et al.* (2020)
Malaysia	Pulau Burung sanitary landfill, Nibong Tebal	>10	Shadi *et al.* (2020)
Ghana	Tema landfill site, Tema	<10	Sackey *et al.* (2020)
China	Landfill in Qingdao	n.a.	Li *et al.* (2019)
Colombia	Landfill in Riohacha	<1	Galindo Montero *et al.* (2019)
China	Sanitary landfill in Wuhan	n.a.	Wang *et al.* (2019)
Colombia	Sanitary landfill in Tubarà city	n.a.	Rebolledo *et al.* (2019)
Malaysia	Jeram Sanitary Landfill, Kuala Lumpur	n.a.	Aziz *et al.* (2018)
Malaysia	Tanjung Langsat Landfill Site, Pasir Gudang	n.a.	Ismail *et al.* (2018)
Ghana	Otil landfill site, Kumasi	13	Boateng *et al.* (2018)
Malaysia	Jeram Sanitary Landfill, Kuala Lumpur	n.a.	Erabee *et al.* (2018)
Egypt	El-hammam sanitary landfill, Alexandria	n.a.	Al-Wasify *et al.* (2018)
India	Hyderabad landfil, Hyderabad	17	Somani *et al.* (2019)

Table 1.2 General information about the dumpsites analyzed.

Country	Site	Age of the Site (years)	Source
Ethiopia	• Arba Minch illegal dumpsite 1 • Arba Minch illegal dumpsite 2 • Arba Minch illegal dumpsite 3 • Arba Minch illegal dumpsite 4	n.a.	Tesseme *et al.* (2022)
India	• Okhla landfill, Delhi • Ghazipur landfill, Delhi • Gurgaon landfil, Gurgaon	24 34 <10	Somani *et al.* (2019)
Thailand	• Dong Mafai controlled dump • Nong Lat controlled dump • Sawang Daen Din controlled dump • Wanon Niwat controlled dump • Bong Tai open dump • Nong Luang open dump • Song Dao open dump • Ban Phon open dump • Kut Bak open dump	n.a.	Ruengruehan *et al.* (2021)
Sri Lanka	Waste dumping site A, Kesbewa	≈20	Koliyabandara *et al.* (2022)
India	Perungudi dumpsite (LS1), Chennai	20	Kuchelar *et al.* (2022)
India	Urali-Devachi landfill, Pune	n.a.	Ingle (2022)
Malaysia	Alor Pongsu landfill site, Perak	>10	Aziz *et al.* (2021)
Lebanon	Naameh municipal solid waste landfill, Naameh village	>10	Sawaya *et al.* (2021)
Iran	Saravan landfill, Rasht	>10	Farhangi *et al.* (2021)
Malaysia	• Ulu Maasop landfill, Senaling • Kampung Keru landfill, Tampin	>10 >10	Hussein *et al.* (2019)
Ethiopia	Landfill site in Mekelle	n.a.	Alemayehu *et al.* (2019)
India	Ramna MSW landfill, Varanasi city	<10	Mishra *et al.* (2019)
India	Dapha landfill, Kolkata	n.a.	De *et al.* (2019)
Sri Lanka	Dumping site in Karadiyana	n.a.	Nayanthika *et al.* (2018)

1.3.3 Leachate quality by site

High levels of Fe and Cr, as well as BOD_5 and COD were found in both the landfills (Figure 20.3), and the dumpsites (Figure 20.4). Relatively high concentrations of Ni and Zn were also found in the landfills. Aluminum (Al) was not included in Figures 1.3 and 1.4 because of the paucity of available data. In particular, studies about it were not found for dumpsites. In both figures, values are expressed in milligram/liter.

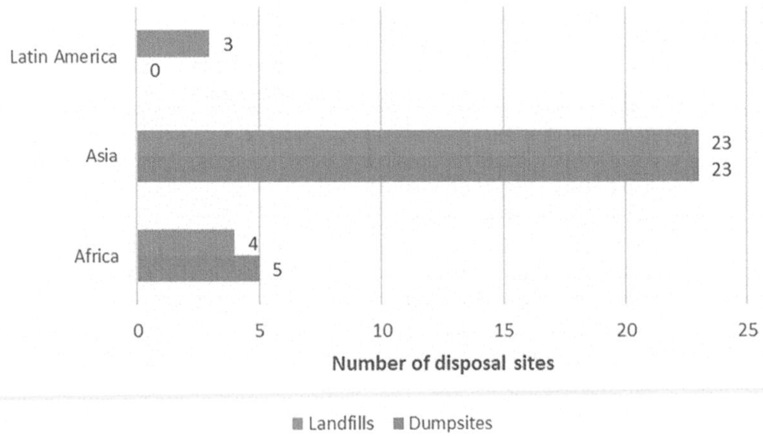

Figure 1.1 Number of waste disposal sites analyzed according to their typology and continent.

1.3.4 Leachate quality by region

Figure 1.5 illustrates that landfills in Africa had higher Fe concentrations, but lower COD levels compared to the other two regions.

Figure 1.6 shows that COD, BOD_5, NH_3-N, Fe, and Zn concentrations were higher in Asian dumpsites compared to African dumpsites. As in Figure 20.1, no studies about Latin America were found. Al does not appear in the figure for the same reasons discussed in Figure 1.3.

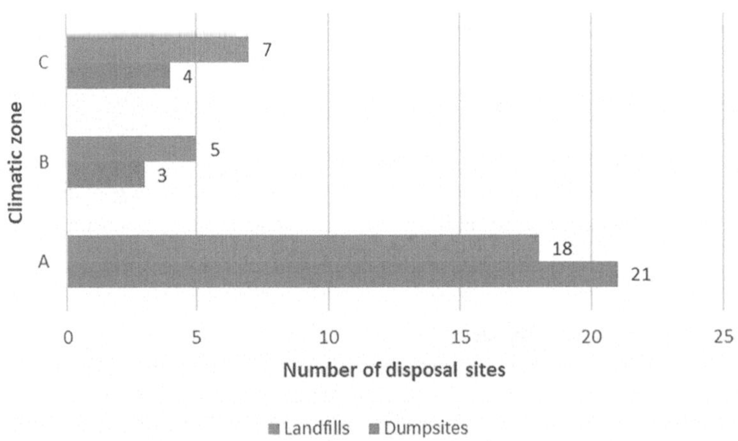

Figure 1.2 Number of waste disposal sites analyzed according to their typology and climatic zone.

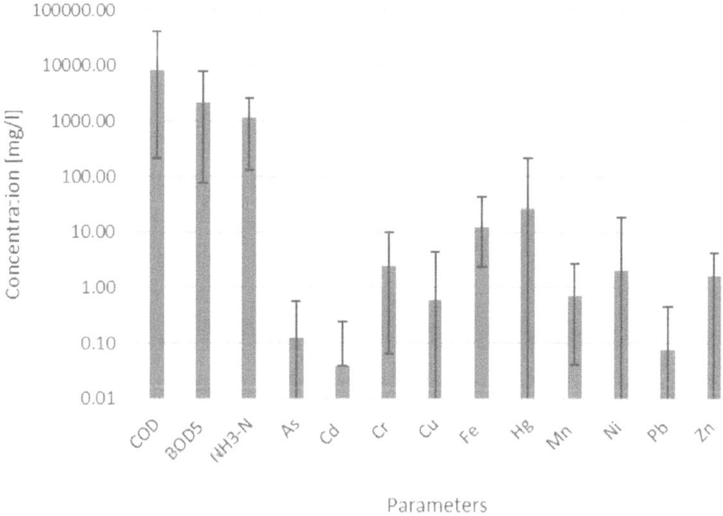

Figure 1.3 Biochemical parameters in all landfills.

1.3.5 Pollutant levels by site type

In the previous study (Vaccari *et al.*, 2019a), almost all pollutant levels were higher in dumpsites than landfills. In this study, the values were higher in dumpsites for some parameters (see Table 1.3). In particular, for COD, BOD_5, As, Cd, Fe, Mn, and Pb. However, there was a significant difference between landfills and dumpsites ($p < 0.05$) only for Cd and Pb.

In dumpsites, only the average concentrations of BOD_5 and COD were higher compared to the values obtained in the previous review. All of the other

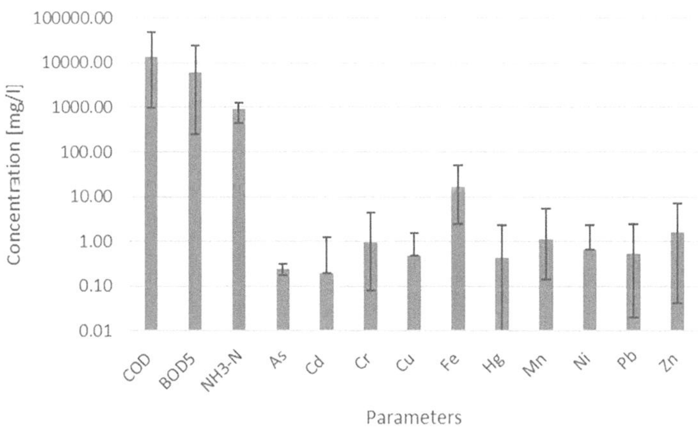

Figure 1.4 Biochemical parameters in all dumpsites.

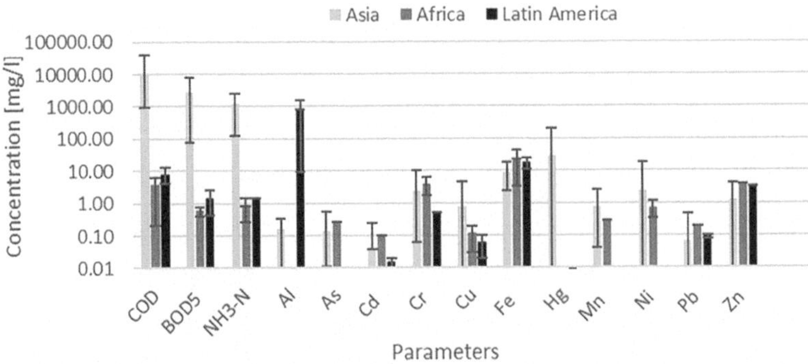

Figure 1.5 Biochemical parameters in landfills from Asia, Africa, and Latin America.

parameters investigated were higher in the study published in 2019. Focusing on landfills, most of the parameters (COD, As, Cd, Cr, Cu, Ni, Zn) were higher in this review than in the previous one. Further investigations would be necessary. Indeed, heavy metals and metalloids such as As, Cd, and Cr may pose significant risks for human health and the environment (Vaccari *et al.*, 2019b). However, the previous review (Vaccari *et al.*, 2019a) gathered more data, with studies covering about two decades (from 1998). In this review, the influence of hot spots, such as the study of Siddiqi *et al.* (2022), could have influenced the results.

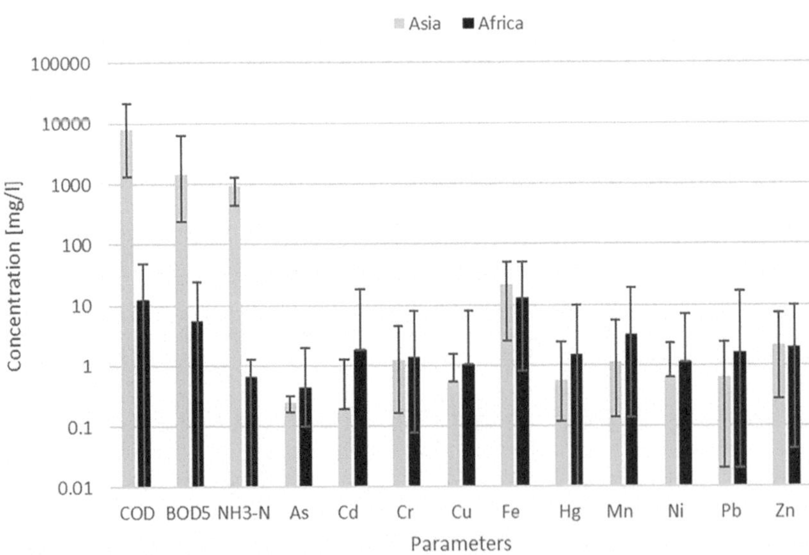

Figure 1.6 Biochemical parameters in dumpsites from Asia and Africa.

Table 1.3 Comparison of the biochemical parameters in the landfills and dumpsites (values are expressed in mg/l).

	Mean		Standard Deviation		Coefficient of Variation		Median	
	Landfills	**Dumpsites**	**Landfills**	**Dumpsites**	**Landfills**	**Dumpsites**	**Landfills**	**Dumpsites**
COD	8,416.01	13,813.57	9,161.58	13,493.01	1.09	0.98	5,198.00	10,700.00
BOD_5	2,202.07	6,057.57	2,651.93	7,969.41	1.20	1.32	1,177.00	1,335.00
NH_3-N	1,156.54	921.55	978.65	394.18	0.85	0.43	986.95	986.50
Al	0.71	–	0.76	–	1.06	–	0.48	–
As	12.10	0.24	11.38	0.10	0.94	0.41	9.43	0.24
Cd	0.04	0.19	0.07	0.30	1.81	1.54	0.01	0.11
Cr	0.60	0.96	1.20	1.28	1.99	1.34	0.10	0.37
Cu	0.08	0.48	0.12	0.48	1.54	1.01	0.02	0.33
Fe	401.27	17.00	795.83	15.93	1.98	0.94	5.04	11.08
Hg	2.05	0.43	5.35	0.65	2.61	1.50	0.36	0.20
Mn	2.42	1.12	3.56	1.26	1.47	1.13	0.51	0.52
Ni	1.63	0.65	1.80	0.69	1.11	1.05	0.50	0.55
Pb	0.13	0.54	0.19	0.62	1.48	1.15	0.06	0.29
Zn	26.40	1.60	74.59	1.99	2.83	1.24	0.01	0.75

Furthermore, both here and in the previous review, few studies investigated Al concentration in leachate from landfills and dumpsites.

Comparing the findings with the previous review (Vaccari *et al.*, 2019a) confirms the differences previously found in the levels of Cd and Pb between landfills and dumpsites. Evidently, Cd and Pb are heavy metals that seem to be often found in the leachate from dumpsites. The presence of these two heavy metals maybe as a result of the dumping of components of electrical and electronic goods and batteries. The breakdown of these metals in the dumpsites would be exacerbated by the climatic conditions. In turn, higher heavy metal concentrations would also lead to greater public and environmental health concerns within the urban poor populations who live on and near to the dumpsites and poorly maintained landfills.

1.3.6 Pollutant levels by region

In climatic zone A, the pollutants concentration was higher in dumpsites than landfills, except for Cr, Fe, Ni, Zn. However, a significant difference was found for BOD_5, and Pb. In climatic zone B, the pollutants concentration was higher in dumpsites, except for Cr, Cu, Ni. Though, a significant difference was found in climatic zone B in Pb concentration from landfills and dumpsites. The most likely reason for the differences in these two regions was most probably due to the influence of climatic conditions (i.e., higher temperatures) on leachate characteristics (Ma *et al.*, 2022). In climatic zone C, given the paucity of data, the analysis was only conducted for COD, BOD_5, Cr and Fe. There were no statistically significant differences found.

1.4 CONCLUSIONS

The presence of heavy metals in dumpsites and landfills, particularly Cd and Pb, poses a significant risk to the environment and to public health, especially of the urban poor. Evidently, the risks are particularly high in Global South countries in Asia, Africa, and Latin America. To mitigate these risks, it is vital for there to be investments in improving the waste management infrastructure and systems in these countries. In addition, there should be improved governance structures to enhance the enforcement of the existing policies. Finally, there should also be attention paid to facilitating stakeholder engagement and co-design with those in the communities who are most at risk from the landfills and dumpsites, as a means of enabling the development of more effective initiatives on the ground. Future research should include case studies from all over the world. Thus, North America, Oceania, and Europe should also be investigated and data could also be aggregated according to the income of the country.

REFERENCES

Alemayehu T., Mebrahtu G., Hadera A. and Bekele D. (2019). Assessment of the impact of landfill leachate on groundwater and surrounding surface water: a case study of Mekelle city, Northern Ethiopia. *Sustainable Water Resource Management*, **5**, 1641–1649, https://doi.org/10.1007/s40899-019-00328-z

Al-Wasify R., Ali M. and Hamed S. (2018). Application of different magnetic intensities for the treatment of landfill leachate in Egypt. *Cogent Engineering*, **5**, 1436114, https://doi.org/10.1080/23311916.2018.1436114

Ashraf M., Zeshan M., Hafeez S., Hussain R., Qadir A., Majid M., Ahmad F. and Ahmad S. (2022). Temporal variation in leachate composition of a newly constructed landfill site in Lahore in context to environmental pollution and risks. *Environmental Science Pollution Research*, **29**, 37129–37143, https://doi.org/10.1007/s11356-022-18646-9

Aziz A., Agamuthu P. and Fauziah S. (2018). Removal of bisphenol A and 2,4-Di-tert-butylphenol from landfill leachate using plant-based coagulant. *Waste Management Research*, **36**, 975–984, https://doi.org/10.1177/0734242X18790360

Aziz H., Rahmat N. and Alazaiza M. (2021). The potential use of *Nephelium lappaceum* seed as coagulant-coagulant aid in the treatment of semi-aerobic landfill leachate. *International Journal of Environmental Research Public Health*, **19**, 420, https://doi.org/10.3390/ijerph19010420

Boateng T., Opoku F. and Akoto O. (2018). Quality of leachate from the Oti landfill site and its effects on groundwater: a case history. *Environmental Earth Science*, **77**, 435, https://doi.org/10.1007/s12665-018-7626-9

Chen D. M., Bodirsky B. L., Krueger T., Mishra A. and Popp A. (2020). The world's growing municipal solid waste: trends and impacts. *Environmental Research Letters*, **15**, 074021, https://doi.org/10.1088/1748-9326/ab8659

De S., Hazra T. and Dutta A. (2019). Treatment of landfill leachate by integrated sequence of air stripping, coagulation–flocculation and adsorption. *Environment and Development Sustainable*, **21**, 657–677, https://doi.org/10.1007/s10668-017-0053-3

Deng M., Kuo D., Wu Q., Zhang Y., Liu X., Liu S., Hu X., Mai B., Liu Z. and Zhang H. (2018). Organophosphorus flame retardants and heavy metals in municipal land fill leachate treatment system in Guangzhou, China. *Environmental Pollution*, **236**, 137–145, https://doi.org/10.1016/j.envpol.2018.01.042

Deng Y., Chen N., Hu W., Wang H., Kuang P., Chen F. and Feng C. (2021). Treatment of old landfill leachate by persulfate enhanced electro-coagulation system: improving organic matters removal and precipitates settling performance. *Chemical Engineering Journal*, **424**, 130262, https://doi.org/10.1016/j.cej.2021.130262

Desai N., Soraganvi V. and Madabhavi V. (2020). Solar photocatalytic degradation of organic contaminants in landfill leachate using TiO_2 nanoparticles by RSM and ANN. *Nature Environment and Pollution Technology*, **19**, 651–662, https://doi.org/10.46488/NEPT.2020.v19i02.019

El Mrabet I., Ihssane B., Valdés H. and Zaitan H. (2022). Optimization of Fenton process operating conditions for the treatment of the landfill leachate of Fez city (Morocco). *International Journal of Environmental Science and Technology*, **19**, 3323–3336, https://doi.org/10.1007/s13762-021-03393-0

Erabee I., Ahsan A., Jose B., Aziz M., Ng A., Idrus S. and Daud N. (2018). Adsorptive treatment of landfill leachate using activated carbon modified with three different methods. *KSCE Journal of Civil Engineering*, **22**, 1083–1095, https://doi.org/10.1007/s12205-017-1430-z

Farhangi M., Ghasemzadeh Z., Ghorbanzadeh N., Khalilirad M. and Unc A. (2021). Phosphate removal from landfill leachate using ferric iron bioremediation under anaerobic condition. *Journal of Material Cycles and Waste Management*, **23**, 1576–1587, https://doi.org/10.1007/s10163-021-01239-y

Ferronato N. and Torretta V. (2019). Waste mismanagement in developing countries: a review of global issues. *International Journal of Environmental Research Public Health*, **16**, 1060, https://doi.org/10.3390/ijerph16061060

Galindo Montero A., Pimienta Serrano E. and Montiel J. (2019). Sanitary landfill leachate treatment with double chamber anaerobic reactor in series with constructed wetland. *Environmental Processes*, **6**, 695–712, https://doi.org/10.1007/s40710-019-00380-z

Gómez-Sanabria A., Kiesewetter G., Klimont Z. and Haberl H. (2022). Potential for future reductions of global GHG and air pollutants from circular waste management systems. *Natural Communication*, **13**, 106, https://doi.org/10.1038/s41467-021-27624-7

Hoai S., Nguyen Lan H., Thi Viet N., Nguyen Hoang G. and Kawamoto K. (2021). Characterizing seasonal variation in landfill leachate using leachate pollution index (LPI) at Nam Son solid waste landfill in Hanoi, Vietnam. *Environments*, **8**, 17, https://doi.org/10.3390/environments8030017

Hussein M., Yoneda K., Zaki Z., Othman N. and Amir A. (2019). Leachate characterizations and pollution indices of active and closed unlined landfills in Malaysia. *Environmental Nanotechnology, Monitoring Management*, **12**, 100232, https://doi.org/10.1016/j.enmm.2019.100232

Hussein M., Yoneda K., Mohd-Zaki Z., Amir A. and Othman N. (2021). Heavy metals in leachate, impacted soils and natural soils of different landfills in Malaysia: an alarming threat. *Chemosphere*, **267**, 128874, https://doi.org/10.1016/j.chemosphere.2020.128874

Ingle G. (2022). Study of soil properties affected by leachate – A case study at Urali-Devachi, Pune, India. *Materials Today Proceedings*, **60**, 588–594, https://doi.org/10.1016/j.matpr.2022.02.118

Ismail Y., Ngadi N., Hassim M., Kamaruddin M., Johari A. and Aziz M. (2018). Preparation of activated carbon from oil palm empty fruit bunch by physical activation for treatment of landfill leachate. *IOP Conference Series Materials Science Engineering*, **458**, 012036–206, https://doi.org/10.1016/j.biteb.2019.03.003

Kaza S., Yao L. C., Bhada-Tata P. and Van Woerden F. (2018). *What a waste 2.0: a global snapshot of solid waste management to 2050*. Washington, DC, USA.

Koliyabandara P., Cooray A., Liyanage S. and Siriwardhana C. (2022). Characterization of landfill leachate at the Karadiyana open dumpsite, Sri Lanka, and assessment of water pollution in its vicinity. *Journal of the National Science Foundation Sri Lanka*, **50**, 111–124.

Kow S.-H., Fahmi M., Abidin C., Rashid N., Ibrahim N., Ibrahim A., Ong S.-A., Wikurendra E. and Handayani D. (2022). Influence of leachate matrix on oxidation performance of ozonation and AOPs. *Biointerface Research Application Chemical*, **12**, 7903–7913.

Kuchelar P., Ramesh S., Ramasamy A. and Sugirthamani A. (2022). Utilization of rice husk ash for the treatment of leachate generated from Perungudi semi-urban solid waste dumping site of Chennai city, Tamil Nadu, India. *Journal of Materials Cycles and Waste Management*, **24**, 1565–1578.

Li J., Li P., Jin C., Zhao Y., Guo L. and Ji J. (2019). Treatment of municipal landfill leachate by electrochemical oxidation: assessing operating variables and oxidation products. *Environmental Engineer Science*, **36**.

Ma S., Zhou C., Pan J., Yang G., Sun C., Liu Y., Chen X. and Zhao J. (2022). Leachate from municipal solid waste landfills in a global perspective: characteristics, influential factors and environmental risks. *Journal of Cleaner Production*, **333**, 130234.

Mao W., Yang R., Shi H., Feng H., Chen S. and Wang X. (2022). Identification of key water parameters and microbiological compositions triggering intensive N_2O emissions during landfill leachate treatment process. *Science of the Total Environment*, **833**, 155135.

Mishra S., Tiwary D., Ohri A. and Agnihotri A. K. (2019). Impact of municipal solid waste landfill leachate on groundwater quality in Varanasi, India. *Groundwater for Sustainable Development*, **9**, 100230. https://doi.org/10.1016/j.gsd.2019.100230

Nayanthika I., Jayawardana D., Bandara N., Manage P. and Madushanka R. (2018). Effective use of iron-aluminum rich laterite based soil mixture for treatment of landfill leachate. *Waste Management*, **74**, 347–361.

Pujara Y., Pathak P., Sharma A. and Govani J. (2019). Review on Indian municipal solid waste management practices for reduction of environmental impacts to achieve sustainable development goals. *Journal of Environmental Management*, **248**, 109238. https://doi.org/10.1016/j.jenvman.2019.07.009

Rebolledo L., Arana V., Trilleras J., Barros G., González-Solano A. and Maury-Ardila H. (2019). Efficiency of combined processes coagulation/solar photo Fenton in the treatment of landfill leachate. *Water*, **11**, 1351.

Reis B., Silveira A., Lebron Y., Moreira V., Teixeira L., Okuma A., Amaral M. and Lange L. (2020). Comprehensive investigation of landfill leachate treatment by integrated Fenton/microfiltration and aerobic membrane bioreactor with. *Process Safety and Environmental Protection*, **143**, 121–128.

Rubel F., Brugger K., Haslinger K. and Auer I. (2017). The climate of the European Alps: shift of very high resolution Köppen-Geiger climate zones 1800–2100. *Meteorologische Zeitschrift*, **26**(2), 115–125.

Ruengruehan K., Junggoth R., Suttibak S., Sirikoon C. and Sanphoti N. (2021). Contamination of cadmium, lead, mercury and manganese in leachate from open dump, controlled dump and sanitary landfill sites in rural Thailand: a case study in Sakon Nakhon province. *Nature Environment and Pollution Technology*, **20**, 1257–1261.

Sackey L., Kočí V. and van Gestel C. (2020). Ecotoxicological effects on *Lemna minor* and *Daphnia magna* of leachates from differently aged landfills of Ghana. *Science of the Total Environment*, **698**, 134295.

Saeed T., Majed N., Miah M. and Yadav A. (2022). A comparative landfill leachate treatment performance in normal and electrodes integrated hybrid constructed wetlands under unstable pollutant loadings. *Science of the Total Environment*, **838**, 155942.

Sawaya R., Halwani J., Bashour I. and Nehme N. (2021). Assessment of the leachate quality from municipal solid waste landfill in Lebanon. *Arabian Journal of Geoscience*, **14**, 2160.

Shadi A., Kamaruddin M., Niza N., Emmanuela M., Shaah M., Yusoff M. and Allafi F. (2020). Characterization of stabilized leachate and evaluation of LPI from sanitary landfill in Penang, Malaysia. *Desalination Water Treatment*, **189**, 152–164.

Siddiqi S., Al-Mamun A., Sana A., Said Baawain M. and Choudhury M. (2022). Characterization and pollution potential of leachate from urban landfills during dry and wet periods in arid regions. *Water Supply*, **22**, 3462.

Somani M., Datta M., Gupta S., Sreekrishnan T. and Ramana G. (2019). Comprehensive assessment of the leachate quality and its pollution potential from six municipal waste dumpsites of India. *Bioresource Technology Reports*, **6**, 198–206.

Tesseme A., Vinti G. and Vaccari M. (2022). Pollution potential of dumping sites on surface water quality in Ethiopia using leachate and comprehensive pollution indices. *Environmental Monitoring and Assessment*, **19**, 545, https://doi.org/10.1007/s10661-022-10217-2

Tulun S. (2020). Treatment of leachate using up-flow anaerobic sludge blanket reactors/vertical flow subsurface constructed wetlands. *Ecological Chemistry and Engineering S*, **27**, 129–137.

Vaccari M., Tudor T. and Vinti G. (2019a). Characteristics of leachate from landfills and dumpsites in Asia, Africa and Latin America: an overview. *Waste Management*, **95**, 416–431, https://doi.org/10.1016/j.wasman.2019.06.032

Vaccari M., Vinti G., Cesaro A., Belgiorno V., Salhofer S., Dias M. I. and Jandric A. (2019b). WEEE treatment in developing countries: environmental pollution and health consequences – An overview. *International Journal of Environmental Research Public Health*, **16**, 1595, https://doi.org/10.3390/ijerph16091595

Vinti G., Bauza V., Clasen T., Medlicott K., Tudor T., Zurbrügg C. and Vaccari M. (2021). Municipal solid waste management and adverse health outcomes: a systematic review. *International Journal of Environmental Research Public Health*, **18**, 4331, https://doi.org/10.3390/ijerph18084331

Wang Z., Li J., Tan W., Wu X., Lin H. and Zhang H. (2019). Removal of COD from landfill leachate by advanced Fenton process combined with electrolysis. *Separation and Purification Technology*, **208**, 3–11, https://doi.org/10.1016/j.seppur.2018.06.048

Wdowczyk A. and Szymańska-Pulikowska A. (2021). Comparison of landfill leachate properties by LPI and phytotoxicity – A case study. *Frontiers in Environmental Science*, **9**, 693112, https://doi.org/10.3389/fenvs.2021.693112

World Bank. (2022). Solid Waste Management [WWW Document]. URL https://www.worldbank.org/en/topic/urbandevelopment/brief/solid-waste-management (accessed 8.22.22).

doi: 10.2166/9781789063318_0015

Chapter 2

Aerobic treatment of landfill leachate

Kaoutar Aboudi[1], *Ankur Rajpal*[2], *Vinay Kumar Tyagi*[3]* *and Ahmed Tawfik*[4]

[1]*Department of Chemical Engineering and Food Technology, Faculty of Sciences, University of Cádiz – International Campus of Excellence (ceiA3), P.O. Box No 40, Cádiz, Puerto Real 11510, Spain*
[2]*Department of Civil Engineering, Indian Institute of Technology, Roorkee 247667, Uttarakhand, India*
[3]*Environmental Hydrology Division, National Institute of Hydrology, Roorkee 247667, Uttarakhand, India*
[4]*National Research Centre, Water Pollution Research Department 12622, Dokki, Cairo, Egypt*
**Corresponding author: vinayiitrp@gmail.com*

ABSTRACT

Landfill leachate contamination is a major risk to natural resources and public health. Sustainable treatment techniques are therefore essential to protect the environment and living beings' health. Biological processes using microorganisms have proven to be successful and cost-effective techniques and could be aerobic or anaerobic processes. In particular, aerobic treatment techniques involve the biodegradation of the organic and nitrogenous material to produce CO_2 and sludge. Aerobic systems have shown to be efficient in treating young and medium-aged leachates but could be limited to the treatment of old highly loaded leachates. Among different aerobic treatment processes, the ones that are commonly used are activated sludge process, sequencing batch reactors, aerated lagoons, and those including suspended biomass growth such as rotating biological contactors, membrane bioreactors, biofilm reactors, and so on. This chapter discussed the most adopted aerobic treatment and from the findings of this document, it was deduced that the effectiveness of each technology is influenced by several factors such as the leachate characteristics, the process variables, and even the site/area in which the technology is implemented.

Keywords: Aerated lagoons, activated sludge process, sequencing batch reactors, biofilm reactor.

2.1 INTRODUCTION

The infiltration of rainwater mainly generates leachate through the landfill, chemical and biological processes within the landfill, and the water content of the waste itself. As a result of these processes, leachate is typically a very complex mixture (Bilgili *et al.*, 2007; Ritzkowski *et al.*, 2006). Factors that influence the composition and concentration of these contaminants are the type of waste, the age of the landfill and the quality of refuse (Shi *et al.*, 2007; Zhao *et al.*, 2006).

Research has proven that leachate contains toxic and sometimes even carcinogenic compounds. Thus, the unchecked release of leachate into the environment poses significant risks to the human population and the surrounding flora and fauna (Saviour, 2012). The main characteristics of the leachate are the high chemical oxygen demand (COD), biochemical oxygen demand (BOD), variable pH values, varied amount of ammonia, nitrogen, and diversity in heavy metals (Table 2.1). It is a colored liquid with a foul odor. The sum of all factors plays a role in the treatment complexity of landfill leachates (Maiti *et al.*, 2010).

Table 2.1 Leachate characteristics of varied aged landfills of different countries.

Landfill Site	Age	pH	BOD	COD	SS	TKN	NH_3-N
Canada	Y	5.8	9,660	13,800	–	212	42
	MA	6.9–9.0	–	3,210–9,190	–	–	–
China	Y	6.6	4,200	15,700	–	–	2,260
	MA	8.2	1,436	7,439	784	–	–
Greece	Y	6.2	26,800	70,900	950	3,400	3,100
	MA	7.9	1,050	5,350	480	1,100	-
Italy	Y	8	4,000	19,900	–	–	3,917
	MA	8.3	1,270	5,050	–	1,670	–
South Korea	Y	7.3	10,800	24,400	2,400	1,766	1,682
	O	8.5	62	1,409	404	141	1,522
Turkey	Y	7.3–7.8	10,800–11,000	16,200–20,000	–	–	1,120–2,500
	MA	8.1	–	9,500	–	1,450	1,270
	O	8.6	–	10,000	1,600	1,680	1,590
Germany	MA	–	1,060	3,180	–	1,135	–
Poland	MA	8	331	1,180	–	–	743
Taiwan	MA	8.1	500	6,500	–	–	5,500
Brazil	O	8.2	150	3,460	–	–	800
Estonia	O	11.5	800	2,170	–	–	–
Finland	O	–	62	556	–	192	159
France	O	7.5	7.1	500	130	540	430
Malaysia	O	7.5–9.4	48–105	1,533–2,580	159–233	–	–

Y: young; MA: medium age; O: old; all values except pH and BOD/COD are in mgl⁻; TKN: total Kjeldahl nitrogen; SS: suspended solids.
Source: adapted from Renou *et al.* (2008).

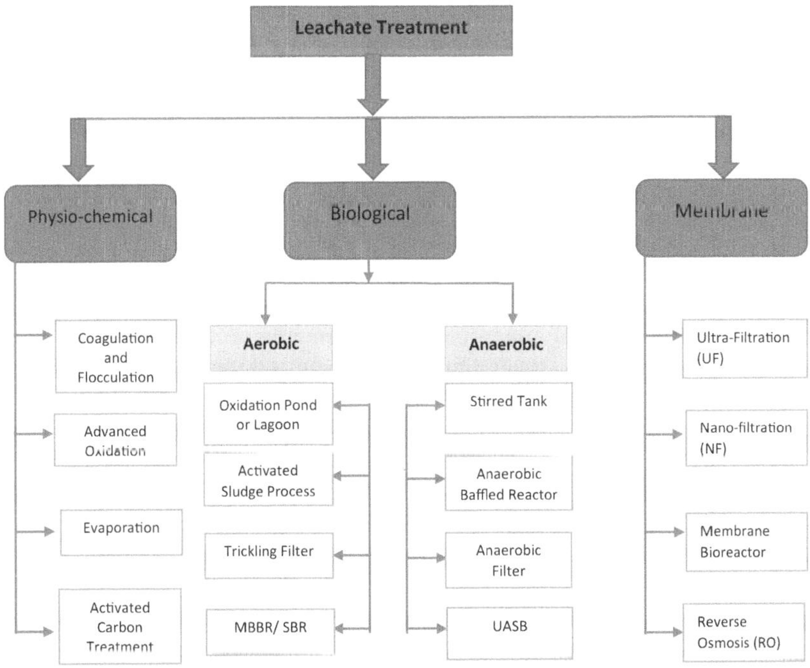

Figure 2.1 Overview of different treatment processes for MSW leachate.

For years, research on landfill leachate treatment has typically been based on wastewater and drinking water technologies (e.g., aerobic and anaerobic biological treatment, coagulation–flocculation, membrane processes, and adsorption processes (Figure 2.1). Leachate treatment plants are coming up at new waste processing facilities and sanitary landfills (SLFs). Mostly, leachate treatment technology is based on a combination of different processes depending upon the discharge conditions (Vij, 2012). The selection of suitable technology for specific leachate problems finalizes after considering various aspects and parameters such as the generation/flow rate of leachate, physical and chemical characteristics, capital investment, and operational expenses.

While analyzing different treatment technologies, it has been perceived that biological methods are suitable for leachate with high organic and nutrients content, which favor the growth of microorganisms, while they are not the best option to remove heavy metals and other toxic compounds from leachates, nor for leachate with low nutrient contents (Salami *et al.*, 2015). Nevertheless, the combination of biological processes with other technologies has also been identified as suitable for leachate heavily loaded with recalcitrant toxicants (Luo *et al.*, 2020; Renou *et al.*, 2008).

Biological treatment processes are reliable, simple, and highly cost-effective for removing high COD and BOD concentrations from leachate (Oller *et al.*, 2011). These processes can be classified into aerobic and anaerobic, depending on oxygen requirement. In aerobic treatment systems, the pollutants in the presence of oxygen, are distorted into CO_2, other solid biological products and sludge (Grady *et al.*, 2011). However, in an anaerobic process, the organic matter is converted to biogas consisting mainly of carbon dioxide, methane, and biological sludge. It has been observed that biological processes are very effective in removing organic and nitrogenous materials from immature (i.e., young) leachate when the BOD/COD ratio has a high value (>0.5) (Naveen *et al.*, 2017). However, the presence of refractory compounds (e.g., humic and fulvic acids) might limit the effectiveness of the process over time (Abbas *et al.*, 2009). Despite being advantageous due to methane generation during anaerobic treatment of leachate, which could be used as an energy source in the plant, some drawbacks of anaerobic systems emerge such as odor generation, the amount of the sludge generated and the required logistics to discharge it, as well as the quality of the final effluent, mainly when the continuous supply of fresh leachate over an old one already treated is carried out. Overall, after the treatment, leachate quality should meet the admissible standardized characteristics for discharge into natural bodies (Lebron *et al.*, 2021).

2.2 AEROBIC TREATMENT OF LANDFILL LEACHATE

The traditional biological process (aerobic) is used as the effective landfill leachate treatment technique in which biodegradable organic pollutants are eliminated and the ammoniacal nitrogen is converted into nitrite through nitrification, resulting in efficient organic material degradation and enabling easy removal of nitrogenous compounds (Ilmasari *et al.*, 2022a, 2022b; Luo *et al.*, 2020).

Among different aerobic treatment processes, the ones that are commonly used are activated sludge process (ASP), sequencing batch reactors (SBRs), aerated lagoons (ALs), and those of suspended biomass growth such as rotating biological contactors (RBCs), membrane bioreactors (MBRs), biofilm reactors, and so on (Amin *et al.*, 2014; Ghaitidak & Yadav, 2013; Jafarinejad, 2017). Table 2.2 depicts the removal efficiencies of organic pollutants and toxics from landfill leachate during aerobic treatment methods.

2.2.1 Aerated lagoons

ALs are kind of basins commonly used efficiently with reduced cost for removing the microbial and organic load through biological oxidation by a continuous air supply. Low maintenance with low operational costs encourages operators and decision-makers to use them for landfill leachate treatment in developing countries (Mara, 2013).

Moreover, the N, P, and Fe removals through this technique were reported to surpass 70% for diluted leachate (Luo *et al.*, 2014). Lagoons' feasibility in treating phenolic compounds together with the organic matter was also reported (Luo *et al.*, 2014). Thus, a 55–64% reduction was achieved for COD and 80–88% for phenol through the AL method (Sil & Kumar, 2017). Similarly, ammonia nitrogen removal of around 75–80% was achieved using AL (Frascari *et al.*, 2004; Mehmood *et al.*, 2009).

Table 2.2 Removal efficiencies of organic material and nutrients contained in landfill leachate using aerobic technologies.

Aerobic Treatment	Removal Efficiency (%)	References
ALs	COD:40%; BOD: 64%; N-NH_3: 77%; TKN: 77%; P: 42%; Fe: 30% COD: 97%; BOD: 99.5%; N-NH_4^+: 92.8%; Fe:98.7%, COD: 75%; Total N: 80%	Frascari *et al.* (2004) Robinson and Grantham (1988) Mehmood *et al.* (2009)
ASP	COD: 85–90%*; BOD: 90–94%; TKN: 77–90%*; N-NH_3: 78–90%*; P: 66–70%. COD: 70%; BOD: 98%;	Boonnorat *et al.* (2021) Bae et al. (1999)
SBRs	COD: 90%; N-NH_4^+: 70% COD: 98%; BOD: 97%; TKN: 80%	Neczaj *et al.* (2005) Neczaj *et al.* (2005)
RBCs	COD: 53–59% NH_3: 70–99%	Castillo *et al.* (2007) Kulikowska *et al.* (2010)
Biofilm reactors	COD: 92–95%; NH_3: 97% COD: 85%; N: 80%; P: 70%	Chen *et al.* (2008) Eldyasti *et al.* (2010)
Membrane reactors	COD:75%; BOD: 90%; NH_3: 90% COD: 54–78%; BOD: 97%	Ahmed and Lan (2012) Sadri *et al.* (2008)
Constructed wetlands	COD: 39–92%; BOD: 61–79%; TN: 34–67%; N-NH_4^+: 30–54% COD: 53%; BOD: 75%; NH_3: 70%; NO_3: 74%; TN: 74%	Akinbile *et al.* (2012) Bakhshoodeh *et al.*, 2020
Fungus	COD: 13–49%; BOD: 48–82%; N-NH_4^+: 71–99% COD: 61%; DOC: 41%	Zegzouti *et al.* (2020) Kalčíková *et al.* (2014)
Phytoremediation	COD: 80%; TN: 70%; P: 95% COD: 19–55%; BOD: 7–60%**; Zn: 80–90%; Fe: 83–87%; Pb: 76–84%	Lavagnolo *et al.* (2016) Abbas *et al.* (2019)
Combined treatments	RBCs and hybrid constructed wetlands = COD: 46%; N-NH_4^+: 95%; P: 41 ASP and Anaerobic process = COD: 94%; Zn: 50%; NH_3: 48–65%	Hu *et al.* (2022) Kheradmand *et al.* (2010)

**According to the contaminant load of leachates, with high removals for moderate loaded leachates.*
***According to the plant species and leachate dilution, being the highest values observed at 75% leachate dilution and water hyacinth specie.*

A significant limitation of this method is temperature dependence, as it mainly affects microbial activity because it was observed that temperatures under 15–20°C decelerate bacterial growth and hence slow down the treatment process (Abbas *et al.*, 2009). Furthermore, retention time is another critical factor in the treatment efficiency as it determines the residence time of microbes in the lagoons and the organic material degradation rates (Frascari *et al.*, 2004; Renou *et al.*, 2008; Robinson & Grantham, 1988).

2.2.2 Activated sludge process

The ASP is a common technology used for the treatment of sludges and wastewater at WWTPs and it mainly consists of using a mixed microbial consortium that degrades organic pollutants contained in wastewater to give rise to gases such as CO_2, water, and new residual biomass. ASP was also used for treating domestic wastewater or the co-treatment of leachate and sewage. For leachate treatment, ASP has shown to be feasible with sufficient treatment performance in terms of removing organic carbon and ammonia, as well as inorganic nutrients and phenolic compounds (Ilmasari *et al.*, 2022a, 2022b).

However, this method has proved inadequate in handling landfill leachate treatment and as a result, treating leachate through the ASP method is scarcely applied (Kamaruddin *et al.*, 2017). Some of the drawbacks of this technique are as follows:

- The need for longer aeration times of about 20 h and inadequate ability of the sludge settlement (Wanner, 2014).
- The excess production of sludge and high demand for energy (Pant *et al.*, 2010).
- The inhibition of microbial growth is due to the high nitrogen content of most leachates (Bernard, 2011).
- The non-compliance of the resulting sludge requires a subsequent treatment technique to fulfil established standards (Ilmasari *et al.*, 2022a, 2022b).

2.2.3 Sequencing batch reactors

SBR technology is a non-steady-state, variable-capacity and suspended-growth biological treatment method that uses the fill and decant-ASP with and without a clarifier. Typically, SBR is divided into five stages: fill, react, settle, draw, and idle (Mahvi, 2008; Rekabi *et al.*, 2007).

Generally, SBR is operated in an intermittent aeration mode where all metabolic reactions and solid–liquid segregation take place in a unit tank through a timed control sequence (Alattabi *et al.*, 2017). The traditional SBR is an integrated nitrification–denitrification process (Duan *et al.*, 2020). It blends both anaerobic and aerobic stages to successfully achieve nitrification, denitrification, and phosphorous removal concurrently.

Many modified approaches are being used in the SBR process for the leachate treatment to enhance biological nitrogen removal, including utterly autotrophic nitrogen removal over nitrite (CANON), anaerobic ammonium oxidation (ANAMMOX), oxygen-limited autotrophic nitrification–denitrification (OLAND), simultaneous nitrification and denitrification (SND) via nitrite, simultaneous nitrification–anammox–denitrification (SNAD), single reactor system for high activity ammonium removal over nitrite (SHARON), and deammonification (DEMON) (Arun *et al.*, 2019). Moreover, combined treatment technologies for leachate treatment using SBR coupled with an MBR were used to treat young leachate. In addition, SBR can be enhanced by the addition of plastic media into the reactor through coagulation to increase the specific surface area of the reactor (Yong *et al.*, 2018).

2.2.4 Rotating biological contactors

An RBC is considered a secondary treatment and it consists of a series of rotating disks containing fixed media filters and aims to remove organic material and ammonia from wastewater and leachates. The rotating disks are partially submerged in wastewater and their rotation allows the degradation of the organic material contained in leachate, 40–50% of their surface area is submerged while the other surface is intermittently in contact with atmospheric air for oxygen absorption. In these conditions, specific microorganisms are continuously growing by forming attached membranes on the disk's surface. The dead cells fall off the disks and together with excess sludge are removed from the system throughout the treatment process (e.g., by a subsequent clarification in a clarifier) (Ilmasari *et al.*, 2022a; Miao *et al.*, 2019). In RBC, the organic material elimination is determined by analyzing BOD and COD, and different forms of nitrogen are analyzed determining the ammonia conversion to nitrates. The RBC efficiency depends mainly on the amount of media surface as well as the leachate's characteristics. Furthermore, operational parameters such as the speed of rotation, hydraulic loading rate, organic loading rate (OLR), dissolved oxygen (DO) level, temperature, and so on are also detrimental to the process (Chen *et al.*, 2006; Ghalehkhondabi *et al.*, 2021; Hassard *et al.*, 2015; Hiras *et al.*, 2004; Maheepala *et al.*, 2022).

RBC systems were used for leachate treatment earlier and currently, but the resulting treatment efficiencies have differed from one study to another, according to the applied operational conditions. Spengel and Dzombak (1991) studied the treatment of high-strength ammonia leachate with low BOD using RBCs. They applied low substrate loading rates to promote nitrification and they obtained complete ammonia oxidation while a low COD removal of 38% was observed. Similarly, Kulikowska *et al.* (2010) investigated the treatment of high-strength ammonia nitrogen leachate using RBC. They found that at OLR of 1.9 g $N\text{-}NH_4/m^2$ d, successful complete nitrification was achieved while increasing the OLR two-fold required the use of two subsequent RBCs for complete nitrification. Analyzing the effect of temperature conditions in RBC systems, a recent study by Fuchigami *et al.* (2021) found that nitrogen removal decreased by 0.1 kg N/m^3 day for each 12°C decrease in temperature. Furthermore, they reported BOD removal up to 90% when the temperature was above 20°C, which decreased to 76% when decreasing temperature below 15°C. In another attempt, Castillo *et al.* (2007) applied different configurations to an RBC system for the treatment of landfill leachate. First, they aimed to select the disk's type considering the criteria of a high COD removal (65%) and biological stability. They used three different disk supports with the same diameter, same OLR, and same rotational speeds (RSs), being the supports corrugated, perforated, or covered with a polyester mesh. Subsequently, they tested different hydraulic retention times (HRTs) in a continuous system, different RS, and different OLRs. Their findings showed that the highest COD removal of 74% and 69% were obtained by using covered supports with polyester mesh and perforated acetate disks, respectively. However, they reported that due to their high mechanical resistance and stability characteristics, the latter ones were selected for the treatment. Regarding the RS parameter, their results showed that COD removals of around 58% were reached at RS of 6 and 9 at the same HRT of 34 h. The lower COD removals of 40% and 14% were observed at RS of 9 rpm at HRTs of 18 and 14 h, respectively. Thus, higher removals were found for high RS and HRT. In another study, Hiras *et al.* (2004) found that a high RS prevents the solids accumulation in the system and led to elevated DO concentrations. However, the suitable RS depends on each system, as elevated RS may also lead to washing out the detached biomass. From these studies, it can be deduced that different operational parameters influence the process performance of RBCs for leachate treatment, which need to be studied separately and in combination to find out the best strategy in each situation. Besides, the leachate age also played an important role when selecting RBC. RBCs are suitable for the treatment of young leachate, and as the leachate age gets mature as these systems show weak treatment efficiency (Kamaruddin *et al.*, 2015).

In general, among RBC's advantages compared to other aerobic systems, the low energy requirement, the simple design and operation, in addition to the low space requirement, are an added advantage. Moreover, the high contact time between biofilms and leachates allows for reaching a high effluent quality and a low excess sludge production. Specifically, RBC systems have been shown to have a high nitrification ability (Ghalehkhondabi *et al.*, 2021; Hassard *et al.*, 2015; Hiras *et al.*, 2004; Maheepala *et al.*, 2022). Nevertheless, the main drawback of this technology is that RBC systems should be protected from extreme weather conditions such as excessive sunlight, wind, rain, and snow (Wang *et al.*, 2022).

2.2.5 Biofilm reactor

As RBC systems, all biological or biochemical reactors involve the activity of specific microorganisms for wastewater or leachate treatment. This function could be boosted by the use of a support media as growth systems inside the reactors, which are commonly referred to as biofilms, manufactured by different materials such as plastics, metals, ceramics, and so on. The biofilm formation occurs naturally through the formation of thick layers of microbial cells, which adhered to the supporting media or form flocs (i.e., granules). Biofilm formation has the advantage of increasing cell density in the reactors, which enhances the performance of any reactor, regardless of its configuration (i.e., CSTR, MBBR, UASB, PBR, RBC, etc.) (Qureshi *et al.*, 2005).

The performance of these reactors also depends on operational parameters. In this line, Ismail *et al.* (2011) studied leachate treatment in a biofilm reactor and observed that increasing OLR led to a decrease in microorganisms' activity and an inefficient treatment. They also studied the effect of pH in the process and reported that pH values close to neutrality led to the formation of a dense biofilm of microbes increasing their proliferation, which conducted in high organic matter and nitrogen removals (60% and 80%, respectively for each parameter). Regarding the role of biofilm thickness on organic pollutants removal, recent research by Sanchez-Huerta *et al.* (2022) found that increasing biofilm thickness and cell density improved the process performance, reaching up to 96% in nitrification and 80% overall COD removal. However, Taşkan *et al.*

(2020) found that an excessive biofilm formation inhibited gas diffusion into the biofilm, which significantly reduced the performance of leachate treatment. Therefore, the control of biofilm formation and thickness is detrimental to a stable process and a successful long-term operation (Lackner *et al.*, 2009). In the same context, Pechaud *et al.* (2022) assessed the effect of different configurations on the biofilm function and effectiveness. They found that increasing HRT in the reactors led to increasing shear stress and a decrease in biofilm thickness. Moreover, they observed that the biofilm density increased by increasing shear stress and OLR and that low HRTs decrease the concentration of suspended biomass, which favors microorganisms agglomeration in thick and dense biofilms. They highlighted that operational conditions strongly affect biological aggregate properties, which influence the stability and performance of the whole process.

2.2.6 Membrane reactors

MBR technology involves the retention of biomass in bioreactors to separate the biosolids from the mixed liquor, which results in a final effluent with much-reduced pollutants content (Lindamulla *et al.*, 2022; Remmas *et al.*, 2017). Membrane technology combines the use of ASP and membrane units for efficient treatment, mainly applied to highly loaded leachates that are difficult to treat using other technologies (Gu *et al.*, 2023). Moreover, membrane technology has proved to be efficient for the treatment of mature landfill leachates with recalcitrant pollutants (Zhang *et al.*, 2020) and offered a stable process with low sludge production (Iorhemen *et al.*, 2016).

Nevertheless, the most challenging aspect of membrane technology is the fouling occurrence due to the use of high organic strength leachate, excessive formation of biosolids, or excessive microbial growth, among other reasons. These challenging aspects substantially increase the operating cost (Abuabdou *et al.*, 2020; Remmas *et al.*, 2017; Werkneh, 2022).

The material used for membrane construction can be different (e.g., polymeric, metallic, ceramic). Some materials are more advantageous due to their possible reutilization through their regeneration, such as ceramic membranes (Wu *et al.*, 2023).

2.2.7 Constructed wetlands

Constructed wetlands (CWs) are implemented to emulate natural wetland ecosystems, in which the following components are involved to depurate leachates and wastewater from pollutants and toxics: Aquatic plants, hydric soils, associated microorganisms, and a filter bed (i.e., sand, gravel, etc.) (Stottmeister *et al.*, 2003; Wdowczyk *et al.*, 2022). Compared to other technologies, CWs have been considered environmentally friendly and of low cost (Siddiqi *et al.*, 2022). In CWs, the extended root system of the plant allowed microorganisms' attachment on the large surface area and increased the decomposition of the organic material. Thus, nitrogen and phosphorous are eliminated via the plant's absorption and ammonia is eliminated via volatilization and nitrification/denitrification (Vymazal, 2007).

CWs have been classified into two categories, free water surface flow CWs (i.e., water flows on the surface of basins) or subsurface flow CWs (i.e., water flows under the surface basins). The latter could be configured for vertical or horizontal flow systems according to the direction of the water flow in the subsurface (Hu *et al.*, 2016; Kivaisi, 2001; Yalcuk and Ugurlu, 2009). Vertical subsurface flow CWs are the most applied configurations in comparison to horizontal systems (Bakhshoodeh *et al.*, 2020).

The substrate used in filling CWs systems is an important parameter for the leachate treatment performance due to its effect on bacterial growth. Wdowczyk *et al.* (2022) studied vertically constructed wetlands for leachate treatment using *Phragmites australis* species, known for their high capacity in nutrient and micropollutants accumulation, and they tested the effect of different filling materials (organic: pine bark; inorganic: zeolite and expanded clay). They observed that depending on the leachate characteristics and charge on micropollutants one material can perform better than the other. Furthermore, they found that inorganic filling material (i.e., zeolite) performed better in terms of high ion exchange capacity, due to their high affinity for ammonium nitrogen ions and their porosity, which increased the growth of microorganisms in the system. They found that for ammonium removal, higher performance with all materials (>96%) could be achieved, while for total nitrogen removal, it ranged between 43% and 67%, with the lowest values for organic filling material, likely due to that these materials can release organic matter, increasing concentrations at the outflow. Furthermore, they found that nitrogen was not completely removed in all systems due to the presence of other cations in the leachate (e.g., K^+, Na^+, Mg^{2+}, etc.), which likely competed for ion exchange sites, inhibiting ammonium nitrogen adsorption. On the other hand, they found that for phosphorous removal, expanded clay and zeolite slightly outperformed bark due to their high adsorption capacity compared to bark pine.

The vegetation used in CWs is another important parameter in the leachate treatment performance as they are responsible for nutrients and heavy metals uptake, hence affecting removal efficiencies (Klomjek & Nitisoravut, 2005; Mbuligwe, 2005). Therefore, selecting the suitable vegetation type according to CWs site include consideration of the inlet strength in terms of organic material and contaminants, as well as climate conditions and the hydrology of the site. The mostly studied vegetation species are *Phragmites australis*, mainly in European countries (O'Connor & Courtney, 2020). However, other plants have shown their effectiveness for specific contaminants such as *Diplachne fusca* Kunth (Kallar grass) for high salinity content (Bakhshoodeh *et al.*, 2020). Other studies tested a combination of different species, which also showed high performance (O'Connor & Courtney, 2020).

One of the challenges of implementing CWs in some countries relies on their poor performance in winter, due to that process performance is affected by temperature conditions (Bove *et al.*, 2015).

2.2.8 Fungal and yeast treatment

Bioremediation of landfill leachate using fungi and yeasts has been widely studied in recent years, mainly for mature leachates, due to its simplicity and efficiency (Collado *et al.*, 2019; Kalčíková *et al.*, 2014). Fungi treatment occurred through the release of extracellular enzymes that allow the decomposition of complex material into simpler substances (Collado *et al.*, 2019). Therefore, fungi utilize the organic material of the leachate as a source of carbon for their growth leading to the removal of the organic matter contained in landfill leachate. Kalčíková *et al.* (2014) evaluated the potential of using the white rot fungus, *Dichomitus squalene*, and its extracellular enzymes for the treatment of a mature landfill leachate from an old closed landfill and reported COD and dissolved organic carbon (DOC) removals of 60%, in addition to a detoxification of the leachate to the bacterial stain *Aliivibrio fischeri*. However, they observed that the used white fungus was inhibited when using leachate from an active landfill, showing COD and DOC removals of only 23%. Besides, they stated that the use of extracellular ligninolytic enzymes in this latter case allowed removing 61% and 44% of COD and DOC, respectively. Their findings highlighted the promising use of fungus as a suitable treatment of mature landfill leachates or as a post-treatment of leachates produced after landfill closure but not for young leachates.

Fungi strain selection is an important parameter for the process and it depends mainly on the characteristics of the treated leachate in terms of contaminants and toxic compound contents (Collado *et al.*, 2019). In this context, Spina *et al.* (2018) compared autochthonous (i.e., from contaminated wastewater) and allochthonous fungi strains (e.g., *Porostereum spadiceum MUT 1585*) performance on landfill leachate treatment and they found that allochthonous fungi strains outperformed autochthonous ones that are supposedly adapted to pollutants and toxic compounds. However, for the decolorization effect, autochthonous strains showed a rapid response compared to the allochthonous ones, although both strains reached a similar removal (around 50%) for some bacterial species. From their results, they concluded that landfill leachates are selective substrates for allochthonous fungal strains. In another study, Islam *et al.* (2019) assessed the effect of lignocellulosic enzymatic activities, which were obtained from six selected fungi species, on the removal efficiency of COD from mature landfill leachate. They found that higher COD removals could be obtained by some fungal strains that were able to produce a high amount of ligninolytic enzymes. They observed that COD removal efficiencies were highly correlated to lignocellulosic extracellular enzyme production.

Some studies suggested the use of co-substrates in the treatment process for the best performance (Bardi *et al.*, 2017; Díaz *et al.*, 2022). In this context, Bardi *et al.* (2017) stated that the use of a co-substrate, such as glucose, for the treatment of old leachate with high recalcitrant content improved COD removals (63% and 53% for total COD and soluble COD, respectively).

Fungi application for leachate treatment is a novel technology that started growing and it could be used in combination with other aerobic treatment methods such as those discussed in previous sections of this chapter, being compliant with many types of bioreactors (del Álamo *et al.*, 2022; Ghosh *et al.*, 2014). Using white rot fungi has shown to be an environmentally friendly and economic treatment strategy for landfill leachates and wastewater. However, the challenge of using fungi for landfill leachate treatment resides in the lack of research on fungi's effect on ammonia removal, as nitrogen is also a concern for the treatment of landfill leachates. In addition, investigating other fungi species than white rot ones should be performed to extend the feasibility and performance of this biological treatment. Furthermore, studies scaling up the processes using fungi are limited.

In the same line as fungi application, leachate treatment using yeast individually or in combination with fungus has been investigated. Brito *et al.* (2012) found that for the treatment of high-strength landfill leachate, the use of yeasts (i.e., *Saccharomyces cerevisiae*) is an efficient strategy, reaching 74%, 82%, and 67% removals for COD, color, and humic substances, respectively. Ghosh *et al.* (2014) treated landfill leachate by using fungi (*Phanerochaete* sp.) and bacteria (*Pseudomonas* sp.), they obtained high removal efficiencies of COD and color of 77% and 45.4%, respectively.

Wichitsathian *et al.* (2004) studied medium-age landfill leachate treatment in bioreactors comparing operation with a mixed bacterial culture and a mixed yeast culture. They found that although both bioreactors showed similar COD removal efficiencies, the one using yeast outperformed the one with bacteria in terms of BOD removal. Furthermore, they observed that the bioreactor operating with yeasts showed a better performance in terms of fouling with lower trans-membrane pressure and longer operating time. They deduced that the high performance of the bioreactor with yeast addition could be related to the yeast cell structure that is larger, in addition to the reduced soluble extracellular polymeric substances (EPS) that cause membrane biofouling. Similarly, Reis *et al.* (2017) analyzed the performance of a commercial bakers' yeast (*Saccharomyces cerevisiae*) versus conventional bacteria-based MBR for the treatment of landfill leachate and reported higher removal efficiencies when using yeasts in bioreactors, reaching values of 69%, 54%, and 34% for COD, color, and ammonia removals, respectively. Thus, conventional bacteria allowed removal efficiencies of 27%, 33%, and 27%, for COD, color, and ammonia, respectively.

In the same way as fungi, yeast allowed the degradation of recalcitrant pollutants that are present in medium and mature-aged leachates. In addition, the use of yeast in membrane reactors is promising due to their low adhesion in surfaces in comparison to conventional bacteria, avoiding fouling occurrence. In addition, some studies found that yeasts are highly resistant to extreme environmental conditions (Buzzini *et al.*, 2018), which make them advantageous in comparison to other technologies that are affected by weather conditions.

2.2.9 Phytoremediation

Phytoremediation technology consists of the use of plants (e.g., trees and grassland) to eliminate pollutants from soil, water, or air. It has been widely studied for leachate treatment due to its relative simplicity with low operational and maintenance costs, as well as being an environment-friendly system (Ilmasari *et al.*, 2022b; Luo *et al.*, 2020).

Leachate treatment through phytoremediation occurred when plants uptake contaminants and accumulate them in their foliar plant tissue, which allows subsequent harvesting and removal of these pollutants. Moreover, contaminants are also uptaken by plant roots and biodegraded by soil microorganisms. Besides, phytoremediation involves the stabilization of inorganic contaminants around the root zone avoiding their leaching into groundwater (Kim & Owens, 2010). The phytoremediation concept is somehow similar to CWs. The vegetation type in phytoremediation is an important factor for the process efficiency. Abbas *et al.* (2019) used aquatic plants for the phytoremediation of landfill leachate and reported high efficiencies in terms of COD, BOD, and heavy metals. They observed that the transport of heavy metals from roots to the aboveground parts of the aquatic plants was low due to their accumulation inside the plants' bodies, without affecting their growth. In another study, Lavagnolo *et al.* (2016) evaluated landfill leachate phytoremediation using three oleaginous species and they obtained high removal efficiencies of COD (>80%), total nitrogen (>70%) and total phosphorous (>95%). Moreover, they observed that the plants irrigated with leachate performed a greater plant mass in comparison to control systems with plants irrigated with water and a synthetic nutrient solution. Furthermore, they found that the soil type also influences the process of reaching high total biomass from oleaginous plants irrigated with leachate and grown on clayey soils, compared to sandy soil. The authors deduced that the differences observed in plant development were likely related to the interaction between the studied plants and the chemical composition of the leachate soils.

Phytoremediation is efficient in eliminating contaminants from landfill leachates, reducing leachate volumes due to evapotranspiration, and even recovering water and nutrient after the treatment (Lavagnolo *et al.*, 2016; Nagendran *et al.*, 2006). Nevertheless, it has been reported to be suitable for weak leachates from old leachates or low or medium-charged leachate, but not for highly contaminated leachates. In this latter case, leachate should be diluted for a better phytoremediation performance.

2.3 CONCLUSIONS

The release of leachate into the environment poses significant risks to the human population and the surrounding flora and fauna. Aerobic biological treatment processes are reliable, simple, and highly cost-effective for removing high COD and BOD concentrations from leachate. The suitable aerobic process for landfill leachate treatment may depend on the leachate characteristics in terms of its age (young, medium, or mature leachate) and its physicochemical composition. Aerobic systems in which suspended biomass supports is included to enhance microorganisms' growth have been identified as advantageous. However, each aerobic process should be analyzed carefully for a better treatment process.

REFERENCES

Abbas A. A., Jingsong G., Ping L. Z., Ya P. Y. and Al-Rekabi W. S. (2009). Review on landfill leachate treatments. *American Journal of Applied Sciences*, **6**(4), 672–684, https://doi.org/10.3844/ajassp.2009.672.684

Abbas Z., Arooj F., Ali S., Zaheer I. E., Rizwan M. and Riaz M. A. (2019). Phytoremediation of landfill leachate waste contaminants through floating bed technique using water hyacinth and water lettuce. *International Journal of Phytoremediation*, **21**, 1356–1367, https://doi.org/10.1080/15226514.2019.1633259

Abuabdou S. M. A., Ahmad W., Aun N. C. and Bashir M. J. K. (2020). A review of anaerobic membrane bioreactors (AnMBR) for the treatment of highly contaminated landfill leachate and biogas production: effectiveness, limitations and future perspectives. *Journal of Cleaner Production*, **255**, 120215, https://doi.org/10.1016/J.JCLEPRO.2020.120215

Ahmed F. N. and Lan C. Q. (2012). Treatment of landfill leachate using membrane bioreactors: a review. *Desalination*, **287**, 41–54, https://doi.org/https://doi.org/10.1016/j.desal.2011.12.012

Akinbile C. O., Yusoff M. S. and Ahmad Zuki A. Z. (2012). Landfill leachate treatment using sub-surface flow constructed wetland by *Cyperus haspan*. *Waste Management*, **32**, 1387–1393, https://doi.org/10.1016/J.WASMAN.2012.03.002

Alattabi A. W., Harris C., Alkhaddar R., Alzeyadi A. and Abdulredha M. (2017). Online Monitoring of a Sequencing Batch Reactor Treating Domestic Wastewater. *Elsevier: Procedia Engineering*, **196**, 800–807, https://doi.org/10.1016/j.proeng.2017.08.010

Al-Rekabi W. S., Qiang H. and Qiang W. W. (2007). Review on sequencing batch reactors. *Nutrition Journal*, **6**, 11–, https://doi.org/10.1186/1475-2891-6-11

Amin M. M., Hashemi H., Bina B., Ebrahimi A., Pourzamani H. R. and Ebrahimi A. (2014). Complementary treatment of leachate using sequencing batch reactor. *International Journal of Health System and Disaster Management*, **2**(4), 216, https://doi.org/10.4103/2347-9019.144403

Arun S., Manikandan N. A., Pakshirajan K. and Pugazhenthi G. (2019). Novel shortcut biological nitrogen removal method using an algae-bacterial consortium in a photo-sequencing batch reactor: process optimization and kinetic modelling. *Journal of Environmental Management*, **250**, 109401, https://doi.org/10.1016/j.jenvman.2019.109401

Bae B. U., Jung E. S., Kim Y. R. and Shin H. S. (1999). Treatment of landfill leachate using activated sludge process and electron-beam radiation. *Water Research*, **33**, 2669–2673, https://doi.org/10.1016/S0043-1354(98)00488-6

Bakhshoodeh R., Alavi N., Oldham C., Santos R. M., Babaei A. A., Vymazal J. and Paydary P. (2020). Constructed wetlands for landfill leachate treatment: a review. *Ecological Engineering*, **146**, 105725, https://doi.org/10.1016/J.ECOLENG.2020.105725

Bardi A., Yuan Q., Siracusa G., Chicca I., Islam M., Spennati F., Tigini V., Di Gregorio S., Levin D. B., Petroni G. and Munz G. (2017). Effect of cellulose as co-substrate on old landfill leachate treatment using white-rot fungi. *Bioresource Technology*, **241**, 1067–1076, https://doi.org/10.1016/J.BIORTECH.2017.06.046

Bernard O. (2011). Hurdles and challenges for modelling and control of microalgae for CO_2 mitigation and biofuel production. *Journal of Process Control*, **21**(10), 1378–1389, https://doi.org/10.1016/j.jprocont.2011.07.012

Bilgili S., Demir A. and Özkaya B. (2007). Influence of leachate recirculation on aerobic and anaerobic decomposition of solid wastes. *Journal of Hazardous Materials*, **143**, 177–183, https://doi.org/10.1016/j.jhazmat.2006.09.012

Boonnorat J., Kanyatrakul A., Prakhongsak A., Ketbubpha K., Phattarapattamawong S., Treesubsuntorn C. and Panichnumsin P. (2021). Biotoxicity of landfill leachate effluent treated by two-stage acclimatized sludge AS system and antioxidant enzyme activity in *Cyprinus carpio*. *Chemosphere*, **263**, 128332, https://doi.org/10.1016/J.CHEMOSPHERE.2020.128332

Bove D., Merello S., Frumento D., Arni S. A., Aliakbarian B. and Converti A. (2015). A critical review of biological processes and technologies for landfill leachate treatment. *Chemical Engineering & Technology*, **38**, 2115–2126, https://doi.org/https://doi.org/10.1002/ceat.201500257

Brito G. C. B., Amaral M. C. S., Lange L. C., Pereira R. C. A., Santos V. L. and Machado M. (2012). Treatment of landfill leachate in membranes bioreactor with yeast (*Saccharomyces cerevisiae*). *Procedia Engineering*, **44**, 934–938, https://doi.org/https://doi.org/10.1016/j.proeng.2012.08.627

Buzzini P., Turchetti B. and Yurkov A. (2018). Extremophilic yeasts: the toughest yeasts around? *Yeast*, **35**, 487–497, https://doi.org/10.1002/yea.3314

Castillo E., Vergara M. and Moreno Y. (2007). Landfill leachate treatment using a rotating biological contactor and an upward-flow anaerobic sludge bed reactor. *Waste Management*, **27**, 720–726, https://doi.org/10.1016/J.WASMAN.2006.08.003

Chen Z., Wen Q., Wang J. and Li F. (2006). Simultaneous removal of carbon and nitrogen from municipal-type synthetic wastewater using net-like rotating biological contactor (NRBC). *Process Biochemistry*, **41**, 2468–2472, https://doi.org/10.1016/J.PROCBIO.2006.06.003

Chen S., Sun D. and Chung J. S. (2008). Simultaneous removal of COD and ammonium from landfill leachate using an anaerobic–aerobic moving-bed biofilm reactor system. *Waste Management*, **28**, 339–346, https://doi.org/10.1016/J.WASMAN.2007.01.004

Collado S., Oulego P., Suárez-Iglesias O. and Díaz M. (2019). Leachates and natural organic matter. A review of their biotreatment using fungi. *Waste Management*, **96**, 108–120, https://doi.org/10.1016/J.WASMAN.2019.07.018

del Álamo A. C., Pariente M. I., Molina R. and Martínez F. (2022). Advanced bio-oxidation of fungal mixed cultures immobilized on rotating biological contactors for the removal of pharmaceutical micropollutants in a real hospital wastewater. *Journal of Hazardous Materials*, **425**, 128002, https://doi.org/10.1016/J.JHAZMAT.2021.128002

Díaz A. I., Laca A. and Díaz M. (2022). Approach to a fungal treatment of a biologically treated landfill leachate. *Journal of Environmental Management*, **322**, 116085, https://doi.org/10.1016/J.JENVMAN.2022.116085

Duan Y., Liu Y., Zhang M., Li Y., Zhu W., Hao M. and Ma S. (2020). Start-up and operational performance of the partial nitrification process in a sequencing batch reactor (SBR) coupled with a micro-aeration system. *Bioresource Technology*, **296**, 122311, https://doi.org/10.1016/j.biortech.2019.122311

Eldyasti A., Chowdhury N., Nakhla G. and Zhu J. (2010). Biological nutrient removal from leachate using a pilot liquid–solid circulating fluidized bed bioreactor (LSCFB). *Journal of Hazardous Materials*, **181**, 289–297, https://doi.org/10.1016/J.JHAZMAT.2010.05.010

Feng L., Chen Y. and Zheng X. (2009). Enhancement of waste activated sludge protein conversion and volatile fatty acids accumulation during waste activated sludge anaerobic fermentation by carbohydrate substrate addition: the effect of pH. *Environmental Science & Technology*, **43**(12), 4373–4380, https://doi.org/10.1021/es8037142

Frascari D., Bronzini F., Giordano G., Tedioli G. and Nocentini M. (2004). Long-term characterization, lagoon treatment and migration potential of landfill leachate: a case study in an active Italian landfill. *Chemosphere*, **54**, 335–343, https://doi.org/10.1016/J.CHEMOSPHERE.2003.08.013

Fuchigami S., Hatamoto M., Takagi R., Akashi T., Watari T. and Yamaguchi T. (2021). Long-term treatment of municipal wastewater using a mesh rotating biological reactor and changes in the biofilm community. *Environmental Technology & Innovation*, **24**, 102074, https://doi.org/10.1016/J.ETI.2021.102074

Ghaitidak D. M. and Yadav K. D. (2013). Characteristics and treatment of greywater – a review. *Environmental Science and Pollution Research*, **20**(5), 2795–2809, https://doi.org/10.1007/s11356-013-1533-0

Ghalehkhondabi V., Fazlali A. and Fallah B. (2021). Performance analysis of four-stage rotating biological contactor in nitrification and COD removal from petroleum refinery wastewater. *Chemical Engineering and Processing – Process Intensification*, **159**, 108214, https://doi.org/10.1016/J.CEP.2020.108214

Ghosh P., Swati and Thakur I. S. (2014). Enhanced removal of COD and color from landfill leachate in a sequential bioreactor. *Bioresource Technology*, **170**, 10–19, https://doi.org/10.1016/J.BIORTECH.2014.07.079

Grady Jr. C. L., Daigger G. T., Love N. G. and Filipe C. D. (2011). Biological Wastewater Treatment. CRC Press.

Gu Z., Bao M., He C. and Chen W. (2023). Transformation of dissolved organic matter in landfill leachate during a membrane bioreactor treatment. *Science of the Total Environment*, **856**, 159066, https://doi.org/10.1016/J.SCITOTENV.2022.159066

Hassard F., Biddle J., Cartmell E., Jefferson B., Tyrrel S. and Stephenson T. (2015). Rotating biological contactors for wastewater treatment – A review. *Process Safety and Environmental Protection*, **94**, 285–306, https://doi.org/10.1016/J.PSEP.2014.07.003

Hiras D. N., Manariotis I. D. and Grigoropoulos S. G. (2004). Organic and nitrogen removal in a two-stage rotating biological contactor treating municipal wastewater. *Bioresource Technology*, **93**, 91–98, https://doi.org/10.1016/J.BIORTECH.2003.06.005

Hoilijoki T. H., Kettunen R. H. and Rintala J. A. (2000). Nitrification of anaerobically pretreated municipal landfill leachate at low temperature. *Water Research*, **34**(5), 1435–1446, https://doi.org/10.1016/S0043-1354(99)00278-X

Hu Y., He F., Ma L., Zhang Y. and Wu Z. (2016). Microbial nitrogen removal pathways in integrated vertical-flow constructed wetland systems. *Bioresource Technology*, **207**, 339–345, https://doi.org/https://doi.org/10.1016/j.biortech.2016.01.106

Hu N., He J., Shi W., He J., Lv B., Liang Y. and Huang L. (2022). Ecological restoration for the Liangtan river by rotating biological contactors combined with hybrid constructed wetlands. *Journal of Cleaner Production*, **375**, 134189, https://doi.org/10.1016/J.JCLEPRO.2022.134189

Ilmasari D., Kamyab H., Yuzir A., Riyadi F. A., Khademi T., Al-Qaim F. F., Kirpichnikova I. and Krishnan S. (2022a). A review of the biological treatment of leachate: available technologies and future requirements for the circular economy implementation. *Biochemical Engineering Journal*, **187**, 108605, https://doi.org/10.1016/J.BEJ.2022.108605

Ilmasari D., Sahabudin E., Riyadi F. A., Abdullah N. and Yuzir A. (2022b). Future trends and patterns in leachate biological treatment research from a bibliometric perspective. *Journal of Environmental Management*, **318**, 115594, https://doi.org/10.1016/J.JENVMAN.2022.115594

Iorhemen O. T., Hamza R. A. and Tay J. H. (2016). Membrane bioreactor (MBR) technology for wastewater treatment and reclamation: membrane fouling. Membranes (Basel), https://doi.org/10.3390/membranes6020033

Islam M., Wai A., Hausner G. and Yuan Q. (2019). Effect of lignocellulosic enzymes on the treatment of mature landfill leachate. *Journal of Environmental Management*, **233**, 400–409, https://doi.org/10.1016/J.JENVMAN.2018.12.045

Ismail T., Tarek D., Mejdi S., Amira B. y., Murano F., Neyla S. and Naceur J. (2011). Cascade bioreactor with submerged biofilm for aerobic treatment of Tunisian landfill leachate. *Bioresource Technology*, **102**, 7700–7706, https://doi.org/10.1016/J.BIORTECH.2011.05.069

Jafarinejad S. (2017). Recent developments in the application of sequencing batch reactor (SBR) technology for the petroleum industry wastewater treatment. *Chemistry International* **3**(3), 241.

Jokela J. P. Y., Kettunen R. H., Sormunen K. M. and Rintala J. A. (2002). Biological nitrogen removal from municipal landfill leachate: low-cost nitrification in biofilters and laboratory scale insitu denitrification. *Water Research*, **36**(16), 4079–4087, https://doi.org/10.1016/S0043-1354(02)00129-X

Kalčíková G., Babič J., Pavko A. and Gotvajn A.Ž. (2014). Fungal and enzymatic treatment of mature municipal landfill leachate. *Waste Management*, **34**, 798–803, https://doi.org/10.1016/J.WASMAN.2013.12.017

Kamaruddin M. A., Yusoff M. S., Rui L. M., Isa A. M., Zawawi M. H. and Alrozi R. (2017). An overview of municipal solid waste management and landfill leachate treatment: Malaysia and Asian perspectives. *Environmental Science and Pollution Research*, **24**(35), 26988–27020, https://doi.org/10.1007/s11356-017-0303-9

Kamaruddin M. A., Yusoff M. S., Aziz H. A. and Hung Y.-T. (2015). Sustainable treatment of landfill leachate. *Applied Water Science*, **5**, 113–126, https://doi.org/10.1007/s13201-014-0177-7

Kheradmand S., Karimi-Jashni A. and Sartaj M. (2010). Treatment of municipal landfill leachate using a combined anaerobic digester and activated sludge system. *Waste Management*, **30**, 1025–1031, https://doi.org/10.1016/J.WASMAN.2010.01.021

Kim K. R. and Owens G. (2010). Potential for enhanced phytoremediation of landfills using biosolids – a review. *Journal of Environmental Management*, **91**, 791–797, https://doi.org/10.1016/J.JENVMAN.2009.10.017

Kivaisi A. K. (2001). The potential for constructed wetlands for wastewater treatment and reuse in developing countries: a review. *Ecological Engineering*, **16**, 545–560, https://doi.org/10.1016/S0925-8574(00)00113-0

Klomjek P. and Nitisoravut S. (2005). Constructed treatment wetland: a study of eight plant species under saline conditions. *Chemosphere*, **58**, 585–593, https://doi.org/https://doi.org/10.1016/j.chemosphere.2004.08.073

Kulikowska D., Jóźwiak T., Kowal P. and Ciesielski S. (2010). Municipal landfill leachate nitrification in RBC biofilm – process efficiency and molecular analysis of microbial structure. *Bioresource Technology*, **101**, 3400 3405, https://doi.org/10.1016/J.BIORTECH.2009.12.050

Lackner S., Holmberg M., Terada A., Kingshott P. and Smets B. F. (2009). Enhancing the formation and shear resistance of nitrifying biofilms on membranes by surface modification. *Water Research*, **43**, 3469–3478, https://doi.org/10.1016/J.WATRES.2009.05.011

La Motta E. J., Silva E., Bustillos A., Padrón H. and Luque J. (2004). Combined anaerobic/aerobic treatment for municipal wastewater. *Journal of Environmental Engineering*, **133**, 4.

Lavagnolo M. C., Malagoli M., Garbo F., Pivato A. and Cossu R. (2016). Lab-scale phytotreatment of old landfill leachate using different energy crops. *Waste Management*, **55**, 265–275, https://doi.org/10.1016/J.WASMAN.2016.06.016

Lebron Y. A. R., Moreira V. R., Brasil Y. L., Silva A. F. R., Santos L. V. d. S., Lange L. C. and Amaral M. C. S. (2021). A survey on experiences in leachate treatment: common practices, differences worldwide and future perspectives. *Journal of Environmental Management*, **288**, 112475, https://doi.org/10.1016/J.JENVMAN.2021.112475

Lindamulla L. M. L. K. B., Jayawardene N. K. R. N., Wijerathne W. S. M. S. K., Othman M., Nanayakkara K. G. N., Jinadasa K. B. S. N., Herath G. B. B. and Jegatheesan V. (2022). Treatment of mature landfill leachate in tropical climate using membrane bioreactors with different configurations. *Chemosphere*, **307**, 136013, https://doi.org/10.1016/J.CHEMOSPHERE.2022.136013

Loukidou M. X. and Zouboulis A. I. (2001). Comparison of two biological treatment processes using attached-growth biomass for sanitary landfill leachate treatment. *Environmental Pollution*, **111**(2), 273–281, https://doi.org/10.1016/S0269-7491(00)00069-5

Luo Y., Guo W., Ngo H. H., Nghiem L. D., Hai F. I., Zhang J. and … Wang X. C. (2014). A review on the occurrence of micropollutants in the aquatic environment and their fate and removal during wastewater treatment. *Science of the Total Environment*, **473–474**, 619–641, https://doi.org/10.1016/j.scitotenv.2013.12.065

Luo H., Zeng Y., Cheng Y., He D. and Pan X. (2020). Recent advances in municipal landfill leachate: a review focusing on its characteristics, treatment, and toxicity assessment. *Science of the Total Environment*, **703**, 135468, https://doi.org/10.1016/J.SCITOTENV.2019.135468

Maheepala S. S., Fuchigami S., Hatamoto M., Akaohi T., Watari T. and Yamaguchi T. (2022). Stable denitrification performance of a mesh rotating biological reactor treating municipal wastewater. *Environmental Technology & Innovation*, **27**, 102543, https://doi.org/10.1016/J.ETI.2022.102543

Mahvi A. H. (2008). Sequencing batch reactor: a promising technology in wastewater treatment. *Iranian Journal of Environmental Health Science and Engineering*, **5**, 79–90.

Maiti S., De S., Hazra T., Debsarkar A. and Dutta A. (2016). Characterisation of leachate and its impact on surface and groundwater quality of a closed dumpsite–a case study at Dhapa, Kolkata, India. *Procedia Environmental Sciences*, **35**, 391–399, https://doi.org/10.1016/j.proenv.2016.07.019

Mara D. (2013). Domestic Wastewater Treatment in Developing Countries, 1st ed. Routledge. ISBN 1-84407-020-4, https://doi.org/https://doi.org/10.4324/9781849771023

Mbuligwe S. E. (2005). Comparative treatment of dye-rich wastewater in engineered wetland systems (EWSs) vegetated with different plants. *Water Research*, **39**, 271–280, https://doi.org/https://doi.org/10.1016/j.watres.2004.09.022

Mehmood M. K., Adetutu E., Nedwell D. B. and Ball A. S. (2009). In situ microbial treatment of landfill leachate using aerated lagoons. *Bioresource Technology*, **100**, 2741–2744, https://doi.org/10.1016/J.BIORTECH.2008.11.031

Miao L., Yang J., Tao T., Peng Y. (2019). Recent advances in nitrogen removal from landfill leachate using biological treatments – a review. *Journal of Environmental Management*, **235**, 178–185, https://doi.org/10.1016/j.jenvman.2019.01.057.

Nagendran R., Selvam A., Joseph K. and Chiemchaisri C. (2006). Phytoremediation and rehabilitation of municipal solid waste landfills and dumpsites: a brief review. *Waste Management*, **26**, 1357–1369, https://doi.org/10.1016/J.WASMAN.2006.05.003

Naveen B. P., Mahapatra D. M., Sitharam T. G., Sivapullaiah P. V. and Ramachandra T. V. (2017). Physico-chemical and biological characterization of urban municipal landfill leachate. *Environmental Pollution*, **220**, 1–12, https://doi.org/10.1016/j.envpol.2016.09.002

Neczaj E., Okoniewska E. and Kacprzak M. (2005). Treatment of landfill leachate by sequencing batch reactor. *Desalination*, **185**, 357–362, https://doi.org/10.1016/J.DESAL.2005.04.044

O'Connor G. and Courtney R. (2020). Constructed wetlands for the treatment of bauxite residue leachate: long term field evidence and implications for management. *Ecological Engineering*, **158**, 106076, https://doi.org/10.1016/J.ECOLENG.2020.106076

Oller I., Malato S. and Sánchez-Pérez J. (2011). Combination of advanced oxidation processes and biological treatments for wastewater decontamination – a review. *Science of the Total Environment*, **409**(20), 4141–4166, https://doi.org/10.1016/j.scitotenv.2010.08.061

Onwosi C. O., Igbokwe V. C., Odimba J. N., Eke I. E., Nwankwoala M. O., Iroh I. N. and Ezeogu L. I. (2017). Composting technology in waste stabilization: on the methods, challenges and future prospects. *Journal of Environmental Management*, **190**, 140–157, https://doi.org/10.1016/j.jenvman.2016.12.051

Pant D., Van Bogaert G., Diels L. and Vanbroekhoven K. (2010). A review of the substrates used in microbial fuel cells (MFCs) for sustainable energy production. *Bioresource Technology*, **101**(6), 1533–1543, https://doi.org/10.1016/j.biortech.2009.10.017

Pechaud Y., Peyre Lavigne M., Bessiere Y., Ochoa J. C., Queinnec I. and Paul E. (2022). Influence of shear stress, organic loading rate and hydraulic retention time on the biofilm structure and on the competition between different biological aggregate morphotypes. *Journal of Environmental Chemical Engineering*, **10**, 107597, https://doi.org/10.1016/J.JECE.2022.107597

Qureshi N., Annous B. A., Ezeji T. C., Karcher P. and Maddox I. S. (2005). Biofilm reactors for industrial bioconversion processes: employing potential of enhanced reaction rates. *Microbial Cell Factories*, **4**, 24, https://doi.org/10.1186/1475-2859-4-24

Reis B. G., Silveira A. L., Tostes Teixeira L. P., Okuma A. A., Lange L. C. and Amaral M. C. S. (2017). Organic compounds removal and toxicity reduction of landfill leachate by commercial bakers' yeast and conventional bacteria based membrane bioreactor integrated with nanofiltration. *Waste Management*, **70**, 170–180, https://doi.org/https://doi.org/10.1016/j.wasman.2017.09.030

Remmas N., Melidis P., Zerva I., Kristoffersen J. B., Nikolaki S., Tsiamis G. and Ntougias S. (2017). Dominance of candidate *Saccharibacteria* in a membrane bioreactor treating medium age landfill leachate: effects of organic load on microbial communities, hydrolytic potential and extracellular polymeric substances. *Bioresource Technology*, **238**, 48–56, https://doi.org/10.1016/J.BIORTECH.2017.04.019

Renou S., Givaudan J. G., Poulain S., Dirassouyan F. and Moulin P. (2008). Landfill leachate treatment: review and opportunity. *Journal of Hazardous Materials*, **150**(3), 468–493, https://doi.org/10.1016/j.jhazmat.2007.09.077

Ritzkowski M., Heyer K.-U. and Stegmann R. (2006). Fundamental processes and implications during in situ aeration of old landfills. *Waste Management*, **26**, 356–372, https://doi.org/10.1016/j.wasman.2005.11.009

Robinson H. D. and Grantham G. (1988). The treatment of landfill leachates in on-site aerated lagoon plants: experience in Britain and Ireland. *Water Research*, **22**, 733–747, https://doi.org/10.1016/0043-1354(88)90184-4

Sadri S., Cicek N. and Van Gulck J. (2008). Aerobic treatment of landfill leachate using a submerged membrane bioreactor – prospects for on-site use. *Environmental Technology*, **29**, 899–907, https://doi.org/10.1080/09593330802075379

Salami L., Fadayini O., Patinvoh R. J. and Koleola O. (2015). Evaluation of leachate contamination potential of Lagos dumpsites using leachate pollution index. Current Journal of Applied Science and Technology, 48–59.

Sanchez-Huerta C., Fortunato L., Leiknes T. and Hong P. Y. (2022). Influence of biofilm thickness on the removal of thirteen different organic micropollutants via a membrane aerated biofilm reactor (MABR). *Journal of Hazardous Materials*, **432**, 128698, https://doi.org/10.1016/J.JHAZMAT.2022.128698

Saviour M. N. (2012). Environmental impact of soil and sand mining: a review. *International Journal of Science, Environment and Technology*, **1**(3), 125–134.

Sheng G. P., Yu H. Q. and Li X. Y. (2010). Extracellular polymeric substances (EPS) of microbial aggregates in biological wastewater treatment systems: a review. *Biotechnology Advances*, **28**(6), 882–894, https://doi.org/10.1016/j.biotechadv.2010.08.001

Shi L., Zhang Q. and Niu D. J. (2007). Microbiological characteristics of aged-refuse based bioreactor for landfill leachate treatment. *Journal of Tongji University (Natural Science)*, **35**(8), 1085–1089.

Siddiqi S. A., Al-Mamun A., Baawain M. S. and Sana A. (2022). A critical review of the recently developed laboratory-scale municipal solid waste landfill leachate treatment technologies. *Sustainable Energy Technologies and Assessments*, **52**, 102011, https://doi.org/10.1016/J.SETA.2022.102011

Sil A. and Kumar S. (2017). Landfill leachate treatment. In: Current developments in biotechnology and bioengineering, pp. 391–406.

Singh M. and Srivastava R. K. (2011). Sequencing batch reactor technology for biological wastewater treatment: a review. *Asia-Pacific Journal of Chemical Engineering*, **6**(1), 3–13, https://doi.org/10.1002/apj.490

Spengel D. B. and Dzombak D. A. (1991). Treatment of landfill leachate with rotating biological contactors: bench-scale experiments. *Research Journal of the Water Pollution Control Federation*, **63**, 971–981.

Spina F., Tigini V., Romagnolo A. and Varese G. C. (2018). Bioremediation of landfill leachate with fungi: autochthonous vs. allochthonous strains. *Life (Basel, Switzerland)*, **8**, https://doi.org/10.3390/life8030027

Stottmeister U., Wießner A., Kuschk P., Kappelmeyer U., Kästner M., Bederski O., Müller R. A. and Moormann H. (2003). Effects of plants and microorganisms in constructed wetlands for wastewater treatment. *Biotechnology Advances*, **22**, 93–117, https://doi.org/https://doi.org/10.1016/j.biotechadv.2003.08.010

Taşkan B., Hasar H. and Lee C.-H. (2020). Effective biofilm control in a membrane biofilm reactor using a quenching bacterium (*Rhodococcus* sp. BH4). *Biotechnology and Bioengineering*, **117**, 1012–1023, https://doi.org/https://doi.org/10.1002/bit.27259

Vij D. (2012). Urbanization and solid waste management in India: present practices and future challenges. *Procedia – Social and Behavioral Sciences*, **37**, 437–447, https://doi.org/10.1016/j.sbspro.2012.03.309

Vymazal J. (2007). Removal of nutrients in various types of constructed wetlands. *Science of the Total Environment*, **380**, 48–65, https://doi.org/10.1016/J.SCITOTENV.2006.09.014

Wang J. H., Zhao X. L., Hu Q., Gao X., Qu B., Cheng Y., Feng D., Shi L. F., Chen W. H., Shen Y. and Chen Y. P. (2022). Effects mechanism of bio-carrier filling rate on rotating biofilms and the reactor performance optimization method. *Chemosphere*, **308**, 136176, https://doi.org/10.1016/J.CHEMOSPHERE.2022.136176

Wanner J. (2014). Activated Sludge: Bulking and Foaming Control. CRC Press.

Wdowczyk A., Szymańska-Pulikowska A. and Gałka B. (2022). Removal of selected pollutants from landfill leachate in constructed wetlands with different filling. *Bioresource Technology*, **353**, 127136, https://doi.org/10.1016/J.BIORTECH.2022.127136

Werkneh A. A. (2022). Application of membrane-aerated biofilm reactor in removing water and wastewater pollutants: current advances, knowledge gaps and research needs – a review. *Environmental Challenges*, **8**, 100529, https://doi.org/10.1016/J.ENVC.2022.100529

Wichitsathian B., Sindhuja S., Visvanathan C. and Ahn K. H. (2004). Landfill leachate treatment by yeast and bacteria based membrane bioreactors. *Journal of Environmental Science and Health Part A, Toxic/Hazardous Substances and Environmental Engineering*, **39**, 2391–2404, https://doi.org/10.1081/ese-200026295

Wiszniowski J., Robert D., Surmacz-Gorska J., Miksch K. and Weber J. V. (2006). Landfill leachate treatment methods: a review. *Environmental Chemistry Letters*, **4**(1), 51–61. 96, https://doi.org/10.1007/s10311-005-0016-z

Wu Z., Qiao W. and Dong R. (2023). Elucidating methanogenesis and membrane filtration performance of a ceramic AnMBR treating municipal waste leachate. *Journal of Water Process Engineering*, **53**, 103905, https://doi.org/10.1016/j.jwpe.2023.103905

Yalcuk A. and Ugurlu A. (2009). Comparison of horizontal and vertical constructed wetland systems for landfill leachate treatment. *Bioresource Technology*, **100**(9), 2521–2526, https://doi.org/10.1016/j.biortech.2008.11.029

Yong Z. J., Bashir M. J. K., Ng C. A., Sethupathi S. and Lim J. W. (2018). A sequential treatment of intermediate tropical landfill leachate using a sequencing batch reactor (SBR) and coagulation. *Journal of Environmental Management*, **205**, 244–252, https://doi.org/10.1016/j.jenvman.2017.09.068

Zegzouti Y., Boutafda A., Ezzariai A., El Fels L., El Hadek M., Hassani L. A. I. and Hafidi M. (2020). Bioremediation of landfill leachate by *Aspergillus flavus* in submerged culture: evaluation of the process efficiency by physicochemical methods and 3D fluorescence spectroscopy. *Journal of Environmental Management*, **255**, 109821, https://doi.org/10.1016/J.JENVMAN.2019.109821

Zhang J., Xiao K. and Huang X. (2020). Full-scale MBR applications for leachate treatment in China: practical, technical, and economic features. *Journal of Hazardous Materials*, **389**, 122138, https://doi.org/10.1016/J.JHAZMAT.2020.122138

Zhao Y. C., Chai X. and Niu D. J. (2006). Characteristics of aged refuse in closed refuse landfill in Shanghai. *Journal of Tongji University (Natural Science)*, **34**(10), 1361–1363.

doi: 10.2166/9781789063318_0037

Chapter 3

Anaerobic treatment of landfill leachate

Ahmed Tawfik[1] and Vinay Kumar Tyagi[2]*

[1]Water Pollution Research Department, National Research Centre, 12622 Dokki, Cairo, Egypt
[2]Environmental Hydrology Division, National Institute of Hydrology, Roorkee 247667, India
**Corresponding author: prof.tawfik.nrc@gmail.com*

ABSTRACT

Anaerobic technologies are the most suitable treatment processes of the landfill leachate, and offer advantages, that is, application of high organic loading rates (OLRs), low operational costs, and biogas productivity. However, the efficiency of anaerobic reactors are highly dependent on the temperature, OLR, ammonia loading rate, sludge loading rate, and reactors configurations. Those factors affect the microbial activities, hydrolysis of leachate and enzyme activities. Moreover, the anaerobic reactors are subjected to biomass washout that impairs the efficiency of the technology. Therefore, intentional sludge discharge is necessary to control the food-to-substrate ratio that highly affects the performance of anaerobic process. The removal of organics is quite high (>90% of BOD) from the leachate by anaerobic digesters where low requirement dose of phosphorous are needed for anaerobes growing, low excess sludge productivity, low energy consumption, and usage. Moreover, the technology produces hydrogen and methane from leachate. Nevertheless, the presence of high concentrations of ammonia in the leachate could highly inhibit the anaerobes and leads to process failure. Therefore, two-stage anaerobic reactors are recommended at low OLR to achieve a good effluent quality complying for discharge and/or reuse.

Keywords: Ammonia toxicity, anaerobic technologies, landfill leachate, operation mode.

3.1 INTRODUCTION

Municipal solid waste is produced in abundance everywhere due to human activities creating environmental damage (Elsamadony & Tawfik, 2018; Farghaly *et al.*, 2017) (Figure 3.1). This waste is mainly organics and inorganics and their microbial biodegradability is acceptable (Tawfik *et al.*, 2013; Tawfik

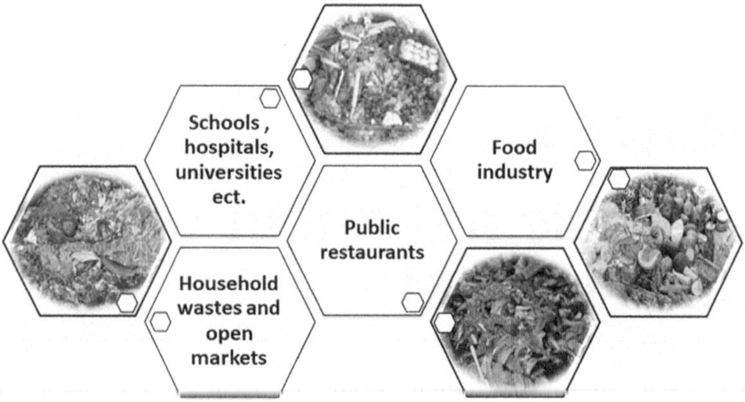

Figure 3.1 Municipal solid waste sources and generation.

& Salem, 2012). Landfilling of municipal solid waste is most common in developing countries (Mahmoud *et al.*, 2018). The process is highly considered the most economical and environmentally acceptable technique all over the world (Ismail *et al.*, 2020; Meky *et al.*, 2019, 2017). However, landfill of wastes generates leachate-rich organics and inorganics due to percolation of rainfall resulting in high strength slurry containing biological and chemical by-products in waste's cells (Figure 3.2). The leachate containing biodegradable and non-biodegradable organics causes serious environmental, health and economic problems due to contamination of surrounding groundwater and soil (Ismail & Tawfik, 2016a; Li *et al.*, 2019). Therefore, treatment of landfill leachate is certainly necessary to avoid the ecosystem failure and overcome the adverse impacts of such dangerous leachate (Meky *et al.*, 2019). Fortunately, landfill leachate contains a huge chunk of organics, ammonia-nitrogen, suspended and dissolved organic matter and high fraction of refractory molecules (El-Bery *et al.*, 2013; Khodary *et al.*, 2018; Meky *et al.*, 2017).

Several biological treatment processes have been proposed for the treatment of landfill leachates, that is, anaerobic and/or aerobic biological degradation

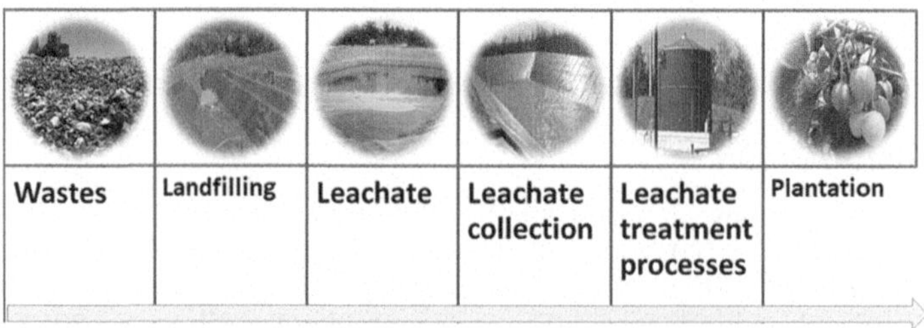

Figure 3.2 Proper landfill leachate management.

(Ismail & Tawfik, 2016a, 2016b), chemical and photo-chemical oxidation (Deng & Englehardt, 2007; El-Gohary *et al.*, 2009; Gar Alalm *et al.*, 2015; Kurniawan *et al.*, 2006; Tauchert *et al.*, 2006), adsorption (Foo & Hameed, 2009), coagulation–precipitation (Vedrenne *et al.*, 2012) and combination of those technologies (Vedrenne *et al.*, 2012). Consequently, an appropriate and low-cost treatment technique of heavily polluted landfill leachate is essential to remove the majority of pollutants.

Biological treatment methods are usually preferred over physicochemical ones due to its high efficiency and do not consume chemicals and intensive energy. However, application of biological treatment alone is not an appropriate process due to the existence of the micro pollutants, refractory compounds, and so on in the leachate. Neither biological process nor chemical treatment method separately achieves high treatment efficiencies of the leachate particularly high variables of contaminants in the slurry stream. There are two reasons for the low removal efficiency of each treatment system: (1) significant presence of high-molecular weight organics that are difficult to remove and (2) inhibitory effects of organics, inorganic salts and heavy metals to activated sludge microorganisms.

Anaerobic treatment of landfill leachate is still the most acceptable technology due to the produced intensive energy and low consumption of chemicals and energies. The anaerobes hydrolyzed the macro-organics into simple amino acids and simple sugars. The acidogenic bacteria convert such hydrolyzed by-products into fatty acids and hydrogen energy. The fast-growing methanogens are converted to short-chain fatty acids and hydrogen into methane via methanogenesis. This process depends on the leachate composition, characteristics, and organic and nutrient contents.

The main objective of this chapter is to emphasize biodegradability of landfill leachate and anaerobic system used for conversion of this leachate into energy in the form of hydrogen and methane.

3.2 LEACHATE GENERATION FROM LANDFILLING OF MUNICIPAL SOLID WASTES AND THEIR CHARACTERISTICS

Leachates are defined as the aqueous slurry rich with organic and inorganic contaminants generated due to dumping of wastes that diluted with natural rainwater (Galeano *et al.*, 2011). The leachate is complex liquid containing high organic, nitrogen, inorganic non-refractory and refractory compounds. The presence of water in the landfill facilitates a good opportunity for combination of physical, chemical, and microbial activities to hydrolyze the solid waste content into the liquid namely leachate-rich high concentrations of ammonia, nitrogen compounds, soluble organic matter, and inorganic ions (Calli *et al.*, 2005a; Li *et al.*, 1999). The major parameter affecting the formation of leachate from landfill is the availability of water (i.e., rainfall, moisture content, etc.) and water distribution in the landfill site (refuse, compaction, age, etc.). The water content in the landfill facilitates the formation of leachate due to an increase in microbial biodegradation. Leachate quantity (volumetric flow rate) and quality (composition) are the main characteristics required to select the proper treatment process and leachate management. The leachate flow rate from landfill sites highly fluctuates from site to another site and seasonally varies at each site. The volumetric flow rate of the leachate is substantially influenced by waste landfilling methods, that is, geotextiles and/or plastics, liner requirements (clay), and waterproof covers, which remain processes to control the water quantity entering the landfill tip and hence decreasing the resultant liquefied waste (Lema *et al.*, 1988).

The local climate conditions have also a great influence on leachate quantity and quality because it affects the precipitation input and liquid losses via evaporation. Finally, leachates from landfill also depend on the liquefaction process, type, nature, quantity, water content, degree of compaction, and composition of the solid waste landfilling (Renou *et al.*, 2008). Furthermore, the leachate composition depends on seasonal rainfall, soil and waste type, and compaction degree and landfill age. The age of landfill reflects stabilization degree of the discharged waste (Kjeldsen *et al.*, 2002). The interaction of these factors produces highly variable leachate compositions. Organic fractions of the discharged wastes in the landfill undergo partial and/or complete anaerobic degradation resulting in leachate containing intermediate by-products with hardly biodegradable soluble pollutants (Mahmud *et al.*, 2012). Landfill leachate is usually characterized by high organics content with large molecules (humic and fulvic acids), inorganic salts, high ammonia-nitrogen, heavy metals and offensive odor (Altin, 2008).

The leachate from landfilling of municipal solid waste are rich in chemical oxygen demand (COD), biochemical oxygen demand (BOD), the BOD/COD ratio, color, pH, alkalinity, oxidation–reduction potential, ammonia nitrogen (NH_4-N), total Kjeldahl nitrogen (TKN), solids (TSS and VSS), and heavy metals. The data presented by Umar *et al.* (2010) show a wide variation of the leachate composition from sanitary landfills of wastes. These results show the high fluctuation of leachate composition resulting from the landfills. Average COD concentrations were 23,878, 5,249, and 2,125 mg/L for young, intermediate, and old landfills respectively, varying from 100 to 70,900 mg/L due to the age of the landfill and the degree of solid waste stabilization. With few exceptions, the pH of leachates lies between 5.8 and 8.5. This can be explained by the biological and microbial degradation activity inside the landfill (Yoon *et al.*, 1998). It is also important to note that the majority of TKN is ammonia-N, than can vary from 0.2 to 13,000 mg/L as NH_3-N. Based on the BOD/COD ratio, landfill leachate is classified into three biodegradation characteristic phases: acid degradation phase (BOD/COD $>$ 0.4); transient transformation phase (0.2 $<$ BOD/COD $<$ 0.4) and the methanogenic conversion phase (BOD/COD $\leq$ 0.2) (Nakashimada *et al.*, 2008). The ratio of BOD/COD highly varies from 0.04 to 0.70 in the older landfills that have the lower values (Lin *et al.*, 2011) due to the release of the large recalcitrant organic molecules from the landfilling of solid waste. Consequently, old landfill leachate that has low BOD/COD ratio is defined as stable methanogen conversion phase and contains hardly biodegradable organics.

3.3 BIOLOGICAL TREATMENT METHODS

Biological treatment processes are mainly proposed to treat the dissolved and colloidal organics (Tawfik *et al.*, 2008). The main goal of the biological treatment processes are to coagulate, agglomerate, remove, or reduce the non-settling organics solids and the dissolved organic load from the leachate using various microbial communities that have the capability to degrade such contaminants via biochemical reaction pathways (Kettunen *et al.*, 1996; Smith, 1995). As a result of its reliability, simplicity, and high cost-effectiveness, biological treatment (suspended/attached growth) is commonly used for reducing the organic content of leachate, particularly when complete onsite treatment is required. Biodegradation is carried out by microorganisms as shown in equation (3.1), which can degrade organics into carbon dioxide and sludge under aerobic conditions and into biogas (CO_2 and CH_4 or H_2) under anaerobic conditions (Loukidou & Zouboulis, 2001; Morgan, 1990). Biological processes were proved to be very effective in removing organics and nitrogenous matter from immature leachates when the BOD/COD ratio exceeding 0.5. However, the presence of refractory compounds (humic and fulvic acids) tends to limit the effectiveness of biological treatment processes.

$$\text{Organic matter} + O_2 + NH_4^{2+} + P \rightarrow \text{New cells} + CO_2 + H_2O \quad (3.1)$$

The performance of biological treatment processes to treat the leachates is highly influenced by hydraulic, organic loadings, and presence of toxic compounds (Smith, 1995).

3.4 ANAEROBIC TREATMENT PROCESS

Anaerobic process is the oldest technology ultimately used for wastewater treatment that was started since the end of nineteenth century. Anaerobic treatment process involves microbial conversion of organic matter and uses inorganic elements such as N, P, S, K, Ca, and Mg for buildup of microorganisms. These microbial anaerobes are active in the absence of molecular oxygen. The anaerobes has the capability to transform the leachate into hydrogen (H_2) and carbon dioxide (CO_2) by acidogenesis process and methane (CH_4) by methanogenesis process (Tawfik & ElBatrawy, 2012) (Figure 3.3).

Various factors would affect the anaerobic technologies treating landfill leachate, that is, temperature, organic loading rate (OLR), ammonia loading rate, sludge loading rate, and hydraulic retention time (HRT). These factors affect the microbial activities, hydrolysis of substrate, and enzymes activities (Sosnowski *et al.*, 2003; Termorshuizen *et al.*, 2003). Most of the anaerobic reactors are subjected to biomass washout that impairs the efficiency of the technology. Therefore, intentional sludge discharge is necessary to control the food-to-substrate ratio that highly affects the performance of anaerobic processes.

The main advantages of anaerobic technology are presented in Figure 3.4. The removal of organics are quite high (>90% of BOD) by anaerobic digesters, low requirement dose of phosphorous are needed for anaerobes growing, low

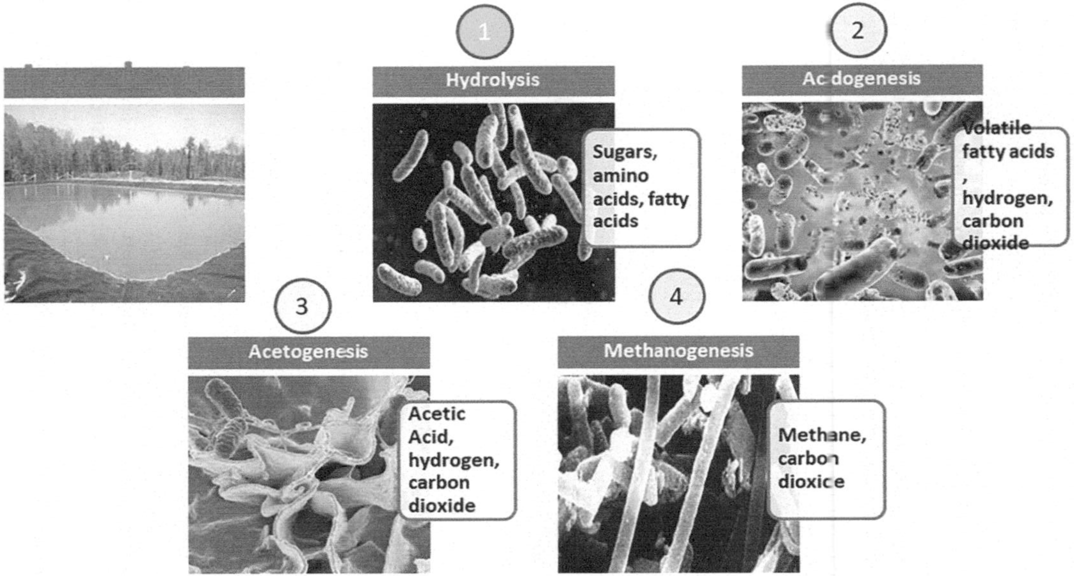

Figure 3.3 Anaerobic degradation pathway of the leachate containing organics.

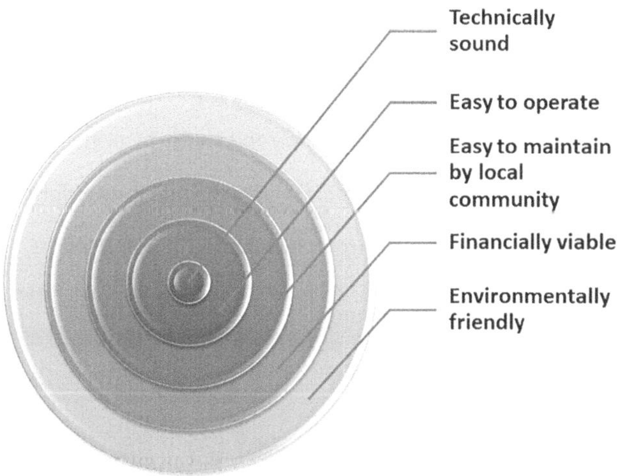

Figure 3.4 Advantages of anaerobic degradation of the leachate containing organics.

excess sludge productivity, low energy consumption and usage. Moreover, the technology produces hydrogen and methane from leachate. However, the presence of high concentrations of ammonia and heavy metals in the leachate could highly inhibit the anaerobes and leads to process failure.

Usually, an anaerobic reactor has commonly been installed in leachate treatment process for treating high loading organic compounds discharged from young landfill sites (Im *et al.*, 2001). Different reactors can be used, such as anaerobic filters (Henry *et al.*, 1987), anaerobic sequencing batch reactor (Timur, 1999), or up-flow anaerobic sludge blanket (UASB) (Kennedy & Lentz, 2000).

3.4.1 Anaerobic technologies

Anaerobic processes are efficient for organics conversion into energy. However, the efficiency of the anaerobic digesters for treatment of landfill leachate depends on its configuration (Figure 3.5). Anaerobic filters, anaerobic sequencing batch reactors, UASBs, anaerobic submerged membrane bioreactors (MBRs) and anaerobic moving-bed biofilm reactors (MBBRs) were employed for treatment of landfill leachate with emphasis on the energy productivity and produced an effluent quality for reuse.

3.4.1.1 Anaerobic filters

The anaerobic filter depends on a bed of packing material to reduce washout of biomass from the reactor at short HRT. This system is a high-rate technology that gathers the advantages of other anaerobic methods and minimizes the disadvantages. The leachate may pass through the anaerobic filter by either up-flow or down-flow mode. Although the up-flow mode appears to be more common and preferable. Even distribution of the leachate across

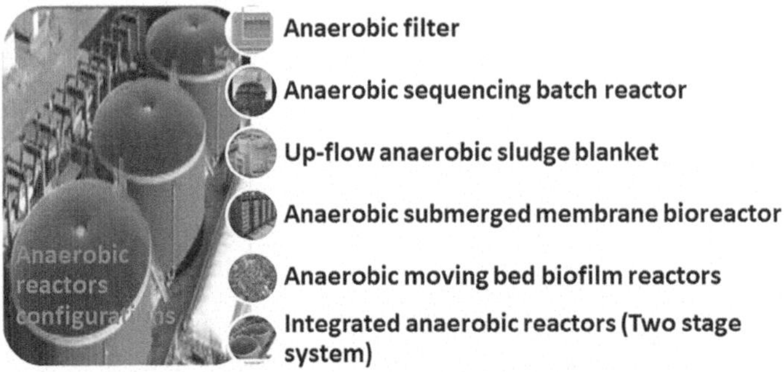

Figure 3.5 Anaerobic treatment technologies treating landfill leachate.

the cross-sectional face of the support media is important to reduce risk of channeling and hence a deterioration of process performance. Henry *et al.* (1987) investigated the organic removal from landfill leachates by up-flow anaerobic filter. Landfill leachates were collected from two different sites. The strong raw leachate from the new landfill had a COD of 14,000 mg/L, a BOD/COD ratio of 0.7 and a COD/P value of 17,900. The partially stabilized leachate from the older landfill had a COD of only 3,750 mg/L, a BOD/COD ratio of 0.3 and a COD/P value of 30,640. The results revealed that the anaerobic filter could reduce the COD of the leachate from landfills of different ages by 90%, at OLRs of 1.26–1.45 kg COD/m^3 d. The total biogas production ranged between 400 and 500 L gas/kg COD and methane content was between 75% and 85%. No addition of phosphorus was required over the loading range studied (Henry *et al.*, 1987).

3.4.1.2 Anaerobic sequencing batch reactor

The anaerobic sequencing batch reactor (AnSBR) is very similar in concept to the aerobic one except the absence of aeration. AnSBR has the capability to achieve solids capture and transform organic matter in one vessel into biogas, eliminating the need for a clarifier. AnSBR has five operating steps: fill, react, settle, draw, and idle process. Timur and Ozturk (1999) studied the anaerobic treatability of municipal landfill leachate using a lab-scale AnSBR. They found that the COD removal efficiencies were in the range of 64–85% depending on volumetric and specific imposed loading rates. The results showed that about 83% of COD were removed during the treatment process and were converted into biomethane.

3.4.1.3 Up-flow anaerobic sludge blanket

The UASB module is considered a breakthrough in the process development and application of high-rate treatment technology. In addition, the UASB offers unique design where liquid, gas, and solid phases can be separated within one vessel. Because of the easy operation, minimal sludge productivity, and high energy efficiencies, the UASB reactor is taken into consideration for leachate treatment. Additionally, during the operation process, methane and/or hydrogen is produced and could be used as fuel. However, pre-treatment of landfill leachate for solids removal might be needed to improve the reactor performance. It is necessary to combine post-treatment with UASB reactor to produce an effluent quality complying for discharge (Tawfik *et al.*, 2006).

Kennedy and Lentz (2000) investigated the treatment of landfill leachate using continuous UASB reactor. Their research study demonstrated the high ability of UASB reactor for the treatment landfill leachate. The continuous UASB reactor achieved a soluble COD removal efficiencies between 77% and 91% at all HRTs of 24, 18, and 12 h and OLRs between 0.6 and 19.7 kg COD/m^3 d. Also, Kettunen and Rintala (1998) demonstrated that leachate can be treated on-site UASB reactor at low temperature. A pilot-scale UASB reactor was used to treat the landfill leachate containing COD of 1.5–3.2 g/L at temperature of 13–23°C. The removal efficiencies by the reactor were 65–75% and up to 95% for COD and BOD_7, respectively at OLRs of 2–4 kg COD/ m^3 d. García *et al.* (1996) concluded that COD rejection efficiency was not affected by temperature between 15 and 35°C. These promising results by UASB showed that high-rate treatment at low temperature could minimize the need for heating the leachate thereby providing an interesting cost-effective option. The main disadvantages of such a treatment are that it stays sensitive to toxic substances such as ammonia and heavy metals (Timur & Özturk, 1997).

3.4.1.4 Anaerobic submerged membrane bioreactor

Recently, membrane technology has obtained a great potential for application in environmental issues particularly landfill leachate treatment (Tawfik *et al.*, 2022a). Membrane technology is efficient for solids separation and produces good water quality complying for discharge standards. A submerged membrane reactor is the most popular process for landfill leachate treatment where the technology is efficient, compact and an energy saving system. The submerged MBR is similar to activated sludge process and/or sequential aerobic batch reactor where the suspended growth bacteria oxidize organics and ammonia into CO_2 and NO_3-N. The submerged membrane separates the bacteria (biomass) from water and keeps high sludge residence time for accomplishment of the biological process. Membrane unit replaces the classical secondary clarifier in the activated sludge process and has some biological metabolism activities. MBRs can be either ultrafiltration or microfiltration unit to reduce the sludge productivity and cells residence time for biological metabolism activities. The diluted leachate with wastewater 5–75% (v/v) was treated by anaerobic submerged membrane bioreactor (Bohdziewicz *et al.*, 2008). Ninety-five percent of the COD was removed by anaerobic submerged membrane bioreactor at dilution ratios of 10% and 20% (v/v). Addition of 25% of leachate to wastewater reduced the COD removal efficiency to 80%. A significant drop in COD removal occurred with addition of leachate to the wastewater by values exceeding 30% due to the microbiological activity inhibition. This indicates that membrane technology is efficient for treatment of diluted landfill leachate and is not recommended for application in the raw or native landfill leachate. However, the technology could be applicable at long HRT and low imposed OLR. Anaerobic membrane bioreactor treating landfill leachate achieved COD removal of 62.2% at ammonia concentration of 3,000 mg/L, COD of 13,000 mg/L and OLR of 4.87 kg COD/m^3 d (Xie *et al.*, 2014). The methane content was 70– 90%, and yield of 0.34 L/g COD was removed. The archaeal microbial community was almost stable while bacterial community was significantly varied, that is, *Alkaliphilus, Petrimonas, Fastidiosipila* and *vadinBC27.* Moreover, phylum TM6 was the most dominant bacterial community at a level of 32.9%. *Methanosarcina* was the dominant Archaea genus.

3.4.1.5 Anaerobic moving-bed biofilm reactors

MBBR is an effective treatment process. The biological process is a typical activated sludge technology but in MBBR the bacteria are attached to the carrier materials to minimize the limitations of suspended growth systems (Tawfik *et al.*, 2012, 2010). The MBBR is the combination of both fluidized-bed reactor and activated sludge process. The process is a completely mixed operational mode. The biomass (biofilm) is attached onto the carrier and carries out their biological and metabolism processes. The carrier materials should not exceed 50% of the total reactor volume and has a lower density than water to be kept in a movement mode in the reactor. The anaerobic MBBR is supplied with mechanical mixer to increase the contact between the substrate and bacteria. The MBBR possesses high biomass retention, high organic loading, tolerance to shock loading, compact reactor, and no sludge washout problems. Anaerobic MBBR system was successfully employed for treatment of landfill leachate by Chen *et al.* (2008). The module achieved COD removal of 91% at OLR of 4.08 kgCOD/m^3 d that was slightly reduced to 86% at OLR of 15.7 kgCOD/m^3 d. Furthermore, the excess sludge yield from the MBBR reactor was only 0.0538 gVS/gCOD$_{removed}$.

3.4.1.6 Integrated anaerobic reactors (two-stage system)

The landfill leachate containing micro- and macro pollutants needs integrated process to accomplish the treatment process and reduces the most dangerous pollutants (Ahmad *et al.*, 2021; Tawfik *et al.*, 2022b). Fortunately, combinations of different anaerobic configuration system, that is, anaerobic filter and membrane bioreactor would offer a unique process for treatment of hazardous landfill leachate (Ismail & Tawfik, 2016a, 2016b). Yiping *et al.* (2008) investigated integrated system consisting of anaerobic filter and membrane bioreactor for treatment of hazardous landfill leachate. The anaerobic filter removed 4-nitrophenol (NP), organochlorine pesticides (OCPs) and polycyclic aromatic hydrocarbons (PAHs), COD, TOC and BOD. The combined anaerobic filter–MBR technology achieved better removal, that is, 99% for BOD, 89% for COD, 87% for TOC, 94% for OCPs, 77% for 4-NP and 59% for PAHs. Ağdağ and Sponza (2005) achieved COD removal of 80% using two-stage UASB reactors treating landfill leachate.

3.5 FACTORS AFFECTING ANAEROBIC DEGRADATION OF LANDFILL LEACHATE

The anaerobic treatment processes are effective for bioenergy productivity from landfill leachate. However, the biodegradation of organics by anaerobes are highly affected by several factors, that is, temperature, OLR, HRT, and substrate concentration (Figure 3.6).

3.5.1 Effect of temperature

The temperature could affect the anaerobic degradation of landfill leachate. The enzymatic and microbial degradation activities of the leachate could be highly affected by temperature. Anaerobic degradation of leachate was feasible at low temperature of 11°C (Kettunen *et al.*, 1996) where 60% COD was removed by UASB and the hybrid reactor at loading rate of 0.7–1.5 kg COD/m^3 d and HRT of 1.5–2 days. This removal efficiency was increased up to 75% by the same reactors at higher temperature of 24°C and OLR of 10 kg COD/m^3 d. Increasing the temperature from 11 to 24 improved the COD removal efficiency and reduced the HRT from 1.5 to 2 days to 10 h. Anaerobic lagoon treating landfill leachate at temperature of 10°C and HRT of 160 days achieved COD removal efficiency of 80% at initial COD concentration of 13,700 mg/L (Zaloum & Abbott, 1997). Only 22% removal efficiency of COD was achieved in a digester from leachate containing COD 8,300 mg/L at HRT of 12.5 days HRT and OLR of 0.67 kg COD/m^3 d (Boyle & Ham, 1974). The COD removal efficiency from leachate by anaerobic digester was 60–80% at temperature of 20–25°C (Henry *et al.*, 1987). However, the OLR was quite high at 3.0 kg COD/m^3 d. Similar COD removal efficiency by anaerobic digestion of leachate was

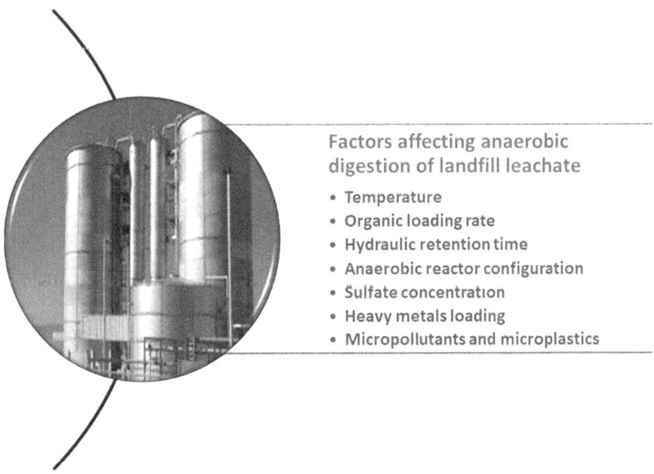

Figure 3.6 Factors affecting anaerobic degradation of landfill leachate.

obtained at temperature of 30–35°C (Ismail & Tawfik, 2016b). This indicates that at low temperature, low OLR is preferable to achieve a high anaerobic degradation of organics in the leachate. Increasing the OLR up to 10.0 kgCOD/m^3 d required a high operational temperature to enhance the microbial degradation activities. Moreover, the short HRT is required for treatment of leachate at high temperatures of 24–35°C.

3.5.2 Effect of reactor operation mode

The operation mode of the anaerobic digesters treating highly fluctuated landfill leachate is important. Completely mixed semi-continuous anaerobic digester treating leachate was investigated by Lin (1991). They found that the digester removed 92% of COD at initial concentration of 22,750 mg COD/L, OLR of 1.1 ± 2.8 kg COD/m^3 d and HRT of 8.0 days. Kennedy *et al.* (1988) compared two reactors namely down flow stationary fixed film and UASB for leachate treatment. Better COD removal efficiency of 95% was achieved by UASB reactor at OLR of 4.8 kg COD/m^3 d and HRTs of 4.0 days. Kennedy and Lentz (2000) found that the performance of the continuous UASB reactor and AnSBR module treating landfill leachate were quite similar at low-to-medium imposed OLRs. COD soluble removal efficiencies by AnSBR was varied from 71% to 92% at HRTs of 24, 18, and 12 h and OLRs of 0.6 –18.4 g COD/L d. The UASB reactor achieved soluble COD removal ranged from 77% to 91% at similar operational conditions. This indicates that reactor operational mode has no effect for treatment of landfill leachate at the same imposed operational conditions. Nevertheless, avoiding the failure of AnSBR treating landfill leachate, OLR should not exceed 3.0 g COD/g VSS d. The efficiency of an UASB reactor, a hybrid-bed reactor, and an anaerobic up-flow filter was compared for treatment of landfill leachate (COD = 18,800–47,800 mg/L) for a period of 230 days (Inanc *et al.*, 2000). The OLR was gradually increased from 1.3 to 8.2 kg COD/m^3 d for all reactors resulting in COD removal efficiencies between 80% and 90%. However, anaerobic filter and hybrid-bed reactor was the best technologies tolerant for ammonia inhibition compared with UASB. This is mainly due to a high imposed sludge residence time in the attached growth anaerobic reactors.

3.5.3 Effect of organic loading rate

The OLR is an important design parameter for application of anaerobic technology for leachate treatment (Ismail *et al.*, 2020; Ismail & Tawfik, 2016a). The OLR is defined as the organics (COD or BOD) multiply the landfill leachate flow rate divided by the digester volume in case of classical reactor or by the digester surface area in case of fixed-bed biofilm module. A hybrid anaerobic digester treating landfill leachate at different OLRs at temperature of 35°C was investigated by Chang (1989). The module removed 92% of COD soluble at OLR of 13 kg COD/m^3 d that was highly reduced to 70% at higher OLR of 21.77 kg COD/m^3 d. This is mainly due to the deterioration of food by microorganisms in the digester (Ismail & Tawfik, 2016a, 2016b). Moreover, increasing the loading rate will increase the accumulation of solids in the reactor reducing the sludge residence time required for accomplishment of the anaerobic degradation process (Meky *et al.*, 2017, 2019). Anaerobic MBBR treating landfill leachate removed COD by 91% at OLR of 4.08 kgCOD/m^3 d, and slightly reduced to 86% at higher OLR of 15.70 kgCOD/m^3 d (Chen *et al.*, 2008). The anaerobic submerged membrane bioreactor treating diluted landfill leachate removed 90% of COD at OLR of 2.5 kg COD/m^3 d (Bohdziewicz *et al.*, 2008). Two anaerobic digesters treating landfill leachate (COD = 48,552–62,150 mg/L) for 8 months was investigated by Kheradmand *et al.* (2010). The digesters showed a COD reduction of 94% at OLR of 2.25 g COD/L d and 93% at OLR of 3.37 g COD/L d. Two-stage UASB reactor operated in series achieved COD removal of 80% from leachate at OLR of 4.3–16 kg/m^3 d with methane content of 64–43% (Ağdağ & Sponza, 2005)

3.5.4 Sulfate concentration

The presence of sulfate ions in the landfill leachate at high concentrations will impair the efficiency of the anaerobic digester. As portions of organics are consumed for the reduction of sulfate into sulfide ions the bioenergy productivity was highly reduced. However, the presence of the sulfide ions in the anaerobic reactors would precipitate the iron to certain extent. Fe accumulated in the anaerobic digester by treating landfill leachate at concentrations of 160–515 mg/L, where the sludge contained Fe concentration of 7,100 mg/L after an operational period of 185 days (Chang, 1989). Furthermore, increasing the sulfate loading of rate from 102 to 683 mg/L d, reduced the energy recovery from 90% to 52% due to the toxicity effect of ions and consumption of big portion of organics in the reduction process.

3.5.5 Ammonia concentration

The ammonia in the feed of the anaerobic reactors increases buffering capacity of methanogenic conditions thus improving the stability of metabolism processes. The anaerobic digestion of landfill leachate containing high ammonia concentration is a bottleneck in the biomethanisation process (Ismail *et al.*, 2019). Microbial communities in anaerobic treatment processes are very sensitive to free ammonia under certain conditions (Ismail *et al.*, 2021). Inhibition and/or toxicity of the microbial communities producing hydrogen/methane could occur due to the presence of free ammonia leading to failures of anaerobic treatment process. Fifty percent reduction in the yield of methane gas in the anaerobic reactor took place at an initial concentration of 11.0 gNH_4-N/L (1.45 gNH_3-N/L) (Nakakubo *et al.*, 2008). UASB reactors performance under high ammonia concentrations was investigated by Calli *et al.* (2005b). Free ammonia nitrogen gradually increased to 800 mg/L in the reactor achieving COD removal efficiencies of 78–96%. However, high COD removal efficiencies were observed at low ammonia concentration in the feed. The authors discovered that acetogenic bacteria degrading propionate were inhibited by free ammonia concentration exceeding 200 mg/L FAN. Furthermore, acetogenic bacteria degrading propionate is more sensitive than methanogenic Archaea at high concentrations of free ammonia in the feed. On the other hand, butyric acid degradation had no problem in the anaerobic reactor at any free ammonia concentrations. This indicates that the resistance of butyrate bacteria degraders to high concentrations of free ammonia.

Procházka *et al.* (2012) found that the anaerobic process was stable at ammonia concentrations of 2.1, 2.6 and 3.1 g/L; however, high ammonia concentration of 4.0 g/L inhibited the biomethane productivity. Likely low methane yield was obtained at low ammonia concentration of 0.5 g/L in the feed due to insufficient nutrients that causes biomass loss, low activity of aceticlastic methanogenic microbes and low buffering capacity. Similar results were recorded by Marchal and Vandecasteele (1993) who found that ammonia concentrations exceeded 4 g N/L inhibited biomethanization process. The methane yield was highly reduced and acetate concentrations increased from 1 to 3 g/L in the medium at high ammonia concentrations. Aceticlastic microbes were highly sensitive to high ammonia concentration compared to hydrogenotrophic methanogens. Therefore, anammox process is much recommended for ammonia and nitrogen removal from landfill leachate in combination with anaerobic pretreatment process for organics removal (Ismail & Tawfik, 2017). Anaerobic up-flow filter, hybrid-bed reactors and sludge blanket system were compared for treatment of landfill leachate containing high ammonia concentrations at OLR of 1.3–23.5 kg COD/m^3 d (Calli *et al.*, 2006). The pH value of the raw leachate was decreased to 4.5 to reduce the inhibition of free ammonia. The performance of hybrid bed and anaerobic filter was stable compared with UASB reactor at high ammonia concentrations. However, acetoclastic methanosaeta species were dominant in all reactors indicating that configuration of the reactors did not have an effect on the microbial diversity.

3.5.6 Carbon/nitrogen (C/N) ratio

The anaerobic treatment process of the landfill leachate could be accomplished at a proper C/N ratio. The balance between nitrogen and organics encourage the growing of anaerobes and subsequently increasing the energy productivity. The imbalance of nitrogen and organic content could lead to the failure of the anaerobic process. COD/N ratios in the feed imposed on an anaerobic reactor were 90, 80, 65, and 50, where reactor performance was assessed by Poggi-Varaldo *et al.* (1997). The syntrophic bacteria in the anaerobic reactors were impaired at increasing ammonia concentration and failed at COD/N exceeding 50 due to accumulation of propionic, butyric, and valeric acids. Wang *et al.* (2014) found that ammonia inhibition of the anaerobic digestion process occurred at C/N ratio of 15 and 20.

3.6 CONCLUSIONS

Usually, an anaerobic reactor has commonly been installed in leachate treatment process for treating high loading organic compounds discharged from young landfill sites. Different reactors can be used, such as anaerobic filters, anaerobic sequencing batch reactor or UASB. Anaerobic up-flow filter, hybrid-bed reactors and sludge blanket system is efficient for treatment of landfill leachate containing high ammonia concentrations at OLR of 1.3–23.5 kg COD/m^3 d. The pH value of the raw leachate could be reduced to 4.5 to mitigate the inhibition of free ammonia. The performance of hybrid bed and anaerobic filter was stable compared with UASB reactor at high ammonia concentrations. However, *Acetoclastic methanosaeta* species were dominant in all reactors indicating that configuration of the reactors did not have any effect on the microbial diversity at high ammonia loading rate. The imbalance of nitrogen and organic content could lead to the failure of the anaerobic process. COD/N ratios in the feed imposed on an anaerobic reactor should be controlled. The syntrophic bacteria in the anaerobic reactors were impaired at increasing ammonia concentration and failed at COD/N exceeding 50 due to accumulation of propionic, butyric and valeric acids. Furthermore, anaerobic digestion process failed at low C/N ratio of 15–20.

ACKNOWLEDGMENTS

The first author acknowledges the support provided by the Academy of Scientific Research and Technology (ASRT)-Egypt and the Natural Science Foundation of China (project no. 9185). The NRC is acknowledged for partially funding this research (project ID: 13040108).

REFERENCES

Ağdağ O. N. and Sponza D. T. (2005). Anaerobic/aerobic treatment of municipal landfill leachate in sequential two-stage up-flow anaerobic sludge blanket reactor (UASB)/completely stirred tank reactor (CSTR) systems. *Process Biochemistry*, **40**, 895–902, https://doi.org/10.1016/j.procbio.2004.02.021

Ahmad H. A., Ahmad S., Cui Q., Wang Z., Wei H., Chen X., Ni S.-Q., Ismail S., Awad H. M. and Tawfik A. (2021). The environmental distribution and removal of emerging pollutants, highlighting the importance of using microbes as a potential degrader: a review. *Science of The Total Environment*, **809**, 151926, https://doi.org/10.1016/j.scitotenv.2021.151926

Altin A. (2008). An alternative type of photoelectro-Fenton process for the treatment of landfill leachate. *Separation and Purification Technology*, **61**, 391–397, https://doi.org/10.1016/j.seppur.2007.12.004

Bohdziewicz J., Neczaj E. and Kwarciak A. (2008). Landfill leachate treatment by means of anaerobic membrane bioreactor. *Desalination*, **221**, 559–565, https://doi.org/10.1016/j.desal.2007.01.117

Boyle W. C. and Ham R. K. (1974). Biological treatability of landfill leachate. *Journal of the Water Pollution Control Federation*, **46**, 860–872.

Calli B., Mertoglu B. and Inanc B. (2005a). Landfill leachate management in Istanbul: applications and alternatives. *Chemosphere*, **59**, 819–829, https://doi.org/10.1016/j.chemosphere.2004.10.064

Calli B., Mertoglu B., Inanc B. and Yenigun O. (2005b). Effects of high free ammonia concentrations on the performances of anaerobic bioreactors. *Process Biochemistry*, **40**, 1285–1292, https://doi.org/10.1016/j.procbio.2004.05.008

Calli B., Mertoglu B., Roest K. and Inanc B. (2006). Comparison of long-term performances and final microbial compositions of anaerobic reactors treating landfill leachate. *Bioresource Technology*, **97**, 641–647, https://doi.org/10.1016/j.biortech.2005.03.021

Chang J. E. (1989). Treatment of landfill leachate with an upflow anaerobic reactor combining a sludge bed and a filter. *Water Science and Technology*, **21**, 133–143, https://doi.org/10.2166/wst.1989.0217

Chen S., Sun D. and Chung J. S. (2008). Simultaneous removal of COD and ammonium from landfill leachate using an anaerobic-aerobic moving-bed biofilm reactor system. *Waste Management*, **28**, 339–346, https://doi.org/10.1016/j.wasman.2007.01.004

Deng Y. and Englehardt J. D. (2007). Electrochemical oxidation for landfill leachate treatment. *Waste Management*, **27**, 380–388, https://doi.org/10.1016/j.wasman.2006.02.004

El-Bery H., Tawfik A., Kumari S. and Bux F. (2013). Effect of thermal pre-treatment on inoculum sludge to enhance bio-hydrogen production from alkali hydrolysed rice straw in a mesophilic anaerobic baffled reactor. *Environmental Technology*, **34**, 1965–1972, https://doi.org/10.1080/09593330.2013.824013

El-Gohary F., Tawfik A., Badawy M. and El-Khateeb M. A. (2009). Potentials of anaerobic treatment for catalytically oxidized olive mill wastewater (OMW). *Bioresource Technology*, **100**, 2147–2154, https://doi.org/10.1016/j.biortech.2008.10.051

Elsamadony M. and Tawfik A. (2018). Maximization of hydrogen fermentative process from delignified water hyacinth using sodium chlorite. *Energy Conversion and Management*, **157**, 257–265, https://doi.org/10.1016/j.enconman.2017.12.013

Farghaly A., Elsamadony M., Ookawara S. and Tawfik A. (2017). Bioethanol production from paperboard mill sludge using acid-catalyzed bio-derived choline acetate ionic liquid pretreatment followed by fermentation process. *Energy Conversion and Management*, **145**, 255–264, https://doi.org/10.1016/j.enconman.2017.05.004

Foo K. Y. and Hameed B. H. (2009). An overview of landfill leachate treatment via activated carbon adsorption process. *Journal of Hazardous Materials*, **171**, 54–60, https://doi.org/10.1016/j.jhazmat.2009.06.038

Galeano L. A., Vicente MÁ and Gil A. (2011). Treatment of municipal leachate of landfill by Fenton-like heterogeneous catalytic wet peroxide oxidation using an Al/Fe-pillared montmorillonite as active catalyst. *Chemical Engineering Journal*, **178**, 146–153, https://doi.org/10.1016/j.cej.2011.10.031

Gar Alalm M., Tawfik A. and Ookawara S. (2015). Comparison of solar TiO_2 photocatalysis and solar photo-Fenton for treatment of pesticides industry wastewater: operational conditions, kinetics, and costs. *Journal of Water Process Engineering*, **8**, 55–63, https://doi.org/10.1016/j.jwpe.2015.09.007

García H., Rico J. and García P. (1996). Comparison of anaerobic treatment of leachates from an urban-solid-waste landfill at ambient temperature and at 35 C. *Bioresource Technology*, **58**, 273–277, https://doi.org/10.1016/S0960-8524(96)00114-9

Henry J. G., Prasad D. and Young H. (1987). Removal of organics from leachates by anaerobic filter. *Water Research*, **21**, 1395–1399, https://doi.org/10.1016/0043-1354(87)90015-7

Im J. H., Woo H. J., Choi M. W., Han K. B. and Kim C. W. (2001). Simultaneous organic and nitrogen removal from municipal landfill leachate using an anaerobic-aerobic system. *Water Research*, **35**, 2403–2410, https://doi.org/10.1016/S0043-1354(00)00519-4

Inanc B., Calli B. and Saatci A. (2000). Characterization and anaerobic treatment of the sanitary landfill leachate in Istanbul. *Water Science and Technology*, **41**, 223–230, https://doi.org/10.2166/wst.2000.0075

Ismail S. and Tawfik A. (2016a). Performance of passive aerated immobilized biomass reactor coupled with Fenton process for treatment of landfill leachate. *International Biodeterioration & Biodegradation*, **111**, 22–30, https://doi.org/10.1016/j.ibiod.2016.04.010

Ismail S. and Tawfik A. (2016b). Treatment of hazardous landfill leachate using Fenton process followed by a combined (UASB/DHS) system. *Water Science and Technology*, **73**, 1700–1708, https://doi.org/10.2166/wst.2015.655

Ismail S. and Tawfik A. (2017). Comprehensive study for Anammox process via multistage anaerobic baffled reactors. *E3S Web of Conferences*, **22**, 00068, https://doi.org/10.1051/e3sconf/20172200068

Ismail S., Elsamadony M., Fujii M. and Tawfik A. (2019). Evaluation and optimization of anammox baffled reactor (AnBR) by artificial neural network modeling and economic analysis. *Bioresource Technology*, **271**, 500–506, https://doi.org/10.1016/j.biortech.2018.09.004

Ismail S., Nasr M., Abdelrazek E., Awad H. M., Zhaof S., Meng F. and Tawfik A. (2020). Techno-economic feasibility of energy-saving self-aerated sponge tower combined with up-flow anaerobic sludge blanket reactor for treatment of hazardous landfill leachate. *Journal of Water Process Engineering*, **37**, 101415, https://doi.org/10.1016/j.jwpe.2020.101415

Ismail S., Elreedy A., Fujii M., Ni S. Q., Tawfik A. and Elsamadony M. (2021). Fatigue of anammox consortia under long-term 1,4-dioxane exposure and recovery potential: N-kinetics and microbial dynamics. *Journal of Hazardous Materials*, **414**, 125533, https://doi.org/10.1016/j.jhazmat.2021.125533

Kennedy K. J. and Lentz E. M. (2000). Treatment of landfill leachate using sequencing batch and continuous flow upflow anaerobic sludge blanket (UASB) reactors. *Water Research*, **34**, 3640–3656, https://doi.org/10.1016/S0043-1354(00)00114-7

Kennedy K. J., Hamoda M. F. and Guiot S. G. (1988). Anaerobic treatment of leachate using fixed film and sludge bed systems. *Journal of The Water Pollution Control Federation*, **60**, 1675–1683.

Kettunen R. H. and Rintala J. A. (1998). Performance of an on-site UASB reactor treating leachate at low temperature. *Water Research*, **32**, 537–546, https://doi.org/10.1016/S0043-1354(97)00319-9

Kettunen R. H., Hoilijoki T. H. and Rintala J. A. (1996). Anaerobic and sequential anaerobic-aerobic treatments of municipal landfill leachate at low temperatures. *Bioresource Technology*, **58**, 31–40, https://doi.org/10.1016/S0960-8524(96)00102-2

Kheradmand S., Karimi-Jashni A. and Sartaj M. (2010). Treatment of municipal landfill leachate using a combined anaerobic digester and activated sludge system. *Waste Management*, **30**, 1025–1031, https://doi.org/10.1016/j.wasman.2010.01.021

Khodary S. M., Negm A. M. and Tawfik A. (2018). Geotechnical properties of the soils contaminated with oils, landfill leachate, and fertilizers. *Arabian Journal of Geosciences*, **11**, 1–17, https://doi.org/10.1007/s12517-017-3372-7

Kjeldsen P., Barlaz M. A., Rooker A. P., Baun A., Ledin A. and Christensen T. H. (2002). Present and long-term composition of MSW landfill leachate: a review. *Critical Reviews in Environmental Science and Technology*, **32**, 297–336, https://doi.org/10.1080/10643380290813462

Kurniawan T. A., Lo W. and Chan G. Y. S. (2006). Physico-chemical treatments for removal of recalcitrant contaminants from landfill leachate. *Journal of Hazardous Materials*, **129**, 80–100, https://doi.org/10.1016/j.jhazmat.2005.08.010

Lema J., Mendez R. and Blazquez R. (1988). Characteristics of landfill leachates and alternatives for their treatment: a review. *Water, Air, and Soil Pollution*, **40**, 223–250, https://doi.org/10.1007/BF00163730

Li X. Z. Z., Zhao Q. L. L. and Hao X. D. D. (1999). Ammonium removal from landfill leachate by chemical precipitation. *Waste Management*, **19**, 409–415, https://doi.org/10.1016/S0956-053X(99)00148-8

Li D., Yang X., Zhou Z., Jiang B., Tawfik A., Zhao S. and Meng F. (2019). Molecular traits of phenolic moieties in dissolved organic matter: linkages with membrane fouling development. *Environment International*, **133**, 105202, https://doi.org/10.1016/j.envint.2019.105202

Lin C. Y. (1991). Anaerobic digestion of landfill leachate. *Water SA*, **17**, 301–306.

Lin Y. H., Juan M. L. and Hsien H. J. (2011). Effects of temperature and initial pH on biohydrogen production from food-processing wastewater using anaerobic mixed cultures. *Biodegradation*, **22**, 551–563, https://doi.org/10.1007/s10532-010-9427-z

Loukidou M. X. and Zouboulis A. I. (2001). Comparison of two biological treatment processes using attached-growth biomass for sanitary landfill leachate treatment. *Environmental Pollution*, **111**, 273–281, https://doi.org/10.1016/S0269-7491(00)00069-5

Mahmoud M., Ismail S. and Tawfik A. (2018). Post-treatment of anaerobic effluent containing 1,4-dioxane and heavy metals via auto-aerated down-flow hanging luffa (ADHL) system. *Process Safety and Environmental Protection*, **117**, 22–32, https://doi.org/10.1016/j.psep.2018.04.008

Mahmud K., Hossain M. D. and Shams S. (2012). Different treatment strategies for highly polluted landfill leachate in developing countries. *Waste Management*, **32**, 2096–2105, https://doi.org/10.1016/j.wasman.2011.10.026

Marchal R. and Vandecasteele J. P. (1993). Thermophilic anaerobic digestion of livestock waste: the effect of ammonia. *Applied Microbiology and Biotechnology*, **38**, 560–564.

Meky N., Fujii M., Yoshimura C. and Tawfik A. (2017). Feasibility of using non-vegetated baffled submerged constructed wetland system for removal of heavy metals, COD and nutrients from hyper-saline hazardous landfill leachate. *Desalination and Water Treatment*, **98**, 254–265, https://doi.org/10.5004/dwt.2017.21738

Meky N., Fujii M. and Tawfik A. (2019). Treatment of hypersaline hazardous landfill leachate using a baffled constructed wetland system: effect of granular packing media and vegetation. *Environmental Technology*, **40**, 518–528, https://doi.org/10.1080/09593330.2017.1397764

Morgan W. E. (1990). Anaerobic/aerobic treatment of hazardous waste leachate. *Journal of Hazardous Materials*, **24**, 279–310, https://doi.org/10.1016/0304-3894(90)87016-B

Nakakubo R., Møller H. B., Nielsen A. M. and Matsuda J. (2008). Ammonia inhibition of methanogenesis and identification of process indicators during anaerobic digestion. *Environmental Engineering Science*, **25**, 1487–1496, https://doi.org/10.1089/ees.2007.0282

Nakashimada Y., Ohshima Y., Minami H., Yabu H., Namba Y. and Nishio N. (2008). Ammonia-methane two-stage anaerobic digestion of dehydrated waste-activated sludge. *Applied Microbiology and Biotechnology*, **79**, 1061–1069, https://doi.org/10.1007/s00253-008-1501-7

Poggi-Varaldo H. M., Rodríguez-Vázquez R., Fernández-Villagómez G. and Esparza-García F. (1997). Inhibition of mesophilic solid-substrate anaerobic digestion by ammonia nitrogen. *Applied Microbiology and Biotechnology*, **47**, 284–291, https://doi.org/10.1007/s002530050928

Procházka J., Dolejš P., MácA J. and Dohányos M. (2012). Stability and inhibition of anaerobic processes caused by insufficiency or excess of ammonia nitrogen. *Applied Microbiology and Biotechnology*, **93**, 439–447, https://doi.org/10.1007/s00253-011-3625-4

Renou S., Givaudan J. G., Poulain S., Dirassouyan F. and Moulin P. (2008). Landfill leachate treatment: review and opportunity. *Journal of Hazardous Materials*, **150**, 468–493, https://doi.org/10.1016/j.jhazmat.2007.09.077

Smith D. (1995). Oxygen flux limitation in aerobic fixed-film biotreatment of a hazardous landfill leachate. *Journal of Hazardous Materials*, **44**, 77–91, https://doi.org/10.1016/0304-3894(95)00050-5

Sosnowski P., Wieczorek A. and Ledakowicz S. (2003). Anaerobic co-digestion of sewage sludge and organic fraction of municipal solid wastes. *Advances in Environmental Research*, **7**, 609–616, https://doi.org/10.1016/S1093-0191(02)00049-7

Tauchert E., Schneider S., de Morais J. L. and Peralta-Zamora P. (2006). Photochemically-assisted electrochemical degradation of landfill leachate. *Chemosphere*, **64**, 1458–1463, https://doi.org/10.1016/j.chemosphere.2005.12.064

Tawfik A. and ElBatrawy O. (2012). Anaerobic biodegradation of personnel care products (PCPs) wastewater in an up-flow anaerobic sludge blanket (UASB) reactor. *Desalination and Water Treatment*, **41**, 232–239, https://doi.org/10.1080/19443994.2012.664719

Tawfik A. and Salem A. (2012). The effect of organic loading rate on bio-hydrogen production from pre-treated rice straw waste via mesophilic up-flow anaerobic reactor. *Bioresource Technology*, **107**, 186–190, https://doi.org/10.1016/j.biortech.2011.11.086

Tawfik A., Ohashi A. and Harada H. (2006). Sewage treatment in a combined up-flow anaerobic sludge blanket (UASB)–down-flow hanging sponge (DHS) system. *Biochemical Engineering Journal*, **29**, 210–219, https://doi.org/10.1016/j.bej.2005.11.018

Tawfik A., El-Gohary F., Ohashi A. and Harada H. (2008). Optimization of the performance of an integrated anaerobic-aerobic system for domestic wastewater treatment. *Water Science and Technology*, **58**, 185–194, https://doi.org/10.2166/wst.2008.320

Tawfik A., El-Gohary F. and Temmink H. (2010). Treatment of domestic wastewater in an up-flow anaerobic sludge blanket reactor followed by moving bed biofilm reactor. *Bioprocess and Biosystems Engineering*, **33**, 267–276, https://doi.org/10.1007/s00449-009-0321-1

Tawfik A., Badr N., Taleb E. A. and El-Senousy W. (2012). Sewage treatment in an up-flow anaerobic sponge reactor followed by moving bed biofilm reactor based on polyurethane carrier material. *Desalination and Water Treatment*, **37**, 350–358, https://doi.org/10.1080/19443994.2012.661291

Tawfik A., Salem A., El-Qelish M., Fahmi A. A. and Moustafa M. E. (2013). Factors affecting hydrogen production from rice straw wastes in a mesophillic up-flow anaerobic staged reactor. *Renewable Energy*, **50**, 402–407, https://doi.org/10.1016/j.renene.2012.06.038

Tawfik A., Al-sayed A., Hassan G. K., Nasr M., El-shafai S. A., Alhajeri N. S., Shariq M., Saeed M., Ahmad Z., Rojas P. and Sanz J. L. (2022a). Electron donor addition for stimulating the microbial degradation of 1, 4 dioxane by sequential batch membrane bioreactor : a techno-economic approach. *Chemosphere*, **306**, 135580, https://doi.org/10.1016/j.chemosphere.2022.135580

Tawfik A., Niaz H., Qadeer K., Qyyum M. A., Liu J. J. and Lee M. (2022b). Valorization of algal cells for biomass and bioenergy production from wastewater: sustainable strategies, challenges, and techno-economic limitations. *Renewable and Sustainable Energy Reviews*, **157**, 112024, https://doi.org/10.1016/j.rser.2021.112024

Termorshuizen A. J., Volker D., Blok W. J., Ten Brummeler E., Hartog B. J., Janse J. D., Knol W. and Wenneker M. (2003). Survival of human and plant pathogens during anaerobic mesophilic digestion of vegetable, fruit, and garden waste. *European Journal of Soil Biology*, **39**, 165–171, https://doi.org/10.1016/S1164-5563(03)00032-3

Timur H. (1999). Anaerobic Sequencing Batch Reactor Treatment of Landfill Leachate 33.

Timur H. and Özturk I. (1997). Anaerobic treatment of leachate using sequencing batch reactor and hybrid bed filter. *Water Science and Technology*, **36**, 501–508, https://doi.org/10.1016/S0273-1223(97)00561-1

Timur H. and Özturk I. (1999). Anaerobic sequencing batch reactor treatment of landfill leachate. *Water Research*, **33**(15), 3225–3230, https://doi.org/10.1016/S0043-1354(99)00048-2

Umar M., Aziz H. A. and Yusoff M. S. (2010). Trends in the use of Fenton, electro-Fenton and photo-Fenton for the treatment of landfill leachate. *Waste Management*, **30**, 2113–2121, https://doi.org/10.1016/j.wasman.2010.07.003

Vedrenne M., Vasquez-medrano R., Prato-garcia D., Frontana-uribe B. A. and Ibanez J. G. (2012). Characterization and detoxification of a mature landfill leachate using a combined coagulation – flocculation/photo Fenton treatment. *Journal of Hazardous Materials*, **205–206**, 208–215, https://doi.org/10.1016/j.jhazmat.2011.12.060

Wang X., Lu X., Li F. and Yang G. (2014). Effects of temperature and carbon-nitrogen (C/N) ratio on the performance of anaerobic co-digestion of dairy manure, chicken manure and rice straw: focusing on ammonia inhibition. *PLoS One*, **9**, 1–7, https://doi.org/10.1371/journal.pone.0097265

Xie Z., Wang Z., Wang Q., Zhu C. and Wu Z. (2014). An anaerobic dynamic membrane bioreactor (AnDMBR) for landfill leachate treatment: performance and microbial community identification. *Bioresource Technology*, **161**, 29–39, https://doi.org/10.1016/j.biortech.2014.03.014

Yiping X. U., Yiqi Z., Donghong W., Shaohua C., Junxin L. I. U. and Zijian W. (2008). Occurrence and removal of organic micropollutants in the treatment of landfill leachate by combined anaerobic-membrane bioreactor technology. *Journal of Environmental Sciences*, **20**, 1281–1287, https://doi.org/10.1016/S1001-0742(08)62222-6

Yoon J., Cho S., Cho Y. and Kim S. (1998). The characteristics of coagulation of Fenton reaction in the removal of landfill leachate organics. *Water Science and Technology*, **38**, 209–214, https://doi.org/10.1016/S0273-1223(98)00481-8

Zaloum R. and Abbott M. (1997). Anaerobic pretreatment improves single sequencing batch reactor treatment of landfill leachates. *Water Science and Technology*, **35**, 207–214, https://doi.org/10.1016/S0273-1223(96)00898-0

doi: 10.2166/9781789063318_0057

Chapter 4

Microbial fuel cell for landfill leachate treatment

Saurabh Dwivedi[1,2], Ashmita Patro[1,2], Yamini Mittal[1,2], Supriya Gupta[1,2] and Pratiksha Srivastava[3*]

[1]CSIR-Institute of Minerals and Materials Technology, Bhubaneswar 751013, Odisha, India
[2]Academy of Scientific and Innovative Research (AcSIR), Ghaziabad, Uttar Pradesh 201002, India
[3]Department of Chemical and Environmental Technology, University Rey Juan Carlos, C. Tulipán, s/n, 28933 Móstoles, Madrid, Spain
*Corresponding author: p.s.immt@gmail.com

ABSTRACT

Solid waste disposal is a major problem arising due to urbanization and development of the society. Secured landfilling is a scientific practice to manage solid waste for its minimum impact on the environment. However, the leachate generated from the landfill is a major concern to the environment. There are several off-site treatment technologies which are energy and cost-intensive. Additionally, transportation of landfill can cause hazardous effects. Bio-electrochemical treatment technologies such as microbial fuel cell (MFC) have the capability to treat the landfill leachate on-site, along with production of bioelectricity and resource recovery. Vast array of pollutant containing leachate act as potential substrate and its high conductivity helps in the electrochemical reactions of MFC. The unstable load of pollutants contained in landfill leachate causes toxicity, which hampers the performance of MFC. Thus, pre-treatment or dilution of leachate could solve this problem and MFC can be used for landfill leachate efficiently at a secondary stage.

Keywords: Microbial fuel cell, landfill, leachate treatment, bio-electricity.

4.1 INTRODUCTION

Rapid urbanization and population growth are generating huge quantity of solid waste. The major contribution is municipal solid waste (MSW) which consists of organic waste, paper waste, plastic waste, glass waste, metal waste, and other wastes. In 2016, the global MSW generation was 2.01 billion tons (BT), which is predicted to be 3.40 BT by 2050 (Sharma & Jain, 2020). The collection, disposal, and treatment of solid waste is performed by incineration, composting, or landfilling (Dantre *et al.*, 2017; Luo *et al.*, 2020). Landfilling is the most suitable method in solid waste management, and it alone disposes more than 95% of MSW globally (Luo *et al.*, 2020).

Landfilling is the most suitable method for managing different solid wastes generated in society than incineration and composting (Dantre *et al.*, 2017; Luo *et al.*, 2020). Landfills are engineered dumpsites for disposal of different solid waste generated from various sections of society. It contains wide range of wastes such as food, plastics, papers, metals, glasses, leathers, chemicals and pharmaceuticals, and so on. The by-products of these secured landfills are gases and highly contaminated leachates (Aziz *et al.*, 2010). The leachates are high organic matter and nitrogenous pollutants containing liquid which is generated from the percolation of the water through solid waste. The leachate generation mainly depends upon the landfill characteristics such as landfill age, geographical hydrology, climatic condition, and rainwater infiltration (Chaudhary *et al.*, 2021; Elmaadawy *et al.*, 2020b, Maturi *et al.*, 2022). The leachate contains phenols, dyes, dissolved solids, nitrite, nitrate, ammonia, phosphates, pesticides, heavy metals, iron, recalcitrant pollutant, and pathogens (Chaudhary *et al.*, 2021; Elmaadawy *et al.*, 2020b). The nature of landfill leachate is mainly basic, but it can also be acidic in pH depending on waste types (Aziz *et al.*, 2010; Greenman *et al.*, 2009; Puig *et al.*, 2011). The concentrations of pollutants also vary in leachate depending on the contributing waste types. It causes toxicity to various ecosystems ranging from groundwater pollution, aquatic toxicity, land pollution to desertification, and also to human health (Dantre *et al.*, 2017; Gupta *et al.*, 2015). The treatment of landfill leachate is important and the most popular techniques to treat leachate is off-site, on-site, and recirculation through landfill (*Critaria for Landfill Management CPCB.Pdf*, n.d.). On the one hand, the recirculation process cannot treat the leachate properly; on the other, off-site treatment is capital intensive and required various materials, infrastructure, and their maintenance since it involves reverse osmosis, activated carbon adsorption, flocculation, and biological treatment (activated sludge, trickling filter, ponding, rotating biological contractors, etc.) (Iskander *et al.*, 2016; Kumar *et al.*, 2013). To replace such intensive technologies, a sustainable and green approach-based bio-electrochemical systems (BES) is highly a recommended technology to address landfill leachate pollution (You *et al.*, 2006; Zhang *et al.*, 2015a). BES systems are microbes-based system which converts chemical energy obtained from organic waste into electrical/chemical energy by performing various electrochemical reactions. BES systems such as microbial fuel cell (MFC), microbial electrolysis cell (MEC) treat leachates by combining biological and electrochemical processes. MFC is a cost-effective BES system to decontaminate the leachate, recovers resource, and produces bioelectricity (Iskander *et al.*, 2016). Due to high organic content and the presence of various inorganic metals, landfill leachate has the potential to be used as a fuel in MFCs for energy recovery (Elmaadawy *et al.*, 2020b). Leachate could also increase power generation due to its high electrical conductivity (Iskander *et al.*, 2016).

4.2 POLLUTANT REMOVAL MECHANISM

MFC works on the redox condition at the anode and cathode chambers, which helps electroactive bacteria (EAB) in their metabolic activities. The higher concentration and conductivity of landfill leachate helps in maintaining redox potential in MFC conducive for electron transfer and oxidation of different pollutants (Deng *et al.*, 2020). The oxidation of pollutants depends on the development of microbial community in MFCs, the microbial biofilm development on electrodes acts as a bio-cathode and bio-anode (Srivastava *et al.*, 2020a, 2020b). The microbial community developed in MFC is known as electrogens/exo-electrogens/EAB. The bio-cathode and bio-anode enhances the electron transfer rate and helps in the bio-electrochemical reactions of MFC systems. However, in some cases such as dual-chambered (DC) MFCs cathodes are generally abiotic rather than biotic. In this section, a brief description about the MFC to address different pollutants present in landfill based on design and mechanism is illustrated.

4.2.1 Biological process for organics/inorganics removal

Landfill leachate is reported as a significant source of nutrient recovery and electricity production using MFC (Deng *et al.*, 2020). Chemically, leachate contains high organic and inorganic matter, which helps to develop a distinguished redox condition in MFC. Exo-electrogenic bacteria typically oxidize these substances (organics and inorganics) at the anode and produces electrons and protons. The electrons instantly transfer from anode to the cathode electrode via an external electric circuit, while protons pass through a cation-exchange membrane to the cathode chamber and interact with terminal electron acceptors (such as oxygen) in the cathode chamber (Srivastava *et al.*, 2020a, 2020b). It results in the production of bio-electricity treating the leachate. The leachate contains higher and complex concentration of organic and inorganic pollutants; which can affect exo-electrogens and power output of MFCs (Wu *et al.*, 2015). The degradability of landfill leachate is very distinct than traditional wastewater due to inability of exo-electrogens to metabolize complex substrate (long-chain fatty acids and aromatic compounds) (Wu *et al.*, 2015). Ozkaya *et al.* (2006) studied that younger landfill leachate (of less than 2 years) is more biodegradable (in terms of (BOD_5/COD) as it contains high COD concentration (10,000 ppm) than older leachate. The MFC has been reported to remove about 45% of COD from younger landfill leachate having 50,000 mg/l of initial COD concentration (Özkaya *et al.*, 2013). However, with increasing the age of landfill, bio-inhibitory agents such as imperishable organic matter (fulvic acid, humic acid), high strength of ammonium increases in leachates, which makes the biodegradability of leachates difficult by decreasing BOD_5/COD, and BOD_5/TN ratios (Hassan *et al.*, 2018). In this way, high concentration of ammonia present in mature leachates could hamper the microbial activity in BES (Elmaadawy *et al.*, 2020b). This could be resolved with pre-treatment of leachate before treating with MFC.

Landfill leachate contains various forms of nitrogen such as nitrite (NO_2), nitrate (NO_3), ammonia (NH_3), ammonium ions (NH_4^+) and organic nitrogen. Some studies have reported that characteristics of landfill leachate includes about 80–90% of ammonia among the total nitrogen composition (Puig *et al.*, 2011; Zhang *et al.*, 2015a). In MFCs, dissimilatory mechanisms utilize ammonia present in landfill leachate for biological nitrogen removal, whereas assimilation removes nitrogen for cell growth of microorganisms and represent about 23% of the total cell weight. Such microbial mechanisms catalyze the degradation of complex form of nitrogen into simpler form in both the chambers of MFCs (Elmaadawy *et al.*, 2020b). In addition to microbial mechanisms several processes such as nitrification and denitrification takes place in aerobic and anaerobic region, respectively (Virdis *et al.*, 2010). The oxidation of ammonia into nitrate/nitrite takes place in the aerobic chamber via nitrification. The resultant product (nitrite/nitrate) travels through the anion-exchange membrane to the anaerobic region where it is further reduced to nitrogen gas by heterotrophic denitrifying bacteria (Figure 4.1) (Elmaadawy *et al.*, 2020b; Feng *et al.*, 2015; Zou *et al.*, 2018). However, limited availability of dissolved oxygen hinders the complete nitrification in MFC (Srivastava *et al.*, 2021b). Sometimes anoxic condition prevails in the cathode chamber favors conversion of ammonium ions into dinitrogen (in presence of nitrite as electron acceptor) by the action of annamox bacterium facilitating enhanced removal of nitrogen in MFC (Lee *et al.*, 2013). Similarly, in anaerobic region, electroactive nitrifying microorganism oxidize the anoxic ammonium into N_2, where hydroxyl-amine serves as substrate for the oxidation reactions (Elmaadawy *et al.*, 2020b; Vilajeliu-Pons *et al.*, 2018). Instead of simultaneous nitrification–denitrification in MFC (Virdis *et al.*, 2010), there exist several other mechanisms such as anaerobic ammonium oxidation, and ammonium volatilization (biological conversion of ammonium ions into non-reactive NH_3^+), which favours nitrogen removal reported in MFC (Deng *et al.*, 2018; Elmaadawy *et al.*, 2020b; Vilajeliu-Pons *et al.*, 2018).

Phosphorus is also an important constituent of landfill leachate. Phosphorus removal takes place through precipitation with cations and biological accumulation with microorganisms (Elmaadawy *et al.*, 2020b). The

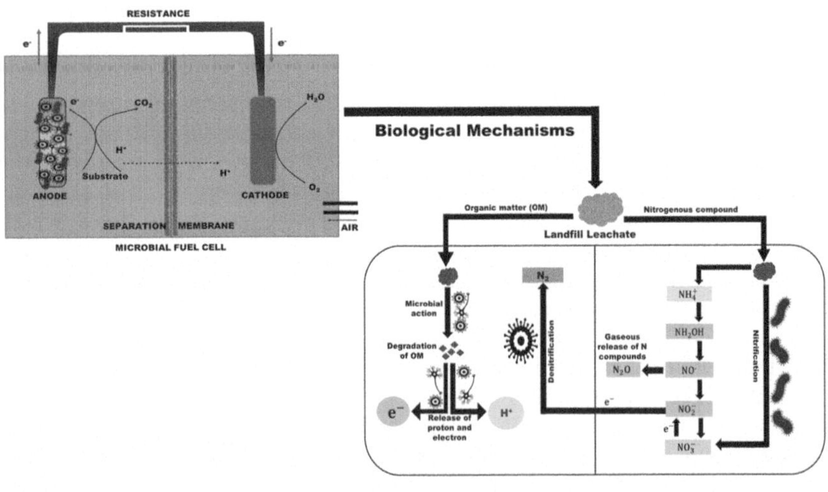

Figure 4.1 Biological mechanism for organics/inorganics removal in MFC based on leachate pollutants.

electrochemical precipitation in the form of salts such as struvite at the cathode also removes the phosphorus in MFC. Several factors such as pH, calcium, and magnesium salts are the limiting factors of struvite formation in MFC. A significant amount of phosphorus has been removed through struvite formation, which could be a slow releasing fertilizer for agricultural systems (Talboys *et al.*, 2016). Phosphorus recovery from landfill leachate hase not been studied so far since it contains very low level of phosphorus (Elmaadawy *et al.*, 2020b). The leachate contains various heavy metals such as Zn, Pb, Cu, Fe, and so on in varied concentration depending upon types of landfill (Wu *et al.*, 2015). These heavy metals are toxic and persistent in the environment, and needs to be removed from the leachate (Ezziat *et al.*, 2019). The deposition and reduction on the surface of the cathode is a dominant mechanism in an abiotic cathode, while chemical precipitation and electrochemical reduction also remove heavy metals in MFC systems. In addition to reduction, accumulation, sorption, and mineralization are the major mechanisms involved at biocathode for heavy metal removal in MFC systems (Colantonio, 2016; Ezziat *et al.*, 2019; Srivastava *et al.*, 2021a, 2021b).

4.2.2 Leachate toxicity and effects on MFC performance

Leachate has been used as a substrate in MFC because of its high organic matter content and nutrient load. The concentration and composition of various pollutants in leachates may vary with the age of the landfill (Özkaya *et al.*, 2013). The older leachate contains less biodegradable and high refractory organics, which causes toxicity to exo-electrogens as they break the biodegradable organic matters. Toxicity in leachate arises due to high organic matter, inorganic constituents, heavy metals, and various xenobiotic compounds. It is also associated with complex organic compounds such as lignin and phenol, heavy metals such as Hg, Pb, Cu, Zn, free ammonia, nitrous acid, and so on, which can alter the electroactive microbial community of MFC (Adekunle *et al.*, 2019b; Bernard *et al.*, 1997). S. Iskander *et al.* (2017) observed poor coulombic efficiency generated in MFC relating to the fact that toxicity present in leachate retards the activity of electron producing microbes. Landfill leachates are rich in ammonium and liberates higher amount of free ammonia and free nitrous acid. The toxicity produced due to nitrous acid and ammonia also inhibits microbial activity in MFC. The ammonia nitrogen present in leachate creates toxic effect on the anodic microbial community which simultaneously decreases the nitrogen removal ability of the system (Hassan *et al.*, 2018). The ammonia nitrogen decreases the nitrogen removal rate, simultaneously hampering the methanogenesis process and generation of bioelectricity (Wu *et al.*, 2015). Additionally, higher toxicity in leachate retards the growth and development of algae used in cathodes of MFCs which subsequently reduces the ammonium removal potential of the system (Nguyen *et al.*, 2017). Toxicity occurring due to dissolved heavy metals of leachate (Mg^{2+} and Fe^{2+}) also interfere with microbial metabolism leading to reduced MFC current (Adekunle *et al.*, 2019a). The higher toxicity could also affect the functioning of proton-exchange membrane (PEM). The chemical modifications might occur on PEM surface which would lead to development of harmful compounds, causing secondary pollution (Feng *et al.*, 2020a). The toxicity present in landfill leachates when treated in MFCs effects the efficiency of the system, for example, the level of toxicity determines removal of ammonium ions (Elmaadawy *et al.*, 2020a). It is highly essential to treat leachate properly as it contains higher amount of ammoniacal nitrogen other than toxic substances (Lakshmidevi *et al.*, 2020). As part of solution to the toxicity of leachate, studies have reported use of sulfate/sulfide MFCs which imitates the natural metabolism of microbes and produces less toxic effluent (Ieropoulos *et al.*, 2013). Using pyrrhotite-cathode in MFC drives bio-Fenton's reaction, which also favors effective reduction of toxicity from landfill leachates (Li *et al.*, 2010). The performance of MFCs can also be enhanced by diluting the leachate being treated, as it will reduce the level of ammonia and other refractory chemicals which deaccelerated bioelectricity generation (Nguyen *et al.*, 2017). There are several types of MFC used for leachate treatment such as single-chambered MFC (SCMFC), column-type MFC, and DCMFC, (Zhang *et al.*, 2015a). In addition, single-chambered series connection MFC (Gálvez *et al.*, 2009), and dual-chambered series and parallel connection MFC have also been reported to treat the landfill leachate (Vázquez-Larios *et al.*, 2014).

4.3 TYPES OF MFC FOR LEACHATE TREATMENT

Various types of MFC configuration have been used for removal of pollutants and bioelectricity generation from landfill leachate. Several designs such as SCMFCs, DCMFCs, up-flow anaerobic sludge blanket (UASB)-MFC, hybrid system and many more have been used for treating landfill leachate (Elmaadawy *et al.*, 2020b). SCMFC comprises of single chamber, where both the anode and cathode electrodes are present. Different electrode material produces different treatment efficiency within the system. For example, Ganesh and Jambeck (2013) have reported the use of SCMFCs for treating landfill leachate with an initial COD concentration of 12,033 ± 3,974 mg/L, where, two different anode material were compared, that is, activated carbon and biochar, which removed 74% and 28% of COD, respectively and no removal of ammonia was observed in the system. Instead, Puig *et al.* (2011) reported use of platinized carbon cloth and graphite rod as cathode and anode material, respectively, in a SCMFC, which was able to remove about 32% of COD and 16% of $NH_4^+ - N$ from diluted landfill leachate having mean COD and $NH_4^+ - N$ concentration of 507 and 802 mg/L, respectively. Whereas undiluted landfill leachate was being treated in the same system, which removed 4% of $NH_4^+ - N$ and 37% of COD, having 5,449 and 3,480 mg/L of initial $NH_4^+ - N$ and COD concentration, respectively. Similarly, studies have also reported use of MFCs for treating various concentrations of landfill leachate in open and closed circuit. Greenman *et al.* (2009) compared the efficiency of open- and closed-circuit MFCs in treating various dilutions of landfill leachate. The influent BOD_5 concentration was in the range of 2,400 mg/L which when diluted in the ratio of 1:8, 1:4, and 1:2 produced 222, 420, and 1,096 mg/L BOD_5 concentration in the effluent of closed-circuit MFC, whereas open-circuit MFC produced 213, 542, and 1,296 mg/L of BOD_5 concentration in the effluent. Another modification in SCMFC was performed by developing up-flow air cathode membrane free MFC using fine powder of C/Pt on carbon cloth, as cathode material, which was wrapped around anode and exposed to air. The characteristics of leachate includes COD (3,400 mg/L) and $NH_4^+ - N$ (744 mg/L) having a treatment efficiency of 89.4% and 23.3%, respectively (Zhang *et al.*, 2008).

The other design aspect to treat landfill leachate is DCMFC, where MFC is made up of two chambers: anode and cathode separated by a cation-exchange membrane. Hassan *et al.* (2018) reported a use of carbon felt as cathode and anode electrode material in a DCMFC separated by a cation-exchange membrane and is used to treat various dilutions of a young landfill leachate. It is observed that the DCMFC is able to remediate about 90% of COD at 60% leachate COD strength (100% COD strength = 16,000 mg/L). The COD removal efficiency of this system decreased to 49% when the influent COD strength was increased to 100%. Similar trend was observed for removal of $NH_4^+ - N$ ions, where 59% and 15–17% of $NH_4^+ - N$ removal is observed at 20% and 100% (>2,000 mg/L) leachate strength. Similar to SCMFC, comparative study of open- and closed circuit for landfill leachate treatment have also been provided for DCMFC. Li *et al.* (2010) reported the use of pyrrhotite-coated graphite as cathode electrode in open- and closed-circuit DCMFC for treating real landfill leachate with initial COD concentration of 1,022 mg/L, where, closed-circuit MFC is found to have better COD removal efficiency (78%) as compared to that of open-circuit DCMFC (28%). On the other hand, Özkaya *et al.* (2013) modified the anode and cathode material with Ti–TiO_2 in DCMFC for treating a young landfill leachate (leachate characteristics: initial COD and sulfate concentration of 50,000 and 3,400 mg/L), which was able to remediate about 40% and 15% of COD and sulfate, respectively, from the influent. Subsequent to change in electrode material and circuit type, several changes were also made in the connection type, such as series and parallel connection, in and outside MFCs, either fluidically or technically, to enhance the power performance of the entire system. Vázquez-Larios *et al.* (2014) reported series and parallel connection of anode within DCMFC, where about 70–86% removal of COD is observed with an initial COD concentration of 12,000 mg/L. Whereas, Gálvez *et al.* (2009) have reported about a hydrodynamic series connection where the fluid passes through three columns serially within the system. The developed system can remove about 79.4% and 81.6% of COD and BOD_5 having 7,050 and 2,962 mg/L of COD and BOD_5 as the initial concentration, respectively. Also, Feng *et al.* (2020b) reported series connection outside MFC, where two MFCs are connected serially to single MEC in the following manner MFC–MFC–MEC and MFC–MEC–MFC, which achieved a maximum COD removal efficiency of about 60% (about 1,000–2,000 mg/L). However, to increase the treatment efficiency of the system, integration of DCMFC had been done to various other treatment systems. For example, Lee *et al.* (2013) developed two integrated ammonium oxidation/MFC reactor and anammox/MFC reactor for removal of nitrogen and bioelectricity generation. The ammonium oxidation/MFC was able to remediate more than 92% of total nitrogen (TN) where the initial TN concentration was 120 mg/L, whereas, anammox/MFC was able to remove 94% TN. Tugtas *et al.* (2013) developed UASB-MFC design where young landfill leachate was pre-treated in UASB and then simultaneously treated with DCMFC. It was observed that the initial $NH_4^+ - N$ concentration was in the range of 4,640–7,020 mg/L which on treatment in UASB decreased to 950–1,270 mg/L and on further treatment in DCMFC, the concentration decreased to 61 mg/L.

Economically, coupling and modifications in MFCs not only enhances power output of the system but also helps in reducing the toxicity of pollutants in a cost-effective and sustainable manner. Moreover, such integrated MFC designs, and configurations can treat non-uniform leachate characteristics (e.g., as reported in some studies with different dilution ratio having different organics load) with least modification and instalment facility, providing enhanced electrical energy and wastewater treatment facility.

4.4 BIOELECTRICITY GENERATION FROM THE LANDFILL LEACHATE

Electricity generation from landfill leachate follows a distinct and complex process in comparison to single substrate such as acetate/glucose (Wu *et al.*, 2015). This is due to the fact that it contains large concentration of dissolved organic substances (i.e., BOD, COD, VFA, humic, and fulvic acids), xenobiotic organic compounds (i.e., phenols, aromatic hydrocarbons, cresols, phthalates, and furans), macro inorganic substances (i.e., Ca^{2+}, Mg^{2+}, SO_4^{2-}, Cl^-, Na^+, etc.) and heavy metals (i.e., Ni^{2+}, Cu^{2+}, Cd^{2+} and $Cr^{3+/6+}$ etc.) (Renou *et al.*, 2008). The volatile fatty acids (VFAs) are easily biodegradable organic fraction of landfill leachate, whereas other humic and fulvic acids can dominate methanogenic phase (Iskander *et al.*, 2016). The presence of heavy metals and recalcitrant pollutants can also impact biodegradability of leachate resulting into low bioelectricity generation. For an instance, Hernández-Flores *et al.* (2017b) compared two substrates, that is, municipal wastewater and mixture of landfill leachate and municipal wastewater. The use of single substrate has performed better than mixture of substrate in terms of COD removal and bioelectricity generation. Municipal wastewater achieved power density, COD removal and CE of 10,380 mW/m^3, 72%, 27% compared to mixed substrate of 8,050 mW/m^3, 53% and 9.5%, respectively. Hence, pre-treatment of landfill leachate is prerequisite to effectively oxidize the complex organic compounds such as long-chain fatty acids and other xenobiotic aromatic compounds by exoelectrogens (Chaudhuri & Lovley, 2003). MFCs fed with complex organic substrates are only able to recover 2–6% of the theoretical voltage (Lee *et al.*, 2008). This indicates that only a limited amount of energy can be harnessed by MFCs (Wu *et al.*, 2015). The detailed mechanisms involved to treat landfill leachate and interrelation with electron transfer is provided in the following section.

4.4.1 Electron transfer mechanism in MFCs to treat landfill leachate

There are several involved mechanisms for electron transfer in MFC. The detailed mechanism is shown in Figure 4.2. The common electron transfer pathway is direct interspecies electron transfer (DIET), which is thermodynamically more favorable in MFC even for treating landfill leachate. Whereas indirect electron transfers, that is, electron transfer through shuttling is also a pathway for exoelectrogens to transfer electron with the help of fermentative microbes such as *Pseudomonas chlororaphis*, *Geothrix fermentans*, and *Clostridium butyricum*.

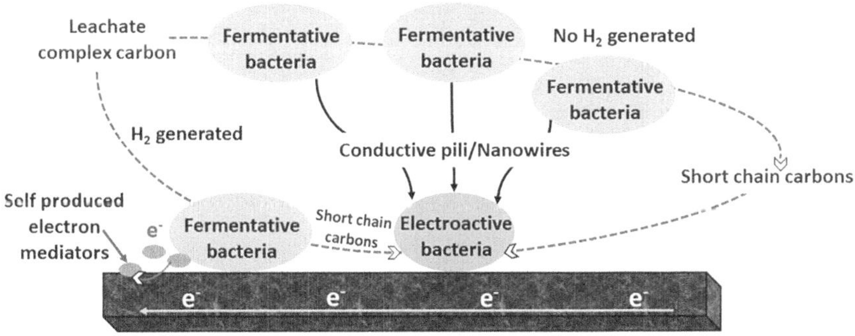

Figure 4.2 Electron transfer mechanism to anode by fermentative bacteria while treating complex carbon of landfill leachate.

These fermentative microbes can help in shuttling of electrons to electrodes with the help of self-produced electron mediators, Fe^{3+} reduction or abiotic oxidation of their products (such as H_2) at the anode (Lovley, 2006; Wu *et al.*, 2015). The syntropy of hydrogen-mediated electron transfer in landfill leachates is well described by Jakobsen *et al.* (1998). Further increased bioelectricity production with the coexistence of *hydrogenophage*, that is, H_2 consuming exoelectrogens and fermentative microorganisms is also mentioned by Kimura and Okabe (2013). In comparison to indirect electron transfer, which is mainly governed by H_2-mediated interspecies electron transfer DIET is more thermodynamically favorable. It is due to the fact that DIET is energy efficient in terms of: (1) H_2 can inhibit organics removal and (2) without H_2 gradient formation interspecies H_2 transfer cannot take place (Wu *et al.*, 2015). However, DIET needs the physical contact between electrodes and microbes for effective electron transfer. Other than hydroge- mediated electron transfer and DIET, mediated electron transfer (MET) is also a frequently observed mechanism for electron transfer in MFC. In MET, electron transfer takes place with the help of redox mediator. In general, landfill leachate can favor MET compared to other type of wastewater since it has high content of accumulated humic substances, that is, around 60% of the dissolved organic matter compared to just <6% in other type of wastewater, which act as a redox mediator for electron transfer (Wu *et al.*, 2015).

Furthermore, fresh and intermediate stage leachate are generally preferred for bioelectricity generation due to its high organic content ranging between 4,000 and 40,000 mg/L (Table 4.1). Among which 60–70% COD of fresh and intermediate leachate is biodegradable (Wu *et al.*, 2015). Whereas, mature or old leachate is avoided due to aging of leachate and the COD/BOD ratio decreases from 0.7 to 0.04, which in turn decreases its biodegradability (Ehrig, 1984). In addition to that, with aging of leachate it changes its characteristics such as elevated levels of toxicants, significantly high levels of ammonium and low dissolved organic matter (Wu *et al.*, 2015). Moreover, fresh and intermediate leachates also cannot be directly utilized in MFCs due to their dynamic COD concentration. In some cases, COD concentration is reported to be too low (e.g.,

Table 4.1 Showcasing literature studies on treatment of landfill leachate through MFCs.

Leachate		MFC Configuration	Mode of Operation	COD/Load of Leachate	COD Removal (%)	Power Output	CE (%)	References
Intermediate	Real	DCMFC	Batch	<100 mg/L 4,900 mg/L	98% 69.54%	2.1 W/m³ 6.8 W/m³	22.1% 3.1%	You *et al.* (2006)
Fresh	Real	SC membrane less MFC	Continuous	3,400 mg/L with 0.65 kg COD/m³ d	89.4%	12.8 W/m³	14.4 (maximum)	Zhang *et al.* (2008)
Intermediate	Real	DCMFC	Continuous mode	7,050 mg/L	79.4%	0.07–0.25 mW/m²	–	Greenman *et al.* (2009)
Mature	Real	Three SCMFC	Continuous (in series)	12,900 mg/L	79.4%	Reactor 1: 1.18 mW/m² Reactor 2: 5.6 mW/m² Reactor 3: 1.7 mW/m²	–	Gálvez *et al.* (2009)
Mature	Real	SCMFC	Continuous	507 mg/L (diluted)	32%	0.06–0.35 W/m³	2	Puig *et al.* (2011)
Intermediate	Real	DCMFC	Continuous (NO_2 acceptor)	250 mg/L	–	12 mW/m²	–	Lee *et al.* (2013)
Fresh	Real	DCMFC	Continuous batch	4,000–4,300 mg/L	98% 90%	158 mW/m² 109 mW/m²	–	Tugtas *et al.* (2013)
Fresh	Real	DCMFC with Ti-TiO_2 electrodes	Continuous	1,000 mg/L 50,000 mg/L	43% at 50,000 mg/L	11 A/m²	57% 1%	Özkaya *et al.* (2013)

–	Real	Single-chambered air- cathode MFC with activated carbon anode and biochar anode	Batch mode	12,033 ± 3,974 mg/L	74.7% ± 5.5% (activated carbon anode) 28.6% ± 8.9% (biochar anode)	699 ± 33 mW/m^3 (activated carbon anode); 575 ± 168 mW/m^3 (biochar anode)	0.58% ± 0.11% (activated carbon anode); 1.27% ± 0.61% (biochar anode)	Ganesh and Jambeck (2013)
Fresh	Real municipal leachate	SCMFC	Batch	12,300 mg/L	72%	1,839 ± 156 mW/m^3	6.7%	Vázquez-Larios *et al.* (2014)
Old/mature	Real	SCMFC	Batch mode	908 to 3,200 mg/L	16%	824 mW/m^3	–	Damiano *et al.* (2014)
Fresh	Real	DCMFC	Continuous	17,500–22,600 mg/L	95.1 ± 1.8%	2.71 ± 0.09 W/m^3	–	Zhang *et al.* (2015a)
–	Real	Dual-chambered membrane less MFC	Continuous	20,100 mg/L	84–89%	2.77 ± 0.26 W m^3	–	Zhang *et al.* (2015b)
Young	Real municipal leachate	SCMFC	Batch mode	12.3 ± 0.5 g/L	71%	13.746 W/m^3	78%	Vázquez-Larios *et al.* (2015)
Young	Real	Single-chambered air cathode MFC	Batch mode	2,000 mg/L	39.3%	20 000 mW/m^3	71.7%	Hernández-Flores *et al.* (2015)
Fresh	Real	Dual-chambered membrane less MFC	Batch mode	20,100 mg/L	86.5%	2.85 W/m^3	–	Zhang *et al.* (2016)

(*Continued*)

Table 4.1 Showcasing literature studies on treatment of landfill leachate through MFCs. (*Continued*)

Leachate		MFC Configuration	Mode of Operation	COD/Load of Leachate	COD Removal (%)	Power Output	CE (%)	References
Fresh	Real	Single-chambered direct air breathing cathode-based MFC	Fed batch mode	–	–	1,513 mW/m^2 with OCV 1.29 V	–	Sonawane *et al.* (2017)
–	Municipal wastewater; Municipal wastewater + landfill; leachate Real (Young)	SCMFC	Batch mode	219 ± 21 mg/L 4,315 ± 304 mg/L 12,305 ± 511 mg/L	80% 86% 53%	315 mW/m^3 82 mW/m^3 489 mW/m^3	47.5% 3.4% 15%	Hernández-Flores *et al.* (2017b)
Young	Real	SCMFC	Batch mode	3,053 ± 588 mg/L	7.4%	10,380 mW/m^3	–	Hernández-Flores *et al.* (2017a)
–	Real (diluted with municipal wastewater)	Dual-chambered algal cathode MFC	Batch	1,938 ± 27 mg/L	97%	300 mV	–	Nguyen *et al.* (2017)
–	Food waste leachate	Air cathode MFC	Batch mode	2,860 mg/L	95.4 ± 0.3%	1.86 W/m^3	11.07 (maximum)	Wang and Lim (2017)
–	Real	Tubular-shaped DCMFC	Batch mode	4,737 mg/L	75.3%	0.123 kWh/m^3	2.8% (maximum)	Iskander *et al.* (2017)
–	Real	DCMFC	–	3,300 mg/L	–	9.15 W/m^3	–	Kumar *et al.* (2017)

	MSW landfill leachate	Dual-chambered algal-assisted MFC	–	2,567.32 mg/L	76.5%	95.63 mW/m²	–	Lakshmidevi *et al.* (2020)
Mature	Real	H type MFC-based bio-electro-Fenton system	Batch mode	2,152 ± 624 mg/L	41.6–49.3	600–700 mA/m²	28–40	Hassan *et al.* (2017)
–	Real	Recirculated DCMFC	Fed batch mode	500–1,250 mg/L	72.27% (maximum)	14.42 mW/m²	10.25%	Moharir and Tembhurkar (2018)
–	Synthetic landfill leachate	Dual-chambered OC MFC; Dual-chambered NA MFC; Dual-chambered A MFC	Fed batch mode	4,000 mg/L	56.82% (OC); 55.17% (NA); 65.12% (A)	105.9 mW/m² (OC); 114.79 mW/m² (NA); 135 mW/m² (A)	–	Huang *et al.* (2018)
Young and old	Real	H type DCMFC	Both continuous and batch	15,000–20,000 mg/L (Young); 800–1,000 (old)	90.0 ± 1.2% (batch, young) 55.5% (continuous, young)	96.8 mW/m² (batch, young); 75 mW/m² (continuous, young)	21.3%	Hassan *et al.* (2018)
–	Simulated leachate	Recirculated DCMFC	Batch mode	3,188 mg/L	90%	230 mW/m²; 9 kWh/m³ (maximum)	20% (maximum)	Li and Chen (2018)
–	Real	Dual-chambered algal cathode MFC	Continuous	–	26%	303 mV	–	Nguyen and Min (2020)

(Continued)

Table 4.1 Showcasing literature studies on treatment of landfill leachate through MFCs. (*Continued*)

Leachate		MFC Configuration	Mode of Operation	COD/Load of Leachate	COD Removal (%)	Power Output	CE (%)	References
Young	Real	Cathodic algal biofilm MFC equipped with biocathode oxygen consuming unit	Fed batch mode	20,055 mg/L	>86%	0.39 V	6.6 ± 0.06%	Elmaadawy *et al.* (2020a)
–	MSW landfill leachate	Dual-chambered algal-assisted MFC	–	2,567.32 mg/L	76.5%	95.63 mW/m^2	–	Lakshmidevi *et al.* (2020)
–	Real	Osmotic MFC	Sequential batch mode	2,638 ± 34 mg/L	–	0.44 W/m^2	–	Jiang *et al.* (2021)
Aged and fresh	Real	Single-chambered air cathode MFC	Batch	12,960, 15,920 mg/L	7.7%~51.4% (aged) 55.8%~61.7% (fresh);	6.61–7.48 W/m^3 (aged); 22.7–25.[illegible] W/m^3 (fresh)	5.9%~11.3% (aged) 4.3%–7.6% (fresh)	Cheng *et al.* (2022)

<150 mg/L), which can be a limiting factor for exoelectrogenic activity; whereas, in some cases COD concentration >1,000 mg/L, which can diminish coulombic efficiency (CE) of treatment systems (You *et al.*, 2006). Özkaya *et al.* (2013) examined young leachate in MFC with initial COD concentration from 1 to 50,000 mg/L, the study reported initial rise followed by significant decrease in power generation and decline in CE from 57% to 1%. Thus, proper pre-stabilization and dilution of landfill leachate stream to an optimum COD range is strongly recommended (Wu *et al.*, 2015). Similar results are also documented by other researchers, where decrease in power density from 2.23 to 1.0 W/m^3 along with decrease in CE from 34% to 20% while treating simulated leachate in MFC is observed (Teng *et al.*, 2010). Furthermore, Puig *et al.* (2011) reported that addition of raw leachate in MFC increases power density from 0.15 to 0.30 W/m^3, whereas, CE decreases to approximately 2%. Similar results were also observed in a recent study, where fresh and matured (aged) leachate was collected from the identical subtropical area of different season and studied in MFC (Cheng *et al.*, 2022). Fresh leachate showed enhanced power generation, that is, 22.5 –25.6 W/m^3 and CE 4.3–7.6% in comparison to 6.61–7.48 W/m^3 from mature leachate and CE of 5.9–11.3% (Cheng *et al.*, 2022). Besides, suitable substrate, configuration, and operating conditions also have higher impact on electricity generation from leachate-fed MFCs (Iskander *et al.*, 2016; Wu *et al.*, 2015), as represented in Table 4.1.

Furthermore, electrode selection also plays a crucial role in electricity production. Carbon cloth is the most applied anode material in leachate-fed MFC systems (Xia *et al.*, 2018). Two studies have achieved different open-circuit voltages (OCV) while operating fresh leachate-fed MFC with similar operating conditions, that is, 534 mV (Damiano *et al.*, 2014) and 433 mV (Ganesh & Jambeck, 2013). Further with identical substrate conditions and different electrodes at the cathode, distinct OCV was achieved by Li *et al.* (2010). The study reported increase in OCV by 120 mV in pyrrhotite cathode compared to graphite-based cathode. Recently, activated carbon ceramic-supported cathodic electrodes were made utilizing three different techniques, that is, wash coat (WC), brush coat (BC) and ultrasound-assisted deposition (SC) and further implemented in MFC treating municipal waste leachate. The highest electricity generation and COD removal was achieved with electrode produced using 20 kHz ultrasound, that is, SC of 10.7 mW/g L and 85%, respectively. Highest performance with SC was reported as a consequence of comparatively low charge transfer and biofilm resistance compared to BC and WC of 23.71 Ω (Pandis *et al.*, 2022). Electrode surface area is important to consider since electrode functions as carrier surface for the growth of microbial communities and also as electron donor/acceptor. Several researchers noted positive impact of electricity generation with increased surface area while treating landfill leachate (Gálvez *et al.*, 2009; Sonawane *et al.*, 2017). Gradual increase in power density from 0.36, 0.84 and upto 1.7 mW/m^3 was reported with increase in cathode electrode surface area from 1, 5 and 36 cm^2, respectively, while treating landfill leachate in SCMFC (Sonawane *et al.*, 2017). Furthermore, Gálvez *et al.* (2009) operated three column MFC, fluidically connected in series with single feed line and recirculation loop and investigated the effect of increasing electrode surface with landfill leachate treatment. The study revealed increase in power output by 264% for 1st column (i.e., from 500.7 to 1,822.6 mW/cm^2), 118% for 2nd column (i.e., from 443.2 to 964.5 mW/cm^2) and 151 for 3rd column (i.e., from 683.5 to 1,714.3 mW/cm^2) with increase in electrode surface area from 360 to 1,080 cm^2. Conclusively, MFCs in different operation conditions and configurations have represented a huge potential towards significant bioelectricity production with distinct forms and stages of highly contaminated landfill leachate wastewater.

4.5 RESOURCE RECOVERY FROM THE LEACHATE

As aforementioned, leachate is characterized as a complex wastewater with high prospects of polluting soil and groundwater and aggravating other environmental problems. Nevertheless, it has been identified as potential source for resource recovery as it contains high concentration of ammonium nitrogen, low levels of phosphorus, and high concentrations of inorganics and metal ions. Thus, the recovery of useful resources, instead of removal is prudent for sustainable leachate treatment and management (Iskander *et al.*, 2016; Nancharaiah *et al.*, 2015).

4.5.1 Metal recovery using MFC from leachate

The leachate is featured with high electrical conductivity which is attributed to the presence of high concentrations of inorganics and metal ions in it. Thus, it favors electricity generation while treating leachate wastewater in MFC (Iskander *et al.*, 2016). Along with, it opportunistically recovers metals present in leachate at the cathode of MFC (Iskander *et al.*, 2016; Nancharaiah *et al.*, 2015). Common metals present in the landfill leachate include iron, manganese, zinc, cadmium, copper, nickel, silver, and lead, however, their prevalence and abundance are specific to the landfill site, source and age (Iskander *et al.*, 2016). In general, the metal concentration is below 1 mg/L which makes metal recovery from leachate non-economical. Nevertheless, much higher concentrations have also been reported in certain cases, for instance, 10 mg/L of copper, 5 mg/L of lead, 13 mg/L of nickel, 1,000 mg/L of zinc, and up to 5,500 mg/L of iron (Baun & Christensen, 2004; Kjeldsen *et al.*, 2002). In MFC, oxidation of organics happens at the anode, resulting in the flow of electrons to the cathode under the effect of redox gradient. Therefore, four strategies have been explored for metal reduction and their recovery at the cathode (Iskander *et al.*, 2016). In the first approach, the MFC is modified, and the cathode is placed anaerobically such that the metals with redox potential higher than the anode potential (e.g., Cu, Fe, etc.) can be reduced on the cathode electrode (Modin *et al.*, 2012; Wang & Ren, 2014). The reaction is thermodynamically favorable and metal acts as terminal electron acceptor for direct recovery (Figure 4.3a). The reduction of metals with redox potential lower than the anode potential, such as cadmium, lead, nickel, zinc and so on, is not feasible in MFC and requires assistance of some external power to drive electrons from an anode electrode to a cathode electrode (MEC mode), so that metal reduction is accomplished (Figure 4.3b) (Modin *et al.*, 2012; Qin *et al.*, 2012). In the third approach, specialized microbes are inoculated

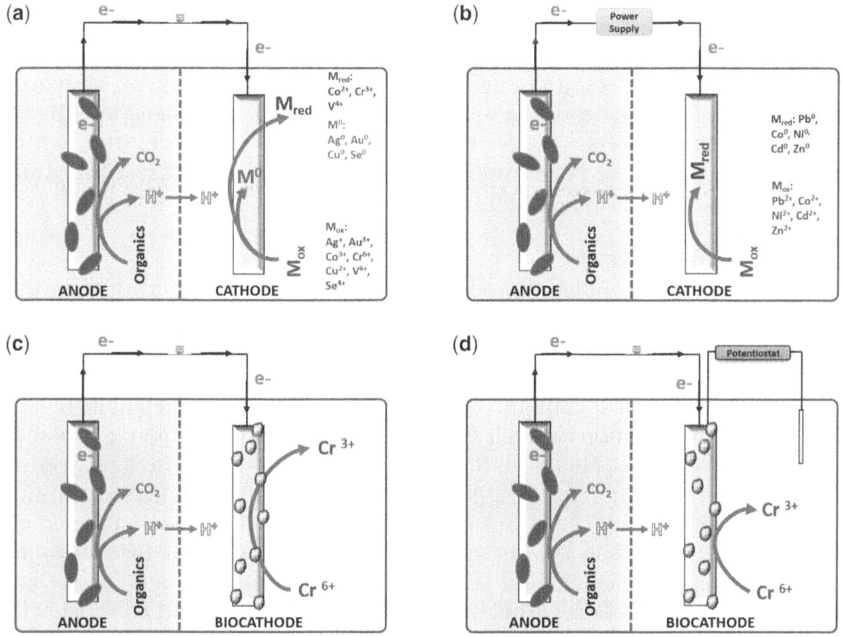

Figure 4.3 Four bio-electrochemical approaches demonstrating heavy metal reduction and recovery by (a) direct reduction at abiotic cathode, (b) abiotic cathode with external power supply, (c) reduction at bio-cathode, (d) reduction at controlled bio-cathode.

in the cathode (bio-cathode) of MFC which selectively reduce metals over the cathode surface which can then be recovered efficiently (Figure 4.3c) (Tandukar *et al.*, 2009). The fourth strategy subject bio-cathode to controlled potential such that the microbes attain required energy to reduce the metals (Figure 4.3d) (Huang *et al.*, 2011). Depending upon the redox speciation the reduced metals get deposited over the cathode surface which can be easily recovered.

The greatest challenge in metal recovery from landfill leachate is its low concentration and form in which it is present. Such as most metals are soluble at low pH with high organic content in acidogenic phase compared to high pH and low organic content in another phase (Harmsen, 1983). Thus, it is important to understand the interlinked factors such as leachate age, available forms of metals and its concentration to recover precious metals economically and strategically in MFC. For this, metals with high concentration are generally targeted for recovery. To intensify the process further, leachate is treated at the anode chamber and the metal ions are allowed to migrate and concentrate in the cathode chamber through a membrane so that they can be directly reduced by the electrons received from the anode. This eliminates the risk of decreased metal reduction and occurrence of undesired secondary reactions in the presence of large amount of organic compounds when fed with anode effluent.

Thus, it is important to investigate the migration of multiple ions and methods for collecting reduced compounds for the successful operation of this process. Recognizing the low concentration of metals, it would be judicious to target the recovery of other resources and couple metals along with them. Moreover, low amount of metals deposited over the cathode could be utilized to enhance other beneficial processes rather than harvesting. For instance, iron can be consumed for cathodic bio-Fenton's reaction that would oxidize recalcitrant pollutants (Rozendal *et al.*, 2009).

4.5.2 Nutrient recovery

Nutrient recovery is of high interest from landfill leachate due to the fact that leachate has high concentration of both ammonium and phosphate, with ammonium concentration higher than phosphate (Shehzad *et al.*, 2016). In BES, recovery of ammonium was tested with various distinct approaches such as (1) ammonium oxidation at cathode region, (2) denitrification in the anode region, (3) ammonium migration through PEM, and (4) ammonium stripping, as shown in Figure 4.4 (Kelly & He, 2014; Shehzad *et al.*, 2016). The ammonia stripped in the process of ammonium stripping can be recovered through its adsorption in diluted sulfuric acid (i.e., H_2SO_4) to form ammonium sulfate (Iskander *et al.*, 2016). This end product is very valuable and has a potential for utilization in agriculture, polymer or food sector as fertilizer, nitrogen synthesizer and source of food production, respectively (Matassa *et al.*, 2015). In contrast, with

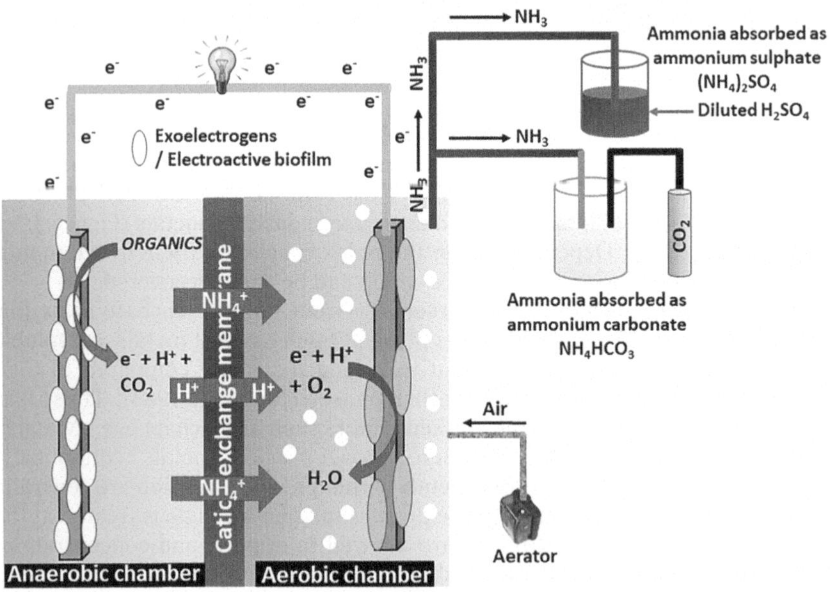

Figure 4.4 Recovery of ammonia in MFC through transport of ammonium to cathode followed by ammonium stripping.

the use of pre-saturated adsorption media, that is, saturated H_2SO_4 solution with ammonium sulfate can result in recovery of pure ammonia in the form of ammonium sulfate crystals which can be further used in laboratory chemical preparation and for production of fertilizers (Tao & Ukwuani, 2015). A study with landfill leachate-fed system revealed enhancement in the recovery of ammonia by approximately 11% with aeration in the cathode chamber. The highest ammonium concentration collected from stripped ammonia with the help of adsorption technique was 0.77 mol/h (Qin *et al.*, 2016). However, aeration can be quite energy intensive, thus development of other alternative methods is required to drive the ammonium out of the cathode in MFCs. Later, it was suggested that in a DCMFC with cation-exchange membrane, phosphate can be removed with the help of microbial adsorption and through chemical precipitation (Ye *et al.*, 2019). However, ammonium can be removed by current generation and recovered through precipitation (Ye *et al.*, 2019).

Phosphorous concentration is generally low in leachate wastewater and its recovery can be generally carried out through precipitation inside the cathode chamber of MFC (Kelly & He, 2014). The formation of struvite/magnesium ammonium sulfate ($NH_4MgPO_4 \cdot 6H_2O$) was practiced at high pH values of 9 ± 0.2 for the recovery of phosphate from mature landfill leachate (Warmadewanthi *et al.*, 2021). Damiano *et al.* (2014) introduced struvite precipitation in DCMFC fed with landfill leachate wastewater and achieved phosphorus removal efficiency of 80%. Although, there are numerous challenges which still needs to be resolved for phosphate recovery in MFCs such as (a) effect of precipitate formation on performance of electrode, (b) electrode regeneration with the purpose of reuse, (c) adjusting optimal stoichiometry for struvite precipitation, (d) amendments in the design of MFC system to recirculate the effluent from anode in order to supply phosphate to cathode region for recovery (Iskander *et al.*, 2016; Shehzad *et al.*, 2016).

4.5.3 Water recovery

Leachate volume reduction is crucial for sustainable landfill leachate management and water recovery/extraction can help in accomplishing this goal. In MFCs, filtration membranes such as nanofiltration, reverse osmosis, ultrafiltration, and microfiltration can be integrated in two ways (1) internally for example: as a separator between the anode and cathode, and (2) externally, where membrane unit is separated from MFC and connected hydraulically (as shown in Figure 4.5) (Yuan & He, 2015). Li *et al.* (2014) revealed enhancement in terms of water recovery, organics removal and power production with the application of ultrafiltration membrane in aerated cathodic chamber of MFC. This synergistic cooperation was named as membrane bio-electrochemical reactor (MBER) and revealed 90% of COD removal, 69% of total inorganic nitrogen removal along with generation of membrane permeate having turbidity <2 NTU (Li *et al.*, 2014). The system is also advantageous in terms of low energy requirement, that is, 0.09 kWh/m^3, which is substantially lower than energy requirement in membrane bioreactors (MBRs) and simultaneously MBER generates 0.032–0.064 kWh/m^3 of bioelectricity. Less fouling rate was also obtained in MBER resulting from aeration in the cathodic chamber.

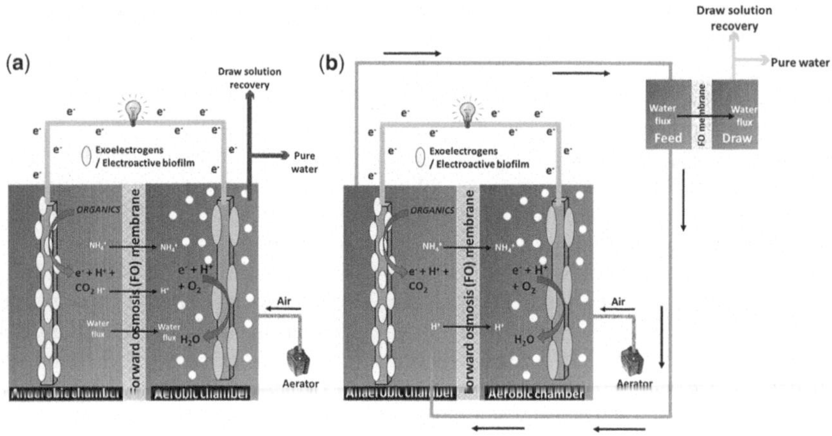

Figure 4.5 Representing (a) internal integration and (b) external integration of forward osmosis membrane in MFCs for water recovery.

An emerging membrane technology, that is, forward osmosis (FO) is investigated with BES to minimize fouling conditions while operating (Zhao *et al.*, 2012). FO is integrated with BES to form osmotic BES (OsBES), which can simultaneously remove organics and recover energy and water. FO utilizes the naturally developed osmotic gradient pressure between the feed solution and draw solution in order to drive water towards the side of concentrated draw solute (Iskander *et al.*, 2016). Draw solute plays a very crucial role since it is used to draw water across a semi-permeable membrane from anode compartment with high osmotic potential (Lu *et al.*, 2014). Ammonium carbonate functions as a very attractive draw solute for FO due to its easy recovery by moderate heating (McCutcheon *et al.*, 2005). BES-integrated FO system has been investigated for the treatment of leachate and revealed low COD removal and ammonia recovery with cathode aeration of approximately 35% and 65%, respectively. Further, with the treated leachate acting as feed in FO system resulted in extraction of 51% water in the period of 3.5 h with utilization of 2 M of ammonium carbonate as a draw solution (Qin *et al.*, 2016). Furthermore, for the application of BES-FO certain challenges need to be addressed: (a) membrane fouling: FO membrane receives bio-fouling and pollutants scaling due to the presence of large concentration of contaminants in the leachate. Thus, to mitigate fouling issues, BES should function as pre-treatment technology to reduce the organics concentration and decrease biodegradability of leachate, (b) utilization of extracted water should be explored. Since the recovered water is free from most of the contaminants, but still quality of water should be monitored and accordingly its application should be determined such as agricultural irrigation/gardening, landscaping, automobile washing, discharge into ground water for water table recharging (c) disposal of brine solution used as draw solution during FO operation should be studied further. FO will remove the excess water and concentrate the leachate but further this brine solution also needs disposal. The one solution can be mixing of this solution into leachate again but its impact on biological BES degradation should be investigated, and (d) successful functioning of FO is critically dependable on selection of appropriate draw solution. Sodium chloride/brine solution is widely applied in FO studies but it is challenging in terms of reverse salt flux and regeneration issues, thus limiting its use in FO studies. Use of ammonium carbonate is gaining interest as draw solution, particularly for removing of ammonia from leachate (Iskander *et al.*, 2016; Lu *et al.*, 2014; Shehzad *et al.*, 2016). FO is generally considered advantageous over other membrane filtration techniques such as micro filtration and ultrafiltration because it is capable of rejecting wide range of contaminants including heavy metals. Besides, reverse osmosis and nanofiltration can also recover water to a great extent, but there is no report on using BES in combination with these technologies for treatment of leachate (Iskander *et al.*, 2016).

4.6 RECENT DEVELOPMENTS IN LEACHATE TREATMENT USING MFC

Since the last two decades, MFCs have proven to be a reliable and sustainable treatment technology for treating various types of wastewaters. The enhancement of bioelectricity generation in MFCs, brought to attention the use of landfill leachate to be used as substrate in MFCs. Till now, the studies reported that landfill leachate in MFCs has proved to be a viable option for treating various kinds of organic (Özkaya *et al.*, 2013) and inorganic pollutants (Greenman *et al.*, 2009; You *et al.*, 2006). Whereas, decreased power output and higher installation costs limit the upscaling of this technology. Keeping in view the applicability of this technology, several strategies have been developed to improve efficiency of MFCs such as using hybridized systems, that is, integrating MFC to other treatment technologies such as UASB, anammox reactor, and so on. Also, cheaper electrode materials, extending surface area of electrode, and optimizing operational parameters helps in improving the functioning of MFCs using landfill leachate. Various studies have reported the use of algal cells in MFCs for treatment of high refractory organic material leachate associated with generation of bioelectricity (Elmaadawy *et al.*, 2020a; Lakshmidevi *et al.*, 2020; Nguyen & Min, 2020; Nguyen *et al.*, 2017). Lakshmidevi *et al.* (2020) used algal-assisted DCMFC to treat MSW leachate. It is observed that the system was able to treat the maximum COD (76.5%) at 50% dilution (initial COD = 2,567.32 mg/L). In addition, when such cathodic algal-assisted MFC were integrated with a bioactive oxygen consuming unit, the hybridized system remediates 86% of biorefractory organic compounds along with 89.4% of ammoniacal nitrogen (Elmaadawy *et al.*, 2020a). Additionally, cathodic algal-assisted MFCs also effectively removed various types of nutrients present in landfill leachate. The extractions of nitrogenous and phosphatic nutrients from leachate make MFCs a viable option with various field-scale application (Nguyen *et al.*, 2017; Nguyen & Min, 2020). MFCs are also integrated with bioelectro-Fenton systems for treating the recalcitrant organic compounds through biochemical processes such as partial nitritation anammox (Hassan *et al.*, 2017). Whereas, tubular MFC have also been reported in remediating UV quenchers from landfill leachate to ease UV disinfection in wastewater treatment facility (Iskander *et al.*, 2017). The performance of MFC is also improved by coupling the technology with various types of physical and chemical processes. When MFC is hybridized to osmotic process, that is, osmotic MFC, the technology exhibited higher pollutant removal capacity with a power output of 0.44 W/m^2 (Jiang *et al.*, 2021). Coupling of coagulation process with MFC, that is, addition of coagulants to anodic compartment treated 99% COD from landfill leachate with a higher power generation capacity (Kumar *et al.* 2019). Apart from this, modifications are also done in electrolytes being used in the MFC. For example, Elmaadawy *et al.* (2020b) used combination of landfill leachate and aerobic sludge for treating wastewater of shale gas fracturing unit associated with an enhanced bioelectricity generation capacity. Till now very limited studies have been reported regarding developments in MFCs for treating various concentrations of landfill leachates. Further, extensive research studies should be carried out with different modifications in MFC design and configurations which will enhance the treatment performance of the system with simultaneous bioelectricity generation.

4.7 CONCLUSION

The handling and management of solid waste is challenging due to its bulk generation. Landfilling is generally sought as an effective solution for solid waste management. BES have shown their potential for leachate treatment with simultaneous energy and resource recovery, utilizing the synergy of biological and electrochemical processes. Several aspects of MFCs such as MFC configuration, electrode materials, integration of several MFCs either electrically or hydraulically and hybrid MFC systems have significantly affected the MFC performance for leachate treatment. The mechanisms underlying the leachate treatment using MFCs are sophisticated and greatly depend on the leachate age and biodegradability, thus, needs to be well understood for better application. Bio-electricity generation is an additional leachate treatment using MFC which eventually minimizes the cost incurred in its operation. Moreover, bio-electrochemical leachate treatment facilitates resource recovery while treating it. The ammonium and phosphate present in leachate have been successfully recovered as valuable ammonium sulfate and struvite, respectively, that can be utilized as fertilizers. Iron, manganese, zinc, cadmium, copper, nickel, silver, and lead have been recovered. Good quality water reclaimed after treatment of leachate can be used for recreational and other suitable purposes. Thus, the treatment of leachate in MFCs is viable and sustainable. However, few challenges, such as low power output, leachate toxicity, low concentration of recoverable pollutants and occurrence of undesirable secondary reactions limit its large-scale application. Considering these challenges and upscaling of this technology several strategies have been developed such as hybridized systems integrating UASB, anammox reactor and algal bioreactor that complemented each other and overcome respective limitations. Also, integration of MFC with membrane has ensured high-quality water reclamation and metal recovery.

REFERENCES

Adekunle A., Raghavan V. and Tartakovsky B. (2019a). A comparison of microbial fuel cell and microbial electrolysis cell biosensors for real-time environmental monitoring. *Bioelectrochemistry*, **126**, 105–112, https://doi.org/10.1016/j.bioelechem.2018.11.007

Adekunle A., Raghavan V. and Tartakovsky B. (2019b). On-line monitoring of heavy metals-related toxicity with a microbial fuel cell biosensor. *Biosensors and Bioelectronics*, **132**, 382–390, https://doi.org/10.1016/j.bios.2019.03.011

Aziz S. Q., Aziz H. A., Yusoff M. S., Bashir M. J. K. and Umar M. (2010). Leachate characterization in semi-aerobic and anaerobic sanitary landfills: a comparative study. *Journal of Environmental Management*, **91**(12), 2608–2614, https://doi.org/10.1016/j.jenvman.2010.07.042

Baun D. L. and Christensen T. H. (2004). Speciation of heavy metals in landfill leachate: a review. *Waste Management and Research*, **22**(1), 3–23, https://doi.org/10.1177/0734242X04042146

Bernard C., Colin J. R. and Anne L. D. D. (1997). Estimation of the hazard of landfills through toxicity testing of leachates. Comparison of physico-chemical characteristics of landfill leachates with their toxicity determined with a battery of tests. *Chemosphere*, **35**(11), 2783–2796, https://doi.org/10.1016/S0045-6535(97)00332-9

Chaudhary R., Nain P. and Kumar A. (2021). Temporal variation of leachate pollution index of Indian landfill sites and associated human health risk. *Environmental Science and Pollution Research*, **28**(22), 28391–28406, https://doi.org/10.1007/s11356-021-12383-1

Chaudhuri S. K. and Lovley D. R. (2003). Electricity generation by direct oxidation of glucose in mediatorless microbial fuel cells. *Nature Biotechnology*, **21**(10), 1229–1232, https://doi.org/10.1038/nbt867

Cheng P., Shan R., Yuan H.-R., Tan X., Chen Y. and Wu J. (2022). Synchronous bio-degradation and bio-electricity generation in a microbial fuel cell with aged and fresh leachate from the identical subtropical area. *Journal of Environmental Management*, **316**, 115017, https://doi.org/10.1016/j.jenvman.2022.115017

Colantonio N. (2016). Heavy Metal Removal From Wastewater Using Microbial Electrolysis Cells. McMaster University, 78.

Criteria for landfill management CPCB.pdf. (n.d.).

Damiano L., Jambeck J. R. and Ringelberg D. B. (2014). Municipal solid waste landfill leachate treatment and electricity production using microbial fuel cells. *Applied Biochemistry and Biotechnology*, **173**(2), 472–485, https://doi.org/10.1007/s12010-014-0854-x

Dantre A., Gupta S. and Com W. W. W. O. (2017). Characterization of leachate from selected landfills. *Open Access International Journal of Science & Engineering*, **2**(9), 69–74.

Deng L., Ngo H.-H., Guo W., Wang J. and Zhang H. (2018). Evaluation of a new sponge addition-microbial fuel cell system for removing nutrient from low C/N ratio wastewater. *Chemical Engineering Journal*, **338**, 166–175, https://doi.org/10.1016/j.cej.2018.01.028

Deng Y., Zhu X., Chen N., Feng C., Wang H., Kuang P. and Hu W. (2020). Review on electrochemical system for landfill leachate treatment: performance, mechanism, application, shortcoming, and improvement scheme. *Science of the Total Environment*, **745**, 140768, https://doi.org/10.1016/j.scitotenv.2020.140768

Ehrig H.-J. (1984). Treatment of sanitary landfill leachate: biological treatment. *Waste Management & Research*, **2**(2), 131–152, https://doi.org/10.1016/0734-242X(84)90136-8

Elmaadawy K., Hu J., Guo S., Hou H. and Xu J. (2020a). Enhanced treatment of land fill leachate with cathodic algal biofilm and oxygen-consuming unit in a hybrid microbial fuel cell system. *Bioresource Technology*, **310**, 123420, https://doi.org/10.1016/j.biortech.2020.123420

Elmaadawy K., Liu B., Hu J., Hou H. and Yang J. (2020b). Performance evaluation of microbial fuel cell for landfill leachate treatment: research updates and synergistic effects of hybrid systems. *Journal of Environmental Sciences*, **96**, 1–20, https://doi.org/10.1016/j.jes.2020.05.005

Ezziat L., Elabed A., Ibnsouda S. and El Abed S. (2019). Challenges of microbial fuel cell architecture on heavy metal recovery and removal from wastewater. *Frontiers in Energy Research*, **7**(January), 1–13, https://doi.org/10.3389/fenrg.2019.00001

Feng C., Huang L., Yu H., Yi X. and Wei C. (2015). Simultaneous phenol removal, nitrification and denitrification using microbial fuel cell technology. *Water Research*, **76**, 160–170, https://doi.org/10.1016/j.watres.2015.03.001

Feng Q., Xu L., Liu C., Wang H., Jiang Z., Xie Z., Liu Y., Yang Z. and Qin Y. (2020a). Treatment of shale gas fracturing wastewater using microbial fuel cells: mixture of aging landfill leachate and traditional aerobic sludge as catholyte. *Journal of Cleaner Production*, **269**, 121776, https://doi.org/10.1016/j.jclepro.2020.121776

Feng Q., Xu L., Xu Y., Liu C., Lu Y. and Wang H. (2020b). Treatment of aged landfill leachate by a self-sustained microbial fuel cell-microbial electrolysis cell system. *International Journal of Electrochemical Science*, **15**, 1022–1033, https://doi.org/10.20964/2020.01.19

Gálvez A., Greenman J. and Ieropoulos I. (2009). Landfill leachate treatment with microbial fuel cells; scale-up through plurality. *Bioresource Technology*, **100**(21), 5085–5091, https://doi.org/10.1016/j.biortech.2009.05.061

Ganesh K. and Jambeck J. R. (2013). Treatment of landfill leachate using microbial fuel cells: alternative anodes and semi-continuous operation. *Bioresource Technology*, **139**, 383–387, https://doi.org/10.1016/j.biortech.2013.04.013

Greenman J., Gálvez A., Giusti L. and Ieropoulos I. (2009). Electricity from landfill leachate using microbial fuel cells: comparison with a biological aerated filter. *Enzyme and Microbial Technology*, **44**(2), 112–119, https://doi.org/10.1016/j.enzmictec.2008.09.012

Gupta N., Yadav K. K. and Kumar V. (2015). A review on current status of municipal solid waste management in India. *Journal of Environmental Sciences*, **37**, 206–217, https://doi.org/10.1016/j.jes.2015.01.034

Harmsen J. (1983). Identification of organic compounds in leachate from a waste tip. *Water Research*, **17**(6), 699–705, https://doi.org/10.1016/0043-1354(83)90239-7

Hassan M., Pous N., Xie B., Colprim J., Balaguer M. D. and Puig S. (2017). Influence of iron species on integrated microbial fuel cell and electro-Fenton LEQUiA, Institute of the Environment, University of Girona, C/Maria Aurelia Type of contribution: research article. *Chemical Engineering Journal*, **328**, 57–65, https://doi.org/10.1016/j.cej.2017.07.025

Hassan M., Wei H., Qiu H., Jaafry S. W. H., Su Y. and Xie B. (2018). Power generation and pollutants removal from landfill leachate in microbial fuel cell: variation and influence of anodic microbiomes. *Bioresource Technology*, **247**(July), 434–442, https://doi.org/10.1016/j.biortech.2017.09.124

Hernández-Flores G., Poggi-Varaldo H. M., Solorza-Feria O., Romero-Castañón T., Ríos-Leal E., Galíndez-Mayer J. and Esparza-García F. (2015). Batch operation of a microbial fuel cell equipped with alternative proton exchange membrane. *International Journal of Hydrogen Energy*, Special Issue on XIV International Congress of the Mexican Hydrogen Society, 30 September–4 October 2014, Cancun, Mexico, **40**(48), 17323–17331, https://doi.org/10.1016/j.ijhydene.2015.06.057

Hernández-Flores G., Poggi-Varaldo H. M., Romero-Castañón T., Solorza-Feria O. and Rinderknecht-Seijas N. (2017a). Harvesting energy from leachates in microbial fuel cells using an anion exchange membrane. *International Journal of Hydrogen Energy*, **42**(51), 30374–30382, https://doi.org/10.1016/j.ijhydene.2017.08.201

Hernández-Flores G., Solorza-Feria O. and Poggi-Varaldo H. M. (2017b). Bioelectricity generation from wastewater and actual landfill leachates: a multivariate analysis using principal component analysis. *International Journal of Hydrogen Energy*, **42**(32), 20772–20782, https://doi.org/10.1016/j.ijhydene.2017.01.021

Huang L., Chai X., Chen G. and Logan B. E. (2011). Effect of set potential on hexavalent chromium reduction and electricity generation from biocathode microbial fuel cells. *Environmental Science and Technology*, **45**(11), 5025–5031, https://doi.org/10.1021/es103875d

Huang L., Li X., Cai T. and Huang M. (2018). Electrochemical performance and community structure in three microbial fuel cells treating landfill leachate. *Process Safety and Environmental Protection*, **113**, 378–387, https://doi.org/10.1016/j.psep.2017.11.008

Ieropoulos I., Gálvez A. and Greenman J. (2013). Effects of sulphate addition and sulphide inhibition on microbial fuel cells. *Enzyme and Microbial Technology*, **52**(1), 32–37, https://doi.org/10.1016/j.enzmictec.2012.10.002

Iskander S. M., Brazil B., Novak J. T. and He Z. (2016). Resource recovery from landfill leachate using bioelectrochemical systems: opportunities, challenges, and perspectives. *Bioresource Technology*, **201**, 347–354, https://doi.org/10.1016/j.biortech.2015.11.051

Iskander S., Novak J. T., Brazil B. and He Z. (2017). Simultaneous energy generation and UV quencher removal from landfill leachate using a microbial fuel cell. *Environmental Science and Pollution Research*, **24**, 26040–26048, https://doi.org/10.1007/s11356-017-0231-8

Jakobsen R., Albrechtsen H.-J., Rasmussen M., Bay H., Bjerg P. L. and Christensen T. H. (1998). H2 concentrations in a landfill leachate plume (Grindsted, Denmark): in situ energetics of terminal electron acceptor processes. *Environmental Science & Technology*, **32**(14), 2142–2148, https://doi.org/10.1021/es970858x

Jiang N., Huang L., Huang M., Cai T., Song J., Zheng S., Guo J., Kong Z. and Chen L. (2021). Electricity generation and pollutants removal of landfill leachate by osmotic microbial fuel cells with different forward osmosis membranes. *Sustainable Environment Research*, **31**(1), 22, https://doi.org/10.21203/rs.3.rs-66054/v3

Kelly P. T. and He Z. (2014). Nutrients removal and recovery in bioelectrochemical systems: a review. *Bioresource Technology*, **153**, 351–360, https://doi.org/10.1016/j.biortech.2013.12.046

Kimura Z. and Okabe S. (2013). Acetate oxidation by syntrophic association between *Geobacter sulfurreducens* and a hydrogen-utilizing exoelectrogen. *The ISME Journal*, **7**(8), 1472–1482, https://doi.org/10.1038/ismej.2013.40

Kjeldsen P., Barlaz M. A., Rooker A. P., Baun A., Ledin A. and Christensen T. H. (2002). Present and long-term composition of MSW landfill leachate: a review. *Critical Reviews in Environmental Science and Technology*, **32**(4), 297–336, https://doi.org/10.1080/10643380290813462

Kumar S., Katoria D. and Singh G. (2013). Leachate treatment technologies. *International Journal of Environmental Engineering and Management*, **4**(5), 439–444.

Kumar S. S., Malyan S. K., Basu S. and Bishnoi N. R. (2017). Syntrophic association and performance of *Clostridium*, *Desulfovibrio*, *Aeromonas* and *Tetrathiobacter* as anodic biocatalysts for bioelectricity generation in dual chamber microbial fuel cell. *Environmental Science and Pollution Research*, **24**(19), 16019–16030, https://doi.org/10.1007/s11356-017-9112-4

Kumar S. S., Kumar V., Kumar R., Malyan S. K. and Bishnoi N. R. (2019). Ferrous sulfate as an in-situ anodic coagulant for enhanced bioelectricity generation and COD removal from land fill leachate. *Energy*, **176**, 570–581, https://doi.org/10.1016/j.energy.2019.04.014

Lakshmidevi R., Gandhi N. N. and Muthukumar K. (2020). *Carbon Neutral Electricity Production from Municipal Solid Waste Landfill Leachate Using Algal-Assisted Microbial Fuel Cell.*

Lee H.-S., Parameswaran P., Kato-Marcus A., Torres C. I. and Rittmann B. E. (2008). Evaluation of energy-conversion efficiencies in microbial fuel cells (MFCs) utilizing fermentable and non-fermentable substrates. *Water Research*, **42**(6), 1501–1510, https://doi.org/10.1016/j.watres.2007.10.036

Lee Y., Martin L., Grasel P., Tawfiq K. and Chen G. (2013). Power generation and nitrogen removal of landfill leachate using microbial fuel cell technology. *Environmental Technology*, **34**(19), 2727–2736, https://doi.org/10.1080/09593330.2013.788040

Li S. and Chen G. (2018). Effects of evolving quality of landfill leachate on microbial fuel cell performance. *Waste Management & Research*, **36**(1), 59–67, https://doi.org/10.1177/0734242X17739969

Li Y., Lu A., Ding H., Wang X., Wang C., Zeng C. and Yan Y. (2010). Microbial fuel cells using natural pyrrhotite as the cathodic heterogeneous Fenton catalyst towards the degradation of biorefractory organics in landfill leachate. *Electrochemistry Communications*, **12**(7), 944–947, https://doi.org/10.1016/j.elecom.2010.04.027

Li J., Ge Z. and He Z. (2014). Advancing membrane bioelectrochemical reactor (MBER) with hollow-fiber membranes installed in the cathode compartment. *Journal of Chemical Technology & Biotechnology*, **89**(9), 1330–1336, https://doi.org/10.1002/jctb.4206

Lovley D. R. (2006). Bug juice: harvesting electricity with microorganisms. *Nature Reviews Microbiology*, **4**(7), 497–508, https://doi.org/10.1038/nrmicro1442

Lu Y., Qin M., Yuan H., Abu-Reesh I. M. and He Z. (2014). When bioelectrochemical systems meet forward osmosis: accomplishing wastewater treatment and reuse through synergy. *Water*, **7**(1), 38–50, https://doi.org/10.3390/w7010038

Luo H., Zeng Y., Cheng Y., He D. and Pan X. (2020). Recent advances in municipal landfill leachate: a review focusing on its characteristics, treatment, and toxicity assessment. *Science of the Total Environment*, **703**, 135468, https://doi.org/10.1016/j.scitotenv.2019.135468

Matassa S., Batstone D. J., Hülsen T., Schnoor J. and Verstraete W. (2015). Can direct conversion of used nitrogen to new feed and protein help feed the world? *Environmental Science & Technology*, **49**(9), 5247–5254, https://doi.org/10.1021/es505432w

Maturi K. C., Gupta A., Haq I. and Kalamdhad A. S. (2022). Chapter 7 – A glance over current status of waste management and landfills across the globe: a review. In: I. Haq, A. S. Kalamdhad and M. P. B. T.-B. and D. of M. in I. W. Shah (eds.), Elsevier, pp. 131–144, https://doi.org/10.1016/B978-0-323-88507-2.00001-4

McCutcheon J. R., McGinnis R. L. and Elimelech M. (2005). A novel ammonia–carbon dioxide forward (direct) osmosis desalination process. *Desalination*, **174**(1), 1–11, https://doi.org/10.1016/j.desal.2004.11.002

Modin O., Wang X., Wu X., Rauch S. and Fedje K. K. (2012). Bioelectrochemical recovery of Cu, Pb, Cd, and Zn from dilute solutions. *Journal of Hazardous Materials*, **235–236**, 291–297, https://doi.org/10.1016/j.jhazmat.2012.07.058

Moharir P. V. and Tembhurkar A. R. (2018). Effect of recirculation on bioelectricity generation using microbial fuel cell with food waste leachate as substrate. *International Journal of Hydrogen Energy*, **43**(21), 10061–10069, https://doi.org/10.1016/j.ijhydene.2018.04.072

Nancharaiah Y. V., Venkata Mohan S. and Lens P. N. L. (2015). Metals removal and recovery in bioelectrochemical systems: a review. *Bioresource Technology*, **195**, 102–114, https://doi.org/10.1016/j.biortech.2015.06.058

Nguyen H. T. H. and Min B. (2020). Leachate treatment and electricity generation using an algae-cathode microbial fuel cell with continuous fl ow through the chambers in series. *Science of the Total Environment*, **723**, 138054, https://doi.org/10.1016/j.scitotenv.2020.138054

Nguyen H. T. H., Kakarla R. and Min B. (2017). Algae cathode microbial fuel cells for electricity generation and nutrient removal from landfill leachate wastewater. *International Journal of Hydrogen Energy*, **42**(49), 29433–29442, https://doi.org/10.1016/j.ijhydene.2017.10.011

Ozkaya B., Demir A. and Bilgili M. S. (2006). Soluble substrate concentrations in leachate from field scale MSW test cells. *Journal of Hazardous Materials*, **134**(1–3), 19–26, https://doi.org/10.1016/j.jhazmat.2005.11.046

Özkaya B., Cetinkaya A. Y., Cakmakci M., Karadağ D. and Sahinkaya E. (2013). Electricity generation from young landfill leachate in a microbial fuel cell with a new electrode material. *Bioprocess and Biosystems Engineering*, **36**(4), 399–405, https://doi.org/10.1007/s00449-012-0796-z

Pandis P. K., Kamperidis T., Bariamis K., Vlachos I., Argirusis C., Stathopoulos V. N., Lyberatos G. and Tremouli A. (2022). Comparative study of different production methods of activated carbon cathodic electrodes in single chamber MFC treating municipal landfill leachate. *Applied Sciences*, **12**(6), 2991, https://doi.org/10.3390/app12062991

Puig S., Serra M., Coma M., Cabré M., Dolors Balaguer M. and Colprim J. (2011). Microbial fuel cell application in landfill leachate treatment. *Journal of Hazardous Materials*, **185**(2), 763–767, https://doi.org/10.1016/j.jhazmat.2010.09.086

Qin B., Luo H., Liu G., Zhang R., Chen S., Hou Y. and Luo Y. (2012). Nickel ion removal from wastewater using the microbial electrolysis cell. *Bioresource Technology*, **121**, 458–461 https://doi.org/10.1016/j.biortech.2012.06.068

Qin M., Molitor H., Brazil B., Novak J. T. and He Z. (2016). Recovery of nitrogen and water from landfill leachate by a microbial electrolysis cell–forward osmosis system. *Bioresource Technology*, **200**, 485–492, https://doi.org/10.1016/j.biortech.2015.10.066

Renou S., Givaudan J. G., Poulain S., Dirassouyan F. and Moulin P. (2008). Landfill leachate treatment: review and opportunity. *Journal of Hazardous Materials*, **150**(3), 468–493, https://doi.org/10.1016/j.jhazmat.2007.09.077

Rozendal R. A., Leone E., Keller J. and Rabaey K. (2009). Efficient hydrogen peroxide generation from organic matter in a bioelectrochemical system. *Electrochemistry Communications*, **11**(9), 1752–1755, https://doi.org/10.1016/j.elecom.2009.07.008

Sharma K. D. and Jain S. (2020). Municipal solid waste generation, composition, and management: the global scenario. *Social Responsibility Journal*, **16**(6), 917–948, https://doi.org/10.1108/SRJ-06-2019-0210

Shehzad A., Bashir M. J. K., Sethupathi S., Lim J.-W. and Younas M. (2016). Bioelectrochemical system for landfill leachate treatment–challenges, opportunities, and recommendations. *Geosystem Engineering*, **19**(6), 337–345, https://doi.org/10.1080/12269328.2016.1188029

Sonawane J. M., Adeloju S. B. and Ghosh P. C. (2017). Landfill leachate: a promising substrate for microbial fuel cells. *International Journal of Hydrogen Energy, Special Issue on The 1st* International Conference on Advanced Energy Materials (AEM2016), 12–14 September 2016, Surrey, England, **42**(37), 23794–23798, https://doi.org/10.1016/j.ijhydene.2017.03.137

Srivastava P., Abbassi R., Yadav A. K., Garaniya V. and Asadnia M. (2020a). A review on the contribution of electron flow in electroactive wetlands: electricity generation and enhanced wastewater treatment. *Chemosphere*, **254**, 126926, https://doi.org/10.1016/j.chemosphere.2020.126926

Srivastava P., Abbassi R., Yadav A. K., Garaniya V. and Khan F. (2020b). Microbial fuel cell-integrated wastewater treatment systems. In: Integrated Microbial Fuel Cells for Wastewater Treatment, Butterworth-Heinemann, Oxford, UK; Cambridge, US, pp. 29–46, https://doi.org/10.1016/B978-0-12-817493-7.00002-3

Srivastava P., Abbassi R., Yadav A., Garaniya V., Asadnia M., Lewis T. and Khan S. J. (2021a). Influence of applied potential on treatment performance and clogging behaviour of hybrid constructed wetland-microbial electrochemical technologies. *Chemosphere*, **284**, 131296, https://doi.org/10.1016/j.chemosphere.2021.131296

Srivastava P., Belford A., Abbassi R., Asadnia M., Garaniya V. and Yadav A. K. (2021b). Low-power energy harvester from constructed wetland-microbial fuel cells for initiating a self-sustainable treatment process. *Sustainable Energy Technologies and Assessments*, **46**, 101282, https://doi.org/10.1016/j.seta.2021.101282

Talboys P. J., Heppell J., Roose T., Healey J. R., Jones D. L. and Withers P. J. A. (2016). Struvite: a slow-release fertiliser for sustainable phosphorus management? *Plant and Soil*, **401**(1–2), 109–123, https://doi.org/10.1007/s11104-015-2747-3

Tandukar M., Huber S. J., Onodera T. and Pavlostathis S. G. (2009). Biological chromium(VI) reduction in the cathode of a microbial fuel cell. *Environmental Science and Technology*, **43**(21), 8159–8165, https://doi.org/10.1021/es9014184

Tao W. and Ukwuani A. T. (2015). Coupling thermal stripping and acid absorption for ammonia recovery from dairy manure: ammonia volatilization kinetics and effects of temperature, pH and dissolved solids content. *Chemical Engineering Journal*, **280**, 188–196, https://doi.org/10.1016/j.cej.2015.05.119

Teng S.-X., Tong Z.-H., Li W.-W., Wang S.-G., Sheng G.-P., Shi X.-Y., Liu X.-W. and Yu H.-Q. (2010). Electricity generation from mixed volatile fatty acids using microbial fuel cells. *Applied Microbiology and Biotechnology*, **87**(6), 2365–2372, https://doi.org/10.1007/s00253-010-2746-5

Tugtas A. E., Cavdar P. and Calli B. (2013). Bio-electrochemical post-treatment of anaerobically treated landfill leachate. *Bioresource Technology*, **128**, 266–272, https://doi.org/10.1016/j.biortech.2012.10.035

Vázquez-Larios A. L., Solorza-Feria O., Poggi-Varaldo H. M., De Guadalupe González-Huerta R., Ponce-Noyola M. T., Ríos-Leal E. and Rinderknecht-Seijas N. (2014). Bioelectricity production from municipal leachate in a microbial fuel cell: effect of two cathodic catalysts. *International Journal of Hydrogen Energy*, **39**(29), 16667–16675, https://doi.org/10.1016/j.ijhydene.2014.05.178

Vázquez-Larios A. L., Poggi-Varaldo H. M., Solorza-Feria O. and Rinderknecht-Seijas N. (2015). Effect of type of inoculum on microbial fuel cell performance that used RuxMoySez as cathodic catalyst. *International Journal of Hydrogen Energy*, **40**(48), 17402–17412, https://doi.org/10.1016/j.ijhydene.2015.09.143

Vilajeliu-Pons A., Koch C., Balaguer M. D., Colprim J., Harnisch F. and Puig S. (2018). Microbial electricity driven anoxic ammonium removal. *Water Research*, **130**, 168–175, https://doi.org/10.1016/j.watres.2017.11.059

Virdis B., Rabaey K., Rozendal R. A., Yuan Z. and Keller J. (2010). Simultaneous nitrification, denitrification and carbon removal in microbial fuel cells. *Water Research*, **44**(9), 2970–2980, https://doi.org/10.1016/j.watres.2010.02.022

Wang Z. J. and Lim B. S. (2017). Electric power generation from treatment of food waste leachate using microbial fuel cell. *Environmental Engineering Research*, **22**(2), 157–161, https://doi.org/10.4491/eer.2016.061

Wang H. and Ren Z. J. (2014). Bioelectrochemical metal recovery from wastewater: a review. *Water Research*, **66**, 219–232, https://doi.org/10.1016/j.watres.2014.08.013

Warmadewanthi I., Zulkarnain M. A., Ikhlas N., Kurniawan S. B. and Abdullah S. R. S. (2021). Struvite precipitation as pretreatment method of mature landfill leachate. *Bioresource Technology Reports*, **15**, 100792, https://doi.org/10.1016/j.biteb.2021.100792

Wu D., Wang T., Huang X., Dolfing J. and Xie B. (2015). Perspective of harnessing energy from landfill leachate via microbial fuel cells: novel biofuels and electrogenic physiologies. *Applied Microbiology and Biotechnology*, **99**(19), 7827–7836, https://doi.org/10.1007/s00253-015-6857-x

Xia C., Zhang D., Pedrycz W., Zhu Y. and Guo Y. (2018). Models for microbial fuel cells: a critical review. *Journal of Power Sources*, **373**, 119–131, https://doi.org/10.1016/j.jpowsour.2017.11.001

Ye Y., Ngo H. H., Guo W., Liu Y., Chang S. W., Nguyen D. D., Ren J., Liu Y. and Zhang X. (2019). Feasibility study on a double chamber microbial fuel cell for nutrient recovery from municipal wastewater. *Chemical Engineering Journal*, **358**, 236–242, https://doi.org/10.1016/j.cej.2018.09.215

You S. J., Zhao Q. L., Jiang J. Q., Zhang J. N. and Zhao S. Q. (2006). Sustainable approach for leachate treatment: electricity generation in microbial fuel cell. *Journal of Environmental Science and Health – Part A Toxic/Hazardous Substances and Environmental Engineering*, **41**(12), 2721–2734, https://doi.org/10.1080/10934520600966284

Yuan H. and He Z. (2015). Integrating membrane filtration into bioelectrochemical systems as next generation energy-efficient wastewater treatment technologies for water reclamation: a review. *Microbial Fuel Cells*, **195**, 202–209, https://doi.org/10.1016/j.biortech.2015.05.058

Zhang J. N., Zhao Q. L., You S. J., Jiang J. Q. and Ren N. Q. (2008). Continuous electricity production from leachate in a novel upflow air-cathode membrane-free microbial fuel cell. *Water Science and Technology*, **57**(7), 1017–1021, https://doi.org/10.2166/wst.2008.063

Zhang G., Jiao Y. and Lee D.-J. (2015a). A lab-scale anoxic/oxic-bioelectrochemical reactor for leachate treatments. *Bioresource Technology*, **186**, 97–105, https://doi.org/10.1016/j.biortech.2015.03.022

Zhang G., Jiao Y. and Lee D.-J. (2015b). Transformation of dissolved organic matters in landfill leachate–bioelectrochemical system. *Bioresource Technology*, **191**, 350–354, https://doi.org/10.1016/j.biortech.2015.05.082

Zhang G., Jiao Y. and Lee D.-J. (2016). Leachate treatment using anoxic/oxic-bioelectrochemical reactor with intermittent aeration. *Journal of the Taiwan Institute of Chemical Engineers*, **58**, 401–406, https://doi.org/10.1016/j.jtice.2015.06.019

Zhao S., Zou L., Tang C. Y. and Mulcahy D. (2012). Recent developments in forward osmosis: opportunities and challenges. *Journal of Membrane Science*, **396**, 1–21, https://doi.org/10.1016/j.memsci.2011.12.023

Zou S., Guan L., Taylor D. P., Kuhn D. and He Z. (2018). Nitrogen removal from water of recirculating aquaculture system by a microbial fuel cell. *Aquaculture*, **497**, 74–81, https://doi.org/10.1016/j.aquaculture.2018.07.036

doi: 10.2166/9781789063318_0087

Chapter 5

Nitrogen removal from landfill leachate

*Ankit Singh and Rajesh Roshan Dash**

Environmental Engineering, School of Infrastructure, Indian Institute of Technology Bhubaneswar, Bhubaneswar 752050, Odisha, India

**Corresponding author: rrdash@iitbbs.ac.in*

ABSTRACT

Population growth, urbanization, and industrialization are the significant factors responsible for generating massive waste worldwide. The generation of solid waste leads to an adverse impact on environment security as well as human health. Solid waste management is required to minimize environmental drawbacks. Sanitary landfill is considered an ultimate disposal method, commonly adopted for solid waste management. However, landfill gas and leachate generation have gained social and environmental concerns. The production of leachate from the percolation of rainfall into a landfill is one of the major concerns. To meet discharge permit criteria, reducing the ammonia concentration in landfill leachate is usually necessary. Among the contaminants found in landfill leachate is ammonia-nitrogen (NH_3–N), which is of particular concern due to its high concentration. It is typically essential to treat landfill leachate to eliminate ammonia nitrogen (free ammonia and dissolved organic nitrogen) because large quantities are known to have detrimental effects on waterways and the leachate treatment process. This chapter deals with the technologies adopted for nitrogen removal from landfill leachate. Biological, physical, and chemical treatments can reduce nitrogen content. However, choosing any of these treatment methods is subjective. Therefore, this chapter discusses the treatment methods for nitrogen removal and their drawbacks and suggests that treatment efficiencies, system cost, and environmental impacts are considered while proposing a suitable treatment method or combination of treatment methods for nitrogen removal from landfill leachate.

Keywords. Nitrogen removal, landfill leachate, treatment technologies, leachate treatment.

5.1 INTRODUCTION

Solid wastes were generated from the era of early human beings; since the population was much less in early human existence, the environmental nuisance in the form of air and water pollution was negligible compared to modern human life. As humans settled in one location and adopted farming and craftsmanship, they became socially organized. The permanent settlement had specific unfavorable effects because waste production significantly increased and accumulated for more extended periods of time. The rising accumulation of remains presented the new communities with additional problems. Higher living standards, unsustainable growth of commercials and industries, the rapid development of the economy, and drastically increasing population are some of the major causes of solid waste production worldwide. According to World Bank statistics (Mandal *et al.*, 2017), there are around 2 billion tonnes of solid waste produced annually, and by 2050, that number will rise to 3.4 billion tonnes. Despite having a small population (16% of the world's population), high-income countries produce 34% of the world's garbage. Solid waste creates an adverse impact on environmental security.

Solid waste management system components are – waste reduction, waste collection, handling, transportation, treatment, and finally, safe disposal. After the treatment of solid waste by thermal or biological means, leftover inert material is called solid waste residues. These residues should be safely disposed of in order to reduce the environmental hazard.

Solid waste residues are the waste that is not recycled and remains after the material recovery facility, that is left after the recovery of thermal or biological conversion products and that remains after energy recovery facilities. Solid waste residue disposal is an important component of the integrated solid waste management system. The safe disposal of these wastes is essential to control environmental hazards. The disposal can be thought of as transforming solid waste from one harmful state to an inert state of waste that is less or not harmful to the environment. This transformation can be done by various approaches, that is, landfilling, incineration, pyrolysis, composting, and so on.

Incineration and landfilling is the waste disposal method commonly adopted (Bolyard & Reinhart, 2016). Incineration performs well for waste disposal when waste contains mainly combustible materials. Still, approximately 10–20% of incoming waste remains in the system in the form of ashes and other forms of residues (Wiszniowski *et al.*, 2006). According to a survey, most developed countries chose sanitary landfills as an effective means of solid waste disposal (Das *et al.*, 2019). Out of these means of solid waste elimination, landfilling is considered the most cost-effective and environmentally friendly approach. A landfill is the economical method of ultimate disposal of solid waste, which is practiced worldwide. Capital cost and extent of environmental exploitation of landfill method is minimum amongst other methods of waste management and disposal. Landfill contains a variety of solid wastes and residues generated at the incineration plant. It allows waste decomposition in a slow and controlled manner with minimal environmental pollution.

However, the generation of leachate from the sanitary landfill of solid waste is considered a major concern regarding the negative impact on environmental quality. Landfill leachate is an aqueous effluent generated from landfill. Leachate is composed of internal water in waste, water generated by biodegradation of waste, and rainfall water precipitation over landfill areas.

The composition and contaminants of solid waste depend on the solid waste filled in the landfill, waste composition, age of the landfill, and ambient climatic conditions (Bandala *et al.*, 2021). Leachates mainly contain biodegradable organic matter, refractory organic matter, humic-type substances, chlorinated organic and inorganic salts, heavy metals, and ammonia-nitrogen (Renou *et al.*, 2008).

Among the different pollutants, there are nitrogen compounds that can be inorganic (ammonium ion, nitrite, and nitrate) and organic nitrogen compounds present in the landfill leachate, which is a serious concern regarding leachate pollution. To reduce nutrient emissions, treatment of such leachate has become a significant problem because high ammonia concentration cannot be removed by conventional biological treatments. Generally, ammonia concentration does not decrease with the landfill age (Miao *et al.*, 2019). Hence pre-treatment/post-treatment before biological treatment or modification in biological treatment is required to reduce the ammonia content.

5.2 LANDFILL LEACHATE

Landfill leachate is an aqueous effluent that has percolated through solid waste contained in the landfill. This is the liquid produced by the biodegradation of wastes, precipitation over an uncapped landfill, or the natural moisture in solid waste; it contains dissolved salts, dissolved organic matter (DOM), and other organic substances (Teng *et al.*, 2021). This aqueous effluent is rich in non-biodegradable organic substances, biodegradable organic substances, heavy metals, inorganic salts, and ammonia nitrogen (Haslina *et al.*, 2021).

5.2.1 Formation

The volumetric flow rate and chemical composition of the leachate are the two factors that characterize it (Lema *et al.*, 1988). The volumetric flow rate is related to the quantity of leachate generated, and composition is related to the quality of leachate. The volumetric flow rate depends on climatic conditions and the nature of the waste itself, whereas the composition of leachate depends on the physical, chemical, and biological properties of leachate.

Leachate is considered the most harmful by-product of landfill operation. Moreover, the design of treatment facilities and operation of the leachate treatment plant required data on the leachate amount entering the treatment plant. The quantity of leachate generated in a landfill is directly related to the quantity of infiltrating water through the landfill. Leachate generation mainly depends on the following factors (Abunama *et al.*, 2019; Segweni, 2017):

(a) Composition of solid waste,
(b) The amount of moisture included in waste,
(c) The age and nature of the landfill,
(d) The climatological circumstances,
(e) Compaction rate,
(f) Hydrology of site,
(g) Geology of site.

Based on prediction complexity and result accuracy, several approaches have been developed to quantify the generated amount of leachate. Out of those approaches, the water balance method (WBM) is considered a simpler approach used in various studies (El-Fadel *et al.*, 1997; Fenn *et al.*, 1975).

In addition, the hydrologic evaluation of landfill performance (HELP) model is used for evaluating water balance and other disposal methods for landfill. In developing countries, the unavailability or deficiency of accurate data makes the WBM approach the most reliable leachate quantification model (Abunama *et al.*, 2019).

5.2.2 Composition

In the early stages, the waste of landfill is degraded aerobically due to the presence of oxygen which is held up in the waste. As soon as oxygen is depleted, anaerobic degradation starts. The anaerobic degradation cycle typically consists of three steps: acid fermentation, intermediate anaerobic phase, and anaerobic degradation. In young leachate, acid fermentation happens quickly. At this stage, aerobic processes predominate, producing a significant amount of volatile fatty acids (VFAs). Because oxygen-depleted landfills become more acidic, hence young leachate is highly biodegradable (Rathnayake & Herath, 2018). When VFAs are reduced, and ammonia is released during the intermediate anaerobic phase, methanogenic bacteria begin to multiply. Finally, anaerobic degradation occurs in the old leachate of the mature landfill, but the degradation process is negligible due to the presence of compounds that cannot be treated biologically.

Conventional criteria, such as pH, chemical oxygen demand (COD), biochemical oxygen demand (BOD), total organic carbon (TOC), suspended solids, ammonium-nitrogen ($NH_4^+ - N$), and heavy metal concentrations are typically used to characterize landfill leachate. The percentages of the BOD5/COD ratio and the COD/TOC ratio are two common markers of biodegradability (Teng *et al.*, 2021).

The operational conditions, waste type, climate, hydrogeology, and age of the landfill are some of the elements that affect the quality of the leachate that is produced by landfills (Dabaghian *et al.*, 2018). The age of a landfill is a major factor that can influence the composition and the qualities of landfill leachate.

According to the age of the landfill, the leachate that is collected can be separated into three categories, young leachate (<5 years); medium leachate (5–10 years); and mature leachate (>10 years) (Miao *et al.*, 2019). Young leachate shows higher biodegradability as compared to mature leachate. Another important contaminant in landfill leachate is ammonia. Table 5.1 shows the typical composition of landfill leachate.

Leachate from mature landfills has a high ammonia concentration, which does not often diminish with landfill age (Teng *et al.*, 2021). Because of high concentrations of ammonia and organic compounds, landfill leachate can cause severe contamination. Humic and fulvic acid are refractory molecules that remain after biodegradable organic components have been completely degraded.

It is important to note that concentrations of heavy metals have a tendency to decrease with aging since the higher pH value tends to reduce the solubility of metals. DOM can also be subdivided into fractions that are biodegradable and

Table 5.1 Landfill composition vs. age.

SN.	Type of Leachate	Young Leachate	Medium Leachate	Mature Leachate
1	Age (years)	<5	5–10	>10
2	pH	<6.5	6.5–7.5	>7.5
3	$NH_4^+ - N(mg/L)$	<400	–	>400
4	COD (mg/L)	>15,000	3,000–15,000	<3,000
5	BOD/COD	0.5–1	0.1–0.5	<0.1
6	TOC/COD	<0.3	0.3–0.5	>0.5
7	Biodegradability	High	Medium	Low
8	Organic compounds	80% VFAs	5–30% VFAs + humic and fulvic acid	Humic and fulvic acid

non-biodegradable. Non-biodegradable fractions are predominately composed of humic chemicals (Iskander *et al.*, 2018). Condensation and polymerization of the by-products of microbial degradation are the processes that lead to the formation of these compounds. On the other hand, the vast majority of the organic matter that can be broken down into smaller components by biodegradation is produced during the early phases of the decomposition of organic waste materials. Utilizing biological treatment approaches will allow for the effective elimination of this. Emerging organic contaminants (EOCs), such as endocrine-disrupting chemicals, persistent organic pollutants, pharmaceutical, and personal care products, antibiotic resistance genes, and disinfection by-products, have received a great deal of attention in recent years due to their detrimental effects on human health and the environment (Teng *et al.*, 2021).

5.3 ENVIRONMENTAL IMPACT OF LEACHATE

The sources of landfill emissions are as follows: solid wastes as they are transported to the site; transportation emissions; wastes blown by the wind; surface dust; landfill gas; and leachate created (Vaverková, 2019a). Landfilling is considered an effective disposal method of solid wastes (Chakravarty & Kumar, 2019); however, for many reasons landfilling is placed in the least preferred method for the integrated solid waste management system. These reasons are:

(a) Only a limited amount of waste may be dumped in landfills, and the period of landfill reclamation could run for hundreds of years (Białowiec, 2011).
(b) The adverse effect of biogas and leachate on the environment and human health.
(c) In many countries, unsorted waste is dumped.
(d) The ‘not in my backyard (NIMBY)’ syndrome for the adjacent areas is primarily brought on by adverse odor and air pollution in the majority of landfills across the world (Weng *et al.*, 2015).

The environment is made up of a variety of elements that surround one. It could include the soil, water, and air environments. Human beings are also an integral part of the environment as well. Hence, major environmental damages are discussed below.

5.3.1 Leachate impact on soil

Soil pollution is caused by the movement of leachate first through wastes dumped in the landfill and then the migration of leachate to the soil. Some important points are discussed here (Daniel *et al.*, 2021):

(a) Leachate may contain harmful materials, heavy metals, electronic trash, medicinal drugs, and so on. The soil becomes a contaminant sink as a result of the migration of these toxic compounds (Şchiopu *et al.*, 2009).
(b) Leachate contains heavy metals such as lead and copper and has a detrimental effect on the microbial diversity, population size, and general activity of soil microbial communities (Hong *et al.*, 2002).
(c) Leachate from landfills can harm soil quality and safety, contaminate the food supply, and endanger long-term health.
(d) Leachate modifies the soil composition, promotes abnormal growth of plants, and affects the local economy by having a negative impact on tourists.

5.3.2 Leachate impact on water

Landfill leachate severely damages the quality of groundwater and surface water quality some important points are discussed here (Daniel *et al.*, 2021; Segweni, 2017).

(a) Leachate contains heavy metals, nitrogen species, chlorinated hydrocarbon phenols, cyanides, and bacteria, among others which affect the groundwater quality.
(b) The ecological balance of the aquatic environment may be severely impacted by surface or groundwater contamination, and the diversity of aquatic creatures is constrained as contamination levels rise.
(c) Electronic waste dumping makes water unsuitable for drinking purposes as it contains heavy metals such as cadmium and lead.
(d) Leachate from landfills includes a high ammonia content. Problems associated with a high concentration of ammonia include reduction in dissolved oxygen in water bodies, eutrophication in surrounding water bodies, toxic to aquatic life if a concentration of ammonia is higher than 1 mg/L.

5.3.3 Leachate impact on air

One of the major detrimental reasons for environmental health damage is landfill gases (LFG). A major part of LFG consists of methane (CH_4) and carbon dioxide (CO_2). However, the small concentration of hydrogen sulfide, mercury vapors, and volatile organic compounds (VOCs) are also present in small concentrations (Ozbay *et al.*, 2021).

Some important points are discussed here (Daniel *et al.*, 2021; Segweni, 2017):

(a) LFG extraction is essential since the gases pose a threat of explosion. Additionally, exposure to these gases may endanger the surrounding population.
(b) LFG can increase the global temperature.
(c) The leachate pollutes the air by promoting a bad smell at the site and in the surrounding area.

5.3.4 Leachate impact on human health

Numerous studies have demonstrated the detrimental health effects that uncontrolled landfills have on the surrounding population. It is a combined effect of air, water, and soil pollution. Some of them are discussed here (Lin *et al.*, 2009; Ozbay *et al.*, 2021; Segweni, 2017).

(a) The migration of gases, particles, and chemicals that attach to the air near landfills may be a cause of chemical contamination in the air.
(b) Health concerns like respiratory problems, skin irritations, gastrointestinal issues, exhaustion, headaches, and psychiatric disorders are more common in areas near uncontrolled landfills.
(c) Low birth weight (<2,500 g), fetal and newborn mortality, spontaneous abortion, and the prevalence of congenital disabilities are some of the reproductive impacts associated with landfill sites (Rushton, 2003). High risk of gastrointestinal, esophageal, stomach, colon, and rectal cancers are also encountered (Griffith *et al.*, 1989; Rushton, 2003).
(d) Ammonia inhalation directly may result in a number of respiratory issues, such as tracheobronchitis, bronchiolitis, laryngitis, bronchopneumonia, and pulmonary edema.

5.4 NITROGEN REMOVAL

As far as the treatment of any wastewater is concerned, there are five forms of nitrogen that will be considered for treatment purposes: Organic nitrogen (complex mixture of amino acids, amino sugars, and proteins), ammonia-nitrogen (NH_3–N), ammonium-nitrogen ($NH_4^+ - N$), nitrite nitrogen ($NO_2^- - N$), and nitrate nitrogen ($NO_3^- - N$). Landfill leachate comprises a variety of contaminants, including NH_3–N, which is particularly concerning due to its high concentration in leachate. Due to the high concentrations of COD and nitrogen content, treatment of landfill leachate is extremely difficult and typically necessitates a number of process applications. Nitrogen is found in high percentages in the majority of landfill leachates and over the period of the landfill's life (Drafts, 2021). High COD concentrations can be treated well by anaerobic biological treatment systems; however, biological treatment systems are inhibited by high ammonium concentrations. As a result, ammonium pre-treatment is crucial.

Leachate from landfills is often treated nitrogen using a combination of physical, chemical, and biological methods, with performance varied depending on dump age. Biological treatment is successful for young leachate (<5 years)

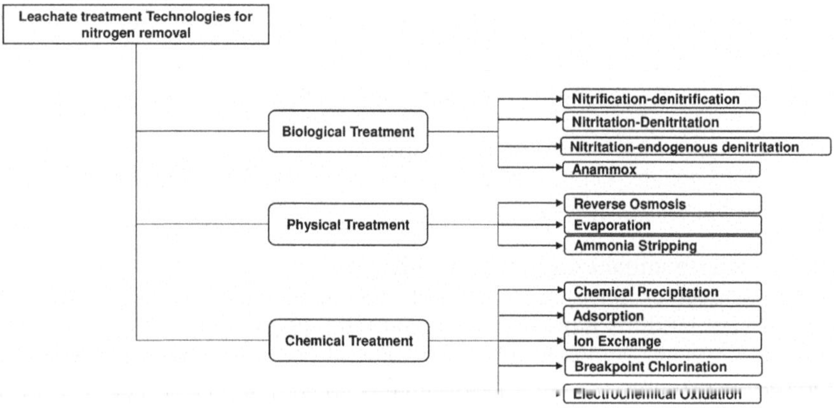

Figure 5.1 Leachate treatment technologies for nitrogen removal.

but not for mature landfill (>10 years). Physico-chemical approaches generally work better in mature landfills (Bandala *et al.*, 2021). Various physical and chemical treatments are used either as a pre-treatment/post-treatment of biological treatment or as a total replacement biological treatment. The COD, BOD5/COD ratio, and age of the landfill are used to assess the types of treatment methods that are most appropriate for landfills leachate. Landfill leachate treatment technologies for nitrogen removal are shown in Figure 5.1. The complexity of leachate treatment makes universal solutions for treatment impossible. Every technology has its own advantages and disadvantages. It is advisable that in addition to analyzing the leachate's composition, volume, and treatment susceptibility, consider the costs of potential treatments also.

5.4.1 Biological treatment

5.4.1.1 Nitrification–denitrification

The conventional biological treatment of ammonia is nitrification–denitrification process. Nitrogen present in landfill leachate is in an aqueous form, that is, ammonium ion (NH_4^+) and the remaining organic nitrogen is converted into NH_4^+ through hydrolysis and ammonification processes. This $NH_4^+ - N$ is converted into $NO_2^- - N$ in the presence of oxygen by ammonia-oxidizing bacteria (AOB). AOB is an aerobic autotrophic bacteria. Further, $NO_2^- - N$ is converted into $NO_3^- - N$ in the presence of oxygen with the help of nitrite-oxidizing bacteria (NOB) (Figure 5.2). The above process can be presented as follows:

$$NH_4^+ + 1.5O_2 \xrightarrow{\text{AOB}} NO_2^- + 2H^+ + H_2O \quad (5.1)$$

$$NO_2^- + 0.5O_2 \xrightarrow{\text{NOB}} NO_3^- \quad (5.2)$$

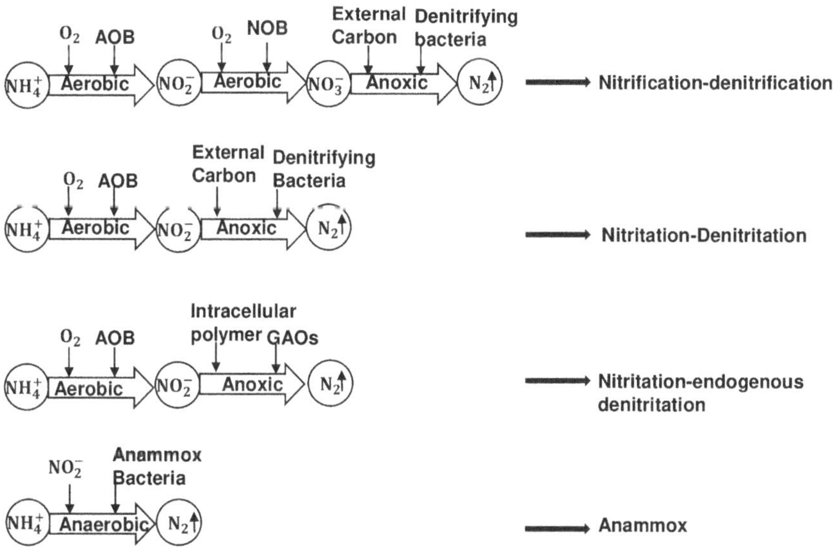

Figure 5.2 Biological treatment technologies for nitrogen removal.

A combination of the above two reactions shown in equations is termed nitrification. The nitrogenous matter is used as an electron donor, and oxygen is used as an electron acceptor. As nitrification is an autotrophic process, nitrifying bacteria have a slow growth rate (Berge *et al.*, 2005). When developing nitrification as a method of leachate treatment, this must be taken into account, as sufficient detention time must be provided to allow for bacterial growth, with average hydraulic residence times for nitrification ranging from 1 to 5 days (Yusof *et al.*, 2010). Various factors, including dissolved oxygen concentration, alkalinity, pH, and temperature, affect the nitrification process.

Denitrification is a process where denitrifying bacteria convert $NO_3^- - N$ to $NO_2^- - N$ and, ultimately, molecular nitrogen gas (N_2). This reaction proceeds under anoxic conditions. Denitrifying bacteria are heterotrophic in nature. During denitrification, the end product of N_2 escapes the leachate, lowering the nitrogen level of the leachate if denitrification progresses to completion. Conversion from nitrite to nitrogen gas is shown in the equations shown below (Berge *et al.*, 2005)

$$NO_3^- + 2e^- + 2H^+ \rightarrow NO_2^- + H_2O \tag{5.3}$$

$$NO_2^- + e^- + 2H^+ \rightarrow NO + H_2O \tag{5.4}$$

$$2NO + 2e^- + 2H^+ \rightarrow N_2O + H_2O \tag{5.5}$$

$$N_2O + 2e^- + 2H^+ \rightarrow N_{2(g)} + H_2O \tag{5.6}$$

As shown in the above equation $NO_3^- - N$ produced from the nitrification process is used as an electron acceptor here, and carbon source as an electron donor. For denitrification to take place, there must be enough organic carbon present. While older leachates with low organic content may need an additional source of organic carbon, young leachates high in VFAs have enough organic carbon to support the growth of denitrifying bacteria (Price *et al.*, 2003). Since denitrification is an anoxic process, excessive dissolved oxygen might inhibit it (Qu *et al.*, 2015). For optimal denitrification, the dissolved oxygen concentration should be between 0.5 and 1.3 mg O_2/L (Medhi *et al.*, 2017).

5.4.1.1.1 Disadvantages of nitrification–denitrification

However, this treatment method has certain operational drawbacks, such as high energy requirement in aeration, high greenhouse gas emission, the expensive cost of an additional carbon source, and a large area requirement, which hinders its widespread adoption. The low COD/N and BOD5/COD ratios in mature landfill leachate are incompatible with the nitrification and denitrification processes (Deng *et al.*, 2022).

Due to the oxygen present during nitrification, autotrophic bacteria can transform the organic molecules in raw wastewater into CO_2. Consequently, there is no carbon source for more denitrification. This disturbs the normal nitrification–denitrification process, leading to the loss of carbon sources and energy as well as high disposal costs. Therefore, mature landfill leachate with a low COD/N ratio and a low BOD5/COD ratio is incompatible with standard nitrification–denitrification procedures.

5.4.1.2 Nitritation–denitritation

In the nitrification–denitrification process, the conversion of $NO_2^- - N$ into $NO_3^- - N$ and $NO_3^- - N$ into $NO_2^- - N$ requires more energy and carbon sources from the landfill leachate. If nitrification could be managed in the first step, when $NH_4^+ - N$ is only oxidized to $NO_2^- - N$ and not further oxidized to $NO_3^- - N$, denitrification process may use nitrite ions (NO_2^-) in place of nitrate ions(NO_3^-). This biochemical reaction is called partial nitrification or nitritation (Figure 5.2). Nitritation or partial nitrification is the mechanism controlling the oxidation of $NH_4^+ - N$ to $NO_2^- - N$ alone, and for denitritation this NO_2^- is used as an electron acceptor. The chemical reactions for nitritation–denitritation process are shown in equations (5.7) and (5.8).

The nitritation–denitritation method provides a number of benefits. First, energy can be conserved by reducing aeration use by 25%. Second, the procedure lowers expenses by 40% by lowering the carbon source utilised for denitrification. Third, both nitritation and denitritation have the potential to significantly reduce sludge production, which will make sludge disposal easier. Fourth, shorter reaction time reduces the volume of reactors and saves space requirements (Miao *et al.*, 2019).

$$NH_4^+ + 1.5O_2 \rightarrow NO_2^- + 2H^+ + H_2O \tag{5.7}$$

$$2NO_2^- + 6H^+ + 6e^- \rightarrow 2OH^- + N_2 + 2H_2O \tag{5.8}$$

Achieving stable nitritation is important for the nitritation–denitritation process. The following aspects should be considered for ammonia to nitrite conversion (Miao *et al.*, 2019).

(a) On a large scale, partial nitrification is accomplished by preventing $NO_2^- - N$ oxidation. From a micro-aspect, the $NO_2^- - N$ accumulation is maintained by accelerating AOB growth and activity and limiting NOB growth and activity. NOB claimed to have a lower rate of growth than AOB. Therefore, limiting SRT could contribute to reducing the number of NOB bacteria (Regmi *et al.*, 2014).
(b) According to some research, AOB has a greater affinity for oxygen than NOB; hence, low dissolved oxygen concentrations could limit NOB activity and growth (Liu *et al.*, 1997).
(c) AOB activity can be accelerated rather than NOB by factors including high pH, high-temperature conditions, and high influent nitrogen content, which is advantageous for nitrite accumulation.

5.4.1.3 Nitritation–endogenous denitritation

The nitritation–endogenous denitritation treatment is based on intracellular polymer storage. It is able to make effective use of organic components in landfill leachate and remove nitrogen without adding external carbon sources. Phosphorous accumulating organisms (PAOs) and glycogen accumulating organisms (GAOs) both work on the same principle that they store carbon and convert it into polyhydroxyalkanoates (PHAs) in the cell and use it when external carbon source is not available. GAOs could play an important role in endogenous denitrification to remove nitrogen content with no need for external carbon sources (Miao *et al.*, 2019). Nitrogen from landfill leachate may be removed through endogenous denitrification by increasing the number of GAOs (Figure 5.2). Using the nitritation–endogenous denitrification process, young and medium landfill leachate, which often contains large amounts of biodegradable organics with a high COD/N ratio and a low P/COD ratio, can be effectively treated (Miao *et al.*, 2019).

To remove nitrogen, the nitritation–endogenous denitritation method may completely use the organics in raw landfill leachate. This results in a longer reaction time than with conventional procedures (Miao *et al.*, 2015), but it can minimize or eliminate the demand for external carbon sources and energy usage for aeration, hence reducing operational expenses. In addition, the generation of sludge could reduce as a result of endogenous metabolism (Miao *et al.*, 2019; Wang *et al.*, 2015).

5.4.1.4 Anammox

Under anaerobic conditions, $NH_4^+ - N$ is oxidized by NO_2^- (as an electron acceptor) to produce N_2 and a small amount of $NO_3^- - N$ (Fig. 5.2), as shown in equation (5.9) (Kuenen, 2008).

$$\begin{aligned} NH_4^+ + 1.32\,NO_2^- + 0.132\,HCO_3^- + 0.512\,H^+ \rightarrow 1.02\,N_2 \\ +0.26\,NO_3^- + CH_2O_{0.5}N_{0.5} + 2.19\,H_2O \end{aligned} \tag{5.9}$$

First, the approach minimizes oxygen uptake by 60% since only around 50% of the $NH_4^+ - N$ in the influent must be converted into $NO_2^- - N$ as opposed to $NO_3^- - N$ (Ma *et al.*, 2016). Second, the technique can save 100% of the carbon source used to extract nitrogen because nitritation and anammox are autotrophic. Third, the slow development of anammox bacteria substantially minimizes sludge generation, hence reducing sludge disposal expenses. Anammox bacteria do not create nitrous oxide (NO_2) during their metabolic processes (Miao *et al.*, 2019). This reduces emissions of greenhouse gases.

However, the anammox technique has downsides as well. The anammox procedure, for instance, involves a lengthy accumulation period and SRT. It delays the start-up of the anammox system (Miao *et al.*, 2019). Oxygen and organic compounds are two additional variables that are unfavorable to anammox. In order to sustain the anammox process, it is necessary to create the appropriate reaction conditions. Difficulty obtaining $NO_2^- - N$ in a stable manner, theoretically, 11% of the $NO_3^- - N$ produced in the anammox process leads to the failure of the effluent total nitrogen (TN), requiring additional treatment. Usually, the concentration of ammonia nitrogen in mature landfill leachate is greater than 2,000 mg/L. However, a very high quantity hinders the activity of anammox microbes. The actual inhibitor is not $NH_4^+ - N$, but rather the free ammonia nitrogen (Deng *et al.*, 2022).

5.4.2 Physical treatment

5.4.2.1 Reverse osmosis

Reverse osmosis (RO) is a membrane filtration method. The membrane filtration process is a separation technique that makes use of a thin semipermeable membrane that, through the pressure gradient, prevents the passage of particles and dissolved components in water or wastewater from influent into effluent streams. Unlike biological approaches, it divides contaminants into two streams: a purified stream known as permeate and a non-purified stream known as concentrate. Membrane procedures separate various components in a solution through a form of filtration; more precisely, they separate them according to the size of the component that needs to be separated.

When a semipermeable membrane separates two solutions, water from the less concentrated side permeates to the most concentrated side. The side that was initially most concentrated experiences additional pressure. Less than 0.0001 μm is the membrane's pore size (Adam *et al.*, 2019). Water can pass through the semipermeable membrane by applying pressure to the side with a higher osmotic pressure, according to the RO principle. Two types of membranes, cellulose acetate (diacetate and triacetate) and aromatic polyamide, are typically employed in the use of RO. Pressure, pH, and membrane type are the most crucial operational variables. A non-volatile compound is created when the ammonia on the shell side combines with the acid on the other interface by diffusing from the bulk solution to the interface between the feed solution and the membrane (Karri *et al.*, 2018).

Thus, the membrane eliminates contaminants such as $NH_4^+ - N$. It is the most promising treatment method because of its exceptional ability to remove heavy metals, suspended and dissolved particles, organic materials, and dissolved inorganic species from leachate. Membrane-based approaches have certain exceptional qualities, including minimum space occupation, large volume loading, good effluent quality, and strong disinfection power. The RO system uses a variety of membrane configurations, including tubular, spiral wound, frame, plate, and so on (Adam *et al.*, 2019). However, there are many disadvantages associated with RO system (Cossu *et al.*, 2018; Teng *et al.*, 2021).

5.4.2.1.1 Disadvantages of RO system

(a) Membrane fouling is produced by the deposition of organic, inorganic, or biological substances on the membrane surface or pores. Membrane fouling can significantly limit the lifetime of membrane modules and result in high operating conditions.
(b) Membrane-concentrated landfill leachate is produced in large quantities.
(c) Retention of small molecules such as ammonia and small halogenated organic compounds (AOX) are not entirely acceptable.
(d) As a result of strong operational pressures, energy consumption is high.
(e) The cost of solid residue storage is high.

5.4.2.2 Evaporation

Evaporation, which can happen naturally or can be accelerated by site operations, is one possible way to reduce the amount of leachate that needs to be handled. By significantly reducing nitrogen concentrations and allowing the recovery of a highly concentrated ammonia distillate solution, evaporation followed by ammonia condensation is considered an effective process for treating landfill leachate. Leachate evaporation is a method for reducing volume, concentrating solids, and promoting ammonia and organic chemical volatilization (Schwarzwälder Sprovieri *et al.*, 2020). $NH_4^+ - N$, which may be converted into $NH_3 - N$ and then recovered by condensation is a crucial component in this process. The amount of leachate that has to be treated has been decreased through leachate evaporation. There are direct processes and indirect mechanisms for evaporation. When using direct methods, the leachate and high-temperature gas come into direct contact. In contrast, energy is transported into the leachate using a heat exchanger when indirect evaporation procedures are used. In ponds, natural evaporation is influenced by a number of variables, including temperature, relative humidity, solar radiation, and pond surface area. Physical systems, such as complex evaporation towers or droplet sprayers, can also improve evaporation. These methods may assist in reducing the amount of leachate, but depending on the type of system, cost of standard treatment, and operational costs connected with each, they may ultimately prove to be more complicated or expensive than typical leachate treatment. Physical technologies include sprinkler/droplet sprayer systems, mechanical vapor recompression, wind-enhanced systems, and heat-based evaporators. To accomplish considerable and high pollutant removal, evaporation is a common treatment procedure that is used in combination with other processes such as reverse osmosis and filtration (Drafts, 2021).

Applications for recovered ammonia from landfill leachates include precipitated magnesium ammonium phosphate (MAP), a multi-nutrient fertilizer (Li & Zhao, 2003). Due to its high refrigerant power and low stratospheric ozone depletion potential, ammonia is used as a source of nitrogen in the production of neutralizing agents for the oil industry and industrial refrigeration gas systems, as well as raw material in the production of explosives, cigarettes, and nitric acid. This allows for the avoidance of the use of organic chlorofluorocarbons in large refrigeration machines (Schwarzwälder Sprovieri *et al.*, 2020). However, this process has some downsides also.

5.4.2.2.1 Disadvantages of evaporation

(a) The process is quite slow.
(b) It occupies a lot of space.
(c) Water temperature and humidity both affect efficiency.
(d) VOCs evaporate and cause air pollution.

5.4.2.3 Ammonia Stripping

Ammonia stripping is one of the important nitrogen removal methods. Using a simple desorption process to reduce the ammonia concentration, ammonia removal using air or steam stripping provides a simpler and more affordable solution. It involves forcing air through a liquid phase to remove one or more volatile components from the liquid phase (Metcalf & Eddy, 2003). This approach relies on encouraging non-ionized ammonia volatilization, which is generally accomplished by increasing the leachate's contact with the air and by increasing pH (Santos *et al.*, 2020).

As shown in equation (5.10), $NH_3 - N$, which is volatile, and $NH_4^+ - N$ are in equilibrium in wastewater.

$$NH_{3(g)} + H_2O \rightleftharpoons NH^+_{4(aq)} + OH^-_{(aq)} \tag{5.10}$$

The effluent pH directly affects this chemical equilibrium. Typically, crude leachate has a pH below 7.5 (Costa *et al.*, 2019). As a result, the pH should be raised by adding an alkalinizing chemical species such as lime, adjusting the pH to 11 units, and raising the temperature to 25°C in order to promote the volatilization of $NH_3 - N$, which at pH 9.25 is 50% ionized (i.e., $NH_4^+ - N$) and 50% free (i.e., $NH_3 - N$). Therefore, the percentage of free ammonia will increase to 99%. According to this characteristic, the air stripping process occurs when a liquid effluent containing free-form ammonia comes into contact with a driven air stream, hence transforming it from a liquid to a gas phase (Metcalf & Eddy, 2003). However, there are some drawbacks to this method as well.

5.4.2.3.1 Disadvantages of ammonia stripping

(a) Demands a large stripping tower due to foaming when $NH_4^+ - N$ in raw leachate is stripped out.
(b) Excessive time and energy use.

(c) Scanning and fouling on packaging
(d) Operates at a particular pH, flow rate, and temperature.
(e) The release of $NH_3 - N$ into the atmosphere could result in severe air pollution if ammonia cannot be adequately absorbed.
(f) The scaling of the stripping tower is formed by calcium carbonate when lime is used to regulate pH. This scaling may cause severe operation and maintenance problems.

5.4.3 Chemical treatment

5.4.3.1 Chemical precipitation

To eliminate high concentrations of $NH_4^+ - N$, chemical precipitation is frequently employed as a pre-treatment method. The removal of heavy metals and $NH_4^+ - N$ is frequently accomplished by the use of chemical precipitation as a pre-treatment method. A highly capable, cost-effective, and simple process of chemical precipitation has been widely used to treat landfill leachate. When treating leachate, struvite, or MAP, and lime are frequently used as the precipitants for $NH_4^+ - N$ precipitation. Chemical reactions take place during chemical precipitation to transform the dissolved ions in the solution into the insoluble solid phase. Eventually, the NH_3 in the water will agglomerate, settle, and form insoluble chemical precipitation of NH_4^+.

Equal molar amounts of magnesium, ammonium, and phosphorus build up the white crystalline substance of MAP. The reaction is shown below (Adam *et al.*, 2019)

$$\begin{aligned} &Mg^{2+} + PO_4^{3-} + NH_4^+ + 6H_2O \rightleftharpoons MgNH_4PO_4 \cdot 6H_2O \\ &\qquad \downarrow pK_s = 12.6(25°C) \end{aligned} \tag{5.11}$$

Equations (5.12), (5.13), and (5.14) can be used to calculate the estimated amount of chemicals required to cause the NH_4^+ to precipitate (Adam *et al.*, 2019).

$$\begin{aligned} &MgCl_2 \cdot 6H_2O + Na_2HPO_4 \cdot 12H_2O + NH_4^+ \\ &\qquad \rightarrow MgNH_4PO_4 \cdot 6H_2O \downarrow + 2NaCl \end{aligned} \tag{5.12}$$

$$MgO + H_3PO_4 + NH_4^+ \rightarrow MgNH_4PO_4 \cdot 6H_2O \downarrow + H_2O \tag{5.13}$$

$$\begin{aligned} &Ca(H_2PO_4)_2 \cdot H_2O + MgSO_4 \cdot 7H_2O \\ &\quad + NH_4^+ \rightarrow MgNH_4PO_4 \cdot 6H_2O \downarrow + CaSO_4 \downarrow \end{aligned} \tag{5.14}$$

MAP is quickly isolated from the aqueous phase and has a low water solubility (0.169 g/L at 25°C). Since MAP has a molecular weight of 245 g/mol, the removal of 1 g of $NH_4^+ - N$ should theoretically result in the precipitation of 17.5 g $MgNH_4PO_4 \cdot 6H_2O$. The molar ratio of $Mg^{2+} : NH_4^+ : PO_4^{3-}$ for the effective removal of $NH_4^+ - N$ was discovered to be 1.15:1:1 (Adam *et al.*, 2019). There are, however, drawbacks to this method (Teng *et al.*, 2021).

5.4.3.1.1 Disadvantages of chemical precipitation

(a) A poor COD removal efficiency.
(b) A high pH dependence.

(c) Large quantities of precipitants are required.
(d) The production of a large amount of sludge.

5.4.3.2 *Adsorption*

The transfer of organic materials from a liquid phase to the surface of a solid phase is known as adsorption. Fundamentally, adsorption is a mass transfer process. Adsorption is the process by which molecules in an aqueous solution are bound to adsorbents with a high internal surface area. The substance that provides the surface for adsorption is known as an adsorbent, whereas the species being adsorbed is known as an adsorbate. Adsorption is one of the most efficient and promising post-treatment methods for removing NH_4^+-N from landfill leachate. In terms of configuration and operation, it is less complicated and more cost-effective.

Physical and chemical adsorption are two types. Weak van der Waals forces of attraction between the adsorbent and adsorbate result from physical adsorption. Chemical absorption, on the other hand, happens when the adsorbent and the adsorbate form a chemical bond.

Adsorption by a solid is influenced by a number of elements, including surface area, the nature of the adsorbent and adsorbate, the pH of the solution, the distribution of the adsorbent's pore size, temperature, and mixed solutes (Segweni, 2017).

Leachate from landfills has been treated using adsorbents having a large surface area, microporous structure, surface reactivity, and thermostability. Adsorbents used in wastewater treatment might come from either industrial production or natural sources. Typical natural adsorbents include natural zeolites, clay minerals, and biopolymers. Activated carbon, chitosan, activated alumina, and silica gel are some of the adsorbents that are manufactured commercially. Activated carbon, biochar, and zeolite are commonly used adsorbents (Teng *et al.*, 2021). Despite these facts, there are a few drawbacks associated with this method.

5.4.3.2.1 Disadvantages of adsorption

(a) The requirement for adsorbent regeneration and
(b) High absorbent consumption levels.

5.4.3.3 *Ion exchange*

Ion exchange procedures are reversible chemical reactions that remove dissolved ions from solutions and replace them with additional ions that have similar charges. Ions are exchanged using this technique between an aqueous solution and a solid known as a resin bed or ion exchanger. It develops a strong affinity for NH_4^+ and a powerful removal capability. It is additionally a simple, inexpensive, and sustainable operation.

The NH_4^+ is intended to take the place of the exchanger cation (A+) through the reversible process. The ion-exchange process between the exchanger and ammonia is as follows:

$$(\text{M-A+})(\text{s}) + (\text{Ammonium+})\ (\text{aq}) \leftrightarrow (\text{M} - \text{Ammonium+})\ (\text{s}) + (\text{A+})\ (\text{aq}) \quad (5.15)$$

Equation (5.16) shows the affinity potential of additional cations relative to NH_4^+, and some of these cations have an impact on the removal of ammonium ions when present in wastewater or landfill leachate (Karri *et al.*, 2018).

$$Cs^+ > Rb^+ > K^+ > NH_4^+ > Sr^{2+} > Na^+ > Ca^{2+} > Fe^{3+} > Al^{3+} > Mg^{2+} \quad (5.16)$$

The two most common types of resins used to remove ammonium ions from wastewater are zeolite and synthetic resins (silica powder) (Prajapati, 2014). Common zeolites found in nature include clinoptilolite, mordenite, chabazite, heulandite, laumonite, analcite, and erionite (Montalvo *et al.*, 2012). In addition to the natural ion exchanger, other investigations have reports on making polymeric ion exchangers that comprise hydrated, scattered metal oxides, like manganese(IV) oxide (MnO_2), hydrated zirconium(IV) oxide (HZrO), iron(III) oxide (HFeO), and hybrid cation exchangers (HCE) and anion exchangers (HAIX) (HCIX) (Adam *et al.*, 2019). This method does have a few drawbacks to it.

5.4.3.3.1 Disadvantages of ion exchange

(a) It is an expensive method due to chemical reagent regeneration costs.
(b) Desorption frequently happens as a result of an equilibrium shift during the ion exchange process for reversible chemical reactions, particularly when the influent ammonia concentration is less.
(c) Suitable for specific pH ranges.
(d) Limited reusability.

5.4.3.4 Breakpoint chlorination

Widely used breakpoint chlorination is a conventional method that converts NH_3 into N_2. This method is only useful for polishing effluent, not for removing large amounts of nitrogen. This procedure involves chlorinating ammonia-containing water, which initially raises the combined chlorine residual. The combined chlorine residual and ammonia concentration then fall together, and finally, the free chlorine residual rises and nearly completely removes the NH_3 as N_2.

NH_3 removal is done in three steps. The first step involves the reaction between chlorine and NH_3 to form monochloramines (NH_2Cl). The second step is associated with the reaction between monochloramine and chlorine, results in an increment in dichloramine ($NHCl_2$) and the disappearance of NH_3. The third step involves the appearance of free chlorine after the breakpoint. A breakpoint is used as a reference to show the complete removal of NH_3. Stoichiometrically NH_3 to N_2 gas requires a Cl_2:NH_3–N weight ratio of 7.6:1 (White, 2010).

The chemistry between chlorine and NH_3 has been confirmed by numerous reviews and research (Wei, 1972; Yoon & Jensen, 1993). According to studies, NH_3 and chlorine generally react to generate chloramine in the following ways:

$$NH_3 + HOCl \rightarrow NH_2Cl + H_2O \text{ (monochloramine)} \quad (5.17)$$

$$NH_2Cl + HOCl \rightarrow NHCl_2 + H_2O \text{ (dichloramine)} \quad (5.18)$$

$$NHCl_2 + HOCl \rightarrow NCl_3 + H_2O \text{ (nitrogen trichloride)} \quad (5.19)$$

The design of a breakpoint chlorination system is quite simple. No unique design elements are required for the initial reaction between NH_3 and chlorine other than ensuring thorough, uniform mixing of the chlorine with the wastewater. A recommended standard is 10 min of minimum interaction time (Stefán *et al.*, 2019). The amount of influent NH_3, as well as the level of wastewater treatment, determines the size of the chlorine-producing and feed device.

5.4.3.4.1 Disadvantages of breakpoint chlorination

(a) If not enough chlorine is added to reach the breakpoint, N_2 gas does not develop, and any generated chloramines must be eliminated before discharge because chloramines are potent disinfectants.
(b) NH_3 removal efficiency will be reduced if there are any chlorine-reducing compounds present in the leachate.

5.4.3.5 Electrochemical oxidation

Electrochemical oxidation (EO) is an advanced oxidation process (AOP). In the EO process, contaminants are eliminated by (a) direct EO, in which organics are oxidized by transferring electrons directly to the anode, or (b) indirect EO, in which electro-active species are created that serve as mediators in the degradation process.

5.4.3.5.1 Direct electrochemical oxidation

Direct electrolysis degrades organic materials in two steps: (1) organic pollutants diffuse from the electrolyte's bulk to the anode surface, where they are adsorbed; and (2) organic contaminants are oxidized at the anode surface by an electron shuttle (Figure 5.3). Chemical reactions are shown in equation (5.20).

$$R \rightarrow P + e^- \quad (5.20)$$

where R is the targeted organic pollutant, P is the oxidized organic pollutant.

In direct EO of contaminants, the electrochemical system is operated at a very low potential (below the 'water discharge' potential) prior to the oxygen evolution process. The reaction rate (pollutant oxidation rate) in direct EO is normally modest but can vary depending on the electrocatalytic activity of the anode materials used. Certain metallic anodes (Pt, Pd, etc.) and metal oxide anodes (IrO_2, Ir-TiO_2, PbO_2, and Ru-O_2) exhibit considerably higher electrocatalytic activity during direct EO, hence accelerating electron transport (Deng & Englehardt, 2007; Gautam *et al.*, 2019; Mandal *et al.*, 2017).

This method may, however, enhance the generation of hydroxyl radicals adsorbed on the anode surface and then oxidize organics by indirect electrolysis.

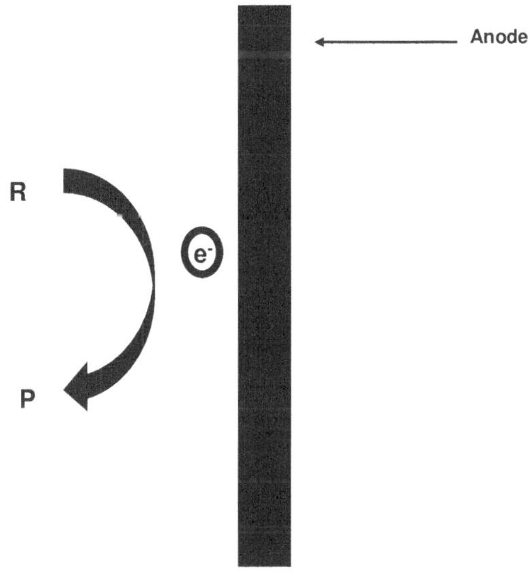

Figure 5.3 Schematic diagram of direct electrochemical oxidation.

5.4.3.5.2 Indirect electrochemical oxidation

Indirect electrolysis takes place in the potential zone where oxygen evolution originates from water oxidation. As a result of oxygen evolution during water discharge, intermediate species are produced, which serve as oxidation spots on the anode surface (Figure 5.4). Consequently, organics can be oxidized in two ways: organic oxidation at the anode surface and organic oxidation in the bulk of the electrolyte solution.

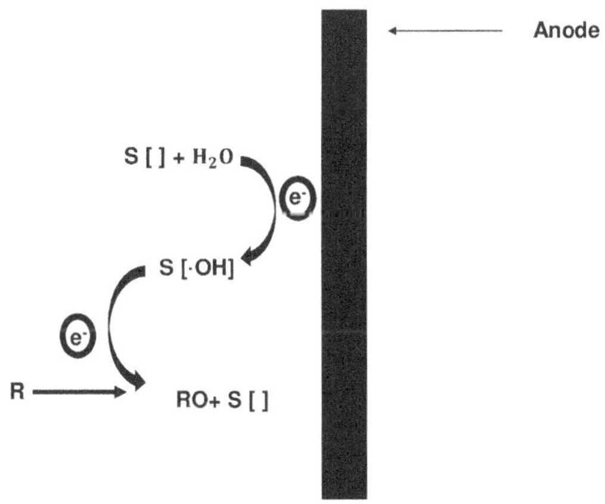

Figure 5.4 Schematic diagram of indirect electrochemical oxidation.

Organic oxidation at the anode surface:

Electrolysis conducted at a potential greater than the stability potential of water results in the generation of hydroxyl radicals. On the anode's surface, these hydroxyl radicals adsorb.

$$S\,[] + H_2O \rightarrow S\,[OH] + H^+ + \tag{5.21}$$

$$S\,[OH] + R \rightarrow S\,[] + RO + H^+ + e^- \tag{5.22}$$

where S [] indicates adsorption sites for generated hydroxyl radical
R indicates organic pollutants and RO indicates oxidized organic pollutants.

A decrease in current efficiency is a challenge related to pollutant removal via indirect hydrolysis, as a part of the given current is used for oxygen generation.

$$S\,[OH] + H_2O \rightarrow S\,[] + O_2 + 3H^+ + 3e^- \tag{5.23}$$

Oxidation of organics in the bulk of electrolyte solution:

Certain oxidizing agents may also be produced during indirect electrolysis, destroying the contaminants in the bulk of the solution. Certain oxidizing agents may also be produced during indirect electrolysis, destroying the contaminants in the bulk of the solution. The most common electrochemically generated oxidants are chlorine (Cl_2), hypochlorite (OCl), hypochlorous acid (HClO), hydrogen peroxide (H_2O_2), peroxodisulfuric acid ($H_2S_2O_8$), HBrO and ozone (O_3), and so on (Mandal *et al.*, 2017).

5.4.3.5.3 Mechanism for ammonium nitrogen removal

EO primarily results in the formation of chlorine/hypochlorite, which removes ammonium nitrogen. There is a competition between COD and ammonium nitrogen to get oxidized, so EO can effectively remove the mature landfills.

$$HOCl + NH_4^+ \rightarrow NH_2Cl + H_2O + H^+ \tag{5.24}$$

$$HOCl + NH_2Cl \rightarrow NHCl_2 + H_2O \tag{5.25}$$

$$H_2O + NHCl_2 \rightarrow NOH + 2H^+ + 2Cl^- \tag{5.26}$$

$$NOH + NHCl_2 \rightarrow N_2 + HOCl + H^+ + Cl^- \tag{5.27}$$

5.4.3.5.4 Disadvantages of electrochemical oxidation

(a) The polymer layer formed on the anode surface, which has a 'poisoning impact' on the electrocatalytic property of the anode material, is the most significant disadvantage of the direct EO process.
(b) High operational cost for electricity required by all electrochemical techniques.
(c) During chlorine-mediated oxidation, the formation of organo-chlorinated molecules is likely.

5.5 CONCLUSION AND PERSPECTIVES

Globally as massive factories and processing industries continue to grow over time, an enormous amount of solid waste will be produced. Landfilling is considered the most cost-effective and environmentally friendly method of ultimate disposal of solid waste, which is practiced worldwide. Treatment efficiency, treatment system cost, and environmental implications are the three most essential criteria for selecting a suitable treatment technology or combination of sustainable treatment technologies for landfill leachate treatment. Biological or physical–chemical therapy approaches on their own have been shown to be ineffectual. Combining two or more physical–chemical treatments would maximize the removal of refractory organic compounds from stabilized leachate, whereas biological and physical–chemical treatments are necessary for the successful removal of $NH_4^+ - N$ and COD. None of the known biodegradation techniques provide a satisfactory solution on their own. Therefore, the integration of these technologies should be given substantial consideration. As both municipal wastewater and landfill leachate include similar contaminants, the novel methods and materials under investigation for wastewater treatment can also be used for leachate treatment. However, leachate contains increased concentrations of pollutants in varying amounts. Therefore, emerging technologies and materials should be handled with caution and applied strategically at each treatment stage. By enhancing the physicochemical properties of traditional adsorbents, effective leachate treatment can be achieved. Enhanced regeneration capabilities would be crucial for application success.

REFERENCES

Abunama T., Othman F., Ansari M. and El-Shafie A. (2019). Leachate generation rate modeling using artificial intelligence algorithms aided by input optimization method for an MSW landfill. *Environmental Science and Pollution Research*, **26**(4), 3368–3381, https://doi.org/10.1007/s11356-018-3749-5

Adam M. R., Othman M. H. D., Abu Samah R., Puteh M. H., Ismail A. F., Mustafa A., Rahman M. and Jaafar J. (2019). Current trends and future prospects of ammonia removal in wastewater: a comprehensive review on adsorptive membrane development. *Separation and Purification Technology*, **213**, 114–132, https://doi.org/10.1016/j.seppur.2018.12.030

Bandala E. R., Liu A., Wijesiri B., Zeidman A. B. and Goonetilleke A. (2021). Emerging materials and technologies for landfill leachate treatment: a critical review. *Environmental Pollution*, **291**, 118133, https://doi.org/10.1016/j.envpol.2021.118133

Berge N. D., Reinhart D. R. and Townsend T. G. (2005). The fate of nitrogen in bioreactor landfills. *Critical Reviews in Environmental Science and Technology*, **35**(4), 365–399, https://doi.org/10.1080/10643380590945003

Białowiec A. (2011). Hazardous emissions from municipal solid waste landfills. *Contemporary Problems of Management and Environmental Protection*, **9**, 7–28.

Bolyard S. C. and Reinhart D. R. (2016). Application of landfill treatment approaches for stabilization of municipal solid waste. *Waste Management*, **55**, 22–30, https://doi.org/10.1016/j.wasman.2016.01.024

Chakravarty P. and Kumar M. (2019). Floral species in pollution remediation and augmentation of micrometeorological conditions and microclimate: an integrated approach. In: Vimal Chandra Pandey & Kuldeep Bauddh (Eds.), Phytomanagement of Polluted Sites. Elsevier, pp. 203–219.

Cossu R., Ehrig H.-J. and Muntoni A. (2018). Physical–chemical leachate treatment. In: Raphael Cossu & Rainer Stegmann (Eds.), *Solid Waste Landfilling*. Elsevier Inc, pp. 575–632, https://doi.org/10.1016/b978-0-12-407721-8.00028-0

Costa A. M., Alfaia R. G. D. S. M. and Campos J. C. (2019). Landfill leachate treatment in Brazil – an overview. *Journal of Environmental Management*, **232**, 110–116, https://doi.org/10.1016/j.jenvman.2018.11.006

Dabaghian Z., Peyravi M., Jahanshahi M. and Rad A. S. (2018). Potential of advanced nano-structured membranes for landfill leachate treatment: a review. *ChemBioEng Reviews*, **5**(2), 119–138, https://doi.org/10.1002/cben.201600020

Daniel A. N., Ekeleme I. K., Onuigbo C. M., Ikpeazu V. O. and Obiekezie S. O. (2021). Review on effect of dumpsite leachate to the environmental and public health implication. *GSC Advanced Research and Reviews*, **7**(2), 051–060, https://doi.org/10.30574/gscarr.2021.7.2.0097

Das S., Lee S. H., Kumar P., Kim K. H., Lee S. S. and Bhattacharya S. S. (2019). Solid waste management: scope and the challenge of sustainability. *Journal of Cleaner Production*, **228**, 658–678, https://doi.org/10.1016/j.jclepro.2019.04.323

Deng Y. and Englehardt J. D. (2007). Electrochemical oxidation for landfill leachate treatment. *Waste Management*, **27**(3), 380–388, https://doi.org/10.1016/j.wasman.2006.02.004

Deng W., Wang L., Cheng L., Yang W. and Gao D. (2022). Nitrogen removal from mature landfill leachate via anammox based processes: a review. *Sustainability (Switzerland)*, **14**(2), 1–17, https://doi.org/10.3390/su14020995

Drafts K. R. (2021). Scholar commons fate of nitrogen and effectiveness of evaporation on managing leachate in collection and storage ponds: a case study at the three rivers solid waste authority regional landfill.

El-Fadel M., Findikakis A. N. and Leckie J. O. (1997). Modeling leachate generation and transport in solid waste landfills. *Environmental Technology*, **18**(7), 669–686, https://doi.org/10.1080/09593331808616586

Fenn D. G., Hanley K. J. and DeGeare T. V. (1975). Use of the Water-Balance Method for Predicting Leachate Generation From Solid-Waste-Disposal Sites (No. PB-87-194643/XAB; EPA/SW-168). Environmental Protection Agency, Washington, DC (USA). Office of Solid Waste.

Gautam P., Kumar S. and Lokhandwala S. (2019). Advanced oxidation processes for treatment of leachate from hazardous waste landfill: a critical review. *Journal of Cleaner Production*, **237**, 117639, https://doi.org/10.1016/j.jclepro.2019.117639

Griffith J., Duncan R. C., Riggan W. B. and Pellom A. C. (1989). Cancer mortality in US counties with hazardous waste sites and ground water pollution. *Archives of Environmental Health: An International Journal*, **44**(2), 69–74, https://doi.org/10.1080/00039896.1989.9934378

Haslina H., NorRuwaida J., Dewika M., Rashid M., Md Ali A. H., Khairunnisa M. P. and Afiq Daniel Azmi M. (2021). Landfill leachate treatment methods and its potential for ammonia removal and recovery – a review. *IOP Conference Series: Materials Science and Engineering*, **1051**(1), 012064, https://doi.org/10.1088/1757-899X/1051/1/012064

Hong K. J., Tokunaga S. and Kajiuchi T. (2002). Evaluation of remediation process with plant-derived biosurfactant for recovery of heavy metals from contaminated soils. *Chemosphere*, **49**(4), 379–387, https://doi.org/10.1088/1757-899x/1051/1/012064

Iskander, S. M., Zhao, R., Pathak, A., Gupta, A., Pruden, A., Novak, J. T. and He, Z. (2018). A review of landfill leachate induced ultraviolet quenching substances: Sources, characteristics, and treatment. *Water Research*, **145**, 297–311.

Karri R. R., Sahu J. N. and Chimmiri V. (2018). Critical review of abatement of ammonia from wastewater. *Journal of Molecular Liquids*, **261**(2017), 21–31, https://doi.org/10.1016/j.molliq.2018.03.120

Kuenen, J. G. (2008). Anammox bacteria: from discovery to application. *Nature Reviews Microbiology*, **6**(4), 320–326, https://doi.org/10.1038/nrmicro1857

Lema J. M., Mendez R. and Blazquez R. (1988). Characteristics of landfill leachates and alternatives for their treatment: a review. *Water, Air, and Soil Pollution*, **40**(3–4), 223–250, https://doi.org/10.1007/BF00163730

Li X. Z. and Zhao Q. L. (2003). Recovery of ammonium-nitrogen from landfill leachate as a multi-nutrient fertilizer. *Ecological Engineering*, **20**(2), 171–181, https://doi.org/10.1016/S0925 8574(03)00012 0

Liu, H., Hull, R. J. and Duff, D. T. (1997). Comparing cultivars of three cool-season turfgrass for soil water NO_3^- concentration and leaching potential. *Crop Science*, **37**(2), 526–534.

Lin A. Y. C., Huang S. T. Y. and Wahlqvist M. L. (2009). Waste management to improve food safety and security for health advancement. *Asia Pacific Journal of Clinical Nutrition*, **18**(4), 538–545.

Ma, B., Wang, S., Cao, S., Miao, Y., Jia, F., Du, R. and Peng, Y. (2016). Biological nitrogen removal from sewage via anammox: recent advances. *Bioresource Technology*, **200**, 981–990, https://doi.org/ 10.1016/j.biortech.2015.10.074

Mandal P., Dubey B. K. and Gupta A. K. (2017). Review on landfill leachate treatment by electrochemical oxidation: drawbacks, challenges and future scope. *Waste Management*, **69**, 250–273, https://doi.org/10.1016/j.wasman.2017.08.034

Medhi K., Singhal A., Chauhan D. K. and Thakur I. S. (2017). Investigating the nitrification and denitrification kinetics under aerobic and anaerobic conditions by *Paracoccus denitrificans* ISTOD1. *Bioresource Technology*, **242**, 334–343, https://doi.org/10.1016/j.biortech.2017.03.084

Metcalf, L., Eddy, H. P. and Tchobanoglous, G. (2003). Wastewater engineering: treatment, disposal, and reuse (Vol. 4). New York: McGraw-Hill.

Miao L., Wang S., Li B., Cao T., Xue T. and Peng Y. (2015). Advanced nitrogen removal via nitrite using stored polymers in a modified sequencing batch reactor treating landfill leachate. *Bioresource Technology*, **192**, 354–360, https://doi.org/10.1016/j.biortech.2015.05.013

Miao L., Yang G., Tao T. and Peng Y. (2019). Recent advances in nitrogen removal from landfill leachate using biological treatments – a review. *Journal of Environmental Management*, **235**, 178–185, https://doi.org/10.1016/j.jenvman.2019.01.057

Montalvo, S., Guerrero, L., Borja, R., Sánchez, E., Milán, Z., Cortés, I. and De La La Rubia, M. A. (2012). Application of natural zeolites in anaerobic digestion processes: A review. *Applied Clay Science*, **58**, 125–133, https://doi.org/10.1016/j.clay.2012.01.013

Ozbay G., Jones M., Gadde M., Isah S. and Attarwala T. (2021). Design and operation of effective landfills with minimal effects on the environment and human health. *Journal of Environmental and Public Health*, **2021**, 1–13, https://doi.org/10.1155/2021/6921607

Prajapati S. (2014). Cation Exchange for Ammonia Removal from Wastewater. Master's thesis.

Price G. A., Barlaz M. A. and Hater G. R. (2003). Nitrogen management in bioreactor landfills. *Waste Management*, **23**, 675–688, https://doi.org/10.1016/S0956-053X(03)00104-1

Qu, D., Wang, C., Wang, Y., Zhou, R. and Ren, H. (2015). Heterotrophic nitrification and aerobic denitrification by a novel groundwater origin cold-adapted bacterium at low temperatures. *Rsc Advances*, **5**(7), 5149–5157, https://doi.org/10.1039/C4RA13141J

Rathnayake W. A. P. P. and Herath G. B. B. (2018). A Review of Leachate Treatment Techniques. The 9th International Conference on Sustainable Built Environment, December, 97–106, https://www.researchgate.net/publication/329915923

Regmi P., Miller M. W., Holgate B., Bunce R., Park H., Chandran K., … and Bott C. B. (2014). Control of aeration, aerobic SRT and COD input for mainstream nitritation/denitritation. *Water Research*, **57**, 162–171, https://doi.org/10.1016/j.watres.2014.03.035

Renou S., Givaudan J. G., Poulain S., Dirassouyan F. and Moulin P. (2008). Landfill leachate treatment: review and opportunity. *Journal of Hazardous Materials*, **150**(3), 468–493, https://doi.org/10.1016/j.jhazmat.2007.09.077

Rushton L. (2003). Health hazards and waste management. *British Medical Bulletin*, **68**(1), 183–197, https://doi.org/10.1093/bmb/ldg034

dos Santos H. A. P., de Castilhos Júnior A. B., Nadaleti W. C. and Lourenço V. A. (2020). Ammonia recovery from air stripping process applied to landfill leachate treatment. *Environmental Science and Pollution Research*, **27**(36), 45108–45120, https://doi.org/10.1007/s11356-020-10397-9

Şchiopu A. M., Apostol I. and Gavrilescu M. (2009). Impact of landfill leachate on soil quality in Iasi County. *Environmental Engineering & Management Journal (EEMJ)*, **8**(5), 1155–1164.

Schwarzwälder Sprovieri J. A., Octavio de Souza T. S. and Contrera R. C. (2020). Ammonia removal and recovery from municipal landfill leachates by heating. *Journal of Environmental Management*, **256**, 109947, https://doi.org/10.1016/j.jenvman.2019.109947

Segweni D. (2017). Options for Treatment of Ammonia in Landfill Leachate. February, 1–198, https://ir.canterbury.ac.nz/bitstream/handle/10092/13626/Segweni%2CDoctor_Master%27sThesis.pdf?sequence=1&isAllowed=y

Stefán D., Erdélyi N., Izsák B., Zaray G. and Vargha M. (2019). Formation of chlorination by-products in drinking water treatment plants using breakpoint chlorination. *Microchemical Journal*, **149**, 104008, https://doi.org/10.1016/j.microc.2019.104008

Teng C., Zhou K., Peng C. and Chen W. (2021). Characterization and treatment of landfill leachate: a review. *Water Research*, **203**, 117525, https://doi.org/10.1016/j.watres.2021.117525

Vaverková M. D. (2019a). Landfill impacts on the environment. *Geosciences*, **9**(10), 431, https://doi.org/10.3390/geosciences9100431

Wang Y., Zhou S., Wang H., Ye L., Qin J. and Lin X. (2015). Comparison of endogenous metabolism during long-term anaerobic starvation of nitrite/nitrate cultivated denitrifying phosphorus removal sludges. *Water Research*, **68**, 374–386, https://doi.org/10.1016/j.watres.2014.09.044

Wei I. W. T. (1972). *Chlorine-ammonia breakpoint reactions: kinetics and mechanism* [Doctoral thesis, HHharvard university Cambridge, Massachusetts] Digital Repository Service, http://hdl.handle.net/2047/D20237252

Weng Y., Fujiwara T., Houng H. J., Sun C., Li W. and Kuo Y. (2015). Management of land fill reclamation with regard to biodiversity preservation, global warming mitigation and land fill mining: experiences from the Asia e Pacific region. *Journal of Cleaner Production*, **104**, 364–373, https://doi.org/10.1016/j.jclepro.2015.05.014

White G. C. (2010). White's Handbook of Chlorination and Alternative Disinfectants. (2011). Germany: Wiley.

Wiszniowski J., Robert D., Surmacz-Gorska J., Miksch K. and Weber J. V. (2006). Landfill leachate treatment methods: a review. *Environmental Chemistry Letters*, **4**(1), 51–61, https://doi.org/10.1007/s10311-005-0016-z

Yoon J. and Jensen J. N. (1993). Distribution of aqueous chlorine with nitrogenous compounds: chlorine transfer from organic chloramines to ammonia. *Environmental Science & Technology*, **27**(2), 403–409, https://doi.org/10.1021/es00039a022

Yusof N., Hassan M. A., Phang L. Y., Tabatabaei M., Othman M. R., Mori M., Wakisaka M., Sakai K. and Shirai Y. (2010). Nitrification of ammonium-rich sanitary landfill leachate. *Waste Management*, **30**, 100–109, https://doi.org/10.1016/j.wasman.2009.08.018

doi: 10.2166/9781789063318_0111

Chapter 6

Constructed wetland for landfill leachate treatment

Sudeep Kumar Mishra[1], Sanket Dey Chowdhury[1], Puspendu Bhunia[2], Arindam Sarkar[3]*

[1]Research Scholar, Environmental Engineering, School of Infrastructure, Indian Institute of Technology Bhubaneswar, Bhubaneswar 752 050, Odisha, India
[2]Professor, Environmental Engineering, School of Infrastructure, Indian Institute of Technology Bhubaneswar, Bhubaneswar 752 050, Odisha, India
[3]Associate Professor, Water Resources Engineering, School of Infrastructure, Indian Institute of Technology Bhubaneswar, Bhubaneswar 752 050, Odisha, India
**Corresponding author: pbhunia@iitbbs.ac.in*

ABSTRACT

Due to complex chemical composition and high recalcitrance, the treatment of landfill leachate becomes challenging. The conventional wastewater treatment processes employed to remediate the landfill leachate necessitate huge operation and maintenance costs. For more than two decades, constructed wetlands (CWs) have been successfully used to treat landfill leachate. These systems mimic the natural wetland process and eliminate pollutants from wastewater. For effective landfill leachate treatment, a comprehensive understanding of its chemical composition and the quantitative and qualitative changes in its characteristics, and the factors that influence them is essential. This chapter reviews the pollutant removal mechanisms, application and performance, and the factors affecting the performance of the CW systems during landfill leachate management. Furthermore, a detailed analysis of the compatibility of various types of CWs for landfill leachate treatment has also been incorporated in this chapter.

Constructed wetland, macrophytes, bed material, biodegradation, phytoremediation.

6.1 INTRODUCTION

Rapid economic growth and rising standards of living have contributed to an increase in municipal solid waste production. There are numerous ways of disposing of municipal solid waste, including landfilling, composting, and incineration. Composting municipal solid waste takes a long time and involves plenty of work. Disposal through incineration requires high energy which makes the process highly expensive. Owing to the economic benefits and easy monitoring, sanitary landfilling has gained widespread acceptance across the globe. However, this technique generates a substantial amount of landfill leachate (Butt *et al.*, 2014).

Landfill leachate is produced by biological degradation, natural humidity and water from residual organic matter, and precipitation entering the landfill sites. A significant amount of inorganic and organic constituents can be found in leachate generated in landfills, posing a threat to both surface water and groundwater resources. Therefore, the exploration of an effective method for treating leachate from landfills is crucial.

The use of constructed wetland (CW) as a leachate treatment method has received a significant amount of attention and effort because of its low cost and potential for on-site application (Akinbile *et al.*, 2012; Dan *et al.*, 2017; Yalçuk & Ugurlu, 2020). CW is an energy-efficient and cost-effective method for treating wastewater with available resources (Langergraber, 2008). These are the man-made wastewater treatment systems that simulate the function of a natural wetland in which contaminants are removed through physical, chemical, and biological processes, which include precipitation, ion exchange, filtration, substrate adsorption, microbial decomposition, and plant uptake due to the combined effects of microorganism, substrate, and plant (He *et al.*, 2017). Adjustment of leachate composition, selection of appropriate media, arrangement of wetland systems, and plant nutrient uptake rate evaluation are the most important factors that must be thoroughly investigated to improve the contaminant removal rate of CWs treating landfill leachate (Camaño Silvestrini *et al.*, 2019; Cano *et al.*, 2019).

6.2 LEACHATE CHARACTERIZATION

As rainwater percolates through landfill waste, it washes out by-products of degraded organic matter and substances deposited in the landfill, creating leachates. The physical and chemical composition of landfill leachate is affected by a variety of factors, including solid waste composition, climate, landfill age, and site hydrogeology. However, leachate typically contains a large number of organic materials, nitrogen, phosphorus, heavy metals, and salts. Generally,

Table 6.1 Landfill leachate characteristics.

Parameters (mg/L)										References
pH*	TSS	COD	BOD_5	NH_3–N	NH_4^+–N	NO_2^-–N	NO_3^-–N	TN	TP	
8.42	685	923.44	686	–	–	–	–	–	–	Akinbile *et al.* (2012)
7.5	458	2037	635	340	–	–	11.4	76	12.3	Bakhshoodeh *et al.* (2020)
8.3	417.50	1275.30	257.10	–	303.4	0.4	11.6	411.2	47.8	Saeed *et al.* (2021)
7.9	2720	2800	204.20	–	198.40	–	1.9		5.5	Lavrova and Koumanova (2010)

*pH is unitless.

the landfill leachate shows neutral pH. The concentration of ammonia nitrogen in municipal solid waste landfill leachate remains as high as 1000 mg/L (Lott *et al.*, 2022). Table 6.1 shows the characteristics of various landfill leachate.

BOD_5/COD and COD/TOC are the standard indices of the biodegradability of organic compounds and the oxidized state of organic carbon, respectively (Teng *et al.*, 2021). BOD_5/COD ratios in landfill leachates, less than three to five years old, can reach as high as 0.7, whereas BOD_5/COD ratios in landfills that are five to 10 years old are significantly lesser, ranging from 0.3 to 0.5 (Bakhshoodeh *et al.*, 2020).

6.3 VARIABILITY IN LEACHATE QUANTITY AND QUALITY

Changes in the qualitative and quantitative characteristics of landfill leachate are influenced by landfill age, type of landfilled waste, landfilling techniques used, and climate (Wojciechowska, 2017). The age of a landfill is a significant element that influences the composition and properties of landfill leachate. Depending on the age of the landfill, leachate can be classified into three groups: young, medium, and old leachate. Young leachate is less than five years old, while medium leachate is five to 10 years old. Leachate older than 10 years is considered an old leachate (Miao *et al.*, 2019). Young landfill leachate is predominantly made of hydrophilic organic matter with low molecular weight, portraying low pH, and high biodegradability index (BOD_5/COD). In contrast, the primary components of the old landfill leachate are humic acid and fulvic acid with high molecular weights, resulting in a lower biodegradability index and elevated pH value. Notably, the concentration of heavy metals tends to decrease with leachate age since the increase in pH diminishes the metals' solubility (Teng *et al.*, 2021).

Organic matter concentration and biodegradability of the landfill leachate depend on the stage of organic matter degradation, which varies with the landfill ages. There are three phases of organic matter degradation: aerobic degradation, anaerobic degradation, and humic phase (Wojciechowska, 2017).

The aerobic degradation of organic waste produces a large quantity of carbon dioxide and water and continues till the depletion of available oxygen, a comparatively short process. Recently, the idea of aerobic landfills with air injections has received major attention from the researchers. Introducing air to the waste tip promotes the aerobic conversion of wastes, which is anticipated to increase the degradation rate, shorten the waste stabilization time, and improve $NH4^+$-N removal (Fadel *et al.*, 2013).

When available oxygen gets exhausted, the anaerobic degradation phase begins, during which sulfates, nitrates, ferric oxide, and manganese oxide are used as electron donors. The anaerobic phase can be further subdivided into the acidic phase and the methanogenic phase. Labile organic compounds like cellulose are broken down during the acidic phase, which results in the formation of volatile fatty acids (VFA) and carbon dioxide, which ultimately causes an increase in the BOD of the landfill leachate. During this time, organic nitrogen undergoes ammonification and the pH shifts slightly towards the acidic range. Landfill leachate is rich in both iron and manganese concentrations (Wojciechowska, 2017).

After 10 years, a landfill enters the phase of methanogenesis when VFA are mineralized and the pH rises. Biogas containing carbon dioxide and methane is produced until the organic matter is depleted. Iron, bound to low-soluble sulfides reduces leachate iron concentration. Typically, the methanogenic phase is regarded as the final phase of organic waste degradation. The landfill reaches the humic phase after approximately 100 years of operation when the ultimate stabilization of organic matter is achieved and the humic substances are produced as the end product. (Bozkurt *et al.*, 2000).

6.4 TYPES OF CWS

According to the functioning of the water flow regime, CW systems can be classified as surface flow constructed wetland (SF CW), subsurface flow constructed wetland (SSF CW), and hybrid systems (Figure 6.1). In SF CW, water slowly flows above the bed media, creating a free water surface and a few-centimeter-deep water column (Figure 6.2a). There are two types of subsurface

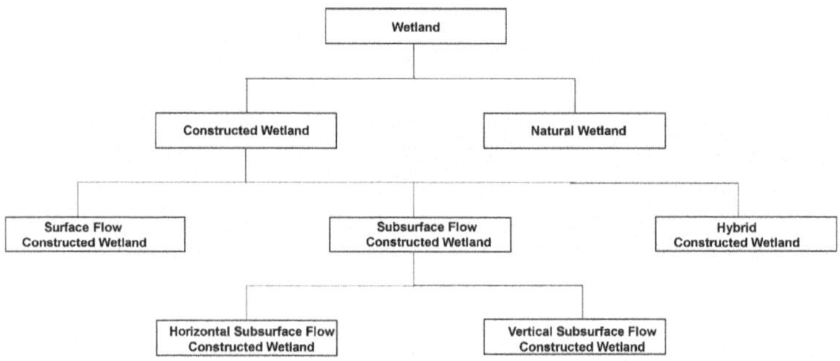

Figure 6.1 Types of constructed wetland.

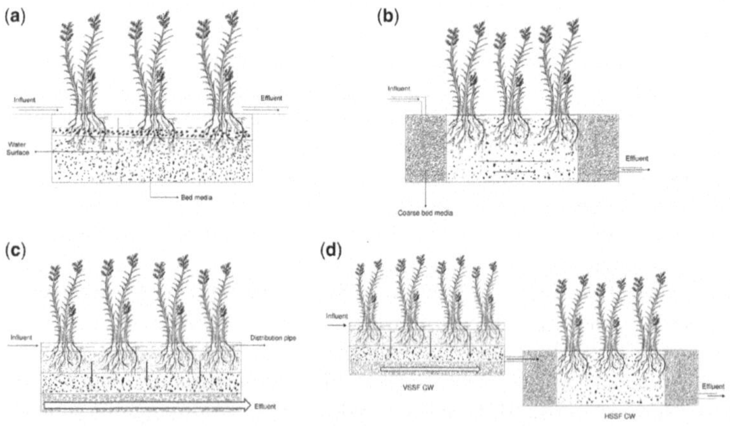

Figure 6.2 Different configurations of CWs: (a) SF CW, (b) HSSF CW, (c) VSSF CW, (d) hybrid CW.

flow CWs: horizontal subsurface flow constructed wetland (HSSF CW) (Figure 6.2b) and vertical subsurface flow constructed wetland (VSSF CW) (Figure 6.2c). Beds are intermittently loaded in the VSSF CW by employing a significant quantity of water throughout the entire bed surface of the wetland. Due to pulse-loading, a satisfactory level of oxygen transfer happens in VSSF CW, which provides better nitrification (Chowdhury *et al.*, 2022; Dey Chowdhury *et al.*, 2022).

Within the HSSF CW, the wastewater is kept below the surface of the media. The wastewater flows horizontally through the bed until it reaches the outlet. The supply of oxygen from plants in the HSSF CW is inadequate which results in anaerobic condition development in the saturated bed, where anaerobic degradation of organic matter and denitrification occurs (Vymazal, 2005). Due to the activity of aerobic microorganisms, VSSF CW removes organic matter, total suspended solids (TSS), and ammonium ions more effectively than HSSF CW, but provides less denitrification (Lavrova & Koumanova, 2010; Yalçuk & Ugurlu, 2020). Various types of CWs (SF, HSSF, and VSSF) may be combined in series to form a hybrid system (Figure 6.2d) to increase the removal efficiency of contaminants, especially nitrogen by providing both aerobic and anaerobic conditions (Dey Chowdhury *et al.*, 2022; Wang *et al.*, 2020).

6.5 MECHANISM OF LEACHATE TREATMENT IN THE CW SYSTEM

CWs have been effectively applied for landfill leachate treatment over the last few years (Akinbile *et al.*, 2012). CWs are natural, easy to operate, and economical wastewater treatment systems. They are also capable of treating wastewater with low biodegradability with significantly greater efficiency than the conventional biological processes (Gajewska *et al.*, 2015).

The removal of pollutants by a CW is accomplished through a combination of physical, chemical, and biological processes. CWs are defined as engineered systems that use natural processes occurring in the presence of macrophytes, microorganisms, and soil for the removal, transformation, and degradation of pollutants from wastewater in a synergistic manner. Macrophytes play a major role in the removal of contaminants (Figure 6.3).

CWs employ a wide range of operations and processes for the removal of contaminants. Biological oxidation, sedimentation, and microbial uptake remove organic material (BOD), while organic contaminants like pesticides are degraded by adsorption, photolysis, volatilization, and biotic processes. Filtration and sedimentation are used to get rid of suspended solids, and plant uptake, nitrification/denitrification, sedimentation, microbial uptake, and volatilization are used for nitrogen removal (Chowdhury *et al.*, 2022; Dey Chowdhury *et al.*, 2022). In the CW treatment system, nitrogen removal follows a complex matrix of biotransformation processes, such as anammox, simultaneous nitrification, and denitrification (Hu *et al.*, 2016). Sedimentation, adsorption, filtration, plant and microbial uptake, and ultraviolet (UV) degradation are all effective ways to remove phosphorus; sedimentation, natural die-off, filtration, UV degradation and adsorption cause elimination of pathogens; and removal of heavy metals is carried out by sedimentation, adsorption by vegetation, and UV degradation (Swarnakar *et al.*, 2022).

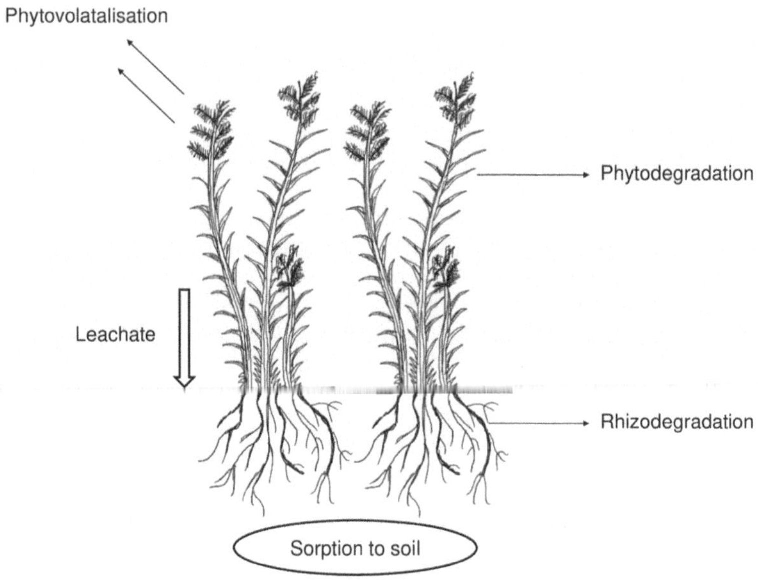

Figure 6.3 Mechanism of contaminant degradation through macrophytes.

CWs designed for landfill leachate treatment must contend with several challenges such as extremely high ammonium concentrations, high concentrations of iron, toxicity to hydrophytes, fluctuations of quantity (caused by precipitation), and low quality of the landfill leachate, resulting from their specific composition, which can inhibit the growth of hydrophytes (Kadlec & Zmarthie, 2010). For a CW to work efficiently, raw leachate is introduced with some dilution with potable water to acclimatize the plants and biofilm to their new conditions. The dilution is increased incrementally over time (Lott *et al.*, 2022).

CW systems used for landfill leachate treatment are mostly surface flow (SF) or sub-surface flow (SSF). An SF system offers high and reliable treatment efficiency and requires a large area and a year-long hydraulic retention time (Wojciechowska, 2017). SSF systems accelerate wetland treatment processes for a longer period over a smaller surface area. Horizontal sub-surface flow (HSSF) systems are easier to run. However, the treatment of the leachate from a landfill often shows trivial nitrogen removal efficiency. A VSSF CW is intended to increase aeration and enhance the nitrification process (Saeed & Sun, 2012).

The efficiency of treatment in the CWs also depends upon the properties and design of the bed (Albuquerque *et al.*, 2009). Mineral substrates have a high adsorption capacity, which allows them to perform well in CW. On the other hand, plant establishment and growth are encouraged by organic substrates due to their enrichment in nutrients and organic content (Wang *et al.*, 2010). pH, along with contact time, is one of the most influential factors, governing the sorption capacity of bed materials. Maintaining an adequate pH level prevents metals trapped on substrates from being released back into the system (Nehrenheim *et al.*, 2008).

6.6 APPLICATION AND PERFORMANCE OF THE CWS IN TREATING LANDFILL LEACHATE

6.6.1 Organic matter

Organic matter in landfill leachate is made up of a complex blend of biopolymers. Some of these components, including carbohydrates, proteins and lipids, are easily destroyed by microbes (i.e., labile), but hemicellulose and lignin are resistant to breakdown (i.e., recalcitrant). A multi-step process degrades the biopolymers. The higher values of the influent COD/BOD_5 ratio indicate the recalcitrant nature of organics present in landfill leachate. In CWs organic matter is degraded by aerobic and anaerobic microbial activity as well as filtration and sedimentation of particulate organic matter. Generally, the CW system follows first-order kinetics during the degradation of the contaminants (Vymazal & Kröpfelová, 2009).

COD removal is observed to be less during the initial stage of treatment because of the non-acclimatization of the leachate sample and the immaturity of the biofilm that plays a major role in the degradation of organic contaminants. Adsorption has no significant role in the degradation of organic matter (He *et al.*, 2017).

6.6.2 Nitrogen

Since the aggravated concentration of nitrogen leads to eutrophication which has detrimental effects on water quality, removing it from landfill leachate is of paramount importance. The predominant nitrogen removal pathways in the CWs include ammonification, volatilization, nitrification, and denitrification (Yalcuk & Ugurlu, 2009).

In subsurface flow wetland systems, ammonification activates the first phase of nitrogen transformation if the wastewater to be treated is rich in organic nitrogen. Ammonification is a complex biochemical process that releases energy in which amino acids undergo oxidative deamination to produce NH_3. The rate of ammonification doubles for every 10°C rise in temperature (Saeed & Sun, 2012). Nitrification is a two-step aerobic process which results in the conversion of NH_4-N into nitrite nitrogen (NO_2-N) which gets converted into nitrate nitrogen (NO_3-N) in the second step. Due to the insufficiency of oxygen, nitrification is inhibited in horizontal subsurface CWs. The aerobic zone is formed around the plant roots in planted CWs, but the region slightly beyond the root remains anoxic. The presence of both the aerobic and anoxic zone is beneficial for the nitrification–denitrification of NH_4^+-N.

Plant rhizosphere aeration may promote aerobic degradation processes, resulting in an increase in nitrification and subsequent gaseous losses of nitrogen via denitrification, and a decrease in relative levels of dissimilatory nitrate reduction to ammonium. Lott *et al.* (2022) studied the removal of

NH_3-N from landfill leachate using various VSSF planted with *Typha latifolia* and *Scirpus californicus* and found the best removal efficiency of 53.2% in the columns planted with *Typha latifolia*. In the experiment, consumption of alkalinity for the nitrification process reduced the pH of the effluent.

Denitrification is dependent on the presence of organic carbon and nitrate nitrogen in the system. Environmental factors such as temperature, pH, microbial attachment surface area, and dissolved oxygen concentration also influence denitrification (Yalcuk & Ugurlu, 2009). The absence of organic substrate can impede denitrification due to the lack of growth of the denitrifiers (Dey Chowdhury & Bhunia, 2021). In addition, plant litter and plant root exudates are reported to continuously add organic carbon to CWs, preventing the complete depletion of the organic carbon inside the system.

It is well known that zeolites are attracted to ammonium ions and other cations. There are high concentrations of ammonia and cationic metals in landfill leachate, and it is advantageous to use zeolite (clinoptilolite) in the CWs used to treat landfill leachate. He *et al.* (2017) observed a rapid reduction in the concentration of ammonia nitrogen and total nitrogen from the initial stage of treatment which is due to the effective adsorption and cation exchange of the substrate.

6.6.3 Total phosphorous and phosphate

In CWs, the removal of phosphorus (P) takes place through chemical and physicochemical processes, including sorption and precipitation (Sim *et al.*, 2008). Filtration and sedimentation also contribute to the removal of P to some extent in CWs (Bakhshoodeh *et al.*, 2020). The biological removal of P from CW is not significant because P is temporarily taken up by microorganisms. Apart from the above-mentioned processes, a large fraction of P is taken up by the macrophytes, if present in the CWs, thereby cutting down the effluent P concentration (Wdowczyk *et al.*, 2022). Variation in media does not affect the P uptake capacity of plants (Saeed *et al.*, 2021).

P removal varies significantly depending on the leachate dilution employed. Wdowczyk *et al.* (2022) observed better phosphorus removal in 100% landfill leachate as compared to 25% diluted leachate. Increased adsorption of P is observed by the substrates rich in aluminium (Al) and calcium (Ca), ensuring more effective removal of P (Johansson Westholm, 2006). Akinbile *et al.* (2012) noted almost 100% removal of total phosphorus from 117 to 0.3 mg/L. Saeed *et al.* (2021) observed that P removal for landfill leachate treatment in hybrid CWs follows a predominantly media-based adsorption route.

6.6.4 Total suspended solids

The presence of very small solid particles in the form of suspended solids increases the turbidity of the leachate. Physical processes, like filtration and sedimentation, are largely responsible for the TSS removal in CW.

Yalçuk and Ugurlu (2020) observed the reactors planted with *Canna indica* and *Typha latifolia* show the best TSS removal. The observed TSS removal after 50 days was, 26–83% in *Canna indica* and 35–88% in *Typha latifolia*.

The presence of *Cyperus haspan* reduced the turbidity by 85% from 140 to 18.7 NTU in the CW treating landfill leachate (Akinbile *et al.*, 2012). Bakhshoodeh *et al.* (2020) observed that the implementation of SF, HSSF, VSSF, and hybrid CWs ensured 59.5 ± 36%, 69.3 ± 17%, 55.5 ± 20%, and 51.8 ± 29% removal of TSS, respectively.

As clogging progresses with time, the filtration bed becomes denser and the rate of removal tends to rise during operation. In HSSF CW, the removal of suspended solids is maximum at the beginning of the filtration bed, and it has been demonstrated that the majority of suspended solids are retained within the starting length of the bed (Wojciechowska *et al.*, 2010).

6.6.5 Heavy metals

Heavy metals are the most toxic pollutants present in landfill leachate. Chromium (Cr) (0.0005–1.6 mg/L), cadmium (Cd) (0.0001–0.13 mg/L), manganese (Mn) (0.01–65 mg/L), iron (Fe) (0.08–2,100 mg/L), nickel (Ni) (0.03–3.2 mg/L), lead (Pb) (0.0005 1.5 mg/L), and zinc (Zn) (0.00005–120 mg/L) are some of the common metals found in landfill leachate. Young (acetogenic) leachate often has higher metal contents than that of older (methanogenic) leachate (Dan *et al.*, 2017).

According to studies, the CWs have the potential to substantially remove the heavy metals from landfill leachate (Dan *et al.*, 2017; Khan *et al.*, 2009; Si *et al.*, 2014). CW primarily removes heavy metals through biological pathways, sorption on the soil surface and plant roots, chemical precipitation, binding to organic matter, and filtration of suspended solids by root systems (Bakhshoodeh *et al.*, 2020). Adsorption is the most important pathway for the removal of heavy metals from leachate in CW, while the contribution of plant uptake is the least toward the same (Mohammed and Babatunde, 2017).

According to a study by Wdowczyk *et al.* (2022), the average reduction of Ni, Zn, Cu, and Cr with sand, gravel, and zeolite as media, in VSSF CW systems was at the level of 41–56%. In the same study, the average reduction of heavy metals in CW without dilution of landfill leachate was 26%, 53%, and 57% using expanded clay, zeolite, and bark substrate, respectively.

Akinbile *et al.* (2012) observed that the landfill leachate embraced the highest concentration of Mg. They reported that the presence of *Cyperus haspan* macrophyte ensured 75.9–89.4%, 51.2–70.5%, 29–75%, and 34.9–59% eradication of zinc (Zn), Mg, manganese (Mn), and iron (Fe), respectively.

Dan *et al.* (2017) reported that the lab-scale VSSF CW facilitated 92–100, 80–100, 75–99, 68–100, and 54–100% removals of Zn, Cr, Ni, Cd, and Pb, respectively while treating the synthetic leachate.

Maine *et al.* (2022) studied the removal of Cr, Ni, and Zn from landfill leachate using VSSF CW with coarse sand and light expanded clay aggregate planted with *Typha domingensis* and *Canna indica*. The result showed better removal of heavy metals in the wetland planted with *Typha domingensis* which exhibits its high capacity to retain metals (Table 6.2). The removal efficiency of heavy metals with higher concentration was higher than the removal efficiency for the metals present in low concentration.

Table 6.2 Removal efficiency of contaminants from landfill leachate in constructed wetland.

Type of Wetland	Vegetation	Substrate	Removal Efficiency (%)			Reference
			Organics	Nutrients	Heavy Metals	
Horizontal SSF CW	*Phragmites australis*	Light expanded clay aggregate (LECA)	–	Ammonia > 62 Nitrate: 90–100	–	Albuquerque *et al.* (2009)
Vertical SSF CW	*Typha latifolia*	Sand, gravel, zeolite	COD: 27.3	NH_4-N: 62.3 PO_4-P: 52.6	Fe: 21	Yalcuk and Ugurlu (2009)
Vertical SSF CW	*Typha latifolia*	Sand, gravel	COD: 30.6	NH_4-N: 48.9 PO_4-P: 51.9	Fe: 40	Yalcuk and Ugurlu (2009)
Horizontal SSF CW	*Typha latifolia*	Gravel	COD: 35.7	NH_4-N: 38.3 PO_4-P: 46.7	Fe: 17	Yalcuk and Ugurlu (2009)
Vertical SSF CW	*Phragmites australis*	Gravel	COD: 96 BOD_5: 92	Ammonia: 100 Total phosphorus: 100	–	Lavrova and Koumanova (2010)
SSF CW	*Cyperus haspan*	Sand and gravel	COD: 39.2–91.8 BOD_5: 60.8–78.7	NH_3-N: 29.8–53.8 TP: 59.8–99.7 TN: 33.8–67.0	Mg: 29.0–75.0 Fe:34.9–59.0 Mn:51.2–70.5 Zn: 75.9–89.4	Akinbile *et al.* (2012)
Two vertical SSF CW with one horizontal SSF CW	*Phragmites australis*	Washed gravel	COD: 47.8–86.6	TN: 68.9–98.5	–	Wojciechowska (2017)
Horizontal SSF CW	*Vetiveria zizanioides*	Fine gravel	BOD_5: 74.5 COD: 53.7	NH_3-N:69.9 NO_3-N:73.5 TN:73.4	–	Bakhshoodeh *et al.* (2017)

Vertical SSF CW	*Juncus effusus*	Pumice stone, loamy soil	–	NH_4-N: 44–73	Cd: 80–96 Cr: 95–99 Fe: 85–100 Mn:73–92 Ni: 88–96 Pb: 54–98 Zn: 96–99	Dan *et al.* (2017)
Horizontal SSF CW	*Reed, cattail*	Zeolite, slag, soil, sand, gravel	COD: 20.5–48.2	NH_3-N: 84.0–99.9 TN: 80.3–92.1	Heavy metals: 90	He *et al.* (2017)
Horizontal SSF CW	*Heliconia psittacovum, Cyperus papyrus*	Calcareous gravel	COD: 40	NH_3-N: 63–81	–	Cano *et al.* (2019)
Vertical SSF CW	*Typha domingensis*	Coarse sand, LECA	–	NH_4^+:34–74 TN:16–48	–	Camaño Silvestrini *et al.* (2019)
Vertical SSF CW	*Canna indica*	Coarse sand, LECA	–	NH_4^+:27–72 TN:18–46	–	Camaño Silvestrini *et al.* (2019)
Vertical SSF CW	*Typha latifolia*	Sand, zeolite, fine gravel, large pebbles	COD:81	PO_4:45 NH_4^+-N:60	–	Yalçuk and Ugurlu (2020)
Vertical SSF CW	*Canna indica*	Sand, zeolite, fine gravel, large pebbles	COD:84	NH_4^+-N:56 PO_4:46	–	Yalçuk and Ugurlu (2020)

(*Continued*)

Table 6.2 Removal efficiency of contaminants from landfill leachate in constructed wetland. (*Continued*)

Type of Wetland	Vegetation	Substrate	Removal Efficiency (%)			Reference
			Organics	Nutrients	Heavy Metals	
Constructed wetland	*Typha* sp. *(cattail)*	Sand, clay, iron powder	BOD: 62.9–76.3 COD: 49.4–70.2 TOC: 60.5–74.7 Organic micropollutant: 67.5–75.4	NH_3:44.4–69.0	–	Witthayaphirom *et al.* (2020)
Hybrid CW	*Phragmites vetiver*	Coal, cocopeat, concrete block, brick, sand	COD: 55–76 BOD: 34–89	N: 50–93 P: 69–100	–	Saeed *et al.* (2021)
Vertical SSF CW	*Typha domingensis*	LECA, coarse sand	–	–	Cr: 60 Zn: 61 Ni: 49	Maine *et al.* (2022)
Vertical SSF CW	*Canna indica*	LECA, coarse sand	–	–	Cr: 54 Zn: 47 Ni: 47	Maine *et al.* (2022)
Three-stage vertical SSF CW	*Canna indica, Phragmites australis, Cyperus involucratus*	Fly ash	–	TN: 91.43 NH_3-N: 94.19 NO_3-N:98.11	–	Wang *et al.* (2022)
Vertical SSF CW	*Phragmites australis*	Gravel, sand, zeolite	–	AN: 96–99 TKN: 75–88.5 TN: 62.5–70 TP:84–88	Heavy metals: 41–56	Wdowczyk *et al.* (2022)

6.7 FACTORS AFFECTING THE PERFORMANCE OF CW

6.7.1 Substrate

The selection of suitable substrate plays a major role in plant growth, microbial adhesion, contaminant storage, and chemical and biological transformations by microorganisms and plants in the CW system (Chowdhury *et al.*, 2022; Liu *et al.*, 2014). Significant research has been performed using sand or gravel as a single substrate material. Both these filter material has good hydraulic properties and pollutant removal capacity, but attributing to the lesser adsorption capacity, these materials can hamper the plant acclimatization and even, clog the system (Wang *et al.*, 2020). Clogging would be a serious challenge to the SSF CW system. The substrate must have adequate porosity to prevent clogging and the ability to filter and absorb contaminants. As the concentration of N and heavy metals is very high in landfill leachate, it is not possible to achieve a significant level of removal efficiency with single filter material. It has been observed that the combination of substrates enhances the adsorption capacity of the bed, which results in greater removal of pollutants (He *et al.*, 2017).

The growth of bacteria carrying out nitrification, ammonification, and denitrification also depends on the properties of the substrate (Cano *et al.*, 2019). Ca- and Mg-rich substrate shows better adsorption of P, resulting in more stable and efficient removal (Liu *et al.*, 2014). Mineral substrates have a high adsorption capacity, which allows them to perform well in CW. On the other hand, plant establishment and growth are encouraged by organic substrates due to their enrichment in organic and nutrient content. Pine bark with a C/N ratio of 104.93 is considered the appropriate material for the treatment of landfill leachate (Wdowczyk *et al.*, 2022). Soluble organic materials and trace amounts of nutrients can be released from bark and other organic substrates. Therefore, while employing organic substrates for treatment in CWs, there is always a possibility of releasing the organic matter back into the system. The other important parameters affecting the selection of the substrate are the universal availability, cost, and the requirement of pretreatment of the materials (Lim *et al.*, 2016).

In comparison to a single substrate system, the combined hybrid substrate shows a better removal of heavy metals (Huong *et al.*, 2020). Due to the higher ion exchange capacity, natural zeolites are one of the most effective substrate materials for the removal of various contaminants. He *et al.* (2017) investigated the efficiency of two HSSF CWs, downflow (F1) and up the flow (F2), with the zeolite–slag hybrid as a substrate for the treatment of rural landfill leachate. The results showed that the CWs were capable of removing NH_3-N, COD, TN, and heavy metals up to 84.0–99.9% (F1) and 93.5–99.2% (F2), 20.5–48.2% (F1) and 18.6–61.2% (F2), 80.3–92.1% (F1) and 80.3–91.2% (F2), and 90% (for both F1 and F2), respectively.

6.7.2 Macrophyte

The treatment performance of the CWs depends on the presence of macrophytes in the system. The root zone of the macrophytes plays an important role in the biodegradation of organic matter and serves as the substrate source for the microbes responsible for the biodegradation of the organics (Brix, 1994).

Photosynthesis by plants generates oxygen which is carried to the root zone where aerobic biodegradation takes place. The oxygen produced is used for the nitrification and mineralization of the organic matter. The ability of wetland plants such *as Typha latifolia, Phragmites mauritianus, Nymphaea spontanea, Cyperus papyrus, Typha angustifola, Limnocharis flava, Eichhornia crassipes,* and *Pistia stratiotes* has been explored by various researchers while treating the domestic as well as the industrial wastewaters (Akinbile & Yusoff, 2012; Sarafraz *et al.*, 2009). *Cyperus haspan* showed a very reliable result in the treatment of landfill leachate in a CW system (Table 8.2) (Akinbile *et al.*, 2012).

It is evident that plants provide numerous benefits to the treatment process, but their role in the overall effectiveness of treatment remains questionable. The plant rhizosphere is thought to be largely responsible for the superior performance of planted wetlands compared to their unplanted controls, as it provides a growing surface for microorganisms on the roots, a source of carbon compounds in the form of root exudates, and a microaerobic environment through the rootzone aeration (Chowdhury *et al.*, 2022; Tanner, 2001). Diffusion of oxygen depends upon the rhizosphere of plants, which further affects the process of nitrification. The rhizosphere of cattail provides favorable conditions for nitrification than those of giant bullrush (Lott *et al.*, 2022). Bakhshoodeh *et al.* (2017) studied the effect of vegetation on compost leachate treatment and found that the unplanted and the vetiver-planted CWs portrayed 26.2 and 53.7, 21.8 and 74.5, 34.1 and 73.5, 17.1 and 69.9, and 35 and 73.4% removals of COD, BOD_5, NO_3^--N, NH_4^+-N, and TN, respectively.

There is a significant effect of vegetation on the removal of heavy metals. The process of metal uptake by plants occurs at a rapid rate. When compared to the uptake of nutrients, the kinetics of metal uptake and the subsequent tissue bioaccumulation of the metals in the macrophytes is significantly faster (Maine *et al.*, 2022).

6.7.3 Temperature

The performance of CW changes with seasonal variations and temperature. The processes involving nitrogen and organics bio-reaction kinetics are highly temperature dependent (Saeed & Sun, 2012). Temperature influences both the direct and indirect uptake of nutrients by plants and microbial activity in wetland ecosystems. In CWs, nitrification is commonly considered a function of temperature. Reportedly, nitrification rates in CWs become inhibited at a water temperature of approximately 10°C and drop precipitously at around 6°C (Werker *et al.*, 2002).

An increase in temperature increases the rate of photosynthesis by the plants in water. Water temperature affects the level of BOD_5 in leachate samples. Water at a higher temperature will have a higher BOD_5 than that of cold water (Akinbile *et al.*, 2012). Nivala *et al.* (2007) noted the removal of organics was higher in the summer (60–97%) than in the winter (44–88%) in an aerated HSSF wetland system. The removal of NH_4^+–N followed the same pattern.

6.7.4 Hydraulic retention time

Hydraulic retention time (HRT) represents the contact time between the wastewater, rhizosphere, substrate, and microorganisms. A longer HRT permits greater contaminant removal, whereas a shorter HRT reduces the treatment efficiency of CW system (Bakhshoodeh *et al.*, 2020; Chowdhury *et al.*, 2022; Dey Chowdhury *et al.*, 2022). The concentrations of bacterial substrates and toxic constituents in a particular CW are dependent on design parameters such as HRT and mass loading rate (Tao *et al.*, 2006). Akinbile *et al.* (2012) observed that as the HRT increased, the COD reduction efficacy was increased. Yalcuk and Ugurlu (2009) compared the performance of VSSF and HSSF CWs for landfill leachate treatment and found the COD removal in VSSF and HSSF to 30.6% and 35.7%, respectively. The greater COD removal in HSSF CW systems in the experiment was due to the higher HRT of HSSF CW than that of VSSF CW systems. Higher HRT also facilitates N removal from wastewater due to longer contact with microorganisms (Saeed & Sun, 2012). It has been reported that an HRT of between 2 and 10 days can improve the amount of N removed by HSSF wetlands (Tunçsiper, 2007). However, if anaerobic conditions prevail in the wetlands, a significant increase in HRT may not help with nitrate removal because it follows a rapid microbiological transformation pathway (Huang *et al.*, 2000).

It is also observed that prolonged HRT results in the remobilization of solids in the CWs, which limits the overall effectiveness and longevity of the CWs (Akinbile *et al.*, 2012). Despite the longer HRT (16.1 days), Headley *et al.* (2005) found that BOD_5 removal was lower in HSSF CW due to the slow decomposition of organic solids accumulated in the substrate interstices.

6.7.5 Mode of feeding

The proper selection of feed mode is essential to maximize the wastewater mixing through the packed media. For SSF wetland systems, different influent feeding modes such as continuous, intermittent, batch, tidal, and step feeding have been reported. The removal of organics and nitrogen can be improved through the use of intermittent loading in VSSF CW (Gervin & Brix, 2001; Saeed & Sun, 2012). Alternating wet and dry periods enhance the diffusion of atmospheric oxygen within packed media. Lott *et al.* (2022) reduced the frequency of loading from 24 to 8 h and found an increase in the NH_3-N removal, which suggests that NH_3-N removal is significantly affected by the instantaneous flow rate and HRT.

Recirculation of effluent helps in adjusting leachate composition, oxygenates the influent, ensures better distribution of wastewater, and enhances the HRT, thereby improving the treatment efficacy of the CWs (Saeed *et al.*, 2021). The method of feeding has been significantly modified in recent years (Saeed & Sun, 2012). The primary goal of this modification was to facilitate N and organic removal in the CWs through the efficient utilization of aerobic conditions within the media.

6.8 CONCLUSIONS AND FUTURE PERSPECTIVES

The potential of the CW's treatment of landfill leachate is significant. With careful design and planning, CWs can treat landfill leachate to the greatest extent possible. In VSSF and HSSF CWs, the prevalence of aerobic and anaerobic conditions, respectively, governs the removal of the organics and nutrients. Both environmental parameters and operational conditions have substantial effects on the removal of nitrogen and organics from landfill leachate in CWs. Future studies should focus on enhancing the life of CW systems by alleviating the clogging of bed material. The researchers should emphasize implementing various modifications to the conventional CWs to mitigate the harmful effects of greenhouse gases. Finally, an in-depth evaluation of the sustainability of the CWs concerning the three major bottom-lines of sustainability, that is, environmental sustainability, economic viability, and social acceptability has to be carried out to reinforce the global acceptability of the CWs as a natural alternative for the conventional landfill leachate management systems.

REFERENCES

Akinbile C. O. and Yusoff M. S. (2012). Assessing water hyacinth (*Eichhornia crassopes*) and lettuce (*Pistia stratiotes*) effectiveness in aquaculture wastewater treatment. *International Journal of Phytoremediation*, **14**(3), 201–211, https://doi.org/10.1080/15226514.2011.587482

Akinbile C. O., Yusoff M. S. and Ahmad Zuki A. Z. (2012). Landfill leachate treatment using sub-surface flow constructed wetland by *Cyperus haspan*. *Waste Management*, **32**(7), 1387–1393, https://doi.org/10.1016/J.WASMAN.2012.03.002

Albuquerque A., Oliveira J., Semitela S. and Amaral L. (2009). Influence of bed media characteristics on ammonia and nitrate removal in shallow horizontal subsurface flow constructed wetlands. *Bioresource Technology*, **100**(24), 6269–6277, https://doi.org/10.1016/j.biortech.2009.07.016

Bakhshoodeh R., Alavi N., Majlesi M. and Paydary P. (2017). Compost leachate treatment by a pilot-scale subsurface horizontal flow constructed wetland. *Ecological Engineering*, **105**, 7–14, https://doi.org/10.1016/j.ecoleng.2017.04.058

Bakhshoodeh R., Alavi N., Oldham C., Santos R. M., Babaei A. A., Vymazal J. and Paydary P. (2020). Constructed wetlands for landfill leachate treatment: a review. *Ecological Engineering*, **146**, 105725, https://doi.org/10.1016/j.ecoleng.2020.105725

Bozkurt S., Moreno L. and Neretnieks I. (2000). Long-term processes in waste deposits. *Science of the Total Environment*, **250**(1–3), 101–121, https://doi.org/10.1016/S0048-9697(00)00370-3

Brix H. (1994). Functions of macrophytes in constructed wetlands. *Water Science and Technology*, **29**(4), 71–78, https://doi.org/10.2166/wst.1994.0160

Butt T. E., Gouda H. M., Baloch M. I., Paul P., Javadi A. A. and Alam A. (2014). Literature review of baseline study for risk analysis – the landfill leachate case. *Environment International*, **63**, 149–162, https://doi.org/10.1016/J.ENVINT.2013.09.015

Camaño Silvestrini N. E., Maine M. A., Hadad H. R., Nocetti E. and Campagnoli M. A. (2019). Effect of feeding strategy on the performance of a pilot scale vertical flow wetland for the treatment of landfill leachate. *Science of the Total Environment*, **648**, 542–549, https://doi.org/10.1016/J.SCITOTENV.2018.08.132

Cano V., Vich D. V., Rousseau D. P. L., Lens P. N. L. and Nolasco M. A. (2019). Influence of recirculation over COD and N-NH_4 removals from landfill leachate by horizontal flow constructed treatment wetland. *International Journal of Phytoremediation*, **21**(10), 998–1004, https://doi.org/10.1080/15226514.2019.1594681

Chowdhury S. D., Surampalli R. Y. and Bhunia P. (2022). Potential of the constructed wetlands and the earthworm-based treatment technologies to remove the emerging contaminants: a review. *Journal of Hazardous, Toxic, and Radioactive Waste*, **26**(2), 04021066, https://doi.org/10.1061/(asce)hz.2153-5515.0000668

Dan A., Oka M., Fujii Y., Soda S., Ishigaki T., Machimura T. and Ike M. (2017). Removal of heavy metals from synthetic landfill leachate in lab-scale vertical flow constructed wetlands. *Science of the Total Environment*, **584–585**, 742–750, https://doi.org/10.1016/j.scitotenv.2017.01.112

Dey Chowdhury S. and Bhunia P. (2021). *Simultaneous Carbon and Nitrogen Removal from Domestic Wastewater using High Rate Vermifilter*, https://doi.org/10.1007/s12088-021-00936-4

Dey Chowdhury S., Bhunia P., Surampalli R. Y. and Zhang T. C. (2022). Nature and characteristics of emerging contaminants as a triggering factor for selection of different configurations and combinations of constructed wetlands: a review. *Journal of Environmental Engineering*, **148**(8), 1, https://doi.org/10.1061/(asce)ee.1943-7870.0002012

Fadel M. E., Fayad W. and Hashisho J. (2013). Enhanced solid waste stabilization in aerobic landfills using low aeration rates and high density compaction. *Waste Management and Research*, **31**(1), 30–40, https://doi.org/10.1177/0734242X12457118

Gajewska M., Jóźwiakowski K., Ghrabi A. and Masi F. (2015). Treatment of Pollution in Constructed Wetlands: from the Fundamental Mechanisms to the Full Scale Applications. WETPOL 2013 Impact of influent wastewater quality on nitrogen removal rates in multistage treatment wetlands, https://doi.org/10.1007/s11356-014-3647-4

Gervin L. and Brix H. (2001). Removal of nutrients from combined sewer overflows and lake water in a vertical-flow constructed wetland system. *Water Science and Technology*, **44**(11–12), 171–176, https://doi.org/10.2166/wst.2001.0825

He H., Duan Z., Wang Z. and Yue B. (2017). The removal efficiency of constructed wetlands filled with the zeolite-slag hybrid substrate for the rural landfill leachate treatment. *Environmental Science and Pollution Research*, **24**(21), 17547–17555, https://doi.org/10.1007/s11356-017-9402-x

Headley T. R., Herity E. and Davison L. (2005). Treatment at different depths and vertical mixing within a 1-m deep horizontal subsurface-flow wetland. *Ecological Engineering*, **25**(5), 567–582, https://doi.org/10.1016/J.ECOLENG.2005.07.012

Hu Y., He F., Ma L., Zhang Y. and Wu Z. (2016). Microbial nitrogen removal pathways in integrated vertical-flow constructed wetland systems. *Bioresource Technology*, **207**, 339–345, https://doi.org/10.1016/J.BIORTECH.2016.01.106

Huang J., Reneau Jr R. B. and Hagedorn C. (2000). Nitrogen removal in constructed wetlands employed to treat domestic wastewater. *Water Research*, **34**(9), 2582–2588, https://doi.org/10.1016/S0043-1354(00)00018-X

Huong M., Costa D. T. and van Hoi B. (2020). Enhanced removal of nutrients and heavy metals from domestic-industrial wastewater in an academic campus of Hanoi using modified hybrid constructed wetlands. *Water Science and Technology*, **82**(10), 1995–2006, https://doi.org/10.2166/wst.2020.468

Johansson Westholm L. (2006). Substrates for phosphorus removal – potential benefits for on-site wastewater treatment? *Water Research*, **40**(1), 23–36, https://doi.org/10.1016/j.watres.2005.11.006

Kadlec R. H. and Zmarthie L. A. (2010). Wetland treatment of leachate from a closed landfill. *Ecological Engineering*, **36**(7), 946–957, https://doi.org/10.1016/J.ECOLENG.2010.04.013

Khan S., Ahmad I., Shah M. T., Rehman S. and Khaliq A. (2009). Use of constructed wetland for the removal of heavy metals from industrial wastewater. *Journal of Environmental Management*, **90**(11), 3451–3457, https://doi.org/10.1016/J.JENVMAN.2009.05.026

Langergraber G. (2008). Modeling of processes in subsurface flow constructed wetlands: a review. *Vadose Zone Journal*, **7**(2), 830–842, https://doi.org/10.2136/vzj2007.0054

Lavrova S. and Koumanova B. (2010). Influence of recirculation in a lab-scale vertical flow constructed wetland on the treatment efficiency of landfill leachate. *Bioresource Technology*, **101**(6), 1756–1761, https://doi.org/10.1016/j.biortech.2009.10.028

Lim C. K., Seow T. W., Neoh C. H., Md Nor M. H., Ibrahim Z., Ware I. and Mat Sarip S. H. (2016). Treatment of landfill leachate using ASBR combined with zeolite adsorption technology. *3 Biotech*, **6**(2), 1–6, https://doi.org/10.1007/s13205-016-0513-8

Liu M., Wu S., Chen L. and Dong R. (2014). How substrate influences nitrogen transformations in tidal flow constructed wetlands treating high ammonium wastewater? *Ecological Engineering*, **73**, 478–486, https://doi.org/10.1016/j.ecoleng.2014.09.111

Lott D. J., Robey N. M., Laux S. and Townsend T. G. (2022). Effect of Loading Frequency on Ammonia-Nitrogen Removal Kinetics in Early-Stage Pilot Scale Vertical Flow Constructed Wetlands Treating Landfill Leachate Effect of Loading Frequency on Ammonia-Nitrogen Removal Kinetics in Early-Stage Pilot Scale Vertical.

Maine M. A., Hadad H. R., Camaño Silvestrini N. E., Nocetti E., Sanchez G. C. and Campagnoli M. A. (2022). Cr, Ni, and Zn removal from landfill leachate using vertical flow wetlands planted with *Typha domingensis* and *Canna indica*. *International Journal of Phytoremediation*, **24**(1), 66–75, https://doi.org/10.1080/15226514.2021.1926909

Miao L., Yang G., Tao T. and Peng Y. (2019). Recent advances in nitrogen removal from landfill leachate using biological treatments – a review. *Journal of Environmental Management*, **235**(January), 178–185, https://doi.org/10.1016/j.jenvman.2019.01.057

Mohammed A. and Babatunde A. O. (2017). Modelling heavy metals transformation in vertical flow constructed wetlands. *Ecological Modelling*, **354**, 62–71, https://doi.org/10.1016/j.ecolmodel.2017.03.012

Nehrenheim E., Waara S. and Johansson Westholm L. (2008). Metal retention on pine bark and blast furnace slag – on-site experiment for treatment of low strength landfill leachate. *Bioresource Technology*, **99**(5), 998–1005, https://doi.org/10.1016/j.biortech.2007.03.006

Nivala J., Hoos M. B., Cross C., Wallace S. and Parkin G. (2007). Treatment of landfill leachate using an aerated, horizontal subsurface-flow constructed wetland. *Science of the Total Environment*, **380**(1–3), 19–27, https://doi.org/10.1016/J.SCITOTENV.2006.12.030

Saeed T. and Sun G. (2012). A review on nitrogen and organics removal mechanisms in subsurface flow constructed wetlands: dependency on environmental parameters, operating conditions and supporting media. *Journal of Environmental Management*, **112**, 429–448, https://doi.org/10.1016/J.JENVMAN.2012.08.011

Saeed T., Miah M. J., Majed N., Alam M. K. and Khan T. (2021). Effect of effluent recirculation on nutrients and organics removal performance of hybrid constructed wetlands: landfill leachate treatment. *Journal of Cleaner Production*, **282**, 125427, https://doi.org/10.1016/J.JCLEPRO.2020.125427

Sarafraz S., Mohammad T. A., Noor M. J. M. M. and Liaghat A. (2009). Wastewater treatment using horizontal subsurface flow constructed wetland. *American Journal of Environmental Sciences*, **5**(1), 99–105, https://doi.org/10.3844/ajessp.2009.99.105

Si W. T., Zhang W. Y., Lv Y., Yang F., Liu J. M. and Zhang Y. M. (2014). Heavy metal removal in a constructed wetland and benefits for the development of the toad *Bufo raddei*. *Polish Journal of Environmental Studies*, **23**(6), 2207–2215, https://doi.org/10.15244/pjoes/26384

Sim C. H., Yusoff M. K., Shutes B., Ho S. C. and Mansor M. (2008). Nutrient removal in a pilot and full scale constructed wetland, Putrajaya city, Malaysia. *Journal of Environmental Management*, **88**(2), 307–317, https://doi.org/10.1016/J.JENVMAN.2007.03.011

Swarnakar A. K., Bajpai S. and Ahmad I. (2022). Various types of constructed wetland for wastewater treatment – a review. *IOP Conference Series: Earth and Environmental Science*, **1032**(1), 012026, https://doi.org/10.1088/1755-1315/1032/1/012026

Tanner C. C. (2001). Growth and nutrient dynamics of soft-stem bulrush in constructed wetlands treating nutrient-rich wastewaters. *Wetlands Ecology and Management*, **9**, 49–73, https://doi.org/10.1023/A:1008454901954

Tao W., Hall K. J. and Duff S. J. B. (2006). Performance evaluation and effects of hydraulic retention time and mass loading rate on treatment of woodwaste leachate in surface-flow constructed wetlands. *Ecological Engineering*, **26**(3), 252–265, https://doi.org/10.1016/J.ECOLENG.2005.10.006

Teng C., Zhou K., Peng C. and Chen W. (2021). Characterization and treatment of landfill leachate: a review. *Water Research*, **203**(March), 117525, https://doi.org/10.1016/j.watres.2021.117525

Tunçsiper B. (2007). Removal of nutrient and bacteria in pilot-scale constructed wetlands. *Journal of Environmental Science and Health – Part A Toxic/Hazardous Substances and Environmental Engineering*, **42**(8), 1117–1124, https://doi.org/10.1080/10934520701418615

Vymazal J. (2005). Horizontal sub-surface flow and hybrid constructed wetlands systems for wastewater treatment. *Ecological Engineering*, **25**(5), 478–490, https://doi.org/10.1016/j.ecoleng.2005.07.010

Vymazal J. and Kröpfelová L. (2009). Removal of organics in constructed wetlands with horizontal sub-surface flow: a review of the field experience. *Science of the Total Environment*, **407**(13), 3911–3922, https://doi.org/10.1016/J.SCITOTENV.2008.08.032

Wang R., Korboulewsky N., Prudent P., Domeizel M., Rolando C. and Bonin G. (2010). Feasibility of using an organic substrate in a wetland system treating sewage sludge: impact of plant species. *Bioresource Technology*, **101**(1), 51–57, https://doi.org/10.1016/j.biortech.2009.07.080

Wang Y., Cai Z., Sheng S., Pan F., Chen F. and Fu J. (2020). Comprehensive evaluation of substrate materials for contaminants removal in constructed wetlands. *Science of the Total Environment*, **701**, 134736, https://doi.org/10.1016/j.scitotenv.2019.134736

Wang H., Yang C., Wang B., He Z. and Fu T. (2022). Environmental technology & innovation nitrogen removal performance and microbiological characteristics for the landfill leachate treatment in a three-stage vertical flow constructed wetlands system. *Environmental Technology & Innovation*, **28**, 102728, https://doi.org/10.1016/j.eti.2022.102728

Wdowczyk A., Szymańska-Pulikowska A. and Gałka B. (2022). Removal of selected pollutants from landfill leachate in constructed wetlands with different filling. *Bioresource Technology*, **353**, 127136, https://doi.org/10.1016/j.biortech.2022.127136

Werker A. G., Dougherty J. M., McHenry J. L. and Van Loon W. A. (2002). Treatment variability for wetland wastewater treatment design in cold climates. *Ecological Engineering*, **19**(1), 1–11, https://doi.org/10.1016/S0925-8574(02)00016-2

Witthayaphirom C., Chiemchaisri C. and Chiemchaisri W. (2020). Optimization of reactive media for removing organic micro-pollutants in constructed wetland treating municipal landfill leachate. *Environmental Science and Pollution Research*, **27**(20), 24627–24638, https://doi.org/10.1007/s11356-019-06010-3

Wojciechowska E. (2017). Potential and limits of landfill leachate treatment in a multi-stage subsurface flow constructed wetland – evaluation of organics and nitrogen removal. *Bioresource Technology*, **236**, 146–154, https://doi.org/10.1016/J.BIORTECH.2017.03.185

Wojciechowska E., Gajewska M. and Obarska-Pempkowiak H. (2010). Treatment of landfill leachate by constructed wetlands: three case studies. *Polish Journal of Environmental Studies*, **19**(3), 643–650.

Yalcuk A. and Ugurlu A. (2009). Comparison of horizontal and vertical constructed wetland systems for landfill leachate treatment. *Bioresource Technology*, **100**(9), 2521–2526, https://doi.org/10.1016/j.biortech.2008.11.029

Yalçuk A. and Ugurlu A. (2020). Treatment of landfill leachate with laboratory scale vertical flow constructed wetlands: plant growth modeling. *International Journal of Phytoremediation*, **22**(2), 157–166, https://doi.org/10.1080/15226514.2019.1652562

doi: 10.2166/9781789063318_0131

Chapter 7

Anaerobic co-digestion of food waste with landfill leachate

Hale Ozgun[1,2], *Ali Izzet Cengiz*[1], *Busra Cicekalan*[1], *Xuedong Zhang*[3] and *Mustafa Evren Ersahin*[1,2*]

[1]*Istanbul Technical University, Civil Engineering Faculty, Environmental Engineering Department, Ayazaga Campus, Maslak, 34469, Istanbul, Turkey*
[2]*National Research Center on Membrane Technologies, Istanbul Technical University, Ayazaga Campus, Maslak, 34469, Istanbul, Turkey*
[3]*Jiangsu Key Laboratory of Anaerobic Biotechnology, School of Environmental and Civil Engineering, Jiangnan University, Wuxi, 214122, China*
**Corresponding author: ersahin@itu.edu.tr*

ABSTRACT

Anaerobic digestion (AD), which is considered to be a viable method of disposing of organic wastes, is a favorable technology to biologically convert organic waste into energy. Landfill leachate, a major organic waste stream, can be reliably treated through the AD process in which both pollution control and energy recovery can be achieved simultaneously. Food waste (FW) is an easily biodegradable substrate for AD with high carbohydrate, lipid, protein, and moderate moisture content. Although FW can be converted into useful bioenergy in the form of methane through the AD process, sole AD of FW has several challenges, including volatile fatty acids inhibition, trace elements deficiency, and poor digestate dewaterability. Compared to mono-digestion of these substrates, co-digestion of FW with landfill leachate provides better treatment performance and is expected to provide better nutrient balance and biogas production, since the high ammonia nitrogen concentration in landfill leachate could enhance the buffering ability of the AD process that ensures the stable operation of AD. Besides, landfill leachate having rich trace elements could improve the nutrient balance for AD bacteria. Co-digestion of FW with landfill leachate could promote refractory organics degradation. Moreover, the divalent cations in the leachate could improve the dewaterability of the digestate and facilitate the subsequent treatment of liquid and solid digestates. Co-digestion of FW with landfill leachate has the potential to greatly increase biogas yield owing to these synergistic effects. Hence, co-digestion of FW with landfill leachate can be considered as a feasible and promising approach.

Keywords: Anaerobic digestion, co-digestion, food waste, landfill leachate, synergistic effects.

7.1 INTRODUCTION

Leachate is a by-product of sanitary and domestic landfills, and it contains wastewater with various compositions and high-level pollutants; ordinarily, it contains four groups of pollutants including dissolved organic matter (DOM), macro inorganic compounds (Ca^{2+}, Mg^{2+}, Na^{+}, K^{+}, NH_4^{+}, Fe^{2+}, Mn^{2+}, HCO_3^{-}), heavy metals (Cd^{2+}, Cr^{3+}, Cu^{2+}, Pb^{2+}, Ni^{2+}, Zn^{2+}, Hg^{2+}), and xenobiotic organic compounds (XOCs) from chemical and urban waste at low concentrations (aromatic hydrocarbons, phenols, and pesticides) and hence, it must be properly treated before being discharged. Globally, around 30 million tons of leachate are generated per year by solid waste landfills (Ma *et al.*, 2022; Moody & Townsend, 2017; Ndao *et al.*, 2022). Discharging leachate into the surrounding environment would lead to severe contamination of groundwater systems. Concentrations of organic components in leachate can be quite high in the early aerobic phase of landfilling, then decrease and stabilize throughout the methanogenic phase (Ndao *et al.*, 2022). Differently, the concentration of ammonia nitrogen in leachate will not be significantly reduced over the various phases of landfilling, which may lead to long-term contamination (Kjeldsen *et al.*, 2002). Moreover, concentrations of heavy metals and XOCs in leachate are modest and vary dramatically among different landfills, posing environmental risks to nearby water and soil (Christensen *et al.*, 2001). Thus, it is should be properly treated to comply with environmental legislation.

Several different methods are currently in use to treat landfill leachate. The majority of these methods can be classified into two main categories: physical/chemical treatments and biological treatments. In the physical treatment technologies used for landfill leachate treatment, air stripping, adsorption, and membrane filtration are the major ones, while coagulation–flocculation, chemical precipitation, and oxidation are the common chemical treatment methods. Biological treatment is classified into two main groups: (1) aerobic biological treatment (i.e. aerated lagoons and activated sludge) and (2) anaerobic biological treatment (i.e. anaerobic lagoons and reactors). Anaerobic biological treatment generally demonstrates better treatment performance than aerobic biological treatment owing to the high chemical oxygen demand (COD) and high biochemical oxygen demand (BOD)/COD ratio of leachates (Azreen & Zahrim, 2018; Raghab *et al.*, 2013). Anaerobic digestion (AD) technology is one of the most effective systems for the biological treatment of organic waste as it has a variety of advantages, such as solid reduction and biogas production, which makes it an attractive technology (Klavon *et al.*, 2013). Anaerobic co-digestion provides potential benefits such as dilution of toxic compounds, improved nutrient balance, development of synergy in the microbiota, an increased biodegradable organic matter loading rate, and improved methane production kinetics and yield, which can lead to higher digestion rates (de Castro *et al.*, 2020). Hence, food waste (FW) can be used as a substrate for anaerobic co-digestion of landfill leachate. FW can be defined as a product from the whole food supply chain, including the production, processing, distribution, storage, sale, preparation, cooking, and serving of food. It is also used specifically for the consumer stage FW, such as the FW generated from homes, restaurants, and schools (Xu *et al.*, 2018).

This chapter intends to bring together the knowledge obtained from different applications of anaerobic co-digestion of FW with landfill leachate. The first part of the chapter covers brief essential information on the fundamentals of landfill leachate treatment and co-substrates used in the digestion of leachate. The remainder of this chapter focuses on the co-digestion of FW with landfill leachate. Examples of applications that reflect the state-of-the-art in co-digestion of FW with landfill leachate are also provided.

7.2 STRATEGIES APPLIED FOR THE IMPROVEMENT OF LANDFILL LEACHATE TREATMENT

Numerous methods have been applied for the effective treatment of landfill leachates involving aerobic/anaerobic biological treatment, physical and chemical treatments. Among them, the AD method has proven to be the most efficient due to faster and higher degradation of toxic substances or pollutants, biogas production, and minimal sludge formation. The types of feedstocks and/or substrates used are also very important in the process of AD. For this reason, the substrate types used in AD and the pretreatment methods applied prior to AD significantly affect both the performance of the AD process and the end products produced.

Pretreatment of the leachate prior to AD is aimed to accelerate the hydrolysis process during digestion and thus, achieving complete degradation of the toxic organic substances. This essentially is a consequence of greater and faster uptake/utilization of smaller organic molecules as substrate by microbial cells (Luste *et al.*, 2009). Moreover, pretreatment of leachate prior to AD offers several advantages including improvement of sludge dewatering, fewer pathogens, reduced foaming, and reduction of bioreactor size (Müller, 2001; Pilli *et al.*, 2011). The pretreatment methods can involve thermal, chemical (acid, alkali, and ozone), physical/mechanical (sonication/ultrasonic, high-pressure homogenization, and other milling processes), and biological techniques resulting in hydrolysis of the large organic molecules in the leachate. Thermal treatment, one of the most studied pretreatment methods, has been successfully applied on an industrial scale (Li *et al.*, 2016). Thermal pretreatment also leads to pathogen removal and enhanced dewatering performance. There is a wide range of temperatures (60–270°C) that are available as thermal treatment to enhance the efficiency of AD. Acids, alkalis, and oxidants in chemical treatment is effective in the breakdown of organic constituents. Among the chemical-based methods, oxidation has been found useful in pretreatment resulting in sludge solubilization. Acid and alkaline methods are mostly applied in combination with other methods for sludge solubilization. Physical/mechanical pretreatment aids in reducing the particles of the organic residues, without generating any products that may have an inhibitory effect. Ultrasonication is efficient in organic waste pretreatment prior to AD. Using a frequency of 20 kHz for ultrasonication at a different time on leachate treatment showed higher organic matter solubilization based on an increment in the ratio of soluble COD (sCOD) to total COD of 63% after 45 min sonication with supplied power of about

600 W/L. AD of the pretreated sample led to 40% higher biogas production compared to control with a methane production rate of 107 m^3 CH_4/day. This demonstrated that low-frequency ultrasonication as a pretreatment step affects the overall performance of the AD process (Oz & Yarimtepe, 2014). Lei *et al.* (2018) showed that magnetite dosing in the AD of leachate enhanced methane production and treatment performance. Methane production rates increased from 3.7 $m^3/(m^3$ day) to 4.8 $m^3/(m^3$ day) at an organic loading rate (OLR) of 18.2 kg COD/(m^3 day) (Amiri *et al.*, 2017; Ariunbaatar *et al.*, 2014; González-Fernández *et al.*, 2012; Nazimudheen *et al.*, 2018).

As an alternative, co-digestion which simultaneously uses more than one organic waste stream as substrates seems promising for improving digestion efficiencies and has a lot of advantages. These include dilution of the toxic substances coming from any of the substrates involved, an improvement of nutrient balance, synergetic effects on microorganisms, and a high digestion rate. In addition, the addition of suitable organic waste increases the digester loads, both hydraulic and organic, favoring more effective stabilization and enhanced biogas production (Cecchi *et al.*, 1996). Anaerobic co-digestion processes require the proper conditions with regard to substrates. FW is an appropriate substrate for the AD process due to the high organic fraction and high moisture content with high biodegradability (Pagliaccia *et al.*, 2019). Moreover, FW is more advantageous in terms of accessibility than other possible co-substrates for the co-digestion process.

Li *et al.* (2009) mention that a mixing ratio of 3:1 was found to be optimum for the co-digestion of cattle manure and kitchen waste, with a methane yield of 233 ml/g volatile solids (VS). For the digestion of FW, it is important to maintain suitable pH during the anaerobic process and maintain the balance between volatile fatty acids (VFA) and methane production. Co-digestion of FW with other organic waste would ensure an improvement of nutrient balance and the development of synergetic effects on microorganisms (Hartmann & Ahring, 2005; Sosnowski & Ledakowicz, 2003). While fresh landfill leachate can be treated by AD due to the high concentration of organic matter, it is hard to treat mature leachate alone; therefore, co-digestion of landfill leachate and co-substrates might promote treatment of mature leachate as well as increase the efficiency of the AD process. Co-digestion of leachate has recently gained interest using different substrates such as wastewater, sewage sludge, and FW.

7.3 HISTORICAL DEVELOPMENT OF LEACHATE CO-DIGESTION

Although the anaerobic co-digestion concept has become more popular in the 21st century, it was initially investigated around 1987 (Mata-Alvarez *et al.*, 2014). In the first study related to anaerobic co-digestion of leachate recorded in the literature conducted by Lin *et al.* (1999), the co-digestion of leachate and septage within a continuous stirred tank reactor (CSTR) in terms of operation parameters and treatment performance was investigated. Thereafter, Lin *et al.* (2000), operated the up-flow anaerobic sludge blanket (UASB) reactor for co-digestion of leachate and septage to determine treatment performance and start-up period compared to CSTR. Anaerobic co-digestion of leachate and FW have been studied recently. Shahriari *et al.* (2012) used untreated leachate to provide supplemental water for the AD of FW. The results showed that although an excess amount of leachate recirculation inhibited the methanogenic activity, a certain amount of leachate recirculation successfully improved the biogas production. In another study on co-digestion of leachate and FW conducted by Liao *et al.* (2014), it was found that landfill leachate addition to the AD of FW can be effective to adjust pH and providing water supplementation. Zhang *et al.* (2015a) investigated the advantages and synergistic effects of co-digestion of FW and landfill leachate. It was found that co-digestion of FW provided better system stability compared to mono-digestion of FW. Moreover, it was found that methane yield was increased with co-digestion compared to mono-digestion. In literature, various studies have been carried out based on such synergistic effects of co-digestion of FW with landfill leachate. Currently, there is one review paper that covers related studies in literature prepared by Lv *et al.* (2021). Figure 7.1 illustrates studies on the co-digestion of leachate and FW that have started to acquire importance in recent years. Obviously, further studies are expected on the co-digestion of leachate and FW.

7.4 CO-SUBSTRATES USED IN THE DIGESTION OF LEACHATE

AD is an appropriate process to treat fresh leachate because it contains a significant amount of biodegradable organic matter. However, mature leachate contains organic matter with low biodegradability such as humic and fulvic substances (Kulikowska & Klimiuk, 2008). Moreover, mono-digestion is a challenging treatment option because of the high ammonia nitrogen and refractory organic compounds present in leachate (Bonu *et al.*, 2023). Therefore, anaerobic co-digestion is applied both to overcome the problems associated with mono-digestion of leachate and take the advantage of co-digestion (Liao *et al.*, 2014). Anaerobic co-digestion can be defined as the simultaneous digestion of two or more substrates. Substrates utilized in the anaerobic co-digestion are called co-substrates.

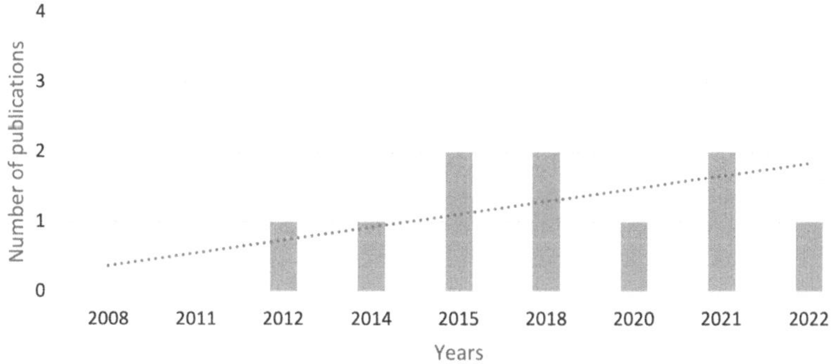

Figure 7.1 Evolution of the number of papers published related to co-digestion of leachate and FW (*source*: Web of Science, 01.08.2022).

In literature, there are several co-substrates used in the AD of leachate. These studies were mainly focused on sewage sludge and FW as co-substrates. Berenjkar *et al.* (2019) investigated the co-digestion of leachate and sewage sludge by applying different mixing ratios. It was reported that low doses of leachate addition (up to 20% leachate out of total mixture volume) enhance biogas production. Montusiewicz and Lebiocka (2011) focused on the co-digestion of intermediate leachate and sewage sludge. It was found that the addition of intermediate landfill leachate enhanced the biogas and methane yields as 13% and 16.9%, respectively. Mesophilic co-digestion of leachate and sewage sludge was found feasible in terms of technical aspects such as methane production by Hombach *et al.* (2003). In this study, an increase of about 28% in daily methane production was observed through a stable operation. Moreover, the dewaterability of the digestate was improved by co-digestion. Recently, Gao *et al.* (2022) investigated the impact of salinity on the co-digestion of landfill leachate and sewage sludge. It was found that when salinity increased from 10 to 30 g/L, biogas production, and methane yield decreased, and treatment performance of the digester deteriorated. In this study, it was reported that the microbial community was not able to adapt to the high salinity and significantly changed, therefore, process stability was adversely affected. Gao *et al.* (2021) expressed that the addition of leachate can improve solubilization, hydrolysis, and acidification of waste activated sludge (WAS). Briefly, a proper amount of leachate addition to WAS can improve process stability and enhance the biogas yield. However, excess amounts of leachate addition can be detrimental to the microbial community. Liao *et al.* (2014) investigated the co-digestion of leachate and FW in a single-phase batch reactor. It was found that leachate addition enhanced biogas production since it stimulated the methanogenic activity and led to a balance between methanogenic and acidogenic microorganisms. Guven *et al.* (2018) investigated the methane yield of the combination of several waste components including the leachate, sewage sludge, reverse osmosis (RO) concentrate, and organic fraction of municipal solid waste (OFMSW), including FW with different mixing ratios. It was determined that the methane yield of mono-digestion of leachate was much lower than co-digestion with other waste components. Moreover, it was indicated that leachate addition can be a good solution to adjust the optimum total solids (TS) content of an anaerobic digester. Jayanth *et al.* (2020) successfully operated an industrial-scale digester for co-digesting of OFMSW and leachate. Zhang *et al.* (2015a) and Zhang *et al.* (2015b) investigated treatment performance and methane yield of co-digestion of leachate and FW. In general, leachate was mixed in certain amounts with mentioned co-substrates to provide a synergistic effect for the AD process in terms of both treatment performance and methane yield. Therefore, studies in the literature mostly have focused on the mixing ratio which can be defined as optimum.

Although sewage sludge and FW are generally considered as main co-substrates for leachate co-digestion in the literature, several studies have also been carried out for different substrates. Han *et al.* (2012) studied the feasibility of co-digestion of leachate and piggery wastewater. Methane yield was examined with the optimization of different parameters such as salinity, mixing ratio, and alkalinity. Based on the results, optimum conditions were indicated as the salinity of 3.4 g/L sodium chloride (NaCl), alkalinity of 2,850 mg $CaCO_3$/L, and mixing ratio of 33:1 (leachate: piggery wastewater in terms of VS, w/w). Zhou *et al.* (2020), successfully operated a UASB reactor for co-digestion of leachate and acid mine drainage. It was reported that the synergistic effect between sulfate-reducing bacteria and the methanogen archaeal community contributed to treatment performance and biogas production. Takeda *et al.* (2022) utilized crude glycerol, which is the main by-product of biodiesel, as a co-substrate in the AD of leachate to overcome the obstacles born with the mono-digestion of leachate. The synergistic effect between the characteristics of leachate and glycerol resulted in stable operation with more than 20 times biogas production compared to mono-digestion of leachate. Kim and Kang (2015), utilized microalgal biomass and sewage sludge as co-substrates of leachate. Methane yield with a value of 176 ± 6.8 CH_4/g VS was obtained from co-digestion of microalgal biomass and sewage sludge for a mixing ratio of 1:1 (V/V). It was observed that methane yield for co-digestion of microalgal biomass and sewage is higher compared to the AD of each substrate individually. Therefore, by considering the energy conversion of waste components, co-digestion was detected as a more feasible way compared to mono-digestion. Leiva *et al.* (2014) evaluated the co-digestion of different waste components, including leachate, WAS, municipal sludge cake, and screen cake obtained from a winery industrial wastewater treatment plant. Darwin *et al.* (2019) utilized starch waste as a co-substrate in leachate co-digestion. It was found that two-stage anaerobic co-digestion provided higher methane yield compared to single-stage anaerobic co-digestion. Moreover, in this study, biofilm was used in order to enhance methane yield with the degradation of organic matter. It was found that the methane yield of two-stage co-digestion was 77.60 mL CH_4/g VS_{added} while methane yield of single-stage co-digestion was 20.57 mL CH_4/g VS_{added}. Methane yield of two stage co digestion that biofilm used was superior with 125.11 mL CH_4/g VS_{added} among the other methods. Many studies showed the potential utilization of different types of co-substrates in leachate digestion such as swine waste (Kim & Ju, 2012), pineapple peel (Jaroenpoj *et al.*, 2015), glycerin (de Castro *et al.*, 2020), domestic wastewater (Moujanni *et al.*, 2019), leather fleshing waste (Lee *et al.*, 2020).

To summarize, there are various co-substrates used in the co-digestion of leachate in literature. However, most of the studies focus on sewage sludge and FW as co substrates because of both technical and economic feasibility. Especially, with the optimization of process parameters, co-digestion of FW and leachate may be a widespread application in the future. Therefore, in the following title, the co-digestion of FW and leachate was investigated in different aspects.

7.5 CO-DIGESTION OF LANDFILL LEACHATE AND FW

7.5.1 Synergistic effect and opportunities

AD is a process that provides biogas generation from organic matter used in leachate management. However, studies in the literature showed that mono-digestion of leachate causes some problems related to process performance, microbial activity, treatment performance, and methane yield. Complementary properties of FW and leachate create a synergistic effect for the AD process. Therefore, co-digestion of FW with landfill leachate is used as a solution to overcome the problems by utilizing the synergistic effects of both substrates. The general synergistic effects of co-digestion of landfill leachate and FW were illustrated in Figure 7.2.

Various studies in the literature show that FW is deficient in trace elements to perform a stable AD process (Ariunbaatar *et al.*, 2016; Zhang & Jahng 2012; Zhang *et al.*, 2012). Trace elements are important for cell growth and thus for the methane formation and stable operation of the AD process (Ma *et al.*, 2018). Zhang *et al.* (2011) evaluated the feasibility of co-digestion of FW and piggery wastewater. It was found that piggery wastewater provided trace elements, in which FW was deficient, to the co-digestion process. Hence, the treatment performance of the system was enhanced for long-term operation. Considering the trace elements in the leachate, it can be concluded that FW and landfill leachate can create a synergistic effect through the co-digestion process. Zhang *et al.* (2015a) investigated this synergistic effect, especially regarding biogas production and process stability. Four different anaerobic reactors (R1, R2, R3, and R4) were operated with different mixing ratios of leachate and FW in three different OLRs (4.0–4.1; 6.1–6.2; 8.1–8.3 g VS/(L day)). Reactors were operated in three phases based on the OLRs. Phase 1 was operated with the OLR 4.0–4.1 g VS/(L day), phase 2 was operated with the OLR 6.1–6.2 g VS/L day and lastly, phase 3 was operated with the OLR 8.1–8.3 g VS/L day. The proportions of fresh leachate basis of VS in the mixed feedstock of R1, R2, R3, and R4 were determined as 0%, 5.8%, 11.6%, and 22.7%, in order. R1 fed only with FW, faced the failure of the process due to the VFAs accumulation and pH decrease in phase 2. However, reactors fed with a mixture of leachate and FW were operated without failure. Moreover, the methane yield of R4 was 452.2–506.3 mL/g VS_{added} while the methane yield of R3 was 419.1–466.4 mL/g

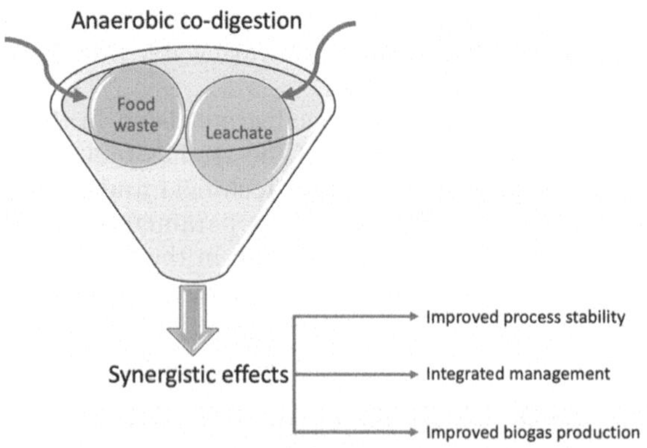

Figure 7.2 Synergistic effects of co-digestion of FW with landfill leachate.

VS_{added}. The results showed that methane yields increased significantly when leachate addition was increased while the process could be operated with stability. According to the results, it was noted that trace elements played a key role in maintaining the stability of the co-digestion process. Especially, high concentration of ferrous (Fe), cobalt (Co), molybdenum (Mo), and nickel (Ni) in the leachate improved process performance and co-digestion stability. The deficiency of trace elements in FW had been overcome with leachate addition and thus, synergistic relation between FW and leachate was constructed. Similarly, Zhang *et al.* (2015b) also investigated the effect of different mixing ratios in the co-digestion of FW and leachate. Four reactors were fed with the feedstocks containing between 0% and 22.1%, respectively, leachate on VS basis. Results showed that removal efficiency of VS in 0% leachate-containing reactor was 72.8%. However, VS removal in 22.1% of the leachate-containing reactor was 77.6%. It was indicated that higher trace element concentration in leachate compared to FW improved AD. Trace elements (Fe, Co, Mo) supplementation was responsible for enhancing methane production by the synthesis of enzyme cofactors. Even if, trace elements can be inhibitory beyond certain limits, Lv *et al.* (2021) indicated that there is no inhibition risk in co-digestion of leachate and FW.

Methane production through AD was significantly improved with co-digestion compared to mono-digestion of leachate. Guven *et al.* (2018) determined methane yield obtained from mono-digestion of leachate as 110 L CH_4/kg VS_{added}. On the other hand, when they investigate the methane yield of co-digestion of leachate and FW, it was found to be 232 L CH_4/kg VS_{added}. Jayanth *et al.* (2020) also compared mono-digestion of leachate and co-digestion of leachate with FW. Three anaerobic gas lift reactors were operated for 47 weeks. The first and second reactors were used to perform mono-digestion of FW and leachate, respectively, whereas the third reactor was used for co-digestion of leachate and FW. Biogas yield of mono-digestion of leachate was determined as 0.39 m^3/kg $VS_{reduced}$. Higher biogas yield was obtained from the third reactor with a value of 0.48 m^3/kg $VS_{reduced}$. Liao *et al.* (2021) showed that FW addition into the AD of the leachate system significantly increased methane production and DOM degradation. It was found that co-digestion increased gas production by 117.69% compared to mono-digestion of leachate. Likely, Guven *et al.* (2018) reported that methane yield was increased by 109% with co-digestion compared to mono-digestion of leachate.

Co-digestion is a method that can be used to improve pH buffering capacity of the anaerobic system (Lin *et al.*, 2011; Wu *et al.*, 2016). Another important benefit of co-digestion is improving the carbon to nitrogen (C/N) balance which is vital for the performance of the AD process (Lee *et al.*, 2019). The C/N ratio is in the range of 13.9–19.6 to properly conduct the AD process, and the optimum C/N ratio was recommended as 15 (Kumar *et al.*, 2010; Zhu, 2007). Zhang *et al.* (2013) reported that co-digestion of cattle manure and FW improved both pH buffering capacity and C/N ratio and thus, process performance. In this study, the C/N of cattle manure and FW were determined as 5.2 and 21.1, respectively. It is known that the alkalinity of leachate is high and the C/N ratio is low. Therefore, a similar synergistic effect can be created between leachate and FW. Ma *et al.* (2018) reported that co-digestion of FW and landfill leachate is an important method to improve the AD of FW since it can improve the pH buffering capacity. Zhang *et al.* (2015b) reported that the high ammonia nitrogen concentration of leachate contributed to the synergistic effect that arose through the co-digestion process by promoting the buffering capacity. Therefore, suitable pH was able to be maintained during the process without an inhibition caused by VFA accumulation. The synergistic effect of FW and leachate on the co-digestion process in terms of buffering capacity was indicated also by Liao *et al.* (2014). In the study C/N ratio of FW was 25.39 while the ammonia nitrogen concentration of leachate was 3,625 mg/L. In the study, different amounts of leachate were added to the FW in six different reactors. Two control reactors were used in order to examine mono-digestion of leachate and FW separately. Reactors introduced with a sufficient amount of leachate (initial value for ammonia nitrogen was between 2,000 and 3,000 mg/L) were able to maintain the pH value in the desired range which is proper for activities of methanogenic microorganisms in the AD process. On the other hand, VFA inhibition was observed in the reactors that are fed with an excess amount of FW. It showed that ammonia nitrogen in leachate provided a buffer effect so that VFA inhibition could be prevented. Results obtained in different studies from the literature indicated that ammonia nitrogen-rich leachate improves the process performance of co-digestion with FW.

High solid concentration of FW causes limitations in using different reactors for the AD process. Although expanded granular sludge bed (EGSB) or UASB reactors provide numerous advantages, CSTR is the main reactor type used generally because of the high solid concentration of FW (Braguglia *et al.*, 2018; Ma *et al.*, 2018). Although the quality and quantity of leachate are significantly variable, suspended solid concentration is very low compared to FW (Yusmartini *et al.*, 2013). Ma *et al.* (2018) suggested using FW as co-substrate in leachate co-digestion to provide dilution for suspended solids and decrease the concentration. By reducing the suspended solids concentration of feedstock, reactors like EGSB and UASB can be used, and thus advantages can be provided such as high OLR and pH buffering capacity. Therefore, dilution of the suspended solids is one of the most important and promising synergistic effects of leachate and FW co-digestion. Preliminary studies conducted by Ma *et al.* (2018) showed that the synergistic effect of dilution made it possible to use EGSB reactors for co-digestion of FW with leachate. It was found that reactors where mono-digestion of FW were applied failed when OLR was increased to 7.2 kg COD/m^3 day. However, the reactor, where co-digestion took place, was able to be operated even at 22.4 kg COD/ m^3 day. Therefore, co-digestion was the proper solution for dilution of FW with high suspended solid concentration. Moreover, toxic compounds can be diluted with the synergistic effect of co-digestion. Jayanth *et al.* (2020) showed that potentially toxic substances in leachate were diluted with the co-digestion of FW and leachate. This increased the stability of the digester and provided process safety.

Hombach *et al.* (2003) showed that co-digestion of sewage sludge with leachate improved the dewaterability of the digestate. This was probably because of the leachate's high calcium (Ca^{2+}) and magnesium (Mg^{2+}) concentrations. Various studies in the literature showed that Ca^{2+} and Mg^{2+} ions improved dewaterability (Liu *et al.*, 2017; Sajjad & Kim, 2015). In parallel, Lv *et al.* (2021) suggested co-digestion of leachate with high Ca^{2+} and Mg^{2+} concentrations and FW to improve the digestate dewaterability compared to mono-digestion.

Co-digestion of FW and leachate provides various advantages because of the complementary characteristics, as shown in Figure 7.3. When two substances are used management of the co-digestion process becomes easier compared to other co-digestion options because of transportation problems.

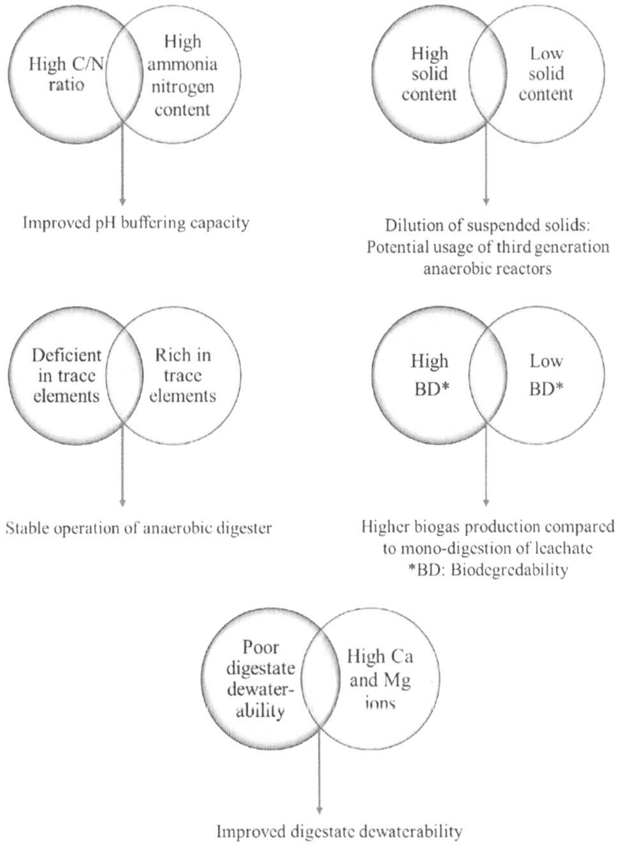

Figure 7.3 Complementary characteristics of leachate (Venn schemes at right) and FW (Venn schemes at left): positive outcomes for anaerobic co-digestion (Adapted from Lv *et al.*, 2021).

In addition to this, strict regulations are applied from day to day regarding the source-separated collection of municipal solid wastes it is therefore likely that OFMSW would contain less impurity in the coming days. Therefore, co-digestion is a promising method for the management of FW and leachate when considering the synergistic effects. Lv *et al.* (2021) indicated that energy conversion efficiency can be 12–18% more in co-digestion compared to mono-digestion. Guven *et al.* (2018) suggested farm-based anaerobic digesters in order to decrease the pollution load of leachate to the environment. Moreover, it was indicated that the leachate can be treated with this integrated management which can be considered as a synergistic effect as well (Figure 7.2).

7.5.2 Challenges and obstacles

Co-digestion of FW and leachate is attracting more and more attention, addressing the challenges of mono-digestion of FW. FW is an ideal substrate for AD treatment due to its high organic matter content (VS/TS: 0.8–0.9) with easy biodegradability; however, AD of FW faces many challenges. VFAs inhibition and absence of trace elements are the main challenges in AD of FW. Moreover, the poor digestate dewaterability is also a challenging issue in engineering, which necessitates extra physical or chemical treatment with additional cost. Besides, in AD of leachate, the major problems are high ammonia nitrogen and low phosphorus concentrations, high organic acid production, unstable characteristics of leachate, and refractory organic matters present in leachate. These challenges might be overcome with the synergistic effects of co-digestion of leachate and FW (see Section 7.5.1). However, ammonia, and VFAs, which are particularly abundant in leachates, are most likely methanogenesis inhibitors linked with co-digestion. The biodegradability and biomethane potential of leachate is relatively low because of the recalcitrant compounds it contains (He *et al.*, 2007). High calcium, ammonia, and heavy metal concentrations of leachate lead to VFA accumulation in the bioreactor, which may limit the process performance. Therefore, it is suggested to consider the inhibition threshold and accordingly decide the mixing ratio of FW and leachate for the co-digestion based on the characteristics of each substrate.

VFA, alkalinity, and pH are typical indicators used to evaluate the stability of an anaerobic digester (Cook *et al.*, 2017). In general, the ideal pH range for methanogens is 6.8–7.2 (Fu *et al.*, 2018). Excessive accumulation of VFAs in an anaerobic digester leads to a rapid pH decrease affecting methanogenic activity (Fisgativa *et al.*, 2016). The AD system is stable when the concentration of VFAs is maintained between 50 and 250 mg/L. But, when it exceeds 8,000 mg/L for total VFAs concentration in an anaerobic digester, it will exert toxic inhibition on methanogens, then lead to reduce the amount of biogas produced (Karthikeyan & Visvanathan, 2013). Besides, the components of VFA are acetic acid, propionic acid, butyric acid, isobutyric acid, valeric acid, isovaleric acid, caproic acid, and heptanoic acid, and the inhibition thresholds of acetic acid, propionic acid and butyric acid in VFAs are 2,000, 1,000, and 3,456 mg/L, respectively (Nayono *et al.*, 2010; Rocamora *et al.*, 2020; Wang *et al.*, 2009). The fast conversion of easily digested FW to VFAs leads to a drastic decrease in pH if no sufficient buffering capacity is present, resulting in the instability of the anaerobic digester. Furthermore, the high protein and lipid contents in FW easily lead to inhibitory levels of ammonia, hydrogen sulfide, and long-chain fatty acids (Chen *et al.*, 2008) or cause digester foaming (Subramanian & Pagilla, 2015). Stable anaerobic digesters usually have alkalinity exceeding 2,000 mg $CaCO_3$/L (Cook *et al.*, 2017). Liu *et al.* (2012) reported that the anaerobic process is considered stable and there is no risk of VFA accumulation when VFA to alkalinity ratio is less than 0.3.

In addition, the digestate, which is the end product of the AD process, can be converted into biofertilizer after applying the additional treatment process, so the whole process achieves zero-solid discharge. Nonetheless, some toxic substances like heavy metals in the leachate may limit the application of this digestate for agricultural purposes. Therefore, it is necessary to conduct a more comprehensive study on the subsequent treatment of the digestate.

7.5.3 Applications

As shown in Table 7.1, only a few studies were performed in co-digestion of FW with landfill leachate, it was found that co-digestion of FW with landfill leachate has a higher digestion efficiency than mono-digestion. The details of the applications related to the co-digestion of landfill leachate and FW in the literature were summarized in Tables 7.1–7.2. Table 7.1 summarizes the results and synergistic effects detected in the applications, and Table 7.2 summarizes the experimental conditions and operating conditions.

7.6 CONCLUSIONS AND FUTURE PERSPECTIVES

Anaerobic treatment is an energy-generating process, in contrast to aerobic systems that often require a significant energy input for aeration reasons. The anaerobic treatment technology is an appealing alternative to existing treatment methods thanks to net energy production from biogas. By treating medium and/or high-strength wastewaters, such as landfill leachate, anaerobic treatment offers great potential for the recovery of biomethane. Fresh leachates have a notable potential for the recovery of biomethane in landfills. Fresh leachates are commonly highly biodegradable compounds and were reported to be more suitable for AD compared to the old leachates. However, these leachates are highly acidic and present a high COD load, which makes them not suitable for the AD process. Thus, AD sometimes tends to be inefficient when some organic waste such as leachate or FW is used as the sole substrate. Co-digestion which simultaneously uses multiple organic waste streams as substrates appear to be a feasible solution for enhancing digestion efficiency.

It has been extensively shown that the co-digestion of several substrates and wastes has a synergistic effect and significantly boosts methane production during the AD process compared to mono-digestion. Co-digestion presents various benefits as it could lower the operational cost of waste management, enhance digestion efficiency, and also could increase the buffer capacity to stabilize the process of AD. Landfill leachate has been proved to be a promising co-substrate for AD with FW since the high ammonia nitrogen concentration in leachate could improve the buffering ability, and leachate is always rich in trace

Table 7.1 Results and synergistic effects of studies related to co-digestion of FW and leachate.

Type of Substrate	Mono/ Co-digestion	Methane Yield (mL/g VS_{added})	VS Reductions Through Co-digestion (%)	Synergistic Effects	References
FW	Mono-digestion	1.1	–	–	Liao *et al.* (2014)
FW + Leachate	Co-digestion	466	N.D	Ammonia nitrogen in leachate formed a buffer system with VFA, acidic inhibition effectively prevented.	
FW	Mono-digestion	425.2 ± 17.0	–	–	Zhang *et al.* (2015a)
FW + Leachate	Co-digestion	479.0 ± 33.0; 452.2 ± 23.4; 506.3 ± 29.0	78.0 ± 0.8; 75.8 ± 0.3; 71.6 ± 1.3	Trace elements in leachate improved the stability of AD process.	
FW + Leachate	Co-digestion	416	77.6	Start-up period was shortened, methane yield was increased with co-digestion.	Zhang *et al.* (2015b)
FW	Mono-digestion	786	–	Biogas potential of the leachate was increased with FW addition.	Guven *et al.* (2018)
FW + Leachate	Co-digestion	232	N.D		
Leachate	Mono-digestion	110	–		
FW + Leachate	Co-digestion	240*	N.D	Acidification and low microorganisms activities were resolved by co-digestion and thus, balance of anaerobic system was improved. System was able to be operated with high OLR.	Ma *et al.* (2018)
FW	Mono-digestion	380–420	–	–	Jayanth *et al.* (2020)
FW + Leachate	Co-digestion	290–390	47	Potentially toxic substances present in leachate were diluted, balance for essential micronutrients was established.	
Leachate	Mono-digestion	230–310	–	–	
FW + Leachate	Co-digestion	N.D	N.D	FW addition improved the degradation of dissolved organic carbon in leachate.	Liao *et al.* (2022)

N.D: no data.
*mL CH_4/g $COD_{removed}$.

Table 7.2 Experimental conditions of studies related to co-digestion of FW and leachate.

Mixing Ratios (FW:Leachate)	OLR (g VS/L day)	Temperature	Reactor Type/ Feeding Pattern/ Configuration	Operation Duration (days)	Scale	References
40 g TS/L:(0–1,200 mL)	41.8	35 + 1°C	Single-stage batch reactor	35	Lab	Liao *et al.* (2014)
0:100%, 5.8:94.2%, 11.6–88.4%, 22.7:77.3%* (VS basis)	4–4.1, 5–6.2, 8.1–8.3	37°C	Schott Duran bottles, semi-continuous reactor	218	Lab	Zhang *et al.* (2015a)
100:0%, 94.4:5.6%, 89.8:11.2%, 77.9:22.1%* (VS basis)	20, 20.1, 20.3, 20.5	37°C	Schott Duran bottles, batch reactor	20	Lab	Zhang *et al.* (2015b)
22:78% (weight basis)	N.D	37°C	Batch reactor	50	Lab	Guven *et al.* (2018)
1:3 (volume basis)	22.4*	N.D	EGSB reactor	74	Lab	Ma *et al.* (2018)
1 kg/day:1 L/day	5.2	24–39°C	Semi-commercial anaerobic gas lift reactor	47**	Industrial	Jayanth *et al.* (2020)
N.D	N.D	35 + 1°C	Sequencing batch reactor	152	Lab	Liao *et al.* (2022)

N.D: no data.
*kg (COD/m^3 day).
**Weeks.

elements which can enhance the stability of anaerobic digester. Thereby, the deficiency of trace elements in FW had been overcome with leachate addition. The mixing ratios of FW and landfill leachate that produces the highest methane yield varied in a large range depending on the properties of FW and leachate. Co-digestion of FW with landfill leachate ensures a lot of advantages thanks to the complementary characteristics involving improved pH buffering capacity, stable operation of the anaerobic digester, and improved digestate dewaterability. The divalent cations in the leachate can effectively enhance the digestate dewaterability and facilitate the subsequent treatment of digestate. Moreover, dilution of the suspended solids is one of the most significant synergistic effects on co-digestion of FW with landfill leachate. By reducing the suspended solids concentration of feedstock, reactors like EGSB and UASB can be used, and thus benefits can be provided like high OLR and pH buffering capacity. Toxic compounds can also be diluted with the synergistic effect of co-digestion and hence, the stability of the digester can be increased. Considering these complementary features of FW and leachate, it appeared to be feasible for co-digestion of FW and landfill leachate.

Co-digestion applications of landfill leachate and FW have the potential to greatly increase biogas yield. Thus, it could be a promising way for energy recovery in the foreseeable future. Nevertheless, it is necessary to choose an appropriate co-digestion proportion in order to promote the synergistic effect of co-digestion for industrial applications. Besides, the digestate from FW can usually be converted into biofertilizer after further treatment, resulting in zero-solid discharge throughout the whole process. Additionally, the concept of co-located treatment facilities in FW treatment is favorable in case FW treatment facilities are close to the landfill leachate treatment plant. In this respect, it can be an economically viable and sustainable, and feasible option that landfill leachate and FW can be co-digested.

REFERENCES

Amiri L., Abdoli M. A., Gitipour S. and Madadian E. (2017). The effects of co-substrate and thermal pretreatment on anaerobic digestion performance. *Environmental Technology*, **38**(18), 2352–2361, https://doi.org/10.1080/09593330.2016.1260643

Ariunbaatar J., Panico A., Esposito G., Pirozzi F. and Lens P. N. (2014). Pretreatment methods to enhance anaerobic digestion of organic solid waste. *Applied Energy*, **123**, 143–156, https://doi.org/10.1016/j.apenergy.2014.02.035

Ariunbaatar J., Esposito G., Yeh D. H. and Lens P. N. L. (2016). Enhanced anaerobic digestion of food waste by supplementing trace elements: role of selenium (vi) and iron (ii). *Frontiers in Environmental Science*, **4**, 8, https://doi.org/10.3389/fenvs.2016.00008

Azreen I. and Zahrim A. Y. (2018). Overview of biologically digested leachate treatment using adsorption. In: Anaerobic Digestion Processes, Horan, N., Yaser, A.Z., and Wid, N. (eds.) Springer, Singapore, pp. 123–148.

Berenjkar P., Islam M. and Yuan Q. (2019). Co-treatment of sewage sludge and mature landfill leachate by anaerobic digestion. *International Journal of Environmental Science and Technology*, **16**(5), 2465–2474, https://doi.org/10.1007/s13762-018-1889-2

Bonu R., Anand N. and Palani S. G. (2023). Impact of thermal pre-treatment on anaerobic co-digestion of sewage sludge and landfill leachate. *Materials Today: Proceedings*, **72**(1), 99–103, https://doi.org/10.1016/j.matpr.2022.06.130.

Braguglia C. M., Gallipoli A., Gianico A. and Pagliaccia P. (2018). Anaerobic bioconversion of food waste into energy: a critical review. *Bioresource Technology*, **248**, 37–56, https://doi.org/10.1016/j.biortech.2017.06.145

Cecchi F., Pavan P. and Mata-Alvarez J. (1996). Anaerobic co-digestion of sewage sludge: application to the macroalgae from the Venice lagoon. *Resources, Conservation and Recycling*, **17**(1), 57–66, https://doi.org/10.1016/0921-3449(96)88182-1

Chen Y., Cheng J. J. and Creamer K. S. (2008). Inhibition of anaerobic digestion process: a review. *Bioresource Technology*, **99**(10), 4044–4064, https://doi.org/10.1016/j.biortech.2007.01.057

Christensen T. H., Kjeldsen P., Bjerg P. L., Jensen D. L., Christensen J. B., Baun A., Albrechtsen H. J. and Heron G. (2001). Biogeochemistry of landfill leachate plumes. *Applied Geochemistry*, **16**(7–8), 659–718, https://doi.org/10.1016/S0883-2927(00)00082-2

Cook S. M., Skerlos S. J., Raskin L. and Love N. G. (2017). A stability assessment tool for anaerobic codigestion. *Water Research*, **112**, 19–28, https://doi.org/10.1016/j.watres.2017.01.027

Darwin Triovanta U. and Rinaldi R. (2019). Two-stage anaerobic co-digestion of landfill leachate and starch wastes using anaerobic biofilm reactor for methane production. *Progress in Agricultural Engineering Sciences*, **15**(1), 53–70, https://doi.org/10.1556/446.15.2019.1.4

de Castro T. M., Torres D. G. B., Arantes E. J., de Carvalho K. Q., Passig F. H., Christ D., Gotardo J. T. and Gomes S. D. (2020). Anaerobic co-digestion of industrial landfill leachate and glycerin: methanogenic potential, organic matter removal and process optimization. *Environmental Technology (United Kingdom)*, **41**(20), 2583–2593.

Fisgativa H., Tremier A. and Dabert P. (2016). Characterizing the variability of food waste quality: a need for efficient valorisation through anaerobic digestion. *Waste Management*, **50**, 264–274, https://doi.org/10.1016/j.wasman.2016.01.041

Fu S. F., Chen K. Q., Zhu R., Sun W. X., Zou H. and Guo R. B. (2018). Improved anaerobic digestion performance of *Miscanthus floridulus* by different pretreatment methods and preliminary economic analysis. *Energy Conversion and Management*, **159**, 121–128, https://doi.org/10.1016/j.enconman.2018.01.014

Gao M., Li S., Zou H., Wen F., Cai A., Zhu R., Tian W., Shi D., Chai H. and Gu L. (2021). Aged landfill leachate enhances anaerobic digestion of waste activated sludge. *Journal of Environmental Management*, **293**, 112853, https://doi.org/10.1016/j.jenvman.2021.112853

Gao M., Yang J., Liu Y., Zhang J., Li J., Liu Y., Wu B. and Gu L. (2022). Deep insights into the anaerobic co-digestion of waste activated sludge with concentrated leachate under different salinity stresses. *Science of the Total Environment*, **838**, 155922, https://doi.org/10.1016/j.scitotenv.2022.155922

González-Fernández C., Sialve B., Bernet N. and Steyer J. P. (2012). Thermal pretreatment to improve methane production of *Scenedesmus* biomass. *Biomass and Bioenergy*, **40**, 105–111, https://doi.org/10.1016/j.biombioe.2012.02.008

Guven H., Akca M. S., Iren E., Keles F., Ozturk I. and Altinbas M. (2018). Co-digestion performance of organic fraction of municipal solid waste with leachate: preliminary studies. *Waste Management*, **71**, 775–784, https://doi.org/10.1016/j.wasman.2017.04.039

Han M. J., Behera S. K. and Park H. S. (2012). Anaerobic co-digestion of food waste leachate and piggery wastewater for methane production: statistical optimization of key process parameters. *Journal of Chemical Technology and Biotechnology*, **87**(11), 1541–1550, https://doi.org/10.1002/jctb.3786

Hartmann H. and Ahring B. K. (2005). Anaerobic digestion of the organic fraction of municipal solid waste: influence of co-digestion with manure. *Water Research*, **39**(8), 1543–1552, https://doi.org/10.1016/j.watres.2005.02.001

He P. J., Qu X., Shao L. M., Li G. J. and Lee D. J. (2007). Leachate pretreatment for enhancing organic matter conversion in landfill bioreactor. *Journal of Hazardous Materials*, **142**(1–2), 288–296.

Hombach S. T., Oleszkiewicz J. A., Lagasse P., Amy L. B., Zaleski A. A. and Smyrski K. (2003). Impact of landfill leachate on anaerobic digestion of sewage sludge. *Environmental Technology (United Kingdom)*, **24**(5), 553–560.

Jaroenpoj S., Yu Q. J. and Ness J. (2015). Development of artificial neural network models for biogas production from co-digestion of leachate and pineapple peel. *The Global EnvironmentalEngineers*,**1**(2),42–47,https://doi.org/10.15377/2410-3624.2014.01.02.2

Jayanth T. A. S., Mamindlapelli N. K., Begum S., Arelli V., Juntupally S., Ahuja S., Dugyala S. K. and Anupoju G. R. (2020). Anaerobic mono and co-digestion of organic fraction of municipal solid waste and landfill leachate at industrial scale: impact of volatile organic loading rate on reaction kinetics, biogas yield and microbial diversity. *Science of the Total Environment*, **748**, 142462, https://doi.org/10.1016/j.scitotenv.2020.142462

Karthikeyan O. P. and Visvanathan C. (2013). Bio-energy recovery from high-solid organic substrates by dry anaerobic bio-conversion processes: a review. *Reviews in Environmental Science and Bio/Technology*, **12**(3), 257–284, https://doi.org/10.1007/s11157-012-9304-9

Kim S. H. and Ju H. J. (2012). Feasibility of co-digestion of sewage sludge, swine waste and food waste leachate. *Journal of Korea Organic Resources Recycling Association*, **20**, 61–70.

Kim J. and Kang C. M. (2015). Increased anaerobic production of methane by co-digestion of sludge with microalgal biomass and food waste leachate. *Bioresource Technology*, **189**, 409–412, https://doi.org/10.1016/j.biortech.2015.04.028

Kjeldsen P., Barlaz M. A., Rooker A. P., Baun A., Ledin A. and Christensen T. H. (2002). Present and long-term composition of MSW landfill leachate: a review. *Critical Reviews in Environmental Science and Technology*, **32**(4), 297–336, https://doi.org/10.1080/10643380290813462

Klavon K. H., Lansing S. A., Mulbry W., Moss A. R. and Felton G. (2013). Economic analysis of small-scale agricultural digesters in the United States. *Biomass and Bioenergy*, **54**, 36–45, https://doi.org/10.1016/j.biombioe.2013.03.009

Kulikowska D. and Klimiuk E. (2008). The effect of landfill age on municipal leachate composition. *Bioresource Technology*, **99**(13), 5981–5985, https://doi.org/10.1016/j.biortech.2007.10.015

Kumar M., Ou Y. L. and Lin J. G. (2010). Co-composting of green waste and food waste at low C/N ratio. *Waste Management*, **30**(4), 602–609, https://doi.org/10.1016/j.wasman.2009.11.023

Lee E., Bittencourt P., Casimir L., Jimenez E., Wang M., Zhang Q. and Ergas S. J. (2019). Biogas production from high solids anaerobic co-digestion of food waste, yard waste and waste activated sludge. *Waste Management*, **95**, 432–439, https://doi.org/10.1016/j.wasman.2019.06.033

Lee J., Hong J., Jeong S., Chandran K. and Park K. Y. (2020). Interactions between substrate characteristics and microbial communities on biogas production yield and rate. *Bioresource Technology*, **303**, 122934, https://doi.org/10.1016/j.biortech.2020.122934

Lei Y., Wei L., Liu T., Xiao Y., Dang Y., Sun D. and Holmes D. E. (2018). Magnetite enhances anaerobic digestion and methanogenesis of fresh leachate from a municipal solid waste incineration plant. *Chemical Engineering Journal*, **348**, 992–999, https://doi.org/10.1016/j.cej.2018.05.060

Leiva M. B., Hosseini Koupaie E. and Eskicioglu C. (2014). Anaerobic co-digestion of wine/fruit-juice production waste with landfill leachate diluted municipal sludge cake under semi-continuous flow operation. *Waste Management*, **34**(10), 1860–1870, https://doi.org/10.1016/j.wasman.2014.06.027

Li R., Chen S., Li X., Saifullah Lar J., He Y. and Zhu B. (2009). Anaerobic codigestion of kitchen waste with cattle manure for biogas production. *Energy & Fuels*, **23**(4), 2225–2228, https://doi.org/10.1021/ef8008772

Li Y., Jin Y., Li J., Li H. and Yu Z. (2016). Effects of thermal pretreatment on the biomethane yield and hydrolysis rate of kitchen waste. *Applied Energy*, **172**, 47–58, https://doi.org/10.1016/j.apenergy.2016.03.080

Liao X., Zhu S., Zhong D., Zhu J. and Liao L. (2014). Anaerobic co-digestion of food waste and landfill leachate in single-phase batch reactors. *Waste Management*, **34**(11), 2278–2284, https://doi.org/10.1016/j.wasman.2014.06.014

Liao X., Zeng F., Akhil G., Zhu J. and Liao L. (2021). Experimental study on the mechanisms involved in the synergetic effect of anaerobic co-digestion of mature leachate with food waste. *Desalination and Water Treatment*, **222**, 137–144, https://doi.org/10.5004/dwt.2021.27119

Liao X., Li H., Chen H., Tang Q., Liao L. and Liu D. (2022). Enhancing the proportion of humic substances in dissolved organic matter via a novel process of anaerobic co-digestion of food waste and young landfill leachate. *Waste and Biomass Valorization*, **1**, 85–92, https://doi.org/10.1007/s12649-022-01868-w

Lin C. Y., Bian F. Y. and Chou J. (1999). Anaerobic co-digestion of septage and landfill leachate. *Bioresource Technology*, **68**(3), 275–282, https://doi.org/10.1016/S0960-8524(98)00151-5

Lin C. Y., Chang F. Y. and Chang C. H. (2000). Co-digestion of leachate with septage using a UASB reactor.

Lin J., Zuo J., Gan L., Li P., Liu F., Wang K., Chen L. and Gan H. (2011). Effects of mixture ratio on anaerobic co-digestion with fruit and vegetable waste and food waste of China. *Journal of Environmental Sciences*, **23**(8), 1403–1408, https://doi.org/10.1016/S1001-0742(10)60572-4

Liu X., Gao X., Wang W., Zheng L., Zhou Y. and Sun Y. (2012). Pilot-scale anaerobic codigestion of municipal biomass waste: focusing on biogas production and GHG reduction. *Renewable Energy*, **44**, 463–468, https://doi.org/10.1016/j.renene.2012.01.092

Liu C. C., Huang C., Sun X. Y., Li R., Li J., Shen J., Han W. and Wang L. (2017). The effect of Mg^{2+} on digestion performance and microbial community structures in sludge digestion systems. *Environmental Science and Pollution Research*, **24**(21), 17474–17484, https://doi.org/10.1007/s11356-017-9320-y

Luste S., Luostarinen S. and Sillanpää M. (2009). Effect of pre-treatments on hydrolysis and methane production potentials of by-products from meat-processing industry. *Journal of Hazardous Materials*, **164**(1), 247–255, https://doi.org/10.1016/j.jhazmat.2008.08.002

Lv Y., Chang N., Li Y. Y. and Liu J. (2021). Anaerobic co-digestion of food waste with municipal solid waste leachate: a review and prospective application with more benefits. *Resources, Conservation and Recycling*, **174**, 105832, https://doi.org/10.1016/j.resconrec.2021.105832

Ma C., Liu J., Ye M., Zou L., Qian G. and Li Y. Y. (2018). Towards utmost bioenergy conversion efficiency of food waste: pretreatment, co-digestion, and reactor type.

Renewable and Sustainable Energy Reviews, **90**, 700–709, https://doi.org/10.1016/j.rser.2018.03.110

Ma S., Zhou C., Pan J., Yang G., Sun C., Liu Y., Chen X. and Zhao Z. (2022). Leachate from municipal solid waste landfills in a global perspective: characteristics, influential factors and environmental risks. *Journal of Cleaner Production*, **333**, 130234, https://doi.org/10.1016/j.jclepro.2021.130234

Mata-Alvarez J., Dosta J., Romero-Güiza M. S., Fonoll X., Peces M. and Astals S. (2014). A critical review on anaerobic co-digestion achievements between 2010 and 2013. *Renewable and Sustainable Energy Reviews*, **36**, 412–427, https://doi.org/10.1016/j.rser.2014.04.039

Montusiewicz A. and Lebiocka M. (2011). Co-digestion of intermediate landfill leachate and sewage sludge as a method of leachate utilization. *Bioresource Technology*, **102**(3), 2563–2571, https://doi.org/10.1016/j.biortech.2010.11.105

Moody C. M. and Townsend T. G. (2017). Comparison of landfill leachates based on waste composition. *Waste Management*, **63**, 267–274, https://doi.org/10.1016/j.wasman.2016.09.020

Moujanni A. E., Qarraey I. and Ouatmane A. (2019). Anaerobic codigestion of urban solid waste fresh leachate and domestic wastewaters: biogas production potential and kinetic. *Environmental Engineering Research*, **24**(1), 38–44, https://doi.org/10.4491/eer.2018.082

Müller J. A. (2001). Prospects and problems of sludge pre-treatment processes. *Water Science and Technology*, **44**(10), 121–128, https://doi.org/10.2166/wst.2001.0598

Nayono S. E., Gallert C. and Winter J. (2010). Co-digestion of press water and food waste in a biowaste digester for improvement of biogas production. *Bioresource Technology*, **101**(18), 6987–6993, https://doi.org/10.1016/j.biortech.2010.03.123

Nazimudheen G., Roy K., Sivasankar T. and Moholkar V. S. (2018). Mechanistic investigations in ultrasonic pretreatment and anaerobic digestion of landfill leachates. *Journal of Environmental Chemical Engineering*, **6**(2), 1690–1701, https://doi.org/10.1016/j.jece.2018.02.001

Ndao A., Drogui P. and Tyagi R. D. (2022). Enhancement of landfill leachate treatment using extracellular polymeric substances as bio-flocculants. *Journal of Environmental Science and Health, Part B*, **57**(1), 62–70, https://doi.org/10.1080/03601234.2021.2024059

Oz N. A. and Yarimtepe C. C. (2014). Ultrasound assisted biogas production from landfill leachate. *Waste Management*, **34**, 1165–1170, https://doi.org/10.1016/j.wasman.2014.03.003

Pagliaccia P., Gallipoli A., Gianico A., Gironi F., Montecchio D., Pastore C., di Bitonto L. and Braguglia C. M. (2019). Variability of food waste chemical composition: impact of thermal pre-treatment on lignocellulosic matrix and anaerobic biodegradability. *Journal of Environmental Management*, **236**, 100–107, https://doi.org/10.1016/j.jenvman.2019.01.084

Pilli S., Bhunia P., Yan S., LeBlanc R. J., Tyagi R. D. and Surampalli R. Y. (2011). Ultrasonic pretreatment of sludge: a review. *Ultrasonics Sonochemistry*, **18**(1), 1–18, https://doi.org/10.1016/j.ultsonch.2010.02.014

Raghab S. M., Abd El Meguid A. M. and Hegazi H. A. (2013). Treatment of leachate from municipal solid waste landfill. *HBRC Journal*, **9**(2), 187–192, https://doi.org/10.1016/j.hbrcj.2013.05.007

Rocamora I., Wagland S. T., Villa R., Simpson E. W., Fernandez O. and Bajon-Fernandez Y. (2020). Dry anaerobic digestion of organic waste: a review of operational parameters and their impact on process performance. *Bioresource Technology*, **299**, 122681, https://doi.org/10.1016/j.biortech.2019.122681

Sajjad M. and Kim K. S. (2015). Studies on the interactions of Ca^{2+} and Mg^{2+} with EPS and their role in determining the physicochemical characteristics of granular sludges in SBR system. *Process Biochemistry*, **50**(6), 966–972, https://doi.org/10.1016/j.procbio.2015.02.020

Shahriari H., Warith M., Hamoda M. and Kennedy K. J. (2012). Effect of leachate recirculation on mesophilic anaerobic digestion of food waste. *Waste Management*, **32**(3), 400–403, https://doi.org/10.1016/j.wasman.2011.10.022

Sosnowski P. and Ledakowicz S. (2003). Energy utilization of sewage sludge and communal wastes in methane fermentation processes. *Przemysl Chemiczny*, **82** (8–9), 1095–1097.

Subramanian B. and Pagilla K. R. (2015). Mechanisms of foam formation in anaerobic digesters. *Colloids and Surfaces B: Biointerfaces*, **126**, 621–630, https://doi.org/10.1016/j.colsurfb.2014.11.032

Takeda P. Y., Gotardo J. T. and Gomes S. D. (2022). Anaerobic co-digestion of leachate and glycerol for renewable energy generation. *Environmental Technology (United Kingdom)*, **43**(8), 1118–1128.

Wang Y., Zhang Y., Wang J. and Meng L. (2009). Effects of volatile fatty acid concentrations on methane yield and methanogenic bacteria. *Biomass Bioenergy*, **33**(5), 848–853, https://doi.org/10.1016/j.biombioe.2009.01.007

Web Of Science. (2022). Retrieved from: https://www.webofscience.com/wos/woscc/summary/7ed1992e-b5ea-4ad8-925d-d6f575422ade-469f193e/relevance/1. Date of access: 01.08.2022.

Wu Q. L., Guo W. Q., Zheng H. S., Luo H. C., Feng X. C., Yin R. L. and Ren N. Q. (2016). Enhancement of volatile fatty acid production by co-fermentation of food waste and excess sludge without pH control: the mechanism and microbial community analyses. *Bioresource Technology*, **216**, 653–660, https://doi.org/10.1016/j.biortech.2016.06.006

Xu F., Li Y., Ge X., Yang L. and Li Y. (2018). Anaerobic digestion of food waste – challenges and opportunities. *Bioresource Technology*, **247**, 1047–1058, https://doi.org/10.1016/j.biortech.2017.09.020

Yusmartini E. S., Setiabudidaya D., Ridwan Marsi and Faizal (2013). Characteristics of leachate at Sukawinatan Landfill, Palembang, Indonesia. *Journal of Physics: Conference Series*, **423**(1), 012048.

Zhang L. and Jahng D. (2012). Long-term anaerobic digestion of food waste stabilized by trace elements. *Waste Management*, **32**(8), 1509–1515, https://doi.org/10.1016/j.wasman.2012.03.015

Zhang L., Lee Y. W. and Jahng D. (2011). Anaerobic co-digestion of food waste and piggery wastewater: focusing on the role of trace elements. *Bioresource Technology*, **102**(8), 5048–5059, https://doi.org/10.1016/j.biortech.2011.01.082

Zhang L., Ouyang W. and Lia A. (2012). Essential role of trace elements in continuous anaerobic digestion of food waste. *Procedia Environmental Sciences*, **16**, 102–111, https://doi.org/10.1016/j.proenv.2012.10.014

Zhang C., Xiao G., Peng L., Su H. and Tan T. (2013). The anaerobic co-digestion of food waste and cattle manure. *Bioresource Technology*, **129**, 170–176, https://doi.org/10.1016/j.biortech.2012.10.138

Zhang W., Zhang L. and Li A. (2015a). Anaerobic co-digestion of food waste with MSW incineration plant fresh leachate: process performance and synergistic effects. *Chemical Engineering Journal*, **259**, 795–805, https://doi.org/10.1016/j.cej.2014.08.039

Zhang W., Zhang L. and Li A. (2015b). The positive effects of waste leachate addition on methane fermentation from food waste in batch trials. *Water Science and Technology*, **72**(3), 429–436, https://doi.org/10.2166/wst.2015.231

Zhou S., Wang J., Peng S., Chen T. and Yue Z. (2020). Anaerobic co-digestion of landfill leachate and acid mine drainage using up-flow anaerobic sludge blanket reactor. *Environmental Science and Pollution Research*, **28**(7), 8498–8506, https://doi.org/10.1007/s11356-020-11207-y

Zhu N. (2007). Effect of low initial C/N ratio on aerobic composting of swine manure with rice straw. *Bioresource Technology*, **98**(1), 9–13, https://doi.org/10.1016/j.biortech.2005.12.003

doi: 10.2166/9781789063318_0153

Chapter 8

Electrochemical treatment of landfill leachate

Brahim Arhoun, Maria Villen-Guzman, Jose Miguel Rodriguez-Maroto and Juan Manuel Paz-Garcia*

Chemical Engineering Department, Faculty of Sciences, University of Malaga, 29071 Malaga, Spain
**Corresponding author: arhoun.b@uma.es*

ABSTRACT

A significant municipal solid waste is managed by disposition in a landfill. This kind of management system makes producing leachate with toxic characteristics unavoidable. Landfill leachate is a potential pollutant source containing several types of organic and inorganic pollutants that can induce risk toward public health and ecosystems if they are not appropriately treated. Among the parameter affecting the landfill leachate, the age of the landfill is considered to be a determinant factor controlling the leachate composition. Conventionally, the leachate has been treated by an individual or combination of biological, physical, and chemical methods. Recently, electrochemical methods have paid particular attention to the treatments of landfill leachate, presenting high efficacy for the degradation of refractory compounds. Electrochemical methods offer several advantages: small space requirement, short treatment time, low requirements of chemical reagents, and easy automation. Electrochemical methods commonly used for the treatment of landfill leachates are: electro-oxidation, electro-coagulation, and electro-Fenton. Each of these methods can be affected by different parameters. Commonly, electrochemical processes are influenced by several operating parameters such as electrode material and design, the inter-distance between electrodes, treatment time, initial pH value and conductivity of the solution, applied current density, mixing and electrolytes added. The electrochemical process can also be combined with biological and physicochemical processes as pre-treatment or post-treatment to enhance the treatment of landfill leachate.

Keywords: Anodic oxidation, electrochemical treatment, electrocoagulation, electro-Fenton, electro-oxidation, landfill leachate.

8.1 INTRODUCTION

Industrialization and the economic development have made the society reach a lifestyle based on a constant increase in consumption. The immediate consequence of this is the generation of a series of urban waste and the problems derived from its inadequate management that can have a detrimental effect on the environment.

Landfilling is one of the main methods of waste management in many countries around the world. In 2020, the world is estimated to generate 2.24 billion tonnes of solid waste, amounting to a footprint of 0.79 kg/person/day (World Bank, 2022). It is estimated that up to 95% of total municipal solid waste collected worldwide is disposed in landfills (Gao *et al.*, 2015). In addition, the decomposition process for 1 ton of solid waste produces approximately 0.2 m^3 of landfill leachate (Christensen *et al.*, 2014).

Landfilled waste undergoes various processes of chemical and biological transformations. A significant drawback of this kind of waste management is that the percolation of rainwater through the landfill produces a heavily polluted effluent loaded with organic and inorganic matter. This effluent, called leachate must be contained in a controlled pond (Luo *et al.*, 2020). Given the negative impact of leachate on the environment, such as an unpleasant odor and high toxicity, which can cause a significant threat, correct management is of utmost importance to avoid contamination of soil, surface water, and groundwater.

Leachate is a liquid produced by the decomposition of organic matter and percolating water (rain, runoff, or the fluid generated in the decomposition of waste). The quantity and composition of the leachate depend on the type of waste, the average rainfall, and the evapotranspiration on the landfill. However, it always has a high content of suspended and dissolved matter that makes it a dangerous contaminant that must be treated to avoid it to cause contamination of aquifers, rivers, and soil, which in turn poses risk to the natural ecosystem and public health (Keyikoglu *et al.*, 2021).

In particular, leacheate's physicochemical and biological characteristics vary from one landfill to another. The age of the landfill strongly influences the composition of pollutants in the leachate. According to the landfill age, the leachate can be classified into three types: young, intermediate, and mature.

To comply with strict discharge standards, the search for the appropriate types of treatments for landfill leachate depending on the age of the landfill is a great urgency to avoid environmental pollution problems. Generally, the leachate characteristics such as chemical oxygen demand (COD), biological oxygen demand (BOD), the BOD_5/COD ratio, and landfill age are crucial determinants for selecting suitable treatment technologies (Luo *et al.*, 2020). Developing new processes for treating landfill leachate with an organic load must be guided by sustainability criteria that bring together economic profitability, social acceptance, and environmental protection.

Traditionally, biological processes such as aerobic and anaerobic treatments are the most used for the treatment of leachates. The main advantage of biological processes is the efficiency of treating biodegradable organic pollutants present in young leachate at a reasonable cost. However, their use is limited with increased leachate ages characterized by a large percentage of refractory compounds. To deal with this limitation, some physicochemical processes have been experimented to treat leachate from sanitary landfills. These processes include physical adsorption, chemical coagulation (Wu *et al.*, 2021), nanofiltration (Meier *et al.*, 2021) and reverse osmosis (Anna Tałałaj *et al.*, 2021). The physical and chemical treatment are limited by the large amount of sludge produced and the need to consume a large amount of coagulant (Verma & Kumar, 2017) or need the combination with other processes (Meier *et al.*, 2021).

Electrochemical processes allow the transformation of organic matter into carbon dioxide and water or its conversion into simpler compounds that can be subsequently treated. The advantages of such processes lie in their non-polluting aspects, ease of automation, reduced treatment times, low reagent requirements, and the possibility of generating electrochemically reactive agents capable of effectively eliminating pollutants (Guo *et al.*, 2022). Electro-oxidation (EO), electrocoagulation (EC), and electro-Fenton (EF) processes are effective solutions for removing and separating a wide range of contaminants such as remove COD, ammonia nitrogen, and heavy metal pollutants (Liu *et al.*, 2022; Tang *et al.*, 2022).

Hence, this chapter focuses on landfill leachate generation, their characteristics, and their treatment by EO, EC, and EF. Meanwhile, it will provide the state of the art, the fundamental principle, and the operating parameters influencing these processes.

8.2 LANDFILL LEACHATE

8.2.1 Generation and characteristics of landfill leachate

Landfill leachate is a liquid produced by decomposing organic matter and percolating water (rain, runoff, or the fluid generated in the decomposition of waste). The quantity and composition of the leachate depend on the type of waste, the average rainfall, and the evapotranspiration on the landfill. However, it always has a high content of suspended and dissolved matter that makes it a dangerous contaminant that must be treated otherwise, and it can cause contamination of aquifers, rivers, and soil, which in turn poses risk to the natural ecosystem and public health (Keyikoglu *et al.*, 2021). Luo *et al.* (2020) reported that approximately 200 hazardous compounds are identified in landfill leachates.

The composition of landfill leachate varies notably with the age of the landfill. According to the landfill age, landfill leachates are commonly classified into three categories: (1) young (<5 years); (2) intermediate (5–10 years); and (3) mature (>10 years) (Luo *et al.*, 2020). As the landfill age increases, the leachate parameters (pH, BOD_5, COD, and BOD_5/COD ratio) change significantly (Fernandes *et al.*, 2015).

8.2.2 Landfill leachate treatment

Generally, the landfill leachate characteristics such as age, COD, BOD_5/COD ratio, salinity, and toxicity are critical factors for the selection of adequate treatment processes (Keyikoglu *et al.*, 2021). Conventional treatments of landfill leachates are generally divided into three major categories: (1) biological processes, (2) physical and chemical processes, and (3) a leachate transfer method. There are two types of landfill leachate transfer methods: (a) combined treatment with domestic sewage, and (b) recycling leachate (Anqi *et al.*, 2020). The effectiveness of leachate treatments methods according to the landfill age is presented in Table 8.1.

In recent years, electrochemical methods have been applied for the treatment of wastewater with a high non-biodegradable organic load. These methods have a promising future because they use readily available robust and compact instrumentation that offers the possibility of easy distribution and can potentially replace sophisticated processes since they require low-capacity content. The advantages of these methods are environmental benefits, compatibility, versatility, energy efficiency, safety, selectivity, ease of process automation, and low costs (Mollah *et al.*, 2004). Also, given its relative ease of operation and its proven effectiveness in removing organic compounds that are difficult to biodegrade, especially in treating industrial wastewater (Yu *et al.*, 2020a).

Table 8.1 Efficiency of process for leachate treatment vs. leachate type

Process	Landfill Classification			Average Removal (%)				
	Young	Medium	Old	BOD	COD	TKN	SS	Turbidity
Ultrafiltration	Poor-Fair	–	–	–	50	60–80	>99	>99
Nanofiltration	Good	Good	Good	80	60–80	60–80	>99	>99
Reverse osmosis	Good	Good	Good	>90	>90	>90	>99	>99
Recycling	Good	Fair	Poor	>90	60–80	–	–	–
Coagulation/ flocculation	Poor	Fair	Fair	–	40–60	>30	>80	>80
Chemical precipitation	Poor	Poor	Fair	–	<30	<30	30–40	>80
Adsorption	Poor	Fair	Good	>80	70–90	–	–	50–70
Oxidation	Poor	Fair	Fair	–	30–90	–	–	>80
Stripping	Poor	Fair	Fair	–	<30	>80	–	30–40
Aerobic processes	Good	Fair	Poor	>80	60–90	>80	60–80	–
Anaerobic processes	Good	Fair	Poor	>80	60–80	>80	60–80	–
Membrane bioreactor	Good	Fair	Fair	>80	>85	>80	>99	40–60

Source: Reprinted from *Journal of Hazardous Materials*, Vol. 150, Issue 3, S. Renou, J.G. Givaudan, S. Poulain, F. Dirassouyan, P. Moulin, Landfill leachate treatment: review and opportunity, pp. 468–493 (2008), with permission from Elsevier.

8.3 EO TREATMENT OF LANDFILL LEACHATE

In a broad sense, electrochemistry includes the study of chemical reactions that produce electrical effects and chemical phenomena caused by the action of currents or voltages. Electrochemical methods for wastewater treatment involve an electrolytic cell and a pair of metallic electrodes whose physical and chemical characteristics are essential variables within the process through which an electric current is circulated. Figure 8.1 shows a conceptual diagram of an electrochemical reactor for wastewater, which includes a power supply, a cathode, an anode, and an electrolyte.

The electrooxidation process is attractive since electrons provide a versatile, efficient, and profitable solution, to which the ease of automation must be added. In addition, it achieves pollutant removal mainly through two mechanisms: direct oxidation (direct transfer of electrons to the oxidizing anode surface) and indirect oxidation (by generating electroactive substances) (Fernandes *et al.*, 2015). The main difference between both mechanisms is related to the electrode material. Direct oxidation and indirect oxidation will destroy the morphology and stability of most pollutants in the electrolyte solution to achieve better removal. The specific electrode reaction is shown in Figure 8.2.

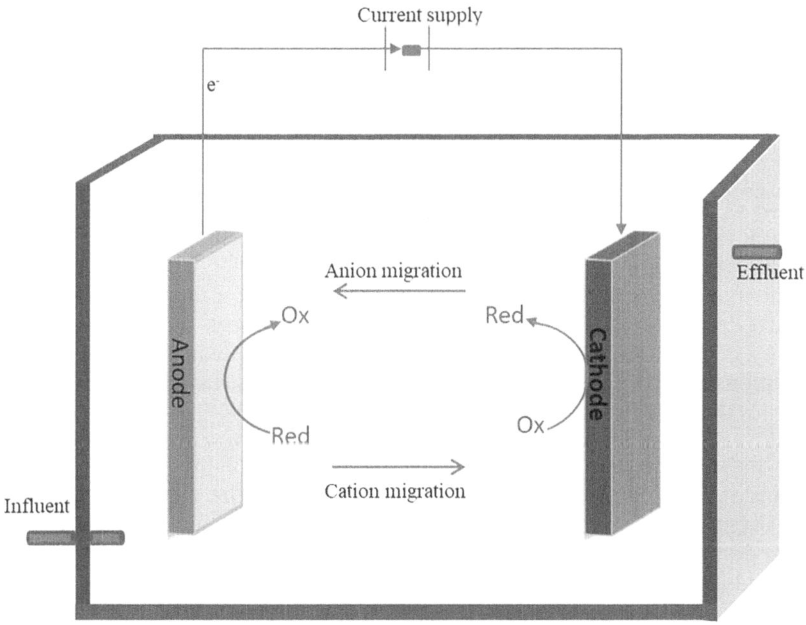

Figure 8.1 Conceptual diagram of an electrochemical reactor.

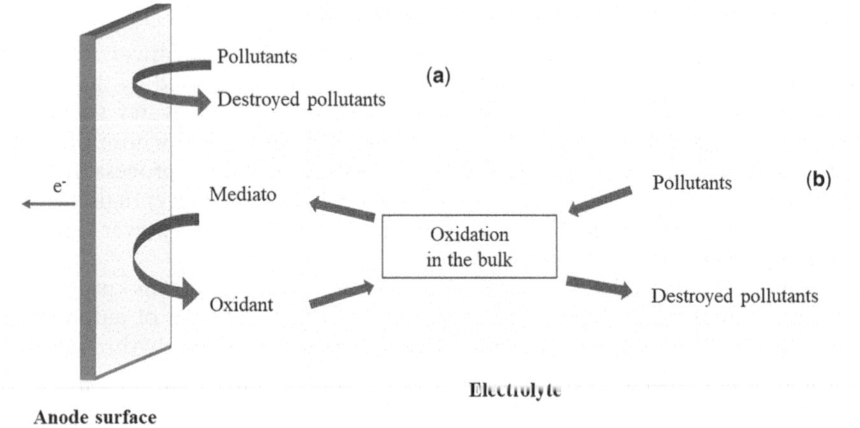

Figure 8.2 Schemes for (a) direct and (b) indirect EO treatment of pollutants. (Adapted from Deng and Englehardt, 2007.)

8.3.1 Direct EO processes

Direct oxidation of pollutants follows two steps: (1) diffusion of pollutants from the bulk solution to the anode surface where they get adsorbed and (2) oxidation of pollutants at the anode surface (equation (8.1)). Therefore, the efficacy of the process depends on the relationship between the mass transfer of the substrate and electron transfer at the electrode surface. In addition, the rate of electron transfer is determined by the electrode activity and current density (Anglada *et al.*, 2009).

$$M + R \rightarrow M(\cdot R) + ne^- \rightarrow P \tag{8.1}$$

where R is an organic pollutant, M is an active site on the anode surface, and P is the product.

The major drawback of this mechanism is forming a layer over the anode surface, 'passivation of the electrode,' which decreases the electrocatalytic activity of the anode material. However, this mechanism may promote the formation of hydroxyl radical adsorbed on the anode surface, further oxidizing organics through indirect electrolysis (Mandal *et al.*, 2017).

8.3.2 Indirect EO processes

The indirect oxidation of wastewater may occur on the electrode surface in two ways: first, the formation of hydroxyl radicals by water oxidation on the electrode surface, and the organic matter is indirectly oxidized by equations (8.2) and (8.3) (Panizza & Cerisola, 2009)

$$M + H_2O \rightarrow H^+ + e^- \tag{8.2}$$

$$M(\cdot OH) + R \rightarrow M + RO + H^+ + e^- \tag{8.3}$$

The efficacy of this degradation mechanism depends upon selected electrode material and experimental conditions (Guo *et al.*, 2022; Mandal *et al.*,

2017). Second, it takes place on the bulk of the solution by other powerful electrogenerated oxidants, such as active chlorine and persulfate. The most common electrochemical oxidant is chlorine, which is formed by the oxidation of chloride at the anode (Anglada *et al.*, 2009). As in the electro-Fenton, metal catalytic mediators (Ag^{+}, Co^{3+}, Fe^{3+}, Ce^{4+}, and Ni^{2+}) are also used to generate hydroxyl radicals. Hence, this approach must operate in highly acidic media and need separation steps to recover the metallic species (Chen, 2004; Martínez-Huitle & Ferro, 2006; Martínez-Huitle *et al.*, 2015).

8.3.3 Influence factors of EO in the treatment of landfill leachate

For the removal rate and current efficiency of pollutants in landfill leachate, the efficiency of the EO process also depends on anode material, reactor design and various operating parameters.

8.3.3.1 Anode material

The choice of electrode material is crucial in the reactor's design, focusing on high activation energies to avoid undesired side reactions. Cathode materials should have high over-voltages for hydrogen evolution, while anode materials should have over voltages for oxygen-evolution (Deng & Englehardt, 2007). The electrode material is of great importance that affects the selectivity and the efficiency of the EO process.

The anode material must have the following properties (Anglada *et al.*, 2009):

- High physical and chemical stability; resistance to erosion, corrosion, and formation of passivation layers.
- High electrical conductivity.
- Catalytic activity and selectivity.
- The use of electrode materials that are inexpensive and durable must be favored (low cost/life ratio).

The anodic activity depends on the value of the overpotential of oxygen evolution. In Table 8.2, several electrode materials are presented. The EO anode materials divided into two categories are as follows:

- Low O_2 overvoltage 'active' anodes are characterized by a high electrochemical activity toward oxygen evolution and low chemical reactivity toward oxidation of organic. Some examples of 'active' anodes are Pt, RuO_2, and graphite. The efficiency of oxidation of pollutants on these anodes occurs at very low current densities or in the presence of high concentrations of chlorides or metallic mediators (Chen, 2004).
- High O_2 overvoltage 'non-active' anodes: higher current densities may be applied with minimal contribution from the oxygen evolution side reaction. Some examples of 'non-active' anodes are boron-doped diamond (BDD), PbO_2, SnO_2 (Mandal *et al.*, 2017) Ti/SnO_2-Sb_2O_5 (Chen *et al.*, 2003). These types of anode are more preferred (Anglada *et al.*, 2009). A 'non-active' electrode does not provide any catalytic active site for the adsorption of reactants and/or products from the aqueous medium (Fernandes *et al.*, 2015).

Table 8.2 Potential of oxygen evolution of different anodes, V vs. NHE.

Anode	Potential (v)	Conditions
Pt	1.3	0.5 mol/L H_2SO_4
Pt	1.6	0.5 mol/L H_2SO_4
IrO_2	1.6	0.5 mol/L H_2SO_4
Graphite	1.7	0.5 mol/L H_2SO_4
PbO_2	1.9	1 mol/L H_2SO_4
SnO_2	1.9	0.5 mol/L H_2SO_4
TiO_2	2.2	1 mol/L H_2SO_4
Si/BDD	2.3	0.5 mol/L H_2SO_4
Ti/BDD	2.7	0.5 mol/L H_2SO_4

Source: Chen (2004).

Several anodes were used on the landfill leachate treatment, such as boron-doped diamond anodes (Agustina *et al.*, 2019; Urtiaga *et al.*, 2022), titanium, ruthenium, iridium, tin oxide-based anodes (DSA) and lead dioxide (PbO_2) anodes (Guo *et al.*, 2022). In particular, boron-doped diamond (BDD) anode is widely used, due to the high organic oxidation rates and greater current efficiencies than other commonly used metal oxides such as PbO_2 and Ti/SnO_2-Sb_2O_5 (Anglada *et al.*, 2009). DSA anodes have high catalytic activity to release active chlorine, and are highly stable and widely used (Turro *et al.*, 2011). For PbO_2 anodes, a large number of hydroxyl radicals can be generated under acidic conditions, and it has a great advantage from the perspective of economic feasibility, so it has been widely used in wastewater treatment (Panizza & Cerisola, 2009).

Chiang *et al.* (1995) reported that the COD removal rate decreased according to this order of different anodes used SPR > DSA > PbO_2/Ti > graphite. COD removal rate of 92% and almost complete removal of ammoniacal nitrogen are obtained with the SPR anode and a current density of 150 mA/cm^2.

8.3.3.2 Reactor design

One of the most critical aspects of electrochemical reactor design is maintaining high mass transfer rates as the primary reaction take place on electrode surfaces (Anglada *et al.*, 2009). In general, the geometry of electrodes can be classified into two-dimensional and three-dimensional constructions (Yu *et al.*, 2020a). Two types of cell configuration, divided and undivided, are used. A divided cell presents better performance in terms of total organic carbon removal using same electrodes (Ti/PbO_2) (Lei *et al.*, 2007).

Two flow characteristics limiting hydrodynamic behavior: plug flow and perfect mixing. The last mentioned presents the advantage of a high electrode surface-to-cell volume ratio value. In the case of flow characteristics within a reactor, two types of limiting hydrodynamic behaviour may be considered: plug flow and perfect mixing (Anglada *et al.*, 2009). Figure 8.3 presents the classification of electrochemical reactors in terms of cell configuration, electrode geometry, and flow type.

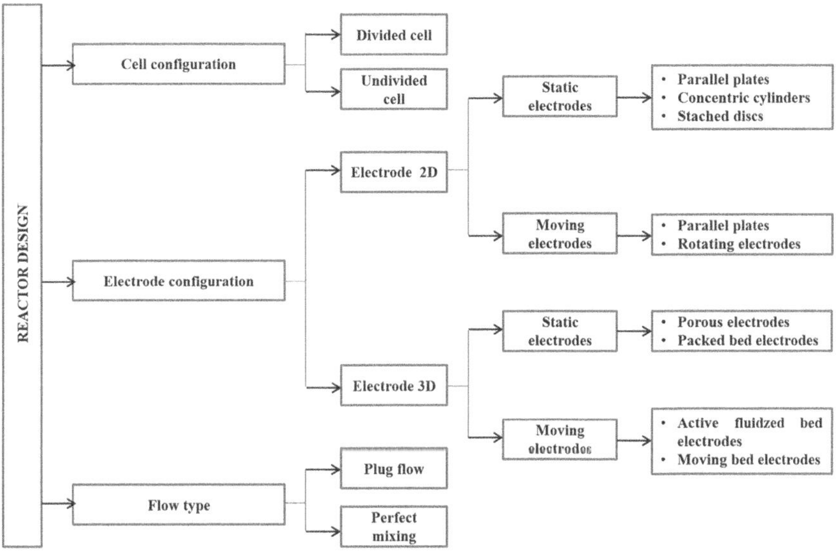

Figure 8.3 Classification of electrochemical reactors in terms of cell configuration, electrode geometry, and flow type. (Adapted from Anglada *et al.*, 2009.)

8.3.4 Operating parameters

In addition to the electrode material and reactor design, pollutant removal, current efficiency, and energy consumption of EO are also affected by different operating parameters. Current density (intensity per unit area of electrode), pH, and electrolytes were added.

8.3.4.1 Current density

Current density is the most crucial parameter in electrochemical oxidation, which is considered determinant of the cost and efficiency of the EO process. Increasing current density during electrolysis yielded enhancement in pollutant removal in many studies, which may be attributed to more hydroxyl radical generation at the anode surface or by the formation of chlorine/hypochlorite. It was further confirmed that the removal of COD and ammonia nitrogen from landfill leachate mainly depends on indirect oxidation under the action of active chlorine (Panizza *et al.*, 2010). The current density can affect the chloride ions during leachate conversion into a large amount of active chlorine, which can improve the landfill leachate treatment efficiency (Guo *et al.*, 2022).

The consequences of applying high current density during EO led to a higher conversion rate of ammonium nitrogen to nitrate nitrogen during ammonium removal (Urtiaga *et al.*, 2009). The greater is the current density, the better will be the oxidation capacity and even better the effect of removal of pollutants. However, the use of higher current densities usually results in higher operating costs due to an increase in energy consumption. It should also be highlighted that an increase in current density does not necessarily increase the process performance. For a given anode material, the effect of current density efficiency depends on the characteristics of the landfill leachate to be treated (Oturan *et al.*, 2015; Papastavrou *et al.*, 2009).

8.3.4.2 pH

COD degradation mainly depends on the large number of hydroxyl radicals generated on the anode surface. Under acidic conditions, it is favorable for the hydroxyl radical's oxidation and advantageous for inhibiting the side reactions of oxygen evolution occurrence (eq (8.4)) (Song-hu & Xiao-hua, 2005). Acidic pH may reduce the concentration of carbonate and bicarbonate ions, known as effective hydroxyl radical scavengers, thus enhancing the oxidation reaction rate. At alkaline pH, more current is used for the oxygen evolution reaction rather than oxidation due to the lower potential requirement for the oxygen evolution reaction equation (8.5), which reduces the current efficiency of the treatment (Mandal *et al.*, 2017).

In acid medium

$$2\,H_2O \rightarrow 4\,H^+ + O_2 + 4\,e^- \tag{8.4}$$

In alkaline medium

$$4\,OH^- \rightarrow 2\,H_2O + O_2 + 4\,e^- \tag{8.5}$$

Under acidic conditions, aqueous chlorine predominates at pH < 3.3 (equation (8.6)). At higher bulk pH values, its diffusion away from the electrode is coupled to its dismutation reaction to form chloric(I) acid (HOCl) at pH < 7.5 (equation (8.7)) and chloric(I) ions (ClO^-) at pH > 7.5 (equation (8.8)). Cl_2, HOCl, and ClO^- ions are called 'active chlorine' (Szpyrkowicz, 2006). According to Pérez *et al.* (2012) ammonium degradation mainly occurs due to indirect oxidation through chlorine/hypochlorite. The highest COD and color removals were obtained under acidic conditions, whereas the lowest removal was obtained in an alkaline medium. In an alkaline medium, the low production of chlorine/hypochlorite ions favors the formation of chlorate or perchlorate. In an acidic medium, the chlorine/chloride present in the solution was in the form of hypochlorous acid, which possesses a high oxidation potential compared with hypochlorite (Mussa *et al.*, 2015; Turro *et al.*, 2011).

$$Cl^- \rightarrow \frac{1}{2}\,Cl_2 + e^- \tag{8.6}$$

$$Cl_2 + H_2O \rightarrow HOCl + H^+ + Cl^- \tag{8.7}$$

$$HOCl \rightarrow OCl^- + H^+ \tag{8.8}$$

The ammonia nitrogen removal efficiency by indirect EO depends on chloride concentration in landfill leachate (Cabeza *et al.*, 2007). As a consequence, chlorinated derivatives of ammonia are generated (equations (8.9) and (8.10)) and further hydrolyzed (equations (8.11) and (8.12)), and finally, the ammonia nitrogen will be completely oxidized (Guo *et al.*, 2022).

$$NH_4^+ + HClO_{(aq)} \rightarrow NH_2Cl + H_2O + H^+ \quad (8.9)$$

$$HClO_{(aq)} + NH_2Cl \rightarrow NHCl_2 + H_2O \quad (8.10)$$

$$NHCl_2 + H_2O \rightarrow NOH + 2H^+ + +2Cl^- \quad (8.11)$$

$$NHCl_2 + NOH \rightarrow N_2 + HClO + H^+ + Cl^- \quad (8.12)$$

8.3.4.3 *Electrolytes*

The conductivity of the electrolyte solution can also be controlled by varying inter-electrode gaps or by the addition of electrolytes (Li *et al.*, 2016). The addition of electrolytes during electrochemical wastewater treatment aims to increase the electrical conductivity of the wastewater to be treated and thus decrease the current density and reduce consumption.

Several electrolytes were added to enhance the conductivity of landfill leachates, such as perchloric acid $HClO_4$, Na_2SO_4, and NaCl. $HClO_4$, as an electrolyte, acts only to reduce energy consumption without involving the evolution of any oxidizing species (Turro *et al.*, 2011). The effect of the addition of Na_2SO_4 as an electrolyte, SO_4^{2-} ion can potentially increase the side reaction of oxygen evolution at the anode surface. Thus, the presence of sulfate may inhibit the electrogeneration of chlorine/hypochlorite resulting in an adverse effect on pollutant removal. However, Na_2SO_4 as an electrolyte had less impact on process performance than NaCl and the formation of $S_2O_8^2$ oxidizing species yielded higher pollutant removal.

Landfill leachate generally contains high conductivity due to its high chloride concentration. The addition of NaCl has been reported as the most influencing parameter for NH_4^+-N removal from landfill leachate (Silveira *et al.*, 2015). In addition, an increase in chloride concentration by the addition of NaCl to landfill leachate enhances process performance and accelerates COD removal up to 5,000 mg L^{-1} of Cl^- through the generation of more chlorine/ hypochlorite (Mussa *et al.*, 2015; Quan *et al.*, 2013; Turro *et al.*, 2011). Then, the chlorine evolution process depends on current density rather than chloride concentration. The addition of NaCl as an electrolyte has been considered more satisfactory than other electrolytes for landfill leachate treatment because it is easily available and low priced (Li *et al.*, 2016).

8.3.5 EO combined with other methods

EO has been used as a pre treatment or post treatment process for the biological treatment of landfill leachate, improving the treatment efficiency (Lei *et al.*, 2007; Tang *et al.*, 2022; Wu *et al.*, 2016; Zolfaghari *et al.*, 2016). Combined EO with biological processes has the advantage of reducing the organic load and energy consumption in the EO process; simultaneously, EO can improve the landfill leachate biodegradability index (Guo *et al.*, 2022).

In addition, other studies have been made on combining EO and physicochemical methods to treat landfill leachate (Misra *et al.*, 2009), such as: chemical coagulation (de Oliveira *et al.*, 2019; Guvenc *et al.*, 2023), EC (Ding *et al.*, 2018; Sato *et al.*, 2021), physical adsorption (Chiang *et al.*, 2001) to treat landfill leachate. The landfill leachate treatment efficiency can also be enhanced by combining ultraviolet radiation to strengthen the EO process (Xiao *et al.*, 2013; Ye *et al.*, 2016).

8.4 EF TREATMENT OF LANDFILL LEACHATE

EF is considered an indirect EO process that combines the EC and Fenton processes intending to increase the degradability of the refractory organic matter present in wastewater. Its advantages over the traditional methods using Fenton's reagent are one or two Fenton's reagents, H_2O_2 and Fe^{2+} can be generated in situ by electrode reactions, the rapid regeneration of Fe^{2+} at the cathode, and lower operating costs (Crispim *et al.*, 2022). In addition, it is easy to achieve automatic control (Wang *et al.*, 2020a). EF process reduces other disadvantages of Fenton processes, such as slow ferrous regeneration rate, sludge generation and increase in solution pH with reaction time (Sruthi *et al.*, 2018). EF is an efficient process in removing COD, NH_3-N, color, and also removing high concentrations of pathogenic microorganisms from stabilized leachate (Aziz *et al.*, 2013).

The EF process can be divided into homogeneous and heterogeneous reactions. Homogeneous process, Fe^{2+} ions can react with H_2O_2 generated in situ. In a heterogeneous process, solid catalysts or integral cathodes were used (Wang *et al.*, 2022). The heterogeneous EF process presents the advantages of regenerating the catalyst, overcoming the main limitation of using ferrous salts in the conventional Fenton process (Sruthi *et al.*, 2018), and reducing iron sludge (Chu *et al.*, 2020; He *et al.*, 2021). In addition, using heterogeneous EF can increase the BOD_5/COD of landfill leachate, effectively removing organic contaminants in mature landfill leachate (Wang *et al.*, 2022). In heterogeneous EF, the catalyst can be reused (He *et al.*, 2021).

8.4.1 Reaction mechanism of EF process

The Fenton process (Fe^{2+}/H_2O_2) was one of the advanced oxidation processes selected for treating of pollutants in wastewater. In this process, the dosage of iron and hydrogen peroxide are the two most important operating variables that determine operating costs and efficiency. Figure 8.4 shows a schematic diagram of an EF cell showing the main reactions involved.

The objective is the production of the hydroxyl radical, which is a strong oxidizing agent, following the reaction between Fe^{2+} and H_2O_2 in acidic conditions $2 < pH < 4$ (eq (8.13)).

$$Fe^{2+} + H_2O_2 \rightarrow Fe^{3+} + OH^- + HO^{\cdot} \tag{8.13}$$

In the electrically assisted Fenton process, there is an increase in the oxidizing power of the H_2O_2 and, therefore, a more significant generation of hydroxyl radicals, which ensures an improvement in the elimination of pollutants. The EF process has four ways to provide Fe^{2+} and/or H_2O_2: in the first, Fe^{2+} are added externally, and H_2O_2 is generated simultaneously at the cathode. In the second, H_2O_2 is added externally, and Fe^{2+} are generated at the sacrificial iron anode; in the third, H_2O_2 is added externally, and Fe^{2+} are generated by reducing

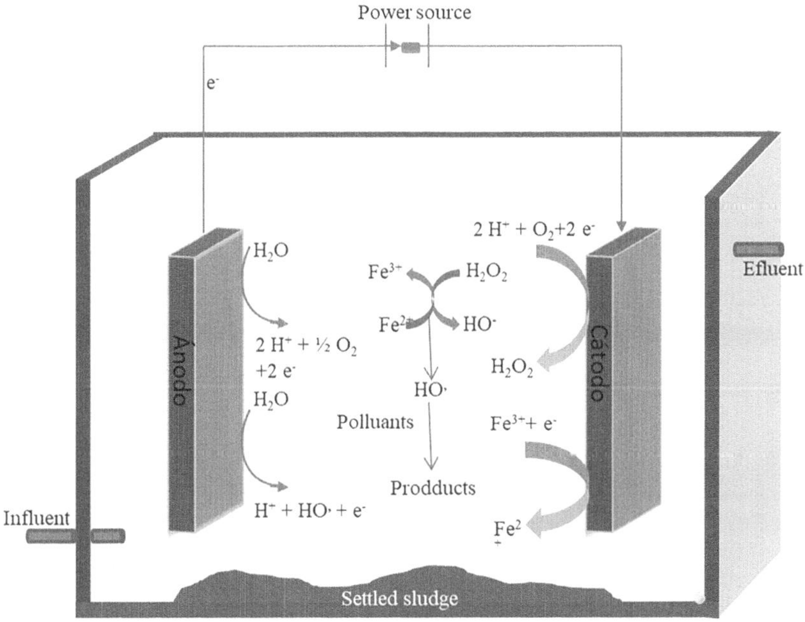

Figure 8.4 Schematic diagram of an EF cell showing the main reactions involved in an EF process. (Adapted from Nidheesh and Gandhimathi, 2012.)

Fe^{3+} or hydroxide sludge. In the fourth both Fe^{2+} and H_2O_2 are electrogenerated at sacrificial electrodes; Fe^{2+} are electrogenerated at the sacrificial iron anode, and H_2O_2 is electrogenerated at the cathode via the two-electron reduction of introduced oxygen gas in the liquid (equations (8.14)) (Zhang *et al.*, 2006).

$$O_2 + 2H^+ + 2e^- \rightarrow H_2O_2 \tag{8.14}$$

In the EF process and when Fe^{2+} ions are electrochemically produced from the sacrificial anode, iron can be regenerated by the following reactions (equations (8.15)–(8.17)):

$$H_2O_2 + Fe^{3+} \rightarrow Fe^{2+} + HO_2 + H^+ \tag{8.15}$$

$$Fe^{3+} + HO_2 \rightarrow Fe^{2+} + H^+ + O_2 \tag{8.16}$$

$$Fe^{3+} + e^- \rightarrow Fe^{2+} \tag{8.17}$$

8.4.2 Operational parameters affecting EF process

In this section, the effects of various operating parameters such as initial pH, dose, treatment time, current density, and distance between the electrodes on treatability of landfill leachate by using EF method are examined. In summary, Table 8.3 lists the works published on applying the EF process to the treatment of landfill leachate at optimal conditions.

Table 8.3 Summary of results of some studies on EF of landfill leachates.

Landfill Leachate Type	Anode/Cathode	Experimental Conditions	Raw Leachate (mg/L)		Removal Efficiency (%)		References
			COD_0	AN_0	COD	AN	
Landfill leachate	Nb/BDD/carbon felt polytetrafluoroethylene	Time: 240 min, c.d.: 90 mA/cm^2, pH: 3	2,684.7	2028.1	66	–	Crispim *et al.* (2022)
Landfill leachate diluted by EF effluent	Ti/RuO_2–IrO_2/Ti/ RuO_2-–IrO_2($FeSO_4$ and H_2O_2)	Time: 50 min, U: 5.5 V, pH: 4, H_2O_2/COD: 1.5, H_2O_2/ Fe^{2+}: 2.5	1,780–1,840	–	93.90	–	Li *et al.* (2022)
Mature membrane concentrated landfill leachate	DSA /iron-loaded needle coke-based titanium mesh composite	Time: 360 min, U: 10 V, pH: 3.6, d:1.4 cm	10,000	–	71.9	–	Gao *et al.* (2022)
Old landfill leachate	Stainless steel/ Stainless steel (Fe_3O_4 NPs@MOF)	Time: 120 min, c.d.: 15 mA/cm^2, pH: 3, Catalyst dosage: 0.75 g/L, d: 1.5 cm	1,600–1,800	–	66.67	–	He *et al.* (2021)
Landfill leachate nanofiltration concentrate	DSA/Fe-loaded, activated C-based Ni-foam composite Fe/ AC/Ni	Time: 300 min, U: 10 V, pH: 2.81, d: 1.53 cm	2,567.39	3276	77.24	70.6	Wang *et al.* (2020a)
Landfill leachate nanofiltration concentrate	DSA/Fe-loaded, activated C-based Fe/ AC/Ti	Time: 280 min, U: 10 V, pH: 7, d: 1cm	1,418.24	–	97.06	–	Wang *et al.* (2020b)

Landfill leachate nanofiltration concentrate	Fe /Fe + (H_2O_2)	Time: 30.3 min, c: 2.72 A, pH: 2.9, H_2O_2/COD; 1.42	5,250	–	60.8	–	Varank *et al.* (2020)
Nanofiltration concentrate of biologically stabilized landfill leachate	Fe /Fe + (H_2O_2)	Time: 30 min, c: 2.75 A, pH: 3.5, H_2O_2/COD: 1.25	4,980	–	69.4	–	Yazici Guvenc *et al.* (2019)
Stabilized landfill leachate	TiO_2/Ti/ Graphite + (FeMoPO nanoparticles)	Time: 90 min, U: 5 V, pH: 2; d: 3 cm Catalyst dose: 50 mg/L	7,184		82		Baiju *et al.* (2018)
Stabilized landfill leachate	Graphite/ Graphite + (iron-manganese binary oxide loaded zeolite (IMZ))	Time: 90 min, U: 4 V, pH: 3; d: 3 cm, Catalyst dose: 25 mg/L	6,160	–	87.5	–	Sruthi *et al.* (2018)
Mature landfill leachate	Ti/RuO_2–IrO_2–SnO_2–TiO_2/Ti mesh + (Fe^{2+} + H_2O_2)	Time: 40 min, c: 1 A, pH: 3; d: 2 cm, H_2O_2/Fe^{2+}: 6	2,720	2,850	72	26	Zhang *et al.* (2012)
Filtered landfill leachate	Fe/Fe + (H_2O_2)	Time: 20 min, c: 2 A, pH: 3; d: 1.8 cm, H_2O_2 dose: 2,000 mg/L	2,350	310	72	28	Atmaca (2009)

8.4.2.1 pH

pH is the main factor on iron speciation and H_2O_2 decomposition and therefore limits the efficiency of EF (Zhang *et al.*, 2006). The optimized range is limited between 2 and 4 because the production of the hydroxyl radical is high in the regeneration of H_2O_2 or increases in the rate of the reaction. The study of hydroxyl radical formation as a function of pH, shows that the highest yield of $^{\cdot}OH$ was observed at pH 3.1 (Lindsey & Tarr, 2000). Figure 8.5 represents total moles of $^{\cdot}OH$ formed as a function of pH and fulvic acid (FA) content.

Fe^{3+} ions mainly control the solubility equilibrium of iron at a pH of less than 3.5. At pH < 3, a significant amount of total iron exists in Fe^{2+} and Fe^{3+} forms, favoring the EF process at pH around 3 (Brillas, 2022; Zhang *et al.*, 2012). At low pH (<2), the reaction of excessive H^+ and H_2O_2 could generate $H_3O_2^+$, competing with the reaction in the formation of $^{\cdot}OH$ (Yang *et al.*, 2022). In addition, at pH > 4, H_2O_2 is unstable and rapidly decomposes to oxygen and water, reducing the efficiency of EF (Yazici Guvenc *et al.*, 2019). In parallel, homogeneous EF, the iron can precipitate at pH > 4, to overcome the drawbacks, solid materials containing iron as catalysts (Gao *et al.*, 2022).

8.4.2.2 Dose of reagents

In the Fenton and EF processes, the Fe^{2+}/H_2O_2 ratio is essential regarding process cost and efficiency. An excess of hydrogen peroxide can give rise to reactions that interfere with and limit the efficiency of the EF process. In these reactions, hydrogen peroxide is consumed for the oxidation of Fe^{2+} ions and produce $HO_2^{\cdot}$ radicals that have one less oxidizing capacity than hydroxyl radicals $^{\cdot}OH$ (equations (8.18) and (8.19)):

$$H_2O_2 + HO^{\cdot} \rightarrow HO_2^{\cdot} + H_2O \tag{8.18}$$

$$Fe^{2+} + HO_2^{\cdot} \rightarrow Fe^{3+} + HO_2^{-} \tag{8.19}$$

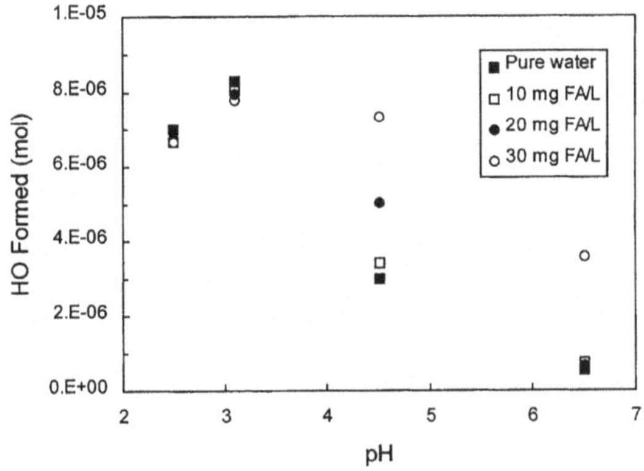

Figure 8.5 Total moles of $^{\cdot}OH$ formed as a function of pH for pure water and aqueous FA. Benzoic acid used as probe at 9 mm. Measurements made 300 s after mixing Fe^{2+} and H_2O_2. (Reprinted from Lindsey and Tarr, 2000, with permission of Elsevier.)

In the case that Fe^{2+} is in excess, it produces an extra amount of sludge and an increase in total dissolved solids. These interfering reactions can be hindered by using the proper Fe^{2+}/H_2O_2 ratio and Fe^{3+}/H_2O_2 ratio and, optimized initial pH.

In parallel, the increase in H_2O_2 concentration leads to an increase in hydroxyl radical concentration and, consequently, high efficacy in removing pollutants from wastewater (Zhang *et al.*, 2012). In addition, H_2O_2/COD ratio is also influenced by COD removal efficiency and, in the literature, ranges between 0.4 and 2 (Varank *et al.*, 2020). At values $H_2O_2/COD > 2$, COD removal efficiency decreased due to ·OH scavenging reactions (Yazici Guvenc *et al.*, 2019).

8.4.2.3 Reagent feed mode

The H_2O_2 feed mode at the beginning, stepwise or continuously during the experiment implies a change in the H_2O_2/COD and H_2O_2/Fe^{2+} ratio, consequently, a change of COD removal rate. The COD removal efficiency increases with dose time, and a high value can be reached continuously. Adding H_2O_2 in a single stage increases the production efficiency of hydroxyl radicals. In the case of the continuous mode, the concentration of peroxide of hydrogen is kept at low levels, thus avoiding scavenging reaction (Zhang *et al.*, 2006). In EF, a batch reactor and a continuous-flow reactor (Crispim *et al.*, 2022; Yazici Guvenc *et al.*, 2019) have been used. In the case of a continuous operation at a steady state, increasing the hydraulic retention time implies that the treatment efficiency increases because leachate has more chance to react with the reactive oxidants (Li *et al.*, 2022).

8.4.2.4 Current density

The applied current is among the most critical factors affecting EF's efficiency. This parameter depends strongly on the conductivity of the wastewater. Applying higher current increases the amount of H_2O_2 produced and means higher electro-regeneration of ferrous ions from ferric ions (Yazici Guvenc *et al.*, 2019). However, a high applied current can inhibit the EF process over a limited value of the current is applied with the following competing reactions (8.20) and (8.21) at the electrodes (Baiju *et al.*, 2018; Yang *et al.*, 2022).

$$2H_2O_2 \rightarrow 4H^+ + O_2 + 4e^- \tag{8.20}$$

$$2H^+ + 2e^- \rightarrow H_2 \tag{8.21}$$

8.4.2.5 Inter-space electrode

In EF reaction, the electrode spacing influences COD removals. When the electrode spacing was too close, concentration polarization could inhibit contaminants transfer in the reaction and then reduce COD removal. It was also observed that reducing electrode spacing increases significantly COD removal efficiency (Atmaca, 2009). In contrast, too long a distance increases reaction resistance and decreases substrate transfer (He *et al.*, 2021). In addition, with the increase of the spacing of the electrode, the current density also decreased, which in turn reduced the electron transfer rate on the cathode surface, therefore, a reduced amount of ·OH generation and mass transfer decreases Fe^{3+}, and consequently decreases COD removal efficiency (Gao *et al.*, 2022). Moreover, increasing the electrode spacing from 3 to 4 cm and mass transfer decrease of Fe^{3+} led to a reduction in COD removal. When the electrode spacing was reduced to 2 cm, the efficiency of COD removal due to the electro-generation of Fe^{2+} to Fe^{3+} is faster at the anode (Baiju *et al.*, 2018). Therefore, the electrode spacing should be optimized for low energy expenses and to avoid short-circuiting (Gao *et al.*, 2022). In addition, the optimization of electrode spacing is of great importance to the highest COD removal when EF operated in batch mode (Zhang *et al.*, 2012).

8.4.2.6 Cathode material

The appropriate cathode material is essential in the heterogeneous EF process (Monteil *et al.*, 2019). Carbonaceous materials are preferred as cathode materials because they can produce H_2O_2 from O_2 reduction. Moreover, the cathode must exhibit high over potential for H_2 evolution and low catalytic activity for H_2O_2 decomposition since the process is optimal in an acidic medium (Brillas *et al.*, 2009).

Currently, three-dimensional carbon materials have been used as cathodes, including carbon felt (Yang *et al.*, 2020), carbon aerogel (Lu *et al.*, 2019), carbon nanotubes (Cheng *et al.*, 2020), carbon sponge (Oturan *et al.*, 2021), and activated carbon fibre (Yu *et al.*, 2019). The cathode base carbon materials are characterized by a high specific surface area and good porosity, which are conductive to loading iron-active species. Nevertheless, the loaded cathode also has some limitations, such as poor conductivity, poor repeatability, and comparative complex preparation (Gao *et al.*, 2022).

More recently, carbon felt modified by anthraquinone 2-sulfonate and polypyrrole (Zhu *et al.*, 2020), carbon felt-polytetrafluoroethylene (CF-PTFE) (Crispim *et al.*, 2022), biochar activated by a molten salt (Sun *et al.*, 2021), microchannel-structured carbon decorated with iron oxides (Yu *et al.*, 2020b) and graphene-modified carbon felt (Yang *et al.*, 2019) iron-loaded needle coke-based titanium mesh composite (Gao *et al.*, 2022), Fe-loaded-activated C-based Ni-foam composite Fe/AC/Ni (Wang *et al.*, 2020a) Fe-loaded, activated C-based Fe/AC/Ti (Wang *et al.*, 2020b) have been investigated as suitable agents for a cathode to generate H_2O_2.

8.4.3 Combination of EF with other methods

Given the complexity of landfill leachate, adopting multiple strategies can improve their treatments. EF has been widely used to treat leachate as a single process, or combined with UV to strengthen the EF.

The combination of EF and the biological process can improve the treatment of mature leachates. The low biodegradability of mature leachates limits their biological treatment. The application of EF prior to the biological process enhances the biodegradability of leachate, which will be beneficial for the subsequent biological process (Baiju *et al.*, 2018; Lin & Chang, 2000).

The integration of UV in EF is called photo-electro-Fenton (PEF) process. It can improve the effectiveness of the process, producing a significant regeneration rate of Fe^{2+} and consequently increasing ·OH in the solution (Altin, 2008; Brillas, 2022). The performance of PEF has some limitations related to wastewater matrix effects and energy requirements (Crispim *et al.*, 2022). In addition, the integration of UV during the combination of EF and biological treatment can improve the performance of the process (Pellenz *et al.*, 2020; Seibert *et al.*, 2019).

Recently, bio-electro-Fenton (BEFs) has been an emerging technology based on coupled anodic microbial metabolisms and electrochemical Fenton reactions. BEFs have paid a potential interest due to multiple advantages such as hydrogen peroxide generation, energy saving, high efficiency, and energy production, low toxicity, and mild operational conditions (Soltani *et al.*, 2021; Wang *et al.*, 2019; Yang *et al.*, 2022). Moreover, there are some disadvantages, including low production of current density, a chemical addition, iron sludge formation, and costly materials used (Soltani *et al.*, 2021).

8.5 EC TREATMENT OF LANDFILL LEACHATE

8.5.1 Definition and operating principle

EC is an alternative technique to conventional effluent treatment systems and is based on the principle of electrolysis since it combines chemical coagulation with electrical principles. EC can be defined as a non-selective process in which the particles are destabilized by coagulants that are suspended, emulsified, or dissolved in different types of wastewaters. Applying an electric current through electrodes immersed in the effluent to be treated gradually dissolves the sacrificial anode in the reaction process, thereby generating many high-charge metal cations in situ (Guo *et al.*, 2022). Figure 8.6 shows the scheme of the EC process.

The application of current produces, in addition to other electrochemical reactions, electrolysis of water with the consequent formation of tiny bubbles of hydrogen at the cathode and of oxygen at the anode. The gas bubbles adhere to the suspended particles and drag them to the surface of the solution, forming a foam. This process is called electroflotation, whose bubbles are very stable, so they retain their large contact surface. In addition, neutralizing the charges of the suspended matter produces the breaking of the emulsions and better and faster flocculation of the particles. To sum up, in EC, pollutants can be removed by adsorption, precipitation, and floatation (Das *et al.*, 2022; Mollah *et al.*, 2004). The principal reactions occurring in an EC cell can be described in the following equations (8.22)–(8.24) (Dia *et al.*, 2017b):

$$\text{At the anode:}\quad M_{(s)} \rightarrow M^{n+}_{(aq)} + ne^- \tag{8.22}$$

$$\text{At the cathode:}\quad 2H_2O + 2e^- \rightarrow 2OH^- + H_2 \tag{8.23}$$

$$\text{In the bulk solution:}\quad M^{n+}_{(aq)} + nOH^- \rightarrow M(OH)_{n(s)} \tag{8.24}$$

where $M_{(s)}$ metal, $M^{n+}_{(aq)}$ metallic ion (Fe or Al ion), $M(OH)_{n(s)}$ metallic hydroxide, and ne^- the number of electrons transferred in the reaction at the electrode.

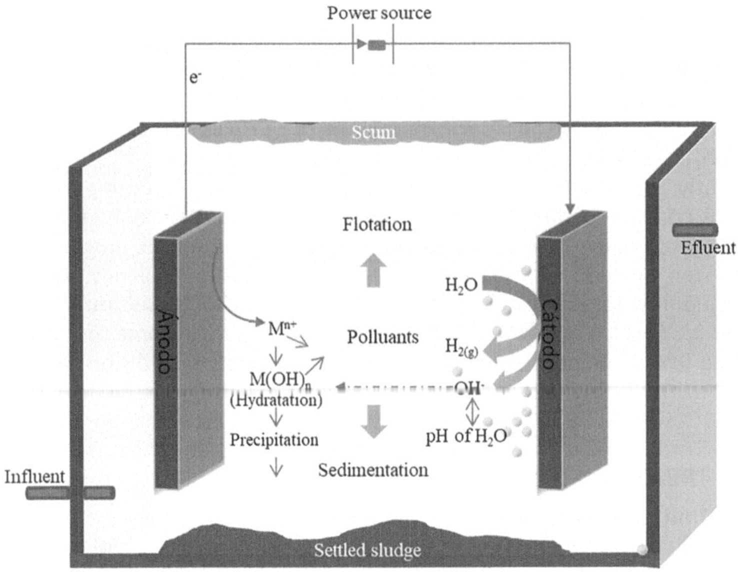

Figure 8.6 Schematic diagram of the EC process.

During an EC process, a sacrificial anodic metal electrode produces the coagulating ions in situ, and three different stages can be identified (Alkhatib *et al.*, 2020; Fernandes *et al.*, 2015):

(i) formation of the coagulants by electrolytic oxidation of the 'sacrificial electrode,'
(ii) destabilization and neutralization of the contaminants and particulate suspension and breaking of emulsions,
(iii) aggregation of the destabilized phases to form flocs that can be settled down by gravity.

The EC process has been successfully applied to landfill leachate treatment due to its versatility and efficiency in removing COD, ammonia, total dissolved solids, total suspended solids, turbidity, color, and other contaminants from landfill leachate (Xu *et al.*, 2020). In addition, EC has gained potential attention over conventional wastewater treatment methods for many advantages, such as ease of implantation and automation and ease in solid–liquid separation. It requires no chemical additives and has a simplicity of equipment, short retention time and low sludge generation and inexpensive, compact facility, and eco-friendly (Alkhatib *et al.*, 2020; Bazrafshan *et al.*, 2015; Naje *et al.*, 2019).

On the other hand, EC presents drawbacks, such as the need to replace periodically sacrificial anodes, conductivity must be high to avoid ohmic drop, the presence of a high concentration of chlorides can lead to the formation of toxic chlorinated organic compounds, an impermeable oxide film formed on the cathode surface (passivation), and a cost of electricity (Bazrafshan *et al.*, 2015).

The advantages of ECs outweigh its drawbacks, which are recommended as an efficient process for wastewater treatment.

8.5.2 EC reactor design and operation

The EC reactor designs, such as high porosity fixed bed, rotating cylinders, and vertical and horizontal plates, have gained significant attention in recent years for wastewater treatment (Das *et al.*, 2022). EC reactor designs can be classified based on three significant distinctions (Holt *et al.*, 2005). The first distinction, whether the reactor operates in a batch-wise or a continuous mode between alternative designs, is whether a reactor was configured as a batch or a continuous system. Batch mode, in which the output depends on operating time working with a fixed wastewater volume, is more suited for pilot plant scale application (Khandegar & Saroha, 2013). The optimized parameters in the batch mode serve as guidelines for operating in the continuous mode. Batch systems present a disadvantage because of conditions within the reactor change with time (Holt *et al.*, 2005). Continuous systems, which are inherently dynamic in the application and operate under steady-state conditions, especially at fixed pollutant concentration and inlet flow rate, are better at the industrial scale for large effluent volumes (Bazrafshan *et al.*, 2015). The outcome is not time-dependent but determined by inlet flow rate and space–time (López-Guzmán *et al.*, 2021). The continuous system of operation is preferred due to being more controlled than the batch mode. Moreover, a steady-state process consumes less energy in a shorter time (Khandegar & Saroha, 2013).

The second distinction is whether the feed flow is horizontal or up flow (Jiang *et al.*, 2002). The third distinction is whether the aggregated pollutants are separated in situ using flotation or sedimentation processes or using a suitable downstream process such as centrifugation or flotation (Al-Qodah & Al-Shannag, 2017). Figure 8.7 presents summary of classification of EC reactor systems.

The wastewater flow through the plate electrodes can be divided into multiple or single channels (Figure 8.8). Multiple channels are simple but offer smaller flow rate, as well as some drawbacks, such as electrode passivation. On the other hand, a simple channel is recommended to increase the flow rate (Chen, 2004).

8.5.3 Operational parameters affecting EC process

The EC process is affected by different factors. Among the most important parameters are pH, conductivity, electrode material, arrangement, electrode spacing, and current density. These parameters determine the efficiency of EC to remove pollutants from landfill leachates by the formation of flocs. Figure 8.9 presents operating parameters influencing the EC process performance.

8.5.3.1 Current density

The current density is one of the most significant factors for controlling the reaction rate in the electrochemical process (Rookesh *et al.*, 2022). The applied current density supply to the EC system determines the amount of Fe^{3+} and Al^{3+} ions released by the respective electrodes, thus affecting the anode

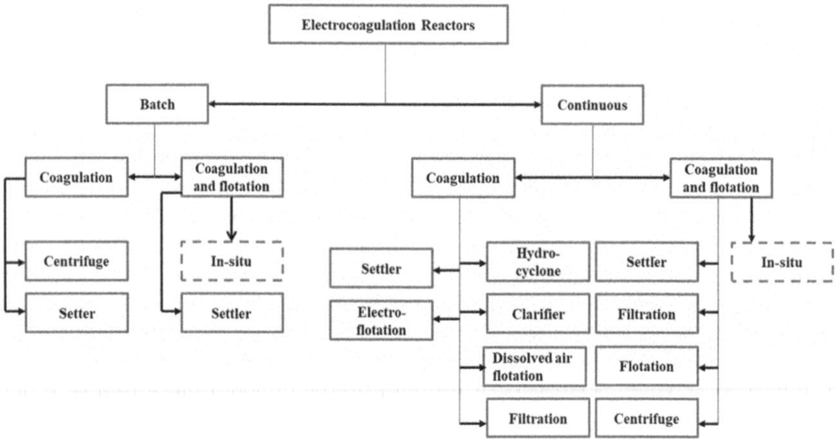

Figure 8.7 Summary of classification of EC reactor systems. (Adapted from Holt *et al.*, 2005, with permission of Elsevier.)

mass transfer process and pollutant removal. Optimizing the current density determines the amount of coagulant, the generation of flotation gas, and the formation of flocs.

According to Faraday's law, which governs the EC process, the number of substances formed at an electrode is proportional to the number of charges passing through the system, and the total number of moles of substance formed at an electrode is stoichiometrically related to the current density put into the system (equation (8.25)).

$$m = \frac{ItM}{nF} \tag{8.25}$$

where m is the amount of electrode material dissolved (g), I is the current (A), t is the electrolysis time (s), M is the molecular mass of the electrode material (g/mol), n is metal valence, and F is the Faraday constant (96,500 C/mol).

In general, an increase in the current density generates an increase in the removal rate of contaminants (Ilhan *et al.*, 2008). Nevertheless, a blocking effect will occur when it exceeds the optimal value. Working rotating electrode EC reactor on landfill leachate treatment, also increases current density COD, total suspended solids and total dissolved solids removal will increase, but, when current density is above 2A, no effect in removal rate was observed. In addition, the same effect was observed on the disinfection by-products produced by EC in the treatment of landfill leachate (Xu *et al.*, 2020). Increasing the working current will increase energy consumption and reduce current efficiency (Naje *et al.*, 2019). In addition, a current density that is too high can produce a significant decrease in efficiency and a loss of energy due to the transformation of electrical energy into heat, increasing the temperature of the water to be treated (Chen, 2004).

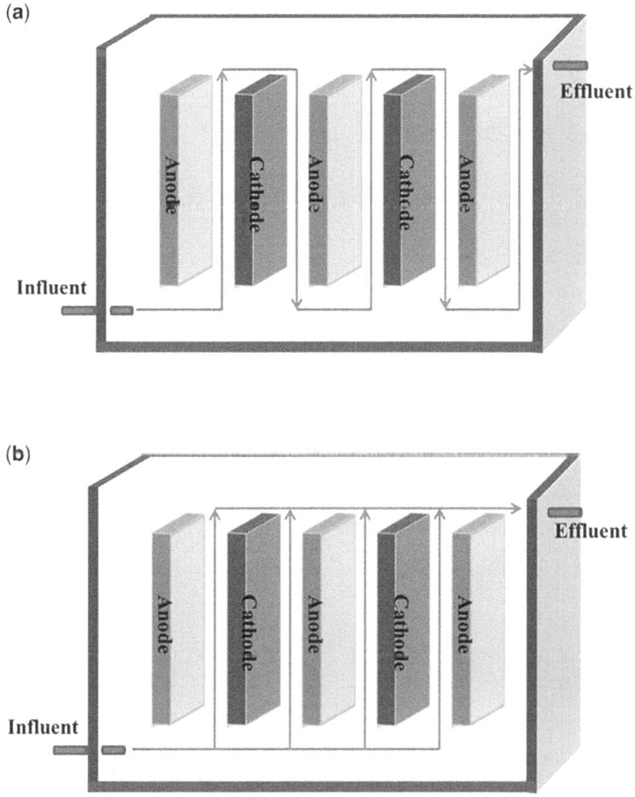

Figure 8.8 Flow types according to the arrangement of the electrodes: (a) single channel; (b) multiple channels. (Adapted from Chen, 2004.)

8.5.3.2 Solution pH

pH value is a crucial factor influencing the EC process performance governing the hydrolyzed metal species. The pH influences the current's efficiency in the metal's solubility process to form hydroxide. The pH varies during EC, depending on the initial pH and the electrode material (Rookesh *et al.*, 2022). Generally, the best removals have been obtained for pH values in the neutral range (Li *et al.*, 2011; Yazici Guvenc *et al.*, 2019). Meanwhile, the EC process performance declined when the pH tended toward acidic either basic values (Naje *et al.*, 2019).

8.5.3.3 Electrode design

In the EC process, the removal of pollutants was attained by sacrificing the anode to form metal ions with high coagulation force. The most common electrode materials used for EC are Al and Fe electrodes, has been widely used to treat landfill leachate (Huda *et al.*, 2017; Naje *et al.*, 2019).

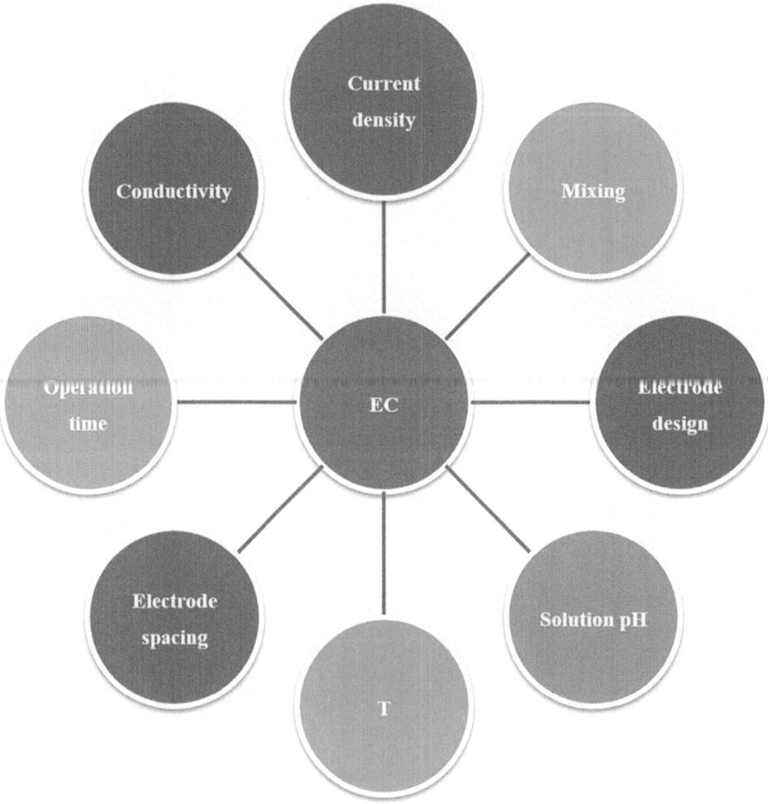

Figure 8.9 Operating parameters influencing the EC process performance.

During the treatment of landfill leachate, the highest removal efficiency was achieved with the Al electrode at the lowest current density of 5 mA/cm^2 and amounts of 22.41%; however, under a higher current density of 25 mA/cm^2, and removal rate of ammonia is 29.58%, the Fe electrode is more efficient than the use of Al electrode (Malinovic *et al.*, 2019). In another study, the Al electrodes' results indicated that the Al electrode's performance was better than that of the Fe electrode in terms of COD and NH_4-N removal, and COD and NH_4-N removal efficiencies were 66% and 63%, respectively (Pirsaheb *et al.*, 2016). However, the Al electrode suffers passivation damage to the reactor's performance, presenting higher toxicity and operating cost than the Fe electrode (Fernandes *et al.*, 2015).

The presence of some anions like Cl^- in an aqueous solution can inhibit the electrode passivation. Some anions, such as HCO_3^- and SO_4^{2-} may precipitate Ca^{2+} and Mg^{2+}, forming a layer on the surface of the electrodes. This effect can significantly be reduced in the presence of chloride ions (Chen, 2004). In addition, several strategies have been proposed to minimize the passivation, including aggressive ion addition (Cl^- or Br^-), alternating current operation, polarity reversal, ultrasonication, and mechanical cleaning of the electrodes (Ingelsson *et al.*, 2020). Table 8.4 summarizes results of some studies on EC of landfill leachates

The shape of the electrodes can also influence the efficiency of EC. Recently, several forms have been designed including, cylindrical electrode (Alkhatib *et al.*, 2020), porous electrode (Ibrahim *et al.*, 2020), novel rotating electrode (Naje *et al.*, 2019), and serpentine anode (Abdel-Aziz *et al.*, 2020). Guo *et al.* (2022) reviewed that changing the shape of the electrode has the advantages of increasing the contact area between the pollutants and the electrode surface, increasing the floc release rate into the bulk solution, and reducing the electrode passivation rate. Consequently, of these benefits, landfill leachate treatment could be improved.

EC cells can present different electrodes' arrangements which can distinguish three types of connexion modes (Mollah *et al.*, 2004; Pretorius *et al.*, 1991):

- Monopolar electrodes with cells in series (Figure 8.10a): In series cell arrangement, the same current intensity flows through all the electrodes and a higher potential difference is required because the cells connected in series have higher resistance. Monopolar connections in series show that n electrodes present $n/2$ resistance in series. Applying a voltage U and a current intensity I, each resistance is traversed by I and subjected to a voltage $U/2n$.
- Bipolar electrodes with cells in series (Figure 8.10b): In bipolar electrodes with cells in parallel arrangement, the sacrificial electrodes are placed between the two parallel electrodes without any electrical connection. This type of connection has characteristics close to the previous connection. The same current passes through the resistance subjected to a voltage of $U/n-1$. For n electrodes, there are $(n-1)$ resistances.
- Monopolar electrodes with cells in parallel (Figure 8.10c): In parallel connections arrangement, the electric current is divided between all the electrodes, in relation to the resistance of the single cells. The monopolar connection is in parallel, by n electrodes that give $n-1$ resistances mounted in parallel. If a voltage U and a current of I are applied. The cells are subjected to the same voltage U. The current that passes through each resistance is $I/n-1$.

Fernandes *et al.* (2015) reviewed the different arrangements used for landfill treatment. They concluded that most of the EC experiments performed were in a batch reactor with monopolar electrodes in parallel connection, varying the number of electrodes between 2 and 14.

8.5.3.4 Inter-electrode distance

The distance between the electrodes affects the performance of the EC process and the electrostatic field effect (Bazrafshan *et al.*, 2015). The inter-distance depends on several parameters such as electrode structure, hydrodynamic conditions, and pollutant nature (Daneshvar *et al.*, 2004; Modirshahla *et al.*,

Table 8.4 Summary of results of some studies on EC of landfill leachates.

Landfill Leachate Type	Anode/Cathode	Experimental Conditions	Raw Leachate (mg/L)		Removal Efficiency (%)		References
			COD_0	AN_0	COD	AN	
Medium-age landfill (pre-treatment with lime)	Fe/Graphite	Time: 120 min, c.d: 64 A/cm^2, pH: 7.5, d: 1.5 cm,	22,371	385	30.63	83.67	Rookesh *et al.* (2022)
Landfill leachate	Al/Fe	Time: 60 min, c: 4 A, pH: 5.4, d: 3.5 cm, T; 40°C	19, 120	–	71.78	–	Sediqi *et al.* (2021)
Landfill leachate	Fe/Fe	Time: 60 min, c.d: 100 mA/cm^2, pH: 6.44, d: 1 cm,	44,900 ± 700	2,000 ± 100	21.8 ± 1.4	20 ± 5	Xu *et al.* (2020)
Landfill leachate	Fe/Fe	Time: 60 min, c.d: 100 mA/cm^2, pH: 7.73, d: 1 cm,	7,800 ± 100	2,200 ± 100	14.1 ± 2.6	20 ± 5	Xu *et al.* (2020)
Landfill leachate	Al/Al	Time: 40 min, c: 2 A, pH: 7, d: 1 cm, speed: 150 rpm	2,130	–	94.5	–	Naje *et al.* (2019)
Leachate nano filtration concentrate	Fe/Fe	Time: 20 min, c: 3.5 A, pH: 7	4,980		57.4		Yazici Guvenc *et al.* (2019)
Raw landfill leachate	Fe/Fe	Time: 60 min, c: 1 A, pH: 7.73, d: 1.16 cm, NaCl: 2 g/L, stirring: 100 rpm	7,230	–	45.1		Huda *et al.* (2017)
Bio-filtrated landfill leachate	Fe/stainless steel	Time: 80 min, c.d.: 10 mA/cm^2, pH: 8.94	596	–	70		Dia *et al.* (2017b)
Bio-filtrated landfill leachate	Al/stainless steel	Time: 80 min, c d.: 8 mA/cm^2, pH: 8.94	596	–	65		Dia *et al.* (2017b)

Landfill leachate	Al/Al	Time: 80 min, c.d.: 20 mA/cm^2, pH: 8	9,800	3, 710	60.5	24.3	Kabuk *et al.* (2014)
Landfill leachate	Fe/Fe	Time: 180 min, c.d.: 30 mA/cm^2, pH: 6.54, d: 0.9 cm	11, 000	–	65.8	–	Orkun and Kuleyin (2012)
Landfill leachate	Fe/Fe	Time: 30 min, U: 5 V, pH: 8.6, d: 2.3 cm	12, 650	–	33	–	Norma *et al.* (2012)
Landfill leachate	Al/Al	Time: 100 min, U: 12 V, pH: 8, d: 6.5 cm	16, 464	–	62.7	–	Jotin *et al.* (2012)
Landfill leachate	Fe/Fe	Time: 100 min, c.d.: 63,1 mA/cm^2, pH: 8.2, d: 6.5 cm	12, 860	2,240	35	11	Ilhan *et al.* (2008)

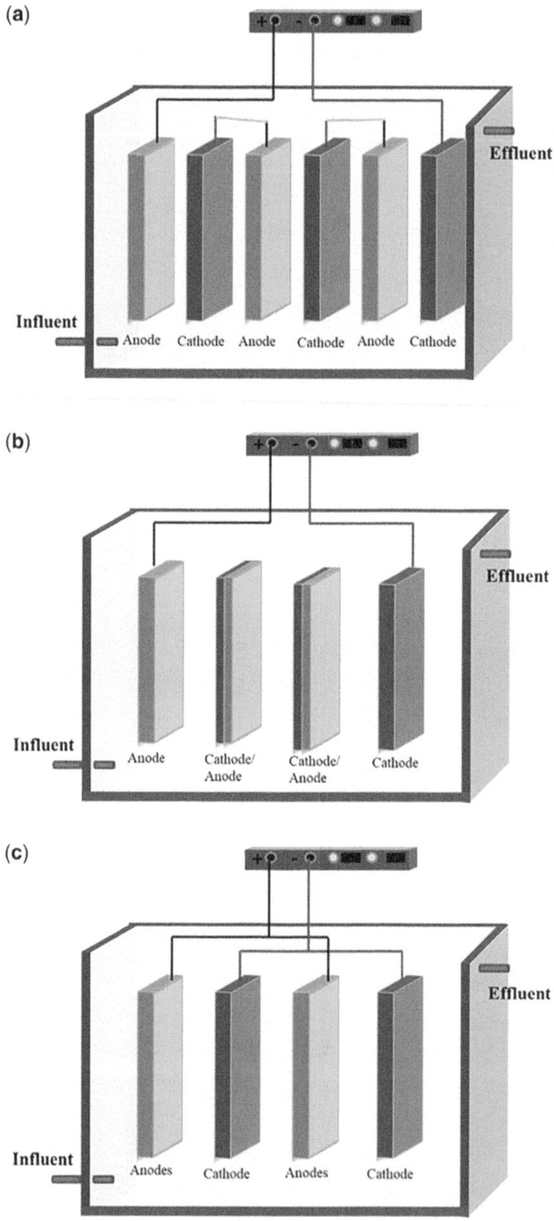

Figure 8.10 Different electrodes' arrangement within the electrochemical cell: (a) monopolar electrodes with cells in series, (b) bipolar electrodes with cells in series, (c) monopolar electrodes with cells in parallel.

2007; Naje *et al.*, 2019). When the inter-distance increases, ohmic drop (IR) increases, as described by the following equation (8.26):

$$IR = \frac{Id}{Ak} \tag{8.26}$$

where I is the current (A), d is the inter-electrode distance (m), A is the active anode surface (m^2), k is the specific conductivity (S/m).

More recently, a novel rotating anode reactor was used for leachate treatment, with the increase of the distance between the electrodes from 10 mm to 15 and 20 mm, the efficiency of COD, TSS, and TDS removal rate decreases. The performance of the parameters mentioned above is reduced because increasing the inter-distance between the electrodes increases the resistance and leads to a decrease in the ion metal dissolution, thereby minimizing the flocs in the bulk solution (Naje *et al.*, 2019). In addition, when the electrode space is wide, the amount of anode dissolution will decrease, and the ions need to travel for a longer distance to form flocs. Consequently, an EC process needs a high voltage due to an increase in the resistance of the solution. In addition, with higher electrode distances, the temperature increases in the EC cell (Rafiee *et al.*, 2020). Oppositely, reducing the distance space increases the probability of contact between metal ions and hydroxyl radicals produced by the anode, improving the performance of the EC process (Abdel-Gawad *et al.*, 2012; Ghosh *et al.*, 2008). Therefore, it seems that the optimum condition of the space electrodes reduces the operational cost and energy consumption of the treatment (Huda *et al.*, 2017; Naje *et al.*, 2019; Rookesh *et al.*, 2022).

8.5.3.5 Stirring speed

Mixing is also one of the parameters that influence the EC efficacy treatment of landfill leachates. Increasing the mixing negatively affected the COD removal rate, and positively increases the ammonia removal, because ammonia is easily stripped by mixing (Ilhan *et al.*, 2008). Other study reports that when the magnetic stirring speed was increased from 50 to 150 rpm, an increase in COD removal and in sludge production was observed (Bouhezila *et al.*, 2011). The optimum rotating speed is 150 rpm and increasing above this value, the rotating speed leads to fracture of the flocs and produces small particle flocs that are difficult to collect after the treatment process and consequently reducing process performance (Naje *et al.*, 2019; Tahreen *et al.*, 2020).

Indeed, the EC process is based on the production of ions applying current density. The movement of the ions is affected by the mixing parameter. The moderate mixing enhances the contact between the contaminant and coagulants and, consequently, the efficiency of the process. On the other hand, high mixing can cause disturbance of ions mobilization and breakup of flocs (Fernandes *et al.*, 2015). Norma *et al.* (2012) showed that stirring increases the time necessary to start the precipitation of the flocs, which can be reduced by an increase in applied current density and, consequently, a higher rate of iron and energy consumption. Another consequence of increasing current density is an increase of the H_2 bubble generation rate according to Faraday's law which can improve solution mixing and mass transfer (Abdel-Aziz *et al.*, 2020).

8.5.3.6 Electrical conductivity

The current density depends significantly on the conductivity of the wastewater (Yazici Guvenc *et al.*, 2019). The wastewater must have some minimum conductivity for the flow of the electric current. With increases in conductivity, the current density efficiency increases, decreasing the treatment time required and reducing energy consumption (Huda *et al.*, 2017; Mussa *et al.*, 2013; Yazici Guvenc *et al.*, 2019). In addition, the costs of the process are directly related to the solution conductivity (Bazrafshan *et al.*, 2015). When the electrical conductivity of the wastewater is low, it needs to be adjusted by the addition of electrolytes some supporting electrolytes such as NaCl (Li *et al.*, 2011). There is an increase in the current density with an increase in the conductivity of the solution at a constant voltage or reduction of voltage at a constant current density. In addition, the removal efficiency also increases (Bazrafshan *et al.*, 2015; Rafiee *et al.*, 2020). The effect of chloride on the color removal of landfill leachate was studied at 0.5–6.25 g/L at optimum conditions. An increase in the concentration of Cl^- from 0.5 to 6.25 g/L yielded increase in the color removal from 24% to 82%. Mussa *et al.* (2013) recommended that for a normal EC process, there should be >20% chloride ions (Chen, 2004).

8.5.3.7 Temperature

The temperature is another independent parameter influencing EC process. An increase in the solution temperature (25–45°C) improves the performance of the EC treatment of landfill leachate (Naje *et al.*, 2019; Sediqi *et al.*, 2021). In the case of the Al electrode, the increase of EC with temperature was attributed to the increased activity of the destruction of the aluminum oxide layer on the electrode surface (El-Ashtoukhy *et al.*, 2013). In addition, when temperature increases to 27°C, the efficiency of color dye removal increases slightly. The reason could be increased mobility and ion collisions with hydroxide polymer (Modirshahla *et al.*, 2007). However, as mentioned, as temperature increases, dye removal efficiency decreases due to the increased movement of the produced ions reducing the formation of the flocs (Daneshvar *et al.*, 2004).

In the other study, the effect of temperatures (20°C, 30°C, and 40°C) was probed on the leachate treatment, and the highest removal rates of COD, NH_4^+, TSS, turbidity, and color were observed at 40°C. However, as mentioned, as temperature increases, the efficiency of EC decreases due to the increased movement of the produced ions reducing the formation of the flocs (Daneshvar *et al.*, 2004). The inhibitory effect when temperature increases can be related to the coagulation reaction and the exothermic reaction of the electrochemical process. During EC, some of the compounds are converted into persistent intermediates (Ambauen *et al.*, 2019). To sum up increasing the temperature results in a very high operational cost of the process.

8.5.3.8 Electrolysis time

Electrolysis time is one of the most important parameters determining the removal rate of pollutants from landfill leachate. Based on Faraday's law (equation (8.11)), the amount of generated coagulant from the anode is linearly proportional to the current density. In addition, hydroxide flocs increase, also the rate of bubble-generation increases. The removal of pollutants from landfill leachate was by the effects of coagulation and flotation (Li *et al.*, 2011). The removal rate of the pollutant also increases with electrolysis time until it reaches an optimum. Above the optimal time it remains constant because, in the medium, the amount of coagulants available is sufficient for pollutant removal (Al-Qodah & Al-Shannag, 2017).

Fernandes *et al.* (2015) reviewed that the optimal reaction time for COD and NH_4^+ removal was obtained at 10–210 min. Low current density requires a higher time for similar efficiency (Rookesh *et al.*, 2022). Consequently, the electrical energy consumption was increased and was calculated by the following equation (8.27) (Rafiee *et al.*, 2020).

$$E = UIt \tag{8.27}$$

where E is the electric energy in Joule (J), U is the electric potential in volts (V), I is the electric current in amperes (A), and t is the time in seconds (s).

8.5.4 EC process combined with other methods

Recently, to achieve a good treatment effect with EC, some technology has been combined with EC, including the introduction of ozone into the EC reactor (Tezcan Un *et al.*, 2018), ultrasonication (Khoramipour *et al.*, 2021) or the combination of the tertiary treatment (Asaithambi *et al.*, 2020), and applying magnetic field enhancement (Ibanez *et al.*, 2012). Other combination focused on the treatment of mature landfill leachate by EC followed by Fenton or UVA-LED photo-Fenton processes (Tejera *et al.*, 2021).

Other studies have investigated the effect of coupling biological treatment with the EC process to improve the efficiency of landfill leachate treatment. EC can be used as a pre-treatment system combined with anaerobic biological treatment (Tezcan Un *et al.*, 2018) and post-treatment system in aeration biofilters (Dia *et al.*, 2017a; Oumar *et al.*, 2016) and activated sludge treatment systems (Djelal *et al.*, 2015).

The combination of physicochemical technology and EC, have been made on the removal combination of EC and membrane filtration process with EC improving COD removal (Ren *et al.*, 2020; Top *et al.*, 2011). The EC process can be used as a pre-treatment for raw leachates before applying other treatment methods (Huda *et al.*, 2017; Tejera *et al.*, 2021)

8.6 CONCLUSIONS

Landfill leachate is considered one of the most heterogeneous and complex wastewater, and the major challenge of modern society is to reduce its hazardous effect on the environment and the choice of adequate methods of treatment. The electrochemical process has proven to be very versatile for broad types of wastewaters, including landfill leachate. Electrochemical methods: EO, EC, and EF are widely used for landfill leachate treatment. In addition, combination of electrochemical process with biological or physical–chemical methods can enhance the treatment of landfill leachate. Different electrode materials and cell configurations have been used. Under optimum conditions, the electrochemical process efficiently removes two major pollutants, COD, and ammonium nitrogen, from landfill leachate. Nevertheless, it is necessary to consider the cost of the process operated with the use of electricity which the use of renewable energy can overcome.

REFERENCES

Abdel-Aziz M. H., El-Ashtoukhy E.-S. Z., Sh. Zoromba M., Bassyouni M. and Sedahmed G. H. (2020). Removal of nitrates from water by electrocoagulation using a cell with horizontally oriented Al serpentine tube anode. *Journal of Industrial and Engineering Chemistry*, **82**, 105–112, https://doi.org/10.1016/j.jiec.2019.10.001

Abdel-Gawad S. A., Baraka A. M., Omran K. A. and Mokhtar M. M. (2012). Removal of some pesticides from the simulated waste water by electrocoagulation method using iron electrodes. *International Journal of Electrochemical Sciences*, **7**, 12.

Agustina F., Bagastyo A. Y. and Nurhayati E. (2019). Electro-oxidation of landfill leachate using boron-doped diamond: role of current density, pH and ions. *Water Science and Technology*, **79**, 921–928, https://doi.org/10.2166/wst.2019.040

Alkhatib A. M., Hawari A. H., Hafiz M. A. and Benamor A. (2020). A novel cylindrical electrode configuration for inducing dielectrophoretic forces during electrocoagulation. *Journal of Water Process Engineering*, **35**, 101195, https://doi.org/10.1016/j.jwpe.2020.101195

Al-Qodah Z. and Al-Shannag M. (2017). Heavy metal ions removal from wastewater using electrocoagulation processes: a comprehensive review. *Separation Science and Technology*, **52**, 2649–2676, https://doi.org/10.1080/01496395.2017.1373677

Altin A. (2008). An alternative type of photoelectro-Fenton process for the treatment of landfill leachate. *Separation and Purification Technology*, **61**, 391–397, https://doi.org/10.1016/j.seppur.2007.12.004

Ambauen N., Muff J., Mai N. L., Hallé C., Trinh T. T. and Meyn T. (2019). Insights into the kinetics of intermediate formation during electrochemical oxidation of the organic model pollutant salicylic acid in chloride electrolyte. *Water*, **11**, 1322, https://doi.org/10.3390/w11071322

Anglada Á, Urtiaga A. and Ortiz I. (2009). Contributions of electrochemical oxidation to waste-water treatment: fundamentals and review of applications. *Journal of Chemical Technology & Biotechnology*, **84**, 1747–1755, https://doi.org/10.1002/jctb.2214

Anna Tałałaj I., Bartkowska I. and Biedka P. (2021). Treatment of young and stabilized landfill leachate by integrated sequencing batch reactor (SBR) and reverse osmosis (RO) process. *Environmental Nanotechnology, Monitoring & Management*, **16**, 100502, https://doi.org/10.1016/j.enmm.2021.100502

Anqi T., Zhang Z., Suhua H. and Xia L. (2020). Review on landfill leachate treatment methods. *IOP Conference Series: Earth and Environmental Science*, **565**, 012038, https://doi.org/10.1088/1755-1315/565/1/012038

Asaithambi P., Govindarajan R., Busier Yesuf M., Selvakumar P. and Alemayehu E. (2020). Enhanced treatment of landfill leachate wastewater using sono(US)-ozone(O3)–electrocoagulation(EC) process: role of process parameters on color, COD and electrical energy consumption. *Process Safety and Environmental Protection*, **142**, 212–218, https://doi.org/10.1016/j.psep.2020.06.024

Atmaca E. (2009). Treatment of landfill leachate by using electro-Fenton method. *Journal of Hazardous Materials*, **163**, 109–114, https://doi.org/10.1016/j.jhazmat.2008.06.067

Aziz H. A., Othman O. M. and Abu Amr S. S. (2013). The performance of electro-Fenton oxidation in the removal of coliform bacteria from landfill leachate. *Waste Management*, **33**, 396–400, https://doi.org/10.1016/j.wasman.2012.10.016

Baiju A., Gandhimathi R., Ramesh S. T. and Nidheesh P. V. (2018). Combined heterogeneous electro-Fenton and biological process for the treatment of stabilized landfill leachate. *Journal of Environmental Management*, **210**, 328–337, https://doi.org/10.1016/j.jenvman.2018.01.019

Bazrafshan E., Mohammadi L., Ansari-Moghaddam A. and Mahvi A. H. (2015). Heavy metals removal from aqueous environments by electrocoagulation process – a systematic review. *Journal of Environmental Health Science and Engineering*, **13**, 74, https://doi.org/10.1186/s40201-015-0233-8

Bouhezila F., Hariti M., Lounici H. and Mameri N. (2011). Treatment of the OUED SMAR town landfill leachate by an electrochemical reactor. *Desalination*, **280**, 347–353, https://doi.org/10.1016/j.desal.2011.07.032

Brillas E. (2022). Fenton, photo-Fenton, electro-Fenton, and their combined treatments for the removal of insecticides from waters and soils. A review. *Separation and Purification Technology*, **284**, 120290, https://doi.org/10.1016/j.seppur.2021.120290

Brillas E., Sirés I. and Oturan M. A. (2009). Electro-Fenton process and related electrochemical technologies based on Fenton's reaction chemistry. *Chemical Reviews*, **109**, 6570–6631, https://doi.org/10.1021/cr900136g

Cabeza A., Urtiaga A., Rivero M.-J. and Ortiz I. (2007). Ammonium removal from landfill leachate by anodic oxidation. "Selected papers of the proceedings of the 5th European Meeting on Chemical Industry and Environment, EMChIE 2006" held in Vienna, Austria, 3–5 May 2006. *Journal of Hazardous Materials*, **144**, 715–719, https://doi.org/10.1016/j.jhazmat.2007.01.106

Chen G. (2004). Electrochemical technologies in wastewater treatment. *Separation and Purification Technology*, **38**, 11–41, https://doi.org/10.1016/j.seppur.2003.10.006

Chen X., Chen G., Gao F. and Yue P. L. (2003). High-performance Ti/BDD electrodes for pollutant oxidation. *Environmental Science and Technology*, **37**, 5021–5026, https://doi.org/10.1021/es026443f

Cheng S., Shen C., Zheng H., Liu F. and Li A. (2020). OCNTS Encapsulating Fe-Co PBA as efficient chainmail-like electrocatalyst for enhanced heterogeneous electro-Fenton reaction. *Applied Catalysis B: Environmental*, **269**, 118785, https://doi.org/10.1016/j.apcatb.2020.118785

Chiang L.-C., Chang J.-E. and Wen T.-C. (1995). Indirect oxidation effect in electrochemical oxidation treatment of landfill leachate. *Water Research*, **29**, 671–678, https://doi.org/10.1016/0043-1354(94)00146-X

Chiang L.-C., Chang J.-E. and Chung C.-T. (2001). Electrochemical oxidation combined with physical–chemical pretreatment processes for the treatment of refractory landfill leachate. *Environmental Engineering Science*, **18**, 369–379, https://doi.org/10.1089/109287501753359609

Christensen TH., Cossu R. and Stegmann R. (Eds.). (2014). Landfilling of Waste: Leachate. CRC Press, London, https://doi.org/10.1201/9781482288933

Chu Y., Miao B., Zhang X. and Lv R. (2020). Heterogeneous electro-Fenton-like oxidation for the degradation of 4-nitrophenol characterized by immobilized Fe(III): performance, mechanism and chlorinated organic compounds formation. *Journal of Water Process Engineering*, **38**, 101662, https://doi.org/10.1016/j.jwpe.2020.101662

Crispim A. C., de Araújo D. M., Martínez-Huitle C. A., Souza F. L. and Dos Santos E. V. (2022). Application of electro-Fenton and photoelectro-Fenton processes for the degradation of contaminants in landfill leachate. *Environmental Research*, **213**, 113552, https://doi.org/10.1016/j.envres.2022.113552

Daneshvar N., Ashassi Sorkhabi H. and Kasiri M. B. (2004). Decolorization of dye solution containing acid Red 14 by electrocoagulation with a comparative investigation of different electrode connections. *Journal of Hazardous Materials*, **112**, 55–62, https://doi.org/10.1016/j.jhazmat.2004.03.021

Das P. P., Sharma M. and Purkait M. K. (2022). Recent progress on electrocoagulation process for wastewater treatment: a review. *Separation and Purification Technology*, **292**, 121058, https://doi.org/10.1016/j.seppur.2022.121058

Deng Y. and Englehardt J. D. (2007). Electrochemical oxidation for landfill leachate treatment. *Waste Management*, **27**, 380–388, https://doi.org/10.1016/j.wasman.2006.02.004

de Oliveira M. S., da Silva L. F., Barbosa A. D., Romualdo L. L., Sadoyama G. and Andrade L. S. (2019). Landfill leachate treatment by combining coagulation and advanced electrochemical oxidation techniques. *ChemElectroChem*, **6**, 1427–1433, https://doi.org/10.1002/celc.201801677

Dia O., Drogui P., Buelna G. and Dubé R. (2017a). Strategical approach to prevent ammonia formation during electrocoagulation of landfill leachate obtained from a biofiltration process. *Separation and Purification Technology*, **189**, 253–259, https://doi.org/10.1016/j.seppur.2017.08.023

Dia O., Drogui P., Buelna G., Dubé R. and Ihsen B. S. (2017b). Electrocoagulation of bio-filtrated landfill leachate: fractionation of organic matter and influence of anode materials. *Chemosphere*, **168**, 1136–1141, https://doi.org/10.1016/j.chemosphere.2016.10.092

Ding J., Wei L., Huang H., Zhao Q., Hou W., Kabutey F. T., Yuan Y. and Dionysiou D. D. (2018). Tertiary treatment of landfill leachate by an integrated electro-oxidation/electro-coagulation/electro-reduction process: performance and mechanism. *Journal of Hazardous Materials*, **351**, 90–97, https://doi.org/10.1016/j.jhazmat.2018.02.038

Djelal H., Lelievre Y. and Ricordel C. (2015). Combination of electro-coagulation and biological treatment by bioaugmentation for landfill leachate. *Desalination and Water Treatment*, **54**, 2986–2993, https://doi.org/10.1080/19443994.2014.908146

El-Ashtoukhy E.-S. Z., El-Taweel Y. A., Abdelwahab O. and Nassef E. M. (2013). Treatment of petrochemical wastewater containing phenolic compounds by electrocoagulation using a fixed bed electrochemical reactor. *International Journal of Electrochemical Sciences*, **8**, 17.

Fernandes A., Pacheco M. J., Ciríaco L. and Lopes A. (2015). Review on the electrochemical processes for the treatment of sanitary landfill leachates: present and future. *Applied Catalysis B: Environmental*, **176–177**, 183–200, https://doi.org/10.1016/j.apcatb.2015.03.052

Gao J., Oloibiri V., Chys M., Audenaert W., Decostere B., He Y., Van Langenhove H., Demeestere K. and Van Hulle S. W. H. (2015). The present status of landfill leachate treatment and its development trend from a technological point of view. *Reviews in Environmental Science and Bio/Technology*, **14**, 93–122, https://doi.org/10.1007/s11157-014-9349-z

Gao Y., Zhu W., Li Y., Zhang Q., Chen H., Zhang J. and Huang T. (2022). Anthraquinone (AQS)/polyaniline (PANI) modified carbon felt (CF) cathode for selective H2O2 generation and efficient pollutant removal in electro-Fenton. *Journal of Environmental Management*, **304**, 114315, https://doi.org/10.1016/j.jenvman.2021.114315

Ghosh D., Solanki H. and Purkait M. K. (2008). Removal of Fe(II) from tap water by electrocoagulation technique. *Journal of Hazardous Materials*, **155**, 135–143, https://doi.org/10.1016/j.jhazmat.2007.11.042

Guo Z., Zhang Y., Jia H., Guo J., Meng X. and Wang J. (2022). Electrochemical methods for landfill leachate treatment: a review on electrocoagulation and electrooxidation. *Science of the Total Environment*, **806**, 150529, https://doi.org/10.1016/j.scitotenv.2021.150529

Guvenc S. Y., Daniser Y., Can-Güven E., Varank G. and Demir A. (2023). Pre-coagulated landfill leachate treatment by electro-oxidation using MMO/Ti, Pt/Ti, and graphite anodes. *Environmental Engineering Research*, **28**, 210419, https://doi.org/10.4491/eer.2021.419

He Z., Liu Y., Wang J., Lv Y., Xu Y. and Jia S. (2021). Enhanced degradation of old landfill leachate in heterogeneous electro-Fenton catalyzed using Fe_3O_4 nano-particles encapsulated by metal organic frameworks. *Journal of Cleaner Production*, **321**, 128947, https://doi.org/10.1016/j.jclepro.2021.128947

Holt P. K., Barton G. W. and Mitchell C. A. (2005). The future for electrocoagulation as a localised water treatment technology. *Chemosphere*, **59**, 355–367, https://doi.org/10.1016/j.chemosphere.2004.10.023

Huda N., Raman A. A. A., Bello M. M. and Ramesh S. (2017). Electrocoagulation treatment of raw landfill leachate using iron-based electrodes: effects of process parameters and optimization. *Journal of Environmental Management*, **204**, 75–81, https://doi.org/10.1016/j.jenvman.2017.08.028

Ibanez J. G., Vazquez-Olavarrieta J. L., Hernandez-Rivera L., Garcia-Sanchez M. A. and Garcia-Pintor E. (2012). A novel combined electrochemical-magnetic method for water treatment. *Water Science and Technology*, **65**, 2079–2083, https://doi.org/10.2166/wst.2012.108

Ibrahim M. H., Moussa D. T., El-Naas M. H. and Nasser M. S. (2020). A perforated electrode design for passivation reduction during the electrochemical treatment of produced water. *Journal of Water Process Engineering*, **33**, 101091, https://doi.org/10.1016/j.jwpe.2019.101091

Ilhan F., Kurt U., Apaydin O. and Gonullu M. T. (2008). Treatment of leachate by electrocoagulation using aluminum and iron electrodes. *Journal of Hazardous Materials*, **154**, 381–389, https://doi.org/10.1016/j.jhazmat.2007.10.035

Ingelsson M., Yasri N. and Roberts E. P. L. (2020). Electrode passivation, faradaic efficiency, and performance enhancement strategies in electrocoagulation – a review. *Water Research*, **187**, 116433, https://doi.org/10.1016/j.watres.2020.116433

Jiang J.-Q., Graham N., André C., Kelsall G. H. and Brandon N. (2002). Laboratory study of electro-coagulation–flotation for water treatment. *Water Research*, **36**, 4064–4078, https://doi.org/10.1016/S0043-1354(02)00118-5

Jotin R., Ibrahim S. and Halimoon N. (2012). Electro coagulation for removal of chemical oxygen demand in sanitary landfill leachate. *International Journal of Environmental Sciences*, **3**, 921–930, https://doi.org/10.6088/ijes.2012030132020

Kabuk H. A., İlhan F., Avsar Y., Kurt U., Apaydin O. and Gonullu M. T. (2014). Investigation of leachate treatment with electrocoagulation and optimization by response surface methodology. *CLEAN – Soil, Air, Water*, **42**, 571–577, https://doi.org/10.1002/clen.201300086

Keyikoglu R., Karatas O., Rezania H., Kobya M., Vatanpour V. and Khataee A. (2021). A review on treatment of membrane concentrates generated from landfill leachate treatment processes. *Separation and Purification Technology*, **259**, 118182, https://doi.org/10.1016/j.seppur.2020.118182

Khandegar V. and Saroha A. K. (2013). Electrocoagulation for the treatment of textile industry effluent – a review. *Journal of Environmental Management*, **128**, 949–963, https://doi.org/10.1016/j.jenvman.2013.06.043

Khoramipour S., Mehralipour J. and Hosseini M. (2021). Optimisation of ultrasonic-electrocoagulation process efficiency in the landfill leachate treatment: a novel advanced oxidation process. *International Journal of Environmental Analytical Chemistry*, **0**, 1–19, https://doi.org/10.1080/03067319.2021.1973449

Lei Y., Shen Z., Huang R. and Wang W. (2007). Treatment of landfill leachate by combined aged-refuse bioreactor and electro-oxidation. *Water Research*, **41**, 2417–2426, https://doi.org/10.1016/j.watres.2007.02.044

Li X., Song J., Guo J., Wang Z. and Feng Q. (2011). Landfill leachate treatment using electrocoagulation. *Procedia Environmental Sciences*, 2011 3rd International Conference on Environmental Science and Information Application Technology ESIAT 2011, **10**, 1159–1164, https://doi.org/10.1016/j.proenv.2011.09.185

Li J., Yang Z., Xu H., Song P., Huang J., Xu R., Zhang Y. and Zhou Y. (2016). Electrochemical treatment of mature landfill leachate using Ti/RuO_2–IrO_2 and Al electrode: optimization and mechanism. *RSC Advances*, **6**, 47509–47519, https://doi.org/10.1039/C6RA05080H

Li M., Zhou M. and Qin X. (2022). A feasible electro-Fenton treatment of landfill leachate diluted by electro-Fenton effluent: evaluation of operational parameters, effect of dilution ratio and assessment of treatment cost. *Journal of Water Process Engineering*, **47**, 102754, https://doi.org/10.1016/j.jwpe.2022.102754

Lin S. H. and Chang C. C. (2000). Treatment of landfill leachate by combined electro-Fenton oxidation and sequencing batch reactor method. *Water Research*, **34**, 4243–4249, https://doi.org/10.1016/S0043-1354(00)00185-8

Lindsey M. E. and Tarr M. A. (2000). Quantitation of hydroxyl radical during Fenton oxidation following a single addition of iron and peroxide. *Chemosphere*, **41**, 409–417, https://doi.org/10.1016/S0045-6535(99)00296-9

Liu D., Yuan Y., Wei Y., Zhang H., Si Y. and Zhang F. (2022). Removal of refractory organics and heavy metals in landfill leachate concentrate by peroxi-coagulation process. *Journal of Environmental Sciences*, **116**, 43–51, https://doi.org/10.1016/j.jes.2021.07.006

López-Guzmán M., Flores-Hidalgo M. A. and Reynoso-Cuevas L. (2021). Electrocoagulation process: an approach to continuous processes, reactors design, pharmaceuticals removal, and hybrid systems – a review. *Processes*, **9**, 1831, https://doi.org/10.3390/pr9101831

Lu X., Zhou M., Li Y., Su P., Cai J. and Pan Y. (2019). Improving the yield of hydrogen peroxide on gas diffusion electrode modified with tert-butyl-anthraquinone on different carbon support. *Electrochimica Acta*, **320**, 134552, https://doi.org/10.1016/j.electacta.2019.07.063

Luo H., Zeng Y., Cheng Y., He D. and Pan X. (2020). Recent advances in municipal landfill leachate: a review focusing on its characteristics, treatment, and toxicity assessment. *Science of the Total Environment*, **703**, 135468, https://doi.org/10.1016/j.scitotenv.2019.135468

Malinovic B. N., Djuricic T. and Zoric D. (2019). The impact of electrode material in the treatment of landfill leachate by electrocoagulation process. *IOP Conference Series: Materials Science and Engineering*, **477**, 012040, https://doi.org/10.1088/1757-899X/477/1/012040

Mandal P., Dubey B. K. and Gupta A. K. (2017). Review on landfill leachate treatment by electrochemical oxidation: drawbacks, challenges and future scope. *Waste Management*, **69**, 250–273, https://doi.org/10.1016/j.wasman.2017.08.034

Martínez-Huitle C. A. and Ferro S. (2006). Electrochemical oxidation of organic pollutants for the wastewater treatment: direct and indirect processes. *Chemical Society Reviews*, **35**, 1324–1340, https://doi.org/10.1039/B517632H

Martínez-Huitle C. A., Rodrigo M. A., Sirés I. and Scialdone O. (2015). Single and coupled electrochemical processes and reactors for the abatement of organic water pollutants: a critical review. *Chemical Reviews*, **115**, 13362–13407, https://doi.org/10.1021/acs.chemrev.5b00361

Meier J., Remmen K., Wintgens T. and Melin T. (2021). Nanofiltration in landfill leachate treatment. In: Nanofiltration, A. I. Schäefer and A. G. Fane (eds), John Wiley & Sons, Ltd, Germany, pp. 663–690, https://doi.org/10.1002/9783527824984.ch15

Misra R., Gedam N., Waghmare S., Masid S. and Neti N. R. (2009). Landfill leachate treatment by the combination of physico-chemical and electrochemical methods. *Journal of Environmental Science and Engineering*, **51**, 315–320.

Modirshahla N., Behnajady M. A. and Kooshaiian S. (2007). Investigation of the effect of different electrode connections on the removal efficiency of Tartrazine from aqueous solutions by electrocoagulation. *Dyes and Pigments*, **74**, 249–257, https://doi.org/10.1016/j.dyepig.2006.02.006

Mollah M. Y. A., Morkovsky P., Gomes J. A. G., Kesmez M., Parga J. and Cocke D. L. (2004). Fundamentals, present and future perspectives of electrocoagulation. *Journal of Hazardous Materials*, **114**, 199–210, https://doi.org/10.1016/j.jhazmat.2004.08.009

Monteil H., Péchaud Y., Oturan N. and Oturan M. A. (2019). A review on efficiency and cost effectiveness of electro- and bio-electro-Fenton processes: application to the treatment of pharmaceutical pollutants in water. *Chemical Engineering Journal*, **376**, 119577, https://doi.org/10.1016/j.cej.2018.07.179

Mussa Z. H., Othman M. R. and Abdullah M. P. (2013). Electrocoagulation and decolorization of landfill leachate. *AIP Conference Proceedings*, **1571**, 829–834, https://doi.org/10.1063/1.4858758

Mussa Z. H., Othman M. R. and Abdullah M. P. (2015). Electrochemical oxidation of landfill leachate: investigation of operational parameters and kinetics using graphite-PVC composite electrode as anode. *Journal of the Brazilian Chemical Society*, **26**, 939–948, https://doi.org/10.5935/0103-5053.20150055

Naje A. S., Ajeel M. A., Ali I. M., Al-Zubaidi H. A. M. and Alaba P. A. (2019). Raw landfill leachate treatment using an electrocoagulation process with a novel rotating electrode reactor. *Water Science and Technology*, **80**, 458–465, https://doi.org/10.2166/wst.2019.289

Nidheesh P. V. and Gandhimathi R. (2012). Trends in electro-Fenton process for water and wastewater treatment: an overview. *Desalination*, **299**, 1–15, https://doi.org/10.1016/j.desal.2012.05.011

Norma D., Fernandes A., Pacheco M. J., Ciriaco L. and Lopes A. (2012). Electrocoagulation and anodic oxidation integrated process to treat leachate from a Portuguese sanitary landfill: port. *Portugaliae Electrochimica Acta*, **30**, 221–234, https://doi.org/10.4152/pea.201203221

Orkun M. O. and Kuleyin A. (2012). Treatment performance evaluation of chemical oxygen demand from landfill leachate by electro-coagulation and electro-fenton technique. *Environmental Progress & Sustainable Energy*, **31**, 59–67, https://doi.org/10.1002/ep.10522

Oturan N., van Hullebusch E. D., Zhang H., Mazeas L., Budzinski H., Le Menach K. and Oturan M. A. (2015). Occurrence and removal of organic micropollutants in landfill leachates treated by electrochemical advanced oxidation processes.

Environmental Science & Technology, **49**, 12187–12196, https://doi.org/10.1021/acs.est.5b02809

Oturan N., Bo J., Trellu C. and Oturan M. A. (2021). Comparative performance of ten electrodes in electro-Fenton process for removal of organic pollutants from water. *ChemElectroChem*, **8**, 3294–3303, https://doi.org/10.1002/celc.202100588

Oumar D., Patrick D., Gerardo B., Rino D. and Ihsen B. S. (2016). Coupling biofiltration process and electrocoagulation using magnesium-based anode for the treatment of landfill leachate. *Journal of Environmental Management*, **181**, 477–483, https://doi.org/10.1016/j.jenvman.2016.06.067

Panizza M. and Cerisola G. (2009). Direct and mediated anodic oxidation of organic pollutants. *Chemical Reviews*, **109**, 6541–6569, https://doi.org/10.1021/cr9001319

Panizza M., Delucchi M. and Sirés I. (2010). Electrochemical process for the treatment of landfill leachate. *Journal of Applied Electrochemistry*, **40**, 1721–1727, https://doi.org/10.1007/s10800-010-0109-7

Papastavrou C., Mantzavinos D. and Diamadopoulos E. (2009). A comparative treatment of stabilized landfill leachate: coagulation and activated carbon adsorption vs. electrochemical oxidation. *Environmental Technology*, **30**, 1547–1553, https://doi.org/10.1080/09593330903252240

Pellenz L., Borba F. H., Daroit D. J., Lassen M. F. M., Baroni S., Zorzo C. F., Guimarães R. E., Espinoza-Quiñones F. R. and Seibert D. (2020). Landfill leachate treatment by a boron-doped diamond-based photo-electro-Fenton system integrated with biological oxidation: a toxicity, genotoxicity and byproducts assessment. *Journal of Environmental Management*, **264**, 110473, https://doi.org/10.1016/j.jenvman.2020.110473

Pérez G., Saiz J., Ibañez R., Urtiaga A. M. and Ortiz I. (2012). Assessment of the formation of inorganic oxidation by-products during the electrocatalytic treatment of ammonium from landfill leachates. *Water Research*, **46**, 2579–2590, https://doi.org/10.1016/j.watres.2012.02.015

Pirsaheb M., Azizi E., Almasi A., Soltanian M., Khosravi T., Ghayebzadeh M. and Sharafi K. (2016). Evaluating the efficiency of electrochemical process in removing COD and NH4-N from landfill leachate. *Desalination and Water Treatment*, **57**, 6644–6651, https://doi.org/10.1080/19443994.2015.1012560

Pretorius W. A., Johannes W. G. and Lempert G. G. (1991). Electrolytic iron flocculant production with a bipolar electrode in series arrangement. *Water SA*, **17**, 133–138, https://doi.org/10.10520/AJA03784738_1579

Quan X., Cheng Z., Chen B. and Zhu X. (2013). Electrochemical oxidation of recalcitrant organic compounds in biologically treated municipal solid waste leachate in a flow reactor. *Journal of Environmental Sciences*, **25**, 2023–2030, https://doi.org/10.1016/S1001-0742(12)60253-8

Rafiee P., Hosseini M. and Ebrahimi S. (2020). The evolution patterns of temperature, pH, and voltage during the removal of chemical oxygen demand from a landfill leachate using electrocoagulation under different conditions. *Reaction Kinetics, Mechanisms and Catalysis*, **131**, 319–334, https://doi.org/10.1007/s11144-020-01846-0

Ren X., Song K., Xiao Y., Zong S. and Liu D. (2020). Effective treatment of spacer tube reverse osmosis membrane concentrated leachate from an incineration power plant using coagulation coupled with electrochemical treatment processes. *Chemosphere*, **244**, 125479, https://doi.org/10.1016/j.chemosphere.2019.125479

Renou S., Givaudan J. G., Poulain S., Dirassouyan F. and Moulin P. (2008). Landfill leachate treatment: review and opportunity. *Journal of Hazardous Materials*, **150**, 468–493, https://doi.org/10.1016/j.jhazmat.2007.09.077

Rookesh T., Samaei M. R., Yousefinejad S., Hashemi H., Derakhshan Z., Abbasi F., Jalili M., Giannakis S. and Bilal M. (2022). Investigating the electrocoagulation treatment of landfill leachate by iron/graphite electrodes: process parameters and efficacy assessment. *Water*, **14**, 205, https://doi.org/10.3390/w14020205

Sato Y., Zeng Q., Meng L. and Chen G. (2021). Importance of combined electrochemical process sequence and electrode arrangements: A lab-scale trial of real reverse osmosis landfill leachate concentrate. *Water Research*, **192**, 116849, https://doi.org/10.1016/j.watres.2021.116849

Sediqi S., Bazargan A. and Mirbagheri S. A. (2021). Consuming the least amount of energy and resources in landfill leachate electrocoagulation. *Environmental Technology & Innovation*, **22**, 101454, https://doi.org/10.1016/j.eti.2021.101454

Seibert D., Henrique Borba F., Bueno F., Inticher J. J., Módenes A. N., Espinoza-Quiñones F. R. and Bergamasco R. (2019). Two-stage integrated system photo-electro-Fenton and biological oxidation process assessment of sanitary landfill leachate treatment: an intermediate products study. *Chemical Engineering Journal*, **372**, 471–482, https://doi.org/10.1016/j.cej.2019.04.162

Silveira J. E., Zazo J. A., Pliego G., Bidóia E. D. and Moraes P. B. (2015). Electrochemical oxidation of landfill leachate in a flow reactor: optimization using response surface methodology. *Environmental Science and Pollution Research*, **22**, 5831–5841, https://doi.org/10.1007/s11356-014-3738-2

Soltani A., Faramarzi M. and Mousavi Parsa S. A. (2021). A review on adsorbent parameters for removal of dye products from industrial wastewater. *Water Quality Research Journal*, **56**, 181–193, https://doi.org/10.2166/wqrj.2021.023

Song-hu Y. and Xiao-hua L. (2005). Comparison treatment of various chlorophenols by electro-Fenton method: relationship between chlorine content and degradation. *Journal of Hazardous Materials*, **118**, 85–92, https://doi.org/10.1016/j.jhazmat.2004.08.025

Sruthi T., Gandhimathi R., Ramesh S. T. and Nidheesh P. V. (2018). Stabilized landfill leachate treatment using heterogeneous Fenton and electro-Fenton processes. *Chemosphere*, **210**, 38–43, https://doi.org/10.1016/j.chemosphere.2018.06.172

Sun C., Chen T., Huang Q., Duan X., Zhan M., Ji L., Li X., Wang S. and Yan J. (2021). Biochar cathode: reinforcing electro-Fenton pathway against four-electron reduction by controlled carbonization and surface chemistry. *Science of the Total Environment*, **754**, 142136, https://doi.org/10.1016/j.scitotenv.2020.142136

Szpyrkowicz L. (2006). Application of electrochemical oxidation for treatment of industrial wastewater – the influence of reactor hydrodynamics on direct and mediated processes. *Journal of Chemical Technology & Biotechnology*, **81**, 1375–1383, https://doi.org/10.1002/jctb.1521

Tahreen A., Jami M. S. and Ali F. (2020). Role of electrocoagulation in wastewater treatment: a developmental review. *Journal of Water Process Engineering*, **37**, 101440, https://doi.org/10.1016/j.jwpe.2020.101440

Tang J., Yao S., Xiao F., Xia J. and Xing X. (2022). Electrochemical oxidation of landfill leachate after biological treatment by electro-Fenton system with corroding electrode of iron. *International Journal of Environmental Research and Public Health*, **19**, 7745, https://doi.org/10.3390/ijerph19137745

Tejera J., Hermosilla D., Gascó A., Miranda R., Alonso V., Negro C. and Blanco Á (2021). Treatment of mature landfill leachate by electrocoagulation followed by Fenton or UVA-LED photo-Fenton processes. *Journal of the Taiwan Institute of Chemical Engineers*, **119**, 33–44, https://doi.org/10.1016/j.jtice.2021.02.018

Tezcan Un U., Filik Iscen C., Oduncu E., Akcal Comoglu B. and Ilhan S. (2018). Treatment of landfill leachate using integrated continuous electrocoagulation and the anaerobic treatment technique. *Environmental Progress & Sustainable Energy*, **37**, 1668–1676, https://doi.org/10.1002/ep.12850

Top S., Sekman E., Hoşver S. and Bilgili M. S. (2011). Characterization and electrocoaggulative treatment of nanofiltration concentrate of a full-scale landfill leachate treatment plant. *Desalination*, **268**, 158–162, https://doi.org/10.1016/j.desal.2010.10.012

Turro E., Giannis A., Cossu R., Gidarakos E., Mantzavinos D. and Katsaounis A. (2011). Electrochemical oxidation of stabilized landfill leachate on DSA electrodes. *Journal of Hazardous Materials*, **190**, 460–465, https://doi.org/10.1016/j.jhazmat.2011.03.085

Urtiaga A., Rueda A., Anglada Á and Ortiz I. (2009). Integrated treatment of landfill leachates including electrooxidation at pilot plant scale. *Journal of Hazardous Materials*, **166**, 1530–1534, https://doi.org/10.1016/j.jhazmat.2008.11.037

Urtiaga A., Gómez-Lavín S. and Soriano A. (2022). Electrochemical treatment of municipal landfill leachates and implications for poly- and perfluoroalkyl substances (PFAS) removal. *Journal of Environmental Chemical Engineering*, **10**, 107900, https://doi.org/10.1016/j.jece.2022.107900

Varank G., Yazici Guvenc S., Dincer K. and Demir A. (2020). Concentrated leachate treatment by electro-Fenton and electro-persulfate processes using central composite design. *International Journal of Environmental Research*, **14**, 439–461, https://doi.org/10.1007/s41742-020-00269-y

Verma M. and Kumar R. N. (2017). Coagulation and electrocoagulation for co-treatment of stabilized landfill leachate and municipal wastewater. *Journal of Water Reuse and Desalination*, **8**, 234–243, https://doi.org/10.2166/wrd.2017.102

Wang D., Hou H., Hu J., Xu J., Huang L., Hu S., Liang S., Xiao K., Liu B. and Yang J. (2019). A bio-electro-Fenton system with a facile anti-biofouling air cathode for efficient degradation of landfill leachate. *Chemosphere*, **215**, 173–181, https://doi.org/10.1016/j.chemosphere.2018.10.018

Wang Y., Meng G., Shan M., Wang D., Bai Z., Zhou X., Lv Y. and Bai J. (2020a). Treatment of high-ammonia-nitrogen landfill leachate nanofiltration concentrate using an Fe-loaded Ni-foam-based electro-Fenton cathode. *Journal of Environmental Chemical Engineering*, **8**, 104243, https://doi.org/10.1016/j.jece.2020.104243

Wang Y., Zhou C., Meng G., Bai J., Geng X., Lv Y. and Chen X. (2020b). Treatment of landfill leachate membrane filtration concentrate by synergistic effect of electrocatalysis and electro-Fenton. *Journal of Water Process Engineering*, **37**, 101458, https://doi.org/10.1016/j.jwpe.2020.101458

Wang Z., Liu M., Xiao F., Postole G., Zhao H. and Zhao G. (2022). Recent advances and trends of heterogeneous electro-Fenton process for wastewater treatment – review. *Chinese Chemical Letters*, **33**, 653–662, https://doi.org/10.1016/j.cclet.2021.07.044

World Bank. (2022). Solid Waste Management [WWW Document]. World Bank. URL https://www.worldbank.org/en/topic/urbandevelopment/brief/solid-waste-management (accessed 9.15.22).

Wu L., Liang D., Xu Y., Liu T., Peng Y. and Zhang J. (2016). A robust and cost-effective integrated process for nitrogen and bio-refractory organics removal from landfill leachate via short-cut nitrification, anaerobic ammonium oxidation in tandem with electrochemical oxidation. *Bioresource Technology*, **212**, 296–301, https://doi.org/10.1016/j.biortech.2016.04.041

Wu C., Chen W., Gu Z. and Li Q. (2021). A review of the characteristics of Fenton and ozonation systems in landfill leachate treatment. *Science of the Total Environment*, **762**, 143131, https://doi.org/10.1016/j.scitotenv.2020.143131

Xiao S., Peng J., Song Y., Zhang D., Liu R. and Zeng P. (2013). Degradation of biologically treated landfill leachate by using electrochemical process combined with UV irradiation. *Separation and Purification Technology*, **117**, 24–29, https://doi.org/10.1016/j.seppur.2013.04.024

Xu B., Iskander S. M. and He Z. (2020). Dominant formation of unregulated disinfection by-products during electrocoagulation treatment of landfill leachate. *Environmental Research*, **182**, 109006, https://doi.org/10.1016/j.envres.2019.109006

Yang W., Zhou M., Oturan N., Li Y. and Oturan M. A. (2019). Electrocatalytic destruction of pharmaceutical imatinib by electro-Fenton process with graphene-based cathode. *Electrochimica Acta*, **305**, 285–294, https://doi.org/10.1016/j.electacta.2019.03.067

Yang W., Zhou M., Oturan N., Bechelany M., Cretin M. and Oturan M. A. (2020). Highly efficient and stable FeIIFeIII LDH carbon felt cathode for removal of pharmaceutical ofloxacin at neutral pH. *Journal of Hazardous Materials*, **393**, 122513, https://doi.org/10.1016/j.jhazmat.2020.122513

Yang Z., Wu S., Sun H., Arhin S. G., Papadakis V. G., Goula M. A., Liu G., Zhang Y., Zhou L. and Wang W. (2022). Efficient degradation of organic compounds in landfill leachate via developing bio-electro-Fenton process. *Journal of Environmental Management*, **319**, 115719, https://doi.org/10.1016/j.jenvman.2022.115719

Yazici Guvenc S., Dincer K. and Varank G. (2019). Performance of electrocoagulation and electro-Fenton processes for treatment of nanofiltration concentrate of biologically stabilized landfill leachate. *Journal of Water Process Engineering*, **31**, 100863, https://doi.org/10.1016/j.jwpe.2019.100863

Ye Z., Zhang H., Zhang X. and Zhou D. (2016). Treatment of landfill leachate using electrochemically assisted UV/chlorine process: effect of operating conditions, molecular weight distribution and fluorescence EEM-PARAFAC analysis. *Chemical Engineering Journal*, **286**, 508–516, https://doi.org/10.1016/j.cej.2015.10.017

Yu F., Wang Y. and Ma H. (2019). Enhancing the yield of H_2O_2 from oxygen reduction reaction performance by hierarchically porous carbon modified active carbon fiber as an effective cathode used in electro-Fenton. *Journal of Electroanalytical Chemistry*, **838**, 57–65, https://doi.org/10.1016/j.jelechem.2019.02.036

Yu D., Cui J., Li X., Zhang H. and Pei Y. (2020a). Electrochemical treatment of organic pollutants in landfill leachate using a three-dimensional electrode system. *Chemosphere*, **243**, 125438, https://doi.org/10.1016/j.chemosphere.2019.125438

Yu X., Fu W., Jiang M., Liu G., Zou Y. and Chen S. (2020b). Automatic microbial electro-Fenton system driven by transpiration for degradation of acid orange 7. *Science of the Total Environment*, **725**, 138508, https://doi.org/10.1016/j.scitotenv.2020.138508

Zhang H., Zhang D. and Zhou J. (2006). Removal of COD from landfill leachate by electro-Fenton method. *Journal of Hazardous Materials*, **135**, 106–111, https://doi.org/10.1016/j.jhazmat.2005.11.025

Zhang H., Ran X. and Wu X. (2012). Electro-Fenton treatment of mature landfill leachate in a continuous flow reactor. *Journal of Hazardous Materials*, **241–242**, 259–266, https://doi.org/10.1016/j.jhazmat.2012.09.040

Zhu W., Li Y., Gao Y., Wang C., Zhang J., Bai H. and Huang T. (2020). A new method to fabricate the cathode by cyclic voltammetric electrodeposition for electro-Fenton application. *Electrochimica Acta*, **349**, 136415, https://doi.org/10.1016/j.electacta.2020.136415

Zolfaghari M., Jardak K., Drogui P., Brar S. K., Buelna G. and Dubé R. (2016). Landfill leachate treatment by sequential membrane bioreactor and electro-oxidation processes. *Journal of Environmental Management*, **184**, 318–326, https://doi.org/10.1016/j.jenvman.2016.10.010

doi: 10.2166/9781789063318_0195

Chapter 9

Treatment of landfill leachate containing emerging micropollutants

Ahmed Tawfik[1] and Vinay Kumar Tyagi[2]*

[1]Water Pollution Research Department, National Research Centre, 12622, Dokki, Cairo, Egypt
[2]Environmental Hydrology Division, National Institute of Hydrology, Roorkee 247667, India
**Corresponding author: prof.tawfik.nrc@gmail.com*

ABSTRACT

Landfills of industrial and municipal solid wastes are popular methods for environmental and ecosystem protection in most of the advanced and developing countries. Nevertheless, dumping of such wastes in landfills creates serious problems particularly hazardous industrial wastes containing recalcitrant compounds. However, landfilling of hazardous industrial wastes is still the appropriate method from an economic point of view. The leachate generated from landfilling of hazardous industrial wastes is of great concern due to its toxic rich and refractory emerging micropollutants (EMs). The hazardous leachate composition is highly variable and depends on the waste composition, density, quantities, temperature, and precipitation. The leachate contains macro and micropollutants that might be potential toxic agents as well as carcinogens. The landfills are recently developed to minimize the leachate; however, recalcitrant micropollutants still are of great concern and remain the main source of pollution in nearby aquatic areas and ecosystems. This highly deteriorates the quality of surface and groundwater that subsequently causes harmful effects for human and animal health. Removal of EMs from the leachate has been a major challenge to overcome their deleterious effects and risks to the environment. Physico-chemical treatment processes such as adsorption, advanced oxidation, and non-thermal plasma are introduced. Biological treatment methods such as constructed wetlands, anaerobic degradation and membrane bioreactors (MBRs) for treatment of landfill leachate are comprehensively reviewed. Integration of MBRs with other biological technologies, that is, anaerobic/aerobic/anoxic is cost effective for degradation of EMs from the leachate. The removal efficiency of 97.4% for di-n-butyl phthalate (DBP) and 98.8% for di (2-ethylhexyl) phthalate (DEHP) was achieved in the integrated system. DBP was removed by microbial degradation and 70.1% of DEHP was eliminated due to filtration processes. The MBR removed phthalic acid esters and phenolic compounds by values of 77–96% via filtration, biodegradation, and adsorption processes. However, their efficiencies highly depend on the biomass retention time and loading of EMs. The MBR module removed alkylphenols by 60–80% and pesticides by 59–74%.

Key words: Emerging micropollutants, hazardous landfill leachate, treatment technologies.

9.1 INTRODUCTION

Landfills of industrial and municipal solid wastes are popular methods for environmental and ecosystem protection in most of the advanced and developing countries (Elsamadony *et al.*, 2021; Luo *et al.*, 2020). Nevertheless, dumping of such wastes in landfills creates serious problems particularly hazardous industrial wastes containing recalcitrant compounds (Antony *et al.*, 2020; Ismail *et al.*, 2019a, 2019b). Recycling of hazardous industrial wastes could be a unique option but it depends on the cost and quality. Landfilling of hazardous industrial wastes is still the proper method from an economic point of view (Abdel-Shafy & Mansour, 2018; Ismail *et al.*, 2019a, 2019b). The leachate generated from landfilling of hazardous industrial wastes is of great concern due to its toxic rich and refractory emerging micropollutants (EMs). The leachate is an hazardous liquid generating as a result of hydrolysis and self-degradation of toxic wastes due to percolation of rainwater (Anastasiou *et al.*, 2014). The composition of hazardous leachate is highly variable and depends on the waste composition, density, quantities, temperature, and precipitation (Al-Yaqout & Hamoda, 2020). The leachate contains macro and micropollutants that might be potentially toxic and carcinogenic (Nurhasanah *et al.*, 2021). The landfills are recently developed to minimize the leachate, however, recalcitrant micropollutants still are of great concern and remain the main source of pollution to nearby aquatic areas and ecosystems (Nouj *et al.*, 2021). This highly deteriorates the quality of surface and groundwater that subsequently causes harmful effects for human and animal health (Parvin & Tareq, 2021).

Studies have focused on the removal of EMs from the leachate in order to overcome their harmful effects and risks to the environment. Adsorption (Ferraz & Yuan, 2020), air stripping (dos Santos *et al.*, 2020), coagulation, biological treatment methods (Miao *et al.*, 2019) such as constructed wetlands (Bakhshoodeh *et al.*, 2020) were employed for treatment of landfill leachate. Biological treatment methods of landfill leachate are still the acceptable technologies from environmental and the economic point of view. However, the presence of recalcitrant substances (PAHs-polyaromatic hydrocarbons, (PCBs-polychlorinated biphenyls, surfactants, humic substances and AOXs-adsorbable organic halogens) in the leachate deteriorates the biological degradation process and should be completely and/or partially removed by physical–chemical processes. The biological treatment processes are not efficient for complete removal of micropollutants particularly pesticides, pharmaceuticals, and personal care products (PCPs) (Essam *et al.*, 2007).

This chapter focuses on the sources and composition of EMs in the hazardous landfill leachate and their impact on health and environment. Types of EMs are comprehensively reviewed. Removal of micro pollutants from landfill leachate using biological and physico-chemical technologies is discussed.

9.2 SOURCES OF EMS IN LANDFILL LEACHATE AND THEIR HEALTH EFFECT

Landfill sites mainly receive huge quantities of wastes from various domestic and industrial sources (Figure 9.1). These wastes are toxic rich and contain

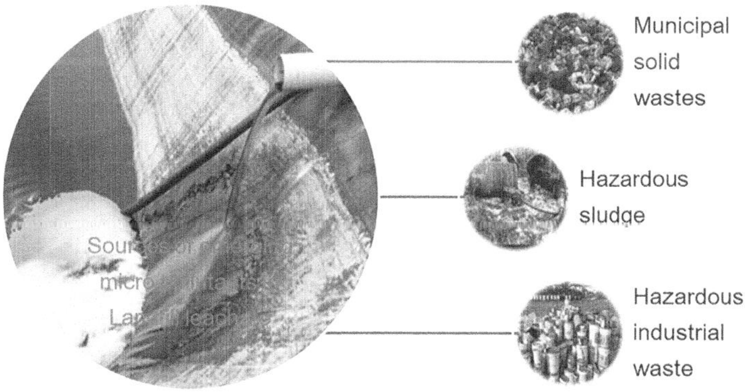

Figure 9.1 Main sources of EMs in landfill leachate.

hazardous chemicals such as EMs. The main sources of EM in the landfill are derived from pharmaceutical and PCPs, pesticides, and flame retardants (Kiefer *et al.*, 2021). A variety of EM in 242 samples include transformation products ($n = 8$), organic compounds ($n = 17$), pesticides ($n = 26$), industrial chemicals (plasticizers, flame retardants, catalysts, and antioxidants) ($n = 82$), food additives ($n = 18$), PCPs ($n = 9$), pharmaceutical intermediates ($n = 18$), and pharmaceuticals ($n = 64$) were identified by Han *et al.* (2022). Xeno organic compounds (XOCs) commonly occur in landfill leachate that includes alkylphenols, alkylbenzenes, and organochlorine compounds, benzene, toluene, ethylbenzene and xylenes (BTEX), phthalic acid esters (PAEs), organophosphate compounds (OPCs) and polyaromatic hydrocarbons (PAHs) (Nabi *et al.*, 2022).

Antibiotics (sulfamethoxazole tetracycline and oxy tetracycline) amounted to 47.9–100.9 μg/kg in municipal solid wastes in China (Song *et al.*, 2016). The concentrations of sulfamethoxazole tetracycline (7.4 mg/kg) and oxy tetracycline (45 mg/kg) in the municipal solid wastes were lower in the United States (Musson & Townsend, 2009). Erythromycin (0.3–20.2 μg/L) and sulfadiazine (0.2–29.2 μg/L) were abundant in the landfill leachate. Pesticides in the landfill leachate cause eye irritation and allergic reactions to the skin (Figure 9.2).

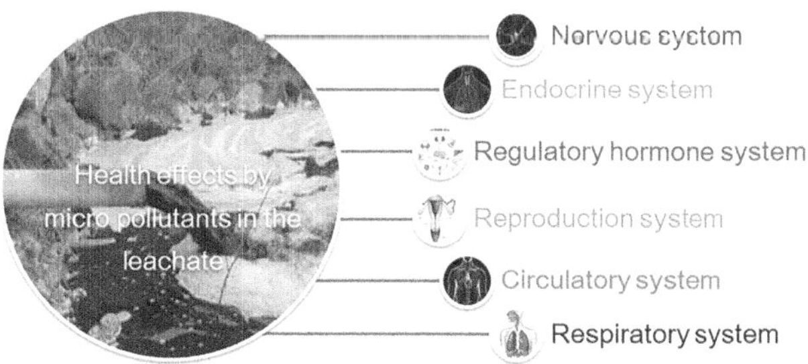

Figure 9.2 Effects of micropollutants on the human body.

Constipation, vomiting, nausea, and heartburn occurred due to the presence of pharmaceuticals in the leachate. Sulfamethazine is the main compound causing dizziness, vomiting, nausea, and headache. Benzene and xylene cause anemia, neurological symptoms, and affect renal and cardiovascular systems

9.3 TYPES OF EMS

The landfill leachate composition comprising micropollutants where extremely toxic chemicals varies based on rainfall, dumped waste, landfill site, and age. Micropollutants such as per/polyfluoroalkyl substances (PFAS), PCPs, fluorochemicals, pharmaceuticals, polycyclic aromatic hydrocarbons (PAHs), and phthalates are the main chemicals in the landfill leachate (Figure 9.3).

9.3.1 Polycyclic aromatic hydrocarbons

PAHs are dangerous pollutants and can be found in the old landfill leachate. These pollutants are carcinogenic and causes severe health problems due to intentional discharge/dumping the leachate onto surface and ground water (Ismail & Tawfik, 2016). It is accounted for ~100–6,000 ng/L in the landfill leachate (Kalmykova *et al.*, 2014; Smol & Włodarczyk-Makuła, 2017). The leachates are rich with phthalates of 0.01–340 μg/L (Baun *et al.*, 2004) and PAHs of 0.03–1220 μg/L (Matejczyk *et al.*, 2011) that highly deteriorated water quality due to discharge into the ecosystem. PAHs are mainly two to six aromatic rings with linear (anthracene and fluorine), clustered (pyrene), or angular arrangements (fluoranthene and phenanthrene) (Sho *et al.*, 2004).

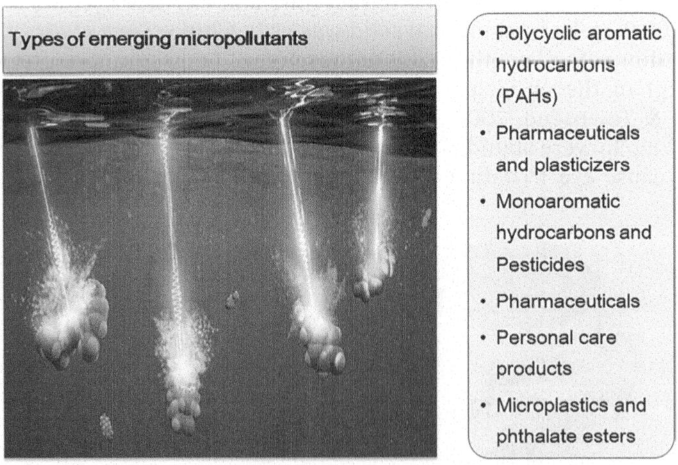

Figure 9.3 Types of EMs in the landfill leachate.

9.3.2 Pharmaceuticals and plasticizers

The majority of hazardous industrial solid wastes are landfilled resulting in high concentrations of emerging contaminants (ECs) in the leachate (Meky *et al.*, 2019). ECs are originally derived from pharmaceutical, plasticizers, and PCP sources. Plasticizers are aromatic contaminants in landfill leachate resulting from landfills of plastic wastes. Bisphenol A (BPA) is one of the plastic wastes, epoxy resins, and thermal papers. Bisphenol S (BPS) is found in food and drugs. Bisphenol F (BPF) is the main component of epoxy resins; 6–17 mg/L of BPA was detected in landfill leachate (Masoner *et al.*, 2016). Pharmaceuticals (anticonvulsants, antibiotics, antihistamines, analgesics, antihypertensive, steroid hormones, diuretics, stimulants) disposal into solid waste is the source of ECs in the leachate. These contaminants varied from 10 to 10,000 ng/L. Ibuprofen (IBP), amphetamine, lidocaine, and carbamazepine (CBZ) compounds were highly detected in the landfill leachate (Clarke *et al.*, 2015). 21,800 ng/L of analgesics acetaminophen (ACT) and 325,000 ng/L of IBP were detected in landfill leachate (Masoner *et al.*, 2016). The CBZ ranged from 165 to 345 ng/L in the landfill leachate (Masoner *et al.*, 2020). 45 ng/L of CBZ and 14,867 ng/L of IBP were detected in the leachate of open cells landfill (Adaryani & Keen, 2022). The authors detected also CBZ, acetaminophen, 17α-ethinylestradiol, doxycycline, and bisphenol.

9.3.3 Monoaromatic hydrocarbons and pesticides

Monoaromatic hydrocarbons such as toluene, benzene, xylene and ethylbenzene are mainly detected in the landfill leachate. These pollutants are derived from petrochemical industries, that is, paints, inks, pharmaceuticals, rubbers, adhesives, cosmetics, and thinners. These hazardous compounds cause severe health problems such as nausea to chromosomal aberrations and can be detected in ground water at a considerable level due to the surrounding landfill leachate (Eganhouse *et al.*, 2001). Pesticide pollutants are carcinogenic and organic persistent. Dumping of polluted vegetables and fruit peels generates pesticide rich leachate. Bentazon (3-isopropyl-1H-21,3-benzothiadiazin-4 (3H)-one 2,2-dioxide) was detected in landfill leachate (Eggen *et al.*, 2010) causing pollution of ground water.

9.3.4 Pharmaceuticals

Antibiotics in landfill leachate cause toxicity and potential health risk. Therefore, it is a matter of concern to the nearby environment, water and the ecosystem. Sulfonamides (SAs) and tetracyclines (TCs) were detected in landfill leachates (Wang *et al.*, 2022). Furthermore, oxytetracycline (OTC), sulfonamide sulfadiazine (SD), doxycycline (DXC), sulfamethazine (SMX), sulfamerazine (SM) and sulfamethoxazole (SMT) were observed in the samples of landfill leachate. SMT concentration was 654 ng/L in 45 samples followed by OTC (219.58 ng/L) in 47 samples and SD (209.98 ng/L) in 49 samples.

Antibiotic resistance genes (ARGs) are presented in landfill leachate. The landfill leachate is rich in organic content, heavy metals, and active microorganisms thereby ARGs exist. 10.0 mobile genetic elements (MGEs) and

203 of ARGs were detected in landfill leachate. 96% of ARGs in the groundwater was originally derived from nearby leachate indicating that ARGs rich leachate penetrates and contaminates the groundwater. Huang *et al.* (2022) found a significant correlation between bacterial communities and ARGs. Twelve bacterial taxa were detected in seven antibiotic-associated ARG classes.

9.3.5 Personal care products

The landfill leachate is rich with PCPs, that is, benzophenone (BPh), 3-(4-methylbenzylidene) camphor (4-MBC), *N,N*-diethyltoluamide, butylparaben (BP), ethylparaben (EP), methylparaben (MP), 1,4-dioxane, BPA and propylparaben (PP). BPA at low levels are typical endocrine disrupting compounds that causes malfunction of hormones in the body. MP, BP, EP, and PP are used as preservatives in PCPs, pharmaceuticals and food products. These chemicals have the ability to dissipate and absorb UV radiation, thereby protecting human skin from radiations of UV. However, these toxic chemicals cause serious health problems. Thus, it is necessary to develop low-cost treatment technologies for removal of PCPs from landfill leachate.

9.3.6 Microplastics and phthalate esters

Phthalate acid esters (PAEs) form a major portion of plasticizers and are present in building materials, PCPs, medical apparatuses and instruments, surfactants, pesticide carriers, and household goods (Mohammadi *et al.*, 2022). More than 30% plastic wastes are dumped into landfills resulting in microplastics (MPs) in the leachate. The size of MPs are <5 mm and generates from medicines, cosmetic, PCPs, and textiles. Plasticizers (plastic debris) are a portion of plastic materials resulting from plastic manufacturing and additive chemicals. Plastic wastes are produced in a large amount everywhere, and are commonly disposed in landfills. The MPs and PAEs in the leachate were 79.16 items/L and 3.27 mg/L, respectively. The PAEs in microplastics were 48.33 μg/g. Microplastics with a size of >1000 μm was abundant. The leachate was rich with di(2-ethylhexyl) phthalate (DEHP) and dibutyl phthalate (DBP). Dioctyl phthalate (DOP), diisobutyl phthalate (DiBP), butyl benzyl phthalate (BBP), and diethyl phthalate (DEP) showed a high risk due to the organisms sensitivity (Mohammadi *et al.*, 2022). The MPs and PAEs are released into the surrounding environment from the landfill leachate sites due to insufficient protection. Thus, constant monitoring and controlling the existing landfills are highly needed to overcome pollution problems.

9.4 TREATMENT PROCESSES OF LANDFILL LEACHATE

Landfill leachate should be treated by low-cost technologies to comply with discharge standards avoiding its environmental risks (Ismail & Tawfik, 2016; Mostafa *et al.*, 2017; Tawfik *et al.*, 2002, 2005). The biological treatment processes are still the acceptable technologies due to its ecofriendly and green smart system as shown in Figure 9.4.

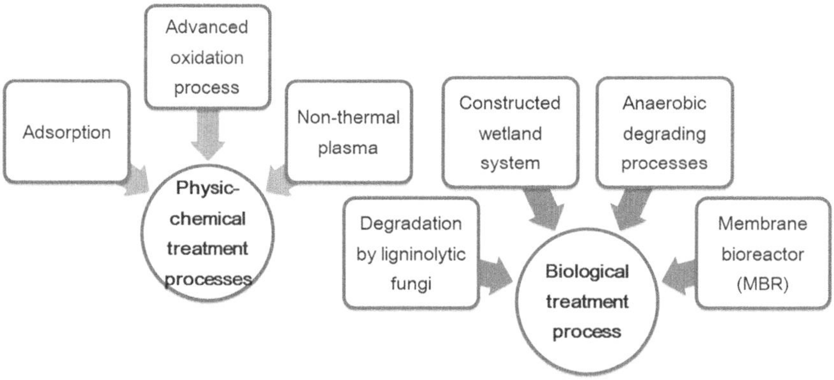

Figure 9.4 Treatment processes of landfill leachate containing EMs.

9.4.1 Physico-chemical treatment processes

9.4.1.1 Adsorption

Adsorption is the main process for removal of micropollutants from landfill leachate. Active sites, surface area and type of adsorbent, retention time, pH value, and substrate types are the main parameters affecting the adsorption process. The process involves transfer of the micropollutants from liquid phase to solid phase. PAHs are removed from the leachate by adsorption process, that is, activated carbon (Eeshwarasinghe *et al.*, 2018), organic waste (Björklund & Li, 2015), and silica gel (Eeshwarasinghe *et al.*, 2018). Biochar from pyrolysis of coconut and orange peels was utilized for PAH adsorption resulting in removal efficiency of 30.33–83.43% for benzo (a)anthracene, 47.09–83.02% for benzo(b)fluoranthene, 24.20–74.25% for benzo(k)fluoranthene, benzo(a) pyrene, and 23.84–84.02% for dibenzo (a,h) anthracene (de Jesus *et al.*, 2017). The removal of 14 PAHs from landfill leachate by adsorption in constructed wetland (sediments or bed substrates) was efficient as reported by Wojciechowska (2013). PAHs in the landfill leachate were highly removed by granulated activated carbon, sand, and peat moss filters (Kalmykova *et al.*, 2014). More than 90% of PAHs (anthracene (ANT), fluoranthene (FLA), fluorene (FLU), phenanthrene (PHN), and pyrene (PYR) was adsorbed on clinoptilolite modified by cationic surfactants–cetylpyridinium chloride, didodecyldimethylammonium bromide, and hexadecyltrimethylammonium bromide due to an increase in hydrophobicity (Hedayati *et al.*, 2021). Hedayati *et al.* (2021) studied the adsorption efficiencies of PAHs from landfill leachate and the environment by surfactant-modified clinoptilolites (MC)–cetylpyridinium chloride (CPC)-MC, didodecyldimethylammonium bromide (DDAB)-MC, and hexadecyltrimethylammonium bromide (HDTMA)-MC. PAH accumulated on CPC-MC, DDAB-MC, and HDTMA-MC with adsorption capacities of anthracene (708, 737, and 750 μg/g), fluorene (973, 1,060, and 1,147 μg/g), fluoranthene (1,355, 1,583, and 1,303 μg/g), pyrene (1,343, 1,569, and 1,269 μg/g) and phenanthrene (844, 1,057, and 989 μg/g). The PAHs were removed from landfill leachate by adsorption values of 97.8% for anthracene, 98.6%, for fluoranthene, 95.7% for fluorene, 97.0% for phenanthrene, and 98.5% for pyrene using CPC-MC and slightly increased to 99.0%, 99.6%, 98.0%, 99.0%, and 99.6% for DDAB-MC, respectively. The hydrophobicity of adsorbents influences the mechanism of PAH adsorption based on π–π stacking.

The PAH compounds tends to form π–π complexes between π-electrons of benzene rings and the active site on the surfactants – cetylpyridinium chloride (CPC) (Eeshwarasinghe *et al.*, 2018). Hydrophobic intermolecular interactions (van der Waal forces) between adsorbent surfaces and PAHs molecules are another way to remove PAHs (Bi *et al.*, 2006). Modification of adsorbents with humic acids will increase the availability of hydrophobic sites for the PAHs removal where the π–π electron–donor–acceptor is the main interaction between adsorbents and PAHs (Pan *et al.*, 2007). Modifications of adsorbents with surfactants can serve as π accepting substrates, and the PAHs are the main π donors (Keiluweit & Kleber, 2009). The clinoptilolite media is a mineral. DDAB and HDTMA are surfactants containing long hydrocarbon chains and nitrogen and PAHs compounds have π bonding. These are the structures needed for high binding and adsorption capacity process resulting in high bonding between PAHs and surfactants. The benzene rings of PAHs provide π electrons and hydrocarbon chains. Further, the surfactant contains quaternary positive ammonium that has a high affinity for electrons.

9.4.1.2 Advanced oxidation process

Advanced oxidation process removes monoaromatic hydrocarbons via oxidation of the organic pollutants by hydroxyl radicals (Figure 9.5). Two reactions are involved in the advanced oxidation process, that is, strong oxidants formation (hydroxyl radicals) and organic pollutants are immediately oxidized by radicals obtained. These radicals are released in the presence of hydrogen

Fenton process	Ozonation process	Electrochemical process	Photocatalysis process
• Enhanced biodegradability • Reduced toxicity • No energy input • Fe in sludge • High chemical consumption • Requirement of acidic pH • No complete mineralization	• No sludge generation • Enhanced biodegradability • Reduced toxicity • Relatively poor • performance on COD removal • High cost	• Easy to handle • Simplicity in operation • In-situ generation of H_2O_2 • Generation of sludge in some cases • Reduction in solution pH at the end of the reaction	• Capability to activate catalyst even in the presence of sunlight • Complete mineralization of pollutant possible • Availability of relatively cheap catalysts, such as TiO_2 • Shading effect by the catalyst, TiO_2 to the light reaching the pollutant • UV light required for activation

Figure 9.5 Advantages and disadvantages of advanced oxidation processes removing EMs (Pisharody *et al.*, 2022).

peroxide (H_2O_2), ultaviolet (UV) light and ozone (O_3). Hydroxyl radicals, are generated due to H_2O_2 photolysis by UV to attack pollutants resulting in their degradation. The H_2O_2/UV at 0.5 ml/L dose, toluene and xylene were removed by 77% (Vakar & Rajan, 2018). Ninety percent mineralization of BTEX was obtained at UV-185/H_2O_2 at an acidic pH value of 3 and reaction time of 6.0 h (Bustillo-Lecompte *et al.*, 2018).

$$H_2O + h\nu 185\,\text{nm} \rightarrow H^{\cdot} + HO^{\cdot} \tag{9.1}$$

$$H_2O_2 + h\nu \rightarrow 2HO^{\cdot} \tag{9.2}$$

$$H_2O_2 + HO^{\cdot} \rightarrow HO^{\cdot 2} + H_2O \tag{9.3}$$

The pesticides can be easily mineralized by advanced oxidation process due to the presence of sulfate and hydroxyl radical in the reaction medium. 100% removal of bentazon was obtained at initial concentration of 15 mg/L by TiO_2 photocatalysis (Pourata *et al.*, 2009). Electro-Fenton using carbon felt cathode or by anodic oxidation process with a BDD anode Pt or boron-doped diamond (BDD) anode was employed for removal of 15 volatile organic compounds (VOCs), 5 organochlorine pesticides (OCPs), 19 PAHs, 7 polychlorobiphenyls (PCBs), 2-polybrominated diphenyl ethers (PBDEs), and 7 alkylphenols from landfill leachate. These electrochemical processes achieved quasi-complete removal efficiency of 98%) of these EMs due to the production of hydroxyl radicals with highly powerful oxidizing species (Oturan *et al.*, 2015). Fenton reaction (1,755 mg Fe^{2+}/L, 26,422 mg H_2O_2/L, pH 3.72, reaction time 99 min removed, anthracene, bis(2-ethylhexyl) phthalate, dieldrin, benzene hexachloride, diuron, diclofenac, and chlorpyriphos by values of 90–99% (Argun *et al.*, 2017). Fenton (reaction time: 20–90 min, 0.51–2.55 g Fe^{2+}/L, 5.1–25.5 g H_2O_2/L, pH 3–5) and ozone (ozonation time: 10–130 min, pH 4–10) oxidation of landfill leachate removed PAHs, phthalates, alkylphenols and pesticides by values of 5–100%. Phthalates were highly removed by Fenton process compared to ozonation. Furthermore, the intermediates were mainly pentanoic acid, benzaldehyde, hydro cinnamic acid, naphthalenediol, and naphthalenone derivatives. Acid ester, naphthalene-based and phenolic compounds resulted from Fenton oxidation process (Ateş & Argun, 2021). Ozonation (0.95 gO_3/g DOC) of BPA and CBZ attained removal efficiency of 99% from landfill leachate. However, only 40% of atrazine and alachlor were removed indicating that ozone dose was not sufficient for complete removal of such contaminants (Yang *et al.*, 2021).

9.4.1.3 Non-thermal plasma

DEHP is the major phthalates in the landfill leachate with high concentrations of 10–400 ng/g waste. Non-thermal plasma (NTP) is one of the advanced oxidation processes utilized for degradation of organics. The NTP is a quasi-neutral medium with ions, electrons, and electrically neutral substances. Plasma is formed by applying voltage in a magnetic field, to produce an electrostatic plasma field. The radicals (O_3 and $O^{\cdot}$) are initially produced in the gas phase to attack the pollutants in the liquid phase. Therefore, the plasma could be used to remove refractory pollutants in a short contact time and alternative process for eliminating EMs from landfill leachates (Joshi & Thagard, 2013). There is no need for chemical oxidants, temperature, and pH control in non-thermal plasma process that makes the technology acceptable from environmental point of view. Detoxification of landfill leachate containing bis (2-ethylhexyl) phthalate using a combined non-thermal plasma and granular sludge sequencing batch reactor was assessed by Seid-mohammadi *et al.* (2022). DEHP concentrations ($\leq$8 mg/L) in the leachate was removed by 100% and reduced to 93% at high influent concentration of 20 mg/L at contact time of 60 min in non-thermal plasma pretreatment unit. In conclusion, non-thermal plasma is a highly competitive technology for treatment of landfill leachate containing EMs.

9.4.2 Biological treatment process

9.4.2.1 Degradation by ligninolytic fungi

The capability of ligninolytic fungi namely white rot fungi for phenolic and pesticides (phenylureas, s-triazines, and phenylamides) in the presence of redox-active mediators' contaminants degradations are mainly due to the secretion of extracellular enzymes, such as laccase, manganese peroxidases and lignin peroxidases. Almond shells, green compost, as adsorbents in combination with ligninolytic fungi (*Stereum hirsutum* and *Pleurotus ostreatus*), were employed to remove ethynilestadiol (EE2), xenoestrogens BPA, 4-n-nonylphenol (NP), linuron (herbicide) and dimethoate (insecticide) at initial concentrations of 10, 1, 1, 10, and 10 mg/L, respectively, from a landfill leachate (Loffredo & Castellana, 2015). The *P. ostreatus* combined with adsorption provided removal efficiencies of 88% for BPA, 96% for EE2, 99% for NP, 58% for linuron, and 46% for dimethoate. *S. hirsutum* achieved an average removal efficiencies of 39% for BPA, 71% for EE2, 100% for NP, 61% for linuron, and 32% for dimethoate. Based on these results, combined adsorption/biodegradation process is efficient and suitable to remove xenobiotic contaminants from the landfill leachate.

9.4.2.2 Constructed wetland system

Hexachlorocyclohexane (HCH) and dichlorodiphenyltrichloroethane (DDT) are the majority of organochlorine pesticides (OCPs) groups in the landfill leachate from nearby agriculture fields. HCH and DDT are highly persistent organic contaminants, stable and carcinogenic (Wang *et al.*, 2020). They cause harmful effects to the human functions and environmental ecosystems (Buah-Kwofie *et al.*, 2018) and therefore, it should be removed from landfill leachate by an effective and low-cost technology. Constructed wetlands (CWs) are a promising technology for removal of such contaminants from landfill leachate. CWs reduced the pesticides from drainage and agricultural runoff (Vymazal & Březinová, 2015). DEHP was removed by adsorption followed by biodegradation in wetland systems with minor removal by plants uptake. CWs remove pollutants via physical, biological, and chemical processes, that is, synergy between adsorption, microorganisms, and plants uptake (Ma *et al.*, 2021). OCPs adsorb and precipitate in the soil of the CWs units. Further decomposition and degradation of OCPs occurred through efficient microbial biodegradation (He *et al.*, 2020). DDT and HCH are mainly converted into CO_2 by microbial mineralization (Huang & Wang, 2013). Multi-stage subsurface flow CWs was employed for treatment of landfill leachate-rich hexachlorocyclohexane (HCH) and dichlorodiphenyltrichloroethane (DDT) (Yang *et al.*, 2022). The contaminants were adsorbed by 50.55% for HCH and 72.74% for DDT and the contribution of microbial activities was 20.38% and 27.89%, respectively. The dominant bacteria were *Betaproteobacteria* (34.37–35.90%) and *Alphaproteobacteria* (21.19–23.84%) in plant roots.

9.4.2.3 Anaerobic degrading processes

A horizontal flow anaerobic reactor inoculated with denitrifying biomass was assessed to remove p-xylene (32.1 mg/L), m-xylene (28.4 mg/L), o-xylene (28.5 mg/L), ethylbenzene (31.1 mg/L), toluene (27.8 mg/L) and hydrocarbons benzene (41.4 mg/L) (Gusmão *et al.*, 2007). The hydrocarbons removal efficiencies were attained at 97%. Propionate-cultured sludge enhanced micro pollutant biodegradation (*N*,*N*-diethyl-*m*-toluamide, 2,6-di-*tert*-butyl-phenol, nicotine, di (ethylhexyl)-phthalate and n-octadecane) and methane productivity from landfill leachate (Elreedy *et al.*, 2015; Pinpatthanapong *et al.*, 2022). The authors found that acclimatized seed sludge was efficient for removal of micropollutants compared to non-acclimatized seed sludge. Moreover, the micro pollutant removal efficiency increased with an increase of the concentration of propionate cultured sludge from 10% to 30%. The concentration of micropollutants (21–51 μg/L) was reduced to the minimum levels of non-toxic to the active microorganisms in the subsequent treatment systems. Two horizontal flow anaerobic reactor equipped with polyurethane foam (5 mm cubes, 95% porosity and 23 kg/m^3 density) was employed for degradation of ethanol, xylenes, toluene and benzene (Ribeiro *et al.*, 2013) at an hydraulic retention time of 13.5 h. The reactors achieved hydrocarbon removal efficiency of 99% at influent concentrations of 6.7 mg/L for benzene, 4.9 mg/L for toluene, 7.2 mg/L for m-xylene and 7.2 mg/L for p-xylene, 3.7 mg/L for ethylbenzene, and 60 mg N/L for nitrate. Occurrence and removal of 16 PAHs, 17 organochlorine pesticides (OCPs), and 4-nonylphenol (4-NP) from landfill leachate using a combined anaerobic filter and membrane bioreactor (MBR) were investigated by Tawfik *et al.* (2008) and Yiping *et al.* (2008). The concentrations of 4-NP, PAHs, and OCPs in the leachate highly fluctuated from 0 to 595.2 ng/L. Removal of 4-NP was accomplished in the MBR process resulting in an overall removal efficiency of 77%. Nevertheless, the 94% of OCPs and 59% of PAHs removals was mainly achieved in the anaerobic filter.

9.4.2.4 Membrane bioreactor

The removal of DBP, phthalic acid diesters (PAEs), and DEHP, from landfill leachates using an integrated system consisting of anaerobic, the aerobic, ultrafiltration membrane, and the reverse osmosis membrane was investigated by Fang *et al.* (2018). The leachate contained DBP and DEHP of 225.8 and 260.91 g/L, respectively, that were highly reduced by the integrated system into 5.8 and 3.2 g/L in the treated effluent. This corresponded to the removal efficiency of 97.4% for DBP and 98.8% for DEHP. DBP was mainly removed by microbial degradation. Membrane processes removed 70.1% of the DEHP due to physical filtration processes (Fang *et al.*, 2018). Two-stage MBR was employed to remove PAEs and phenolic compounds from landfill leachate (Boonyaroj *et al.*, 2012). The MBR removed PAEs and phenolic compounds by values of 77–96%. The main removal mechanism in the membrane units was due to filtration, biodegradation, and adsorption. The removal of organic micropollutants from landfill leachate using MBR modules was substantially improved at high biomass concentration and long sludge age. Aryanti *et al.* (2019) found that ultrafiltration unit removed 80% of xylene and 99% of toluene. The sequential anoxic and oxic units removed 62% of PAHs and 87% of alkylphenols from landfill leachate. The removal of alkylphenols and pesticides was 60–80% and 59–74% in membrane processes. Cu, Ni, and Cr, was eliminated by 92%, 91%, and 51%, respectively, in the bioreactor due to co-precipitation of solids by microbial secreting biopolymers (Argun *et al.*, 2020).

9.5 CONCLUSIONS

Treatment of hazardous landfill leachate is necessary to avoid health-related problems. Integration of MBRs with other biological technologies, that is, anaerobic/aerobic/ anoxic is cost effective for degradation of EMs from the leachate. The leachate contained DBP and DEHP of 225.8 and 260.9l g/L, respectively, that were highly reduced by the integrated system consisting of anaerobic, aerobic, ultrafiltration membrane, and the reverse osmosis membrane into 5.8 and 3.2 g/L in the treated effluent. The overall removal efficiency was 97.4% for DBP and 98.8% for DEHP. DBP was mainly removed by microbial degradation. Membrane processes removed 70.1% of the DEHP due to physical filtration processes. The MBR removed PAEs and phenolic compounds by values of 77–96%. The main removal mechanism in the membrane units was due to filtration, biodegradation, and adsorption. The removal of organic micropollutants from landfill leachate using MBR modules depends on biomass concentration and long sludge age. The removal of alkylphenols and pesticides was 60–80% and 59–74% in membrane processes.

ACKNOWLEDGMENTS

The first author acknowledges the Academy of Scientific Research and Technology (ASRT)-Egypt and the Natural Science Foundation of China. The NRC is acknowledged for partially funding this research (project ID: 13040108).

REFERENCES

Abdel-Shafy H. I. and Mansour M. S. M. (2018). Solid waste issue: sources, composition, disposal, recycling, and valorization. *Egyptian Journal of Petroleum*, **27**(4), 1275–1290, https://doi.org/10.1016/j.ejpe.2018.07.003

Adaryani A. R. and Keen O. (2022). Occurrence of pharmaceuticals and plasticizers in leachate from municipal landfills of different age. *Waste Management*, **141**, 1–7, https://doi.org/10.1016/j.wasman.2022.01.023

Al-Yaqout A. and Hamoda M. F. (2020). Long-term temporal variations in characteristics of leachates from a closed landfill in an arid region. *Water, Air, and Soil Pollution*, **231**(6), 302–319, https://doi.org/10.1007/s11270-020-04688-7

Anastasiou E., Lorentz K. O., Stein G. J. and Mitchell P. D. (2014). Prehistoric schistosomiasis parasite found in the Middle East. *The Lancet Infectious Diseases*, **14**(7), 553–554, https://doi.org/10.1016/S1473-3099(14)70794-7

Antony J., Niveditha S. V., Gandhimathi R., Ramesh S. T. and Nidheesh P. V. (2020). Stabilized landfill leachate treatment by zero valent aluminium-acid system combined with hydrogen peroxide and persulfate based advanced oxidation process. *Waste Management*, **106**, 1–11, https://doi.org/10.1016/j.wasman.2020.03.005

Argun M. E., Alver A. and Karataş M. (2017). Optimization of landfill leachate oxidation at extreme conditions and determination of micropollutants removal. *Desalination and Water Treatment*, **90**, 130–138, https://doi.org/10.5004/dwt.2017.21241

Argun M. E., Akkuş M. and Ateş H. (2020). Investigation of micropollutants removal from landfill leachate in a full-scale advanced treatment plant in Istanbul city, Turkey. *Science of the Total Environment*, **748**, 141423, https://doi.org/10.1016/j.scitotenv.2020.141423

Aryanti N., Kusworo T. D., Oktiawan W. and Wardhani D. H. (2019). Performance of ultrafiltration-ozone combined system for produced water treatment. *Periodica Polytechnica Chemical Engineering*, **63**(3), 438–447, https://doi.org/10.3311/PPch.13491

Ateş H. and Argun M. E. (2021). Advanced oxidation of landfill leachate: removal of micropollutants and identification of by-products. *Journal of Hazardous Materials*, **413**, 125326, https://doi.org/10.1016/j.jhazmat.2021.125326

Bakhshoodeh R., Alavi N., Oldham C., Santos R. M., Babaei A. A., Vymazal J. and Paydary P. (2020). Constructed wetlands for landfill leachate treatment: a review. *Ecological Engineering*, **146**(January), 105725, https://doi.org/10.1016/j.ecoleng.2020.105725

Baun A., Ledin A., Reitzel L. A., Bjerg P. L. and Christensen T. H. (2004). Xenobiotic organic compounds in leachates from ten Danish MSW landfills – chemical analysis and toxicity tests. *Water Research*, **38**(18), 3845–3858, https://doi.org/10.1016/j.watres.2004.07.006

Bi E., Schmidt T. C. and Haderlein S. B. (2006). Sorption of heterocyclic organic compounds to reference soils: column studies for process identification. *Environmental Science and Technology*, **40**(19), 5962–5970, https://doi.org/10.1021/es060470e

Björklund K. and Li L. (2015). Evaluation of low-cost materials for sorption of hydrophobic organic pollutants in stormwater. *Journal of Environmental Management*, **159**, 106–114, https://doi.org/10.1016/j.jenvman.2015.05.005

Boonyaroj V., Chiemchaisri C., Chiemchaisri W., Theepharaksapan S. and Yamamoto K. (2012). Toxic organic micro-pollutants removal mechanisms in long-term operated membrane bioreactor treating municipal solid waste leachate. *Bioresource Technology*, **113**, 174–180, https://doi.org/10.1016/j.biortech.2011.12.127

Buah-Kwofie A., Humphries M. S. and Pillay L. (2018). Bioaccumulation and risk assessment of organochlorine pesticides in fish from a global biodiversity hotspot: iSimangaliso Wetland Park, South Africa. *Science of the Total Environment*, **621**, 273–281, https://doi.org/10.1016/j.scitotenv.2017.11.212

Bustillo-Lecompte C. F., Kakar D. and Mehrvar M. (2018). Photochemical treatment of benzene, toluene, ethylbenzene, and xylenes (BTEX) in aqueous solutions using advanced oxidation processes: towards a cleaner production in the petroleum refining and petrochemical industries. *Journal of Cleaner Production*, **186**, 609–617, https://doi.org/10.1016/j.jclepro.2018.03.135

Clarke B. O., Anumol T., Barlaz M. and Snyder S. A. (2015). Investigating landfill leachate as a source of trace organic pollutants. *Chemosphere*, **127**, 269–275, https://doi.org/10.1016/j.chemosphere.2015.02.030

de Jesus J. H. F., da Cunha G. D., Cardoso E. M. C., Mangrich A. S. and Romão L. P. C. (2017). Evaluation of waste biomasses and their biochars for removal of polycyclic aromatic hydrocarbons. *Journal of Environmental Management*, **200**, 186–195, https://doi.org/10.1016/j.jenvman.2017.05.084

dos Santos H. A. P., de Castilhos Júnior A. B., Nadaleti W. C. and Lourenço V. A. (2020). Ammonia recovery from air stripping process applied to landfill leachate treatment. *Environmental Science and Pollution Research*, **27**(36), 45108–45120, https://doi.org/10.1007/s11356-020-10397-9

Eeshwarasinghe D., Loganathan P., Kalaruban M., Sounthararajah D. P., Kandasamy J. and Vigneswaran S. (2018). Removing polycyclic aromatic hydrocarbons from water using granular activated carbon: kinetic and equilibrium adsorption studies. *Environmental Science and Pollution Research*, **25**(14), 13511–13524, https://doi.org/10.1007/s11356-018-1518-0

Eganhouse R. P., Cozzarelli I. M., Scholl M. A. and Matthews L. L. (2001). Natural attenuation of volatile organic compounds (VOCs) in the leachate plume of a municipal landfill: using alkylbenzenes as process probes. *Ground Water*, **39**(2), 192–202, https://doi.org/10.1111/j.1745-6584.2001.tb02300.x

Eggen T., Moeder M. and Arukwe A. (2010). Municipal landfill leachates: a significant source for new and emerging pollutants. *Science of the Total Environment*, **408**(21), 5147–5157, https://doi.org/10.1016/j.scitotenv.2010.07.049

Elreedy A., Tawfik A., Kubota K., Shimada Y. and Harada H. (2015). Hythane ($H_2 + CH_4$) production from petrochemical wastewater containing mono-ethylene glycol via stepped anaerobic baffled reactor. *International Biodeterioration & Biodegradation*, **105**, 252–261, https://doi.org/10.1016/j.ibiod.2015.09.015

Elsamadony M., Mostafa A., Fujii M. and Tawfik A. (2021). Advances towards understanding long chain fatty acids-induced inhibition and overcoming strategies for efficient anaerobic digestion process. *Water Research*, **190**, 116732, https://doi.org/10.1016/j.watres.2020.116732

Essam T., Aly Amin M., El Tayeb O., Mattiasson B. and Guieysse B. (2007). Solar-based detoxification of phenol and p-nitrophenol by sequential TiO_2 photocatalysis and photosynthetically aerated biological treatment. *Water Research*, **41**(8), 1697–1704, https://doi.org/10.1016/j.watres.2007.01.015

Fang C., Chu Y., Jiang L., Wang H., Long Y. and Shen D. (2018). Removal of phthalic acid diesters through a municipal solid waste landfill leachate treatment process. *Journal of Material Cycles and Waste Management*, **20**(1), 585–591, https://doi.org/10.1007/s10163-017-0625-1

Ferraz F. M. and Yuan Q. (2020). Organic matter removal from landfill leachate by adsorption using spent coffee grounds activated carbon. *Sustainable Materials and Technologies*, **23**, e00141, https://doi.org/10.1016/j.susmat.2019.e00141

Gusmão V. R., Chinalia F. A., Sakamoto I. K. and Varesche M. B. A. (2007). Performance of a reactor containing denitrifying immobilized biomass in removing ethanol and aromatic hydrocarbons (BTEX) in a short operating period. *Journal of Hazardous Materials*, **139**(2), 301–309, https://doi.org/10.1016/j.jhazmat.2006.06.028

Han Y., Hu L., Liu T., Liu J., Wang Y., Zhao J. and Liu Y. (2022). Non-target, suspect and target screening of chemicals of emerging concern in land fi ll leachates and groundwater in Guangzhou, South China. *Science of the Total Environment*, **837**(April), 155705, https://doi.org/10.1016/j.scitotenv.2022.155705

He W., Ye M., He H., Zhu M. and Li Y. (2020). The decomposition and ecological risk of DDTs and HCHs in the soil-water system of the Meijiang River. *Environmental Research*, **180**, 108897, https://doi.org/10.1016/j.envres.2019.108897

Hedayati M. S., Abida O. and Li L. Y. (2021). Adsorption of polycyclic aromatic hydrocarbons by surfactant-modified clinoptilolites for landfill leachate treatment. *Waste Management*, **131**(July), 503–512, https://doi.org/10.1016/j.wasman.2021.06.033

Huang Y. and Wang J. (2013). Degradation and mineralization of DDT by the ectomycorrhizal fungi, *Xerocomus chrysenteron*. *Chemosphere*, **92**(7), 760–764, https://doi.org/10.1016/j.chemosphere.2013.04.002

Huang F., Zhou S., Zhao Y., Zhou X., Li H. and Zhang X. (2022). *Journal of Hazardous Materials*, **440**(1799), 129763, https://doi.org/10.1016/j.jhazmat.2022.129763

Ismail S. and Tawfik A. (2016). Performance of passive aerated immobilized biomass reactor coupled with Fenton process for treatment of landfill leachate. *International Biodeterioration and Biodegradation*, **111**(July), 22–30, https://doi.org/10.1016/j.ibiod.2016.04.010

Ismail S., Elsamadony M., Elreedy A., Fujii M. and Tawfik A. (2019a). Physico-chemical and microbial characterization of compartment-wise profiles in an anammox baffled reactor. *Journal of Environmental Management*, **232**, 875–886, https://doi.org/10.1016/j.jenvman.2018.11.134

Ismail S., Elsamadony M., Fujii M. and Tawfik A. (2019b). Evaluation and optimization of anammox baffled reactor (AnBR) by artificial neural network modeling and economic analysis. *Bioresource Technology*, **271**(September), 500–506, https://doi.org/10.1016/j.biortech.2018.09.004

Joshi R. P. and Thagard S. M. (2013). Streamer-like electrical discharges in water: part II. environmental applications. *Plasma Chemistry and Plasma Processing*, **33**(1), 17–49, https://doi.org/10.1007/s11090-013-9436-x

Kalmykova Y., Moona N., Strömvall A. M. and Björklund K. (2014). Sorption and degradation of petroleum hydrocarbons, polycyclic aromatic hydrocarbons, alkylphenols, bisphenol A and phthalates in landfill leachate using sand, activated carbon and peat filters. *Water Research*, **56**, 246–257, https://doi.org/10.1016/j.watres.2014.03.011

Keiluweit M. and Kleber M. (2009). Molecular-level interactions in soils and sediments: the role of aromatic π-systems. *Environmental Science and Technology*, **43**(10), 3421–3429, https://doi.org/10.1021/es8033044

Kiefer K., Du L., Singer H. and Hollender J. (2021). Identification of LC-HRMS nontarget signals in groundwater after source related prioritization. *Water Research*, **196**, 116994, https://doi.org/10.1016/j.watres.2021.116994

Loffredo E. and Castellana G. (2015). Comparative evaluation of the efficiency of low-cost adsorbents and ligninolytic fungi to remove a combination of xenoestrogens and pesticides from a landfill leachate and abate its phytotoxicity. *Journal of Environmental Science and Health – Part A Toxic/Hazardous Substances and Environmental Engineering*, **50**(9), 958–970, https://doi.org/10.1080/10934529.2015.1030294

Luo H., Zeng Y., Cheng Y., He D. and Pan X. (2020). Recent advances in municipal landfill leachate: a review focusing on its characteristics, treatment, and toxicity assessment. *Science of the Total Environment*, **703**, 135468, https://doi.org/10.1016/j.scitotenv.2019.135468

Ma Y., Huang J., Han T., Yan C., Cao C. and Cao M. (2021). Comprehensive metagenomic and enzyme activity analysis reveals the negatively influential and potentially toxic mechanism of polystyrene nanoparticles on nitrogen transformation in constructed wetlands. *Water Research*, **202**(July), 117420, https://doi.org/10.1016/j.watres.2021.117420

Masoner J. R., Kolpin D. W., Furlong E. T., Cozzarelli I. M. and Gray J. L. (2016). Landfill leachate as a mirror of today's disposable society: pharmaceuticals and other contaminants of emerging concern in final leachate from landfills in the conterminous United States. *Environmental Toxicology and Chemistry*, **35**(4), 906–918, https://doi.org/10.1002/and so on.3219

Masoner J. R., Kolpin D. W., Cozzarelli I. M., Smalling K. L., Bolyard S. C., Field J. A., Furlong E. T., Gray J. L., Lozinski D., Reinhart D., Rodowa A. and Bradley P. M. (2020). Landfill leachate contributes per-/poly-fluoroalkyl substances (PFAS) and pharmaceuticals to municipal wastewater. *Environmental Science: Water Research and Technology*, **6**(5), 1300–1311, https://doi.org/10.1039/d0ew00045k

Matejczyk M., Płaza G. A., Nałcz-Jawecki G., Ulfig K. and Markowska-Szczupak A. (2011). Estimation of the environmental risk posed by landfills using chemical, microbiological and ecotoxicological testing of leachates. *Chemosphere*, **82**(7), 1017–1023, https://doi.org/10.1016/j.chemosphere.2010.10.066

Meky N., Fujii M. and Tawfik A. (2019). Treatment of hypersaline hazardous landfill leachate using a baffled constructed wetland system: effect of granular packing media and vegetation. *Environmental Technology (United Kingdom)*, **40**(4), 518–528, https://doi.org/10.1080/09593330.2017.1397764

Miao L., Yang G., Tao T. and Peng Y. (2019). Recent advances in nitrogen removal from landfill leachate using biological treatments – a review. *Journal of Environmental Management*, **235**(January), 178–185, https://doi.org/10.1016/j.jenvman.2019.01.057

Mohammadi A., Malakootian M. and Dobaradaran S. (2022). Occurrence, seasonal distribution, and ecological risk assessment of microplastics and phthalate esters in leachates of a land fi ll site located near the marine environment: Bushehr port, Iran as a case. *Science of the Total Environment*, **842**(June), 156838, https://doi.org/10.1016/j.scitotenv.2022.156838

Mostafa A., Elsamadony M., El-Dissouky A., Elhusseiny A. and Tawfik A. (2017). Biological H_2 potential harvested from complex gelatinaceous wastewater via attached versus suspended growth culture anaerobes. *Bioresource Technology*, **231**(January), 9–18, https://doi.org/10.1016/j.biortech.2017.01.062

Musson S. E. and Townsend T. G. (2009). Pharmaceutical compound content of municipal solid waste. *Journal of Hazardous Materials*, **162**(2–3), 730–735, https://doi.org/10.1016/j.jhazmat.2008.05.089

Nabi M., Liang H., Cheng L., Yang W. and Gao D. (2022). A comprehensive review on the use of conductive materials to improve anaerobic digestion: focusing on landfill leachate treatment. *Journal of Environmental Management*, **309**(January), 114540, https://doi.org/10.1016/j.jenvman.2022.114540

Nouj N., Heddadi N., Azougarh Y., Hafid N. and El Alem N. (2021). A combined treatment of municipal solid waste landfill leachate using *Opuntia ficus-indica* cactus as coagulant and titaniferous sand as filter material. *Nanotechnology for Environmental Engineering*, **6**(1), 1–14, https://doi.org/10.1007/s41204-021-00105-4

Nurhasanah Cordova M. R. and Riani E. (2021). Micro- and mesoplastics release from the Indonesian municipal solid waste landfill leachate to the aquatic environment: case study in Galuga Landfill Area, Indonesia. *Marine Pollution Bulletin*, **163**(January), 111986, https://doi.org/10.1016/j.marpolbul.2021.111986

Oturan N., Van Hullebusch E. D., Zhang H., Mazeas L., Budzinski H., Le Menach K. and Oturan M. A. (2015). Occurrence and removal of organic micropollutants in landfill leachates treated by electrochemical advanced oxidation processes. *Environmental Science and Technology*, **49**(20), 12187–12196, https://doi.org/10.1021/acs.est.5b02809

Pan B., Ghosh S. and Xing B. (2007). Nonideal binding between dissolved humic acids and polyaromatic hydrocarbons. *Environmental Science and Technology*, **41**(18), 6472–6478, https://doi.org/10.1021/es070790d

Parvin F. and Tareq S. M. (2021). Impact of landfill leachate contamination on surface and groundwater of Bangladesh: a systematic review and possible public health risks assessment. *Applied Water Science*, **11**(6), 1–17, https://doi.org/10.1007/s13201-021-01431-3

Pinpatthanapong K., Panichnumsin P. and Phalakornkule C. (2022). Propionate-cultured sludge bioaugmentation to enhance methane production and micropollutant degradation in landfill leachate treatment. *Bioresource Technology*, **355**(April), 127241, https://doi.org/10.1016/j.biortech.2022.127241

Pisharody L., Gopinath A., Malhotra M., Nidheesh P. V. and Kumar M. S. (2022). Occurrence of organic micropollutants in municipal landfill leachate and its effective treatment by advanced oxidation processes. *Chemosphere*, **287**(P2), 132216, https://doi.org/10.1016/j.chemosphere.2021.132216

Pourata R., Khataee A. R., Aber S. and Daneshvar N. (2009). Removal of the herbicide bentazon from contaminated water in the presence of synthesized nanocrystalline TiO_2 powders under irradiation of UV-C light. *Desalination*, **249**(1), 301–307, https://doi.org/10.1016/j.desal.2008.10.033

Ribeiro R., de Nardi I. R., Fernandes B. S., Foresti E. and Zaiat M. (2013). BTEX removal in a horizontal-flow anaerobic immobilized biomass reactor under denitrifying conditions. *Biodegradation*, **24**(2), 269–278, https://doi.org/10.1007/s10532-012-9585-2

Seid-mohammadi A., Asgari G., Rafiee M., Samadi M. T., Nouri F., Pirsaheb M. and Asadi F. (2022). Fate and inhibition of bis (2-ethylhexyl) phthalate in biophysical reactors for treating real landfill leachate. *Process Safety and Environmental Protection*, **160**, 450–464, https://doi.org/10.1016/j.psep.2022.02.028

Sho M., Hamel C. and Greer C. W. (2004). Two distinct gene clusters encode pyrene degradation in *Mycobacterium* sp. strain S65. *FEMS Microbiology Ecology*, **48**(2), 209–220, https://doi.org/10.1016/j.femsec.2004.01.011

Smol M. and Włodarczyk-Makuła M. (2017). The effectiveness in the removal of PAHs from aqueous solutions in physical and chemical processes: a review. *Polycyclic Aromatic Compounds*, **37**(4), 292–313, https://doi.org/10.1080/10406638.2015.1105828

Song L., Li L., Yang S., Lan J., He H., McElmurry S. P. and Zhao Y. (2016). Sulfamethoxazole, tetracycline and oxytetracycline and related antibiotic resistance genes in a large-scale landfill, China. *Science of the Total Environment*, **551–552**(266), 9–15, https://doi.org/10.1016/j.scitotenv.2016.02.007

Tawfik A., Klapwijk B., El-Gohary F. and Lettinga G. (2002). Treatment of anaerobically pre-treated domestic sewage by a rotating biological contactor. *Water Research*, **36**(1), 147–155, https://doi.org/10.1016/S0043-1354(01)00185-3

Tawfik A., Klapwijk A., El-Gohary F. and Lettinga G. (2005). Potentials of using a rotating biological contactor (RBC) for post-treatment of anaerobically pre-treated domestic wastewater. *Biochemical Engineering Journal*, **25**(1), 89–98, https://doi.org/10.1016/j.bej.2005.03.013

Tawfik A., El-Gohary F., Ohashi A. and Harada H. (2008). Optimization of the performance of an integrated anaerobic-aerobic system for domestic wastewater treatment. *Water Science and Technology*, **58**(1), 185–194, https://doi.org/10.2166/wst.2008.320

Vakar D. and Rajan T. V. (2018). BTEX compounds removal from waste water by using UV&UV/H_2O_2 process. *International Journal of Recent Engineering Science*, **5**(1), 22–25, https://doi.org/10.14445/23497157/ijres-v5i1p105

Vymazal J. and Březinová T. (2015). The use of constructed wetlands for removal of pesticides from agricultural runoff and drainage: a review. *Environment International*, **75**, 11–20, https://doi.org/10.1016/j.envint.2014.10.026

Wang H., Cheng Z., Yuan H., Zhu N., Lou Z. and Otieno P. (2020). Occurrence of banned and commonly used pesticide residues in concentrated leachate: implications for ecological risk assessment. *Science of the Total Environment*, **710**, 136287, https://doi.org/10.1016/j.scitotenv.2019.136287

Wang Y., Lei Y., Liu X., Song L., Hamid N. and Zhang R. (2022). Sulfonamide and tetracycline in land fi ll leachates from seven municipal solid waste (MSW) land fi lls: seasonal variation and risk assessment. *Science of the Total Environment*, **825**, 153936, https://doi.org/10.1016/j.scitotenv.2022.153936

Wojciechowska E. (2013). Removal of persistent organic pollutants from landfill leachates treated in three constructed wetland systems. *Water Science and Technology*, **68**(5), 1164–1172, https://doi.org/10.2166/wst.2013.316

Yang Y., Liu Z., Demeestere K. and Van Hulle S. (2021). Ozonation in view of micropollutant removal from biologically treated landfill leachate: removal efficiency, [rad]OH exposure, and surrogate-based monitoring. *Chemical Engineering Journal*, **410**, 128413, https://doi.org/10.1016/j.cej.2021.128413

Yang C., Wang B., Wang H., He Z., Pi Y., Zhou J., Liang T., Chen M., He T. and Fu T. (2022). Removal of organochlorine pesticides and metagenomic analysis by multi-stage constructed wetland treating landfill leachate. *Chemosphere*, **301**(January), 134761, https://doi.org/10.1016/j.chemosphere.2022.134761

Yiping X. U., Yiqi Z., Donghong W., Shaohua C., Junxin L. I. U. and Zijian W. (2008). Occurrence and removal of organic micropollutants in the treatment of landfill leachate by combined anaerobic-membrane bioreactor technology. *Journal of Environmental Sciences*, **20**(11), 1281–1287, https://doi.org/10.1016/S1001-0742(08)62222-6

doi: 10.2166/9781789063318_0213

Chapter 10

Application of carbon-based adsorbents for landfill leachate treatment

Rahul Mishra and Sunil Kumar*

CSIR-National Environmental Engineering Research Institute, Nagpur 440 020, Maharashtra, India
**Correspondence author: s_kumar@neeri.res.in*

ABSTRACT

Currently, the most popular method for getting rid of municipal solid waste is sanitary landfilling; however, this method has a significant disadvantage in that it produces leachates that are extremely contaminated. The high concentration of heavy metals, halogenated hydrocarbons, poisonous gases, inorganic salts, and organic chemicals are present in landfill leachate, which are frequently highly polluted, poses a major risk to human health and the environment. Therefore, it needs to be treated before being released directly into receiving waters. Selection of leachate treatment techniques can be a difficult task. The efficacy of carbon-based adsorbents such as activated carbon, biochar, graphene oxide, and carbon nanotubes in treating various pollutants, such as organic pollutants, personal care items, pharmaceuticals, and heavy metal ions in leachate is also reported. There are several research directions that might be explored in the relatively carbon-based adsorbents to remediate landfill leachate. When complete life-cycle costs are considered, carbon-based adsorbents are substantially more economical than they originally look because they are produced at a cheaper cost compared to conventional adsorbents. An initial examination of the prevailing synthesis methods reveals a significantly greater environmental impact in comparison to conventional adsorbents, especially in terms of energy use and global warming potential. If environmental effects are considered, the study should transform toward the usage of greener processes and chemicals for carbon-based adsorbents. The objective of this chapter is to explain the utilization of carbon-based adsorbents for landfill leachate treatment.

Keywords: Municipal solid waste, landfill leachate, activated carbon, biochar, graphene oxide.

10.1 INTRODUCTION

The field of solid waste that deals with the production of sludge, garbage, and refuse as a result of human activity is quite important (Leton & Omotosho, 2004). More diverse consumption habits have emerged as a consequence of modifications in consumption habits in some culture, which have significantly increased the production of solid waste (Babayemi & Dauda, 2009; Tiwari *et al.*, 2023). According to estimates, the world produces ~1.3 billion tons of municipal solid waste (MSW) annually (Hoornweg & Bhada-Tata, 2012). Although currently landfills are well-designed and reduce the negative effects arising from MSW, the inevitable side effect of landfilling is the production of leachate (Wiszniowski *et al.*, 2006). The generation of landfill leachate is produced as a result of numerous factors, including moisture content, rainfall percolation, and various physicochemical processes (Rani *et al.*, 2020). MSW composition, weather, and age of the landfill can have a substantial effect on the leachate's composition. High amounts of heavy metals, halogenated hydrocarbons, recalcitrant substances, inorganic salts, organic pollutants, and ammoniacal-nitrogen (NH_3-N) are present in landfill leachate (Ukundimana *et al.*, 2020). Dissolved organic matter (DOM), total organic carbon (TOC), sulfur components, heavy metals, and biological oxygen demand (BOD)/chemical oxygen demand (COD) ratio are some of the fundamental indicators of leachate quality. Other indicators include total suspended solids (TSSs), ammonia nitrogen, total dissolved solids, total kjeldahl nitrogen (TKN), TSSs, and DOM (Bhatt *et al.*, 2017; Kjeldsen *et al.*, 2002). The composition is further complicated by the existence of complex pollutants (such as phthalic esters, alkaline earth metals, Polycyclic aromatic hydrocarbons (PAHs), and Perfluorinated Compounds (PFCs)) (Silva *et al.*, 2013). Due to the possibility of landfill leachate penetrating into groundwater, surface water and soil may become polluted (Cheng *et al.*, 2021). Consequently, the development of treatment processes could be required before discharging leachate into receiving waters to avoid pollution and to comply with stringent discharge criteria (Mishra *et al.*, 2022b; Singh *et al.*, 2021a). Moreover, appropriate treatment strategies should be chosen after completely comprehending the leachate properties (Shadi *et al.*, 2021). One of the problems that need to be solved is choosing leachate treatment techniques to reduce environmental effects. Making general guidelines is challenging with the complexity of the leachate's composition. A suitable treatment method should be affordable in terms of maintenance and operation cost, simple to use, efficient, and ecologically friendly to reduce energy consumption and surplus sludge production (Omran *et al.*, 2021). For landfill leachate remediation, a variety of biological and physicochemical techniques, such as membrane filtering, improved oxidation process, adsorption, air stripping, coagulation–flocculation, and membrane bioreactor have been employed (Djeffal *et al.*, 2021). Leachate from landfills can be effectively and efficiently treated using the adsorption technique. Non-specific forces such as van der Waals, hydrogen bonding, hydrophobic interactions, and ionic binding can cause adsorption (Ismail & Baek, 2020). The fact that the adsorption approach especially using carbon-based adsorbents is easier to use, less expensive, and results in no apparent modifications to the structural arrangement of enzymes is one of its benefits (Reis *et al.*, 2019). Recently, carbon-based adsorbents such as activated carbons (ACs), biochar, graphene oxide (GO), and carbon nanotubes (CNTs) have been gaining more interest by several researchers for removing contaminants.

To lower the application's material cost, scientists are looking for adsorbents created from waste products as replacements to commercial ACs. As part of the waste-to-wealth concept, the development of adsorbents from various waste materials, such as waste carpet (Hassan & Elhadidy, 2017), incineration of fly-ash (Xue *et al.*, 2014), and tires (Lin & Wang, 2017) has been studied. The goal is to turn these waste materials, which in some cases require special permits and cost money, into value-added goods that can be marketed for extra revenue (Egun, 2012). Due to biochar's improved surface functional group, porous structure, high carbon content, and inorganic compounds, it has demonstrated its versatility in a number of disciplines, including waste management, pollution remediation, and wastewater treatment (Kumar *et al.*, 2022a). The application of biochar and its efficacy in removing pollutants are closely related to its characteristics. Due to its distinctive qualities including ion-exchange capacity, high-adsorption limit, and microporosity, biochar has more widespread ecological applications (Kumar *et al.*, 2022a). Due to its outstanding optical, mechanical, and thermal qualities, a two-dimensional carbon substance known as graphene has attracted a lot of interest. GO, which is produced frequently by exfoliating and oxidizing graphite, is the oxidized form of graphene. The phenol functional groups, hydroxyl, carboxyl, carbonyl, and epoxide present on graphene sheets provide the GO with distinctive functional groups for alteration (Mishra *et al.*, 2022a). To develop GO composites that can be utilized as an adsorbent for eliminating contaminants, a variety of composites can be produced by modifying GO with polymers, small molecules, metal oxides/hydroxides, and metal nanoparticles (Mishra *et al.*, 2022a). The chapter aims to explain the utilization of carbon-based adsorbents for landfill leachate treatment. The leachate generation and composition as well as the characterization technique have also been discussed. Finally, different carbon-based adsorbents along with non-carbon adsorbents have been discussed.

10.2 LEACHATE GENERATION AND COMPOSITION

A liquid effluent is also known as landfill leachate that generates from landfill sites due to moisture content and runoff of wastes, groundwater percolation through the landfill, and degradation as well as biochemical reactions of wastes (Oliveira *et al.*, 2014). The successive phases of waste breakdown are the initial, acidogenic, methanogenic, and final stabilizing phases (Mandal *et al.*, 2017). Leachate can also be classified as more than 10 years (old), between 5 and 10 years (medium), or <5 years (young) depending on its age, as illustrated in Table 10.1. During the stabilization phase, old leachate is formed and is identified by BOD/COD having a low ratio. In the first and acidogenic phases, young leachate is often formed with a high ratio of BOD/COD, high volatile fatty acids (VFAs), and high COD (Abuabdou *et al.*, 2020). Quantity and quality are two ways to define leachate characteristics. Heavy metals, alkalinity, phosphorus, chloride, pH, total nitrogen, suspended particles, ammonia, COD, and BOD are a few physicochemical variables that can be used to evaluate the quality of landfill leachate (Luo *et al.*, 2020; Mishra *et al.*, 2022b). The level of compaction within landfills, density and particle size, local meteorological conditions, groundwater

Table 10.1 Leachate characteristics based on landfill age.

Parameters	Young	Intermediate	Stabilized	Old
Age	<5	5–10	10–20	>20
Organic compounds	80% VFAs	5–30% VFAs, HA and FA	HA and FA	HA and FA
Conductivity (μS/cm)	2,000–50,000	1,000–15,000	–	–
Alkalinity	1,000–20,000	5,000–6,000	–	–
pH	3–7	6–8	>7.5	>7.5
BOD (mg/L)	2,000–50,000	500–15,000	50–1,000	<300
COD (mg/L)	4,000–90,000	1,000–30,000	1,000–5,000	<3,000
BOD/COD	0.5–1.0	0.1–0.5	<0.1	<0.1
TOC/COD	<0.3	0.3–0.5	>0.5	>0.5
Heavy metals (mg/L)	>2	–	–	<2
NH_4^+-N (mg/L)	500–4,500	–	–	<1,500
NH_3-N (mg/L)	<400	–	>400	–
TKN (mg/L)	500–4,500	400–2,000	50–2,000	<2,000
Mg^{2+} (mg/L)	300–3,000	200–2,000	50–500	<200
Na^+ (mg/L)	1,000–7,000	200–3,000	50–500	<200
Ca^{2+} (mg/L)	1,000–7,000	200–4,000	100–500	<400
Fe^{3+} (mg/L)	200–3,000	200–2,000	50–500	<200
Zn^{2+} (mg/L)	50–400	20–200	5–50	<20
Cl (mg/L)	500–6,000	200–4,000	50–500	<200
P (mg/L)	50–500	5–200	–	<20
SO_4^{2-} (mg/L)	300–4,000	100–2,000	20–200	<100

Sources: Babaei *et al.* (2021); Mojiri *et al.* (2021); Roshadi *et al.* (2020).

percolation, runoff, and precipitation are the key factors influencing leachate formation (Kamaruddin *et al.*, 2017). Landfill leachate comprised of a variety of constituents and has a dark color (Elmaadawy *et al.*, 2020). The components of leachate may be divided into the following groups: xenobiotics (halogenated and monoaromatic hydrocarbons); heavy metals (zinc, nickel, lead, copper, chromium, and cadmium); dissolved inorganic matter (variety of cations such as sodium, potassium, magnesium, iron, calcium, and ammonium; and anions such as sulfate, phosphate, chloride, and hydrogen carbonate); and DOM (fulvic acid (FA) and humic acid (HA), VFAs, TOC, COD, proteins, and carbohydrates). These values are in the range of 600–7,500 mg/L HCO^{3-}, 10–8,000 mg/L SO_4^{2-}, 150–5,000 mg/L Cl^-, 50–2,200 mg/L NH_4^+-N, 0.2–13,000 mg/L TKN, pH 5.8–8.5, and 100–70,900 mg/L COD with heavy metal content varying between 0.01 and 1.0×10^4 mg/kg as well as a 0.04 and 0.7 BOD/COD ratio (Kjeldsen *et al.*, 2002).

10.3 LANDFILL LEACHATE CHARACTERIZATION METHODS

For the characterization of landfill leachate, several analytical techniques have been employed, which have been explained in the following sections.

10.3.1 Structural analytical methods

Elemental analysis can report various details regarding the landfill leachate's structural makeup and composition (Xiaoli *et al.*, 2008). The ratio of O/C is a measure of the carboxylic acid and carbohydrate levels, whereas the ratio of H/C determines the aromatic condensations and maturity level (Shouliang *et al.*, 2008). The O/C >0.4 indicates oxidation with a significant level, whereas the O/C <0.3 indicates oxidation with a comparatively moderate level (Yuan *et al.*, 2017). The molecular composition of the landfill leachate can be distinguished using the combination of electrospray ionization with Fourier-transform ion cyclotron resonance mass spectrometry (ESI-FT-ICR-MS), which offers ultra-high resolution (Zhang *et al.*, 2019). Therefore, a high level of resolution and sensitivity has been found by ESI-FT-ICR-MS for reliably identifying elements and their combinations. Although ESI-FT-ICR-MS has proven to be a spectacular technology for DOM characterization and has been successfully used for assessing samples, the application of this technology is constrained by high analytical expenses and the need for intricate sample pretreatment. A method dependent on nuclear magnetic induction and quantum mechanics is nuclear magnetic resonance (NMR) spectroscopy. By determining the signal intensity ratios, coupling constants, and chemical shifts of various nuclei, ^{1}H-NMR can identify important structural information. ^{13}C-NMR is more sensitive and provides greater chemical shifts compared to ^{1}H-NMR, and without hydrogen ions, it can determine organic structure (Tong *et al.*, 2015). Additionally, the nitrogen species in landfill leachate have been identified using ^{15}N-NMR. The findings demonstrated that only freshly produced leachate containing proteinaceous nitrogen was readily biodegradable, whereas primarily refractory heterocyclic nitrogen was present in older leachate (Zhao *et al.*, 2017).

10.3.2 Molecular weight-associated techniques

One important characteristic that reflects the contents of leachate DOM is the distribution of molecular weight. Molecular weight-related methods utilized in the leachate's characterization include size-exclusion chromatography (SEC) and membrane filtration. Investigation of the leachate's molecular weight distribution has been carried out using ultrafiltration (UF) membranes (Chen *et al.*, 2019a, 2019b). The COD removal rate was shown to decrease from 92.7% to 72.2% with an increase of 1–8 kDa in molecular weight cutoff, suggesting that removal of macromolecular organic matter should be emphasized throughout UF procedures (Pi *et al.*, 2009). On the basis of various mobile phases, SEC can be classified into two categories: gel permeation chromatography, which uses an organic solvent (i.e., methylbenzene and tetramethylene oxide) and gel filtration chromatography, which uses water solution or a buffer. By using SEC, it is possible to identify and quantify five different sized fractions, which are divided into the following categories: acids (<350 Da), low molecular neutrals, building blocks (300–500 Da), humic compounds (1 kDa), and biopolymers (>10 kDa) (Aftab & Hur, 2019). As there is no reaction and no sample loss during the separation process, as well as longer service life, molecular weight-related methods are better in comparison to conventional techniques. Nevertheless, they are unable to differentiate between substances with comparable molecular sizes, and only >10% of molecular weight differences can be used to segregate substances.

10.3.3 Spectroscopic approaches

Rapidly determining the relative quantity of compounds and chemical constituents in leachate samples is possible with spectroscopic methods. Due to good selectivity, quick analysis, ease, and great sensitivity of spectroscopy approaches, they have developed into one of the most widely utilized, extensive, and effective analytical techniques.

10.3.3.1 Fourier-transform infrared (FTIR) spectroscopy

The underlying structure of leachate comprising both non-fluorescent and fluorescent constituents may be identified by FTIR spectroscopy. When performed in attenuated total reflectance mode, FTIR is a fairly non-destructive and quick method that needs little sample processing (Bolyard *et al.*, 2019). The fingerprint and functional group region of the FTIR spectroscopy, which has a wavelength range of 2.5–25 m (400–4,000/cm) can be used to describe chemical characteristics. Strong bands of absorbance at 866 and 1,460/cm were attributable to the presence of many amide-I groups and carboxylic acid, respectively. The presence of benzene, polysaccharides, esters, and carbohydrates is also indicated by this technique (Liu *et al.*, 2015). On the basis of chemical alterations, FTIR spectroscopy can be used to analyze the breakdown of leachate DOM during different treatment methods (Xiaoli *et al.*, 2013). The use of FTIR spectroscopic methods in the leachate's characterization is not without certain inherent restrictions, though. Due to DOM's high complexity and polyfunctionality in leachate, there is a significant overlap in the peak positions, making it challenging to immediately interpret the FTIR spectroscopy results. Additionally, the FTIR spectroscopic technique for leachate DOM characterization is more of a descriptive method than a quantitative approach. Additionally, because of their great infrared absorption, oxygen–hydrogen interactions in water molecules cause significant measurement disruptions.

10.3.3.2 Fluorescence spectroscopy

Details on the composition of the biogeochemical cycling and the leachate Fluorescent dissolved organic matter (FDOM) are qualitatively and semi-quantitatively provided by fluorescence spectra (Huguet *et al.*, 2009). There are five sections of the fluorescence excitation–emission matrix (EEM) map after separation according to the locations of fluorescent peaks (He *et al.*, 2016). To demonstrate that primarily protein-like species are shown by fluorescent species in early landfill leachate, EEM can characterize the DOM of leachate at various phases of landfill aging (Shouliang *et al.*, 2008). By computing the fluorescence indexes, fluorescence EEM may also evaluate the source and level of humification. Between 435–480/254 and 300–345/254 nm of EM/EX wavelengths, the ratio of fluorescence peak regions is used to compute the humification index (Zhang *et al.*, 2019). Several mathematical techniques, such as self-organizing map analysis, fluorescence regional integration (FRI), and parallel factor analysis (PARAFAC), have been established to find quantitative and qualitative structural data of DOM. By enabling EEM records for landfill leachate to be divided into many separate fluorescence components, which have unique spectrum properties corresponding to diverse compositions, sources, and reactivities, EEM–PARAFAC has expanded the EEM application. The FRI approach, which derives EEM spectra for total and regional fluorescence intensity data to quantitatively determine the configuration and composition of leachate's DOM, may separate the DOM's EEM into regionally specific EEM volumes (Wang *et al.*, 2017). Because the quantity of a fluorescent element is often not significantly linked to fluorescence intensity, applying FRI to determine the leachate's DOM component is not generally acknowledged as an appropriate technique. A best practice for high-sensitivity investigation of DOM's chemical structure and composition is fluorescence spectroscopy. To depict the entire structure of DOM, the utilization of these fluorophores is still difficult because leachate has only a small fraction (1%) of DOM having fluorescence.

10.3.3.3 Ultraviolet–visible (UV–Vis) absorption spectroscopy

Several molecular characteristics, such as the reactivity of compounds, molecular weight distribution, and aromatic content can be obtained using UV–Vis absorption spectroscopy. Leachate DOM is frequently characterized using both the indices produced by data processing and the original UV–Vis absorption spectroscopy data. Carbonyl or conjugated groups, heterocyclic aromatic hydrocarbons, and conjugated unsaturated bonds are indicated by absorption at wavelengths 290–350, 250–290, and <250 nm, respectively (Chen *et al.*, 2019a, 2019b). The spectroscopic properties of various DOM portions differ. There may be macromolecular organic matter and aromatic compounds with conjugated double and carbonyl bonds because the FA and HA fractions are more complicated than Hydrophilic (HPI) and show prominent peaks at wavelengths 210–375 nm (Liu *et al.*, 2015). Specific UV wavelength absorption can offer useful structural data. For instance, absorbance values at 280 and 254 nm (E_{280} and E_{254}, respectively) are frequently utilized as markers of aromaticity and hydrophobicity that are directly related to the quantity of leachate's DOM. The advantages of UV–Vis spectroscopy over other spectral techniques include the ability to simultaneously monitor many parameters, no need for sample preparation, quick analysis time, and easy operational procedures (Chen *et al.*, 2015). Nevertheless, it may be challenging to differentiate distinct peaks due to the high overlap between the absorption peaks of several compounds. Additionally, a number of variables, including the presence of coexisting compounds such as nitrates and metal ions, pH, and temperature might affect the UV–Vis spectra. As a result, interference during the collection of spectral data should be avoided, and extreme caution should be taken when interpreting the DOM properties of leachate.

10.4 ADSORPTION PROCESS FOR LANDFILL TREATMENT

One of the most promising and efficient methods for removing NH_4^+-N and DOM from landfill leachate is the adsorption technique. For the leachate treatment, adsorbents with thermostability, surface reactivity, micropore structures, and surface area have been used. Adsorption is frequently used as a step of pre- or post-treatment in conjunction with other physicochemical techniques such as Fenton reactions, ozonation, and coagulation. The maximum removal rates for NH_4^+-N and COD using a combined adsorption treatment of ozone and AC were 92% and 86%, respectively (Kurniawan *et al.*, 2006). Under ideal circumstances, adsorption and coagulation were used to remediate biologically treated leachate, leading to an 80% rate of COD removal (Papastavrou *et al.*, 2009). Discovering an affordable and efficient adsorbent has received a lot of scientific attention. By adsorbing DOM, ACs created from sewage sludge treated with $ZnCl_2$ have been used to remediate landfill leachate, obtaining an 85.61% COD removal rate under ideal circumstances (Zeng *et al.*, 2020). Additionally, with 30% rate of COD removal, zeolite has been used in leachate post-treatment (Poblete *et al.*, 2019). Currently, biochar has become a viable AC replacement. For the treatment of landfill leachate, rice husk–transformed phosphoric acid–activated biochar has been effective, removing 80% of COD (Luo *et al.*, 2019).

10.5 DIFFERENT TYPES OF ADSORBENTS FOR LANDFILL LEACHATE TREATMENT

10.5.1 Conventional adsorbents

Adsorption is a method that occurs on surfaces commonly and is used to remove both organic and inorganic compounds. As a result of physical and chemical interactions, species are drawn to the surface of highly porous material (Foo & Hameed, 2009). Carbon-based adsorption has dominated adsorption processes all over the globe because it is straightforward in design, widely accessible, and inexpensive to carry out. Adsorption is considered to be an effective and essential method for treating wastewater (Mukherjee *et al.*, 2015). High porosity is one of the most crucial qualities to aim for in an adsorbent because it results in a larger surface area and faster elimination of pollutants (Dąbrowski, 2001). Adsorption has been effective in treating landfill leachate, particularly in the elimination of ammonia nitrogen and organic compounds (Foo & Hameed, 2009). Previously, the most popular adsorbents have been ACs. However, modern porous-based materials are also advised for improved capacities of adsorption and reduced negative environmental effects (Torretta *et al.*, 2016). These materials are available in various geometries, surface structures and chemical forms. Overall, a variety of materials such as biological adsorbents, agricultural, industrial wastes, and by-products as well as synthetic materials such as zeolites and resins, improve natural materials, that is, ACs and natural materials may be utilized as adsorbents (Crini *et al.*, 2019).

10.5.2 Carbon-based materials

In recent literature, a variety of carbon-based materials have been employed as an adsorbent. Utilizing carbon-based substances such as biochar or hydrochar as an alternative to ACs is a novel technique. These materials are economical and environmentally beneficial adsorbents to eliminate the contaminants from the aqueous phase, which has received a lot of research interest (Nadarajah *et al.*, 2021). The key benefit is different biomass can be utilized for their generation (Hong *et al.*, 2020). The potential to alter surface functional groups for the production of materials with an improved capacity of adsorption, appropriate for immobilizing nanoparticles with a variety of chemical activities is another crucial property of both biochar and hydrochar (Kumar *et al.*, 2022a; Singh *et al.*, 2022b). Notwithstanding these benefits, there has been comparatively much less research on the viability of using biochar and hydrochar to remediate landfill leachate, and the majority of research is in a very early stage of development. There is a critical information gap here that needs to be filled right away. The utilization of these materials for different environmental purposes is an alluring concept, which incorporates all of these ideas, receiving lot of attention in recent years for developing a circular economy (Singh *et al.*, 2022b). To achieve a thorough understanding of the difficulties and benefits of using these bio-adsorbents to manage landfill leachate, as well as their potential for reusability or secure final disposition, research on a large scale should be carried out.

10.5.2.1 Activated carbon

The word 'activated carbon' generally refers to carbonaceous materials (Cuhadaroglu & Uygun, 2008), which vary from elemental carbon by the oxidation of carbon atoms. There are two processes, that is, physical and chemical activation to produce ACs. Furthermore, the physical activation is divided into two steps: carbonization and activation (Ioannidou & Zabaniotou, 2007). The temperatures of carbonization and activation are in the range of 400–850 and 600–900°C, respectively (Ioannidou & Zabaniotou, 2007). Carbonization is performed to produce char with more fixed carbon for activation purposes by lowering the volatile component of the raw material (McDougall, 1991). Preoxidation, steam activation, and pyrolysis are the stages of physical activation. Chemical activation, also known as wet oxidation (Al-Qodah & Shawabkah, 2009), produces ACs by impregnating the precursor with a catalyst and washing it afterward (Abdullah *et al.*, 2001). The essential temperature range for this activation is between 300 and 700°C (Giraldo & Moreno-Piraján, 2012). AC has been the most utilized adsorbent due to its potential to eliminate several contaminants, greater pore structure, low cost, strength against acids/bases, thermal stability, good porosity, and larger surface area (Foo & Hameed, 2009; Mukherjee *et al.*, 2015). Granular activated carbon (GAC) and powdered activated carbon are the two kinds of ACs that can be employed. As a feedstock for generating AC, almost all materials with a high carbon concentration, that is, lignite, tobacco, nutshells, peat, timber, and coal are appropriate (Torretta *et al.*, 2016). Other carbonaceous sources are used for the production of ACs

Table 10.2 Removal of different contaminants by AC.

Biomass Source	Target Contaminant	Adsorption Capacity (mg/g)	Reference
Silkworm cocoon	Cr^{6+}	366.3	Sun *et al.* (2017)
Green alga	Ni^{2+}	64.51	Nemchi *et al.* (2017)
Banana peel	Cu^{2+}	14.3	Van Thuan *et al.* (2017)
Melia azedarach fruit	Pb^{2+}	111.11	Sabermahani *et al.* (2018)
Typha angustifolia	Cd^{2+}	48.08	Tang *et al.* (2017)
Peach stone	*p*-Nitrophenol	234.3	Álvarez-Torrellas *et al.* (2017a)
Sawdust	Bisphenol A	334.28	Thue *et al.* (2017)
	Caffeine	256.29	
Date	Ciprofloxacin	2.094	Darweesh and Ahmed (2017)
Waste tea residue	Oxytetracycline	273.7	Kan *et al.* (2017)

such as pith, bamboo, oat hulls, peanut hulls, corn cobs, apricot stones, banana peels, rice straw, rice hulls, pinecones peach stones, corn straw, cotton residues, olive stones, soyabean hulls, sugarcane bagasse, and so on (Chan *et al.*, 2012; Skodras *et al.*, 2007). Table 10.2 presents the removal of different contaminants by AC. A schematic diagram of production and the activation techniques of AC is shown in Figure 10.1. ACs are typically created using an activation process and a continuous two-stage carbonization (Zubair *et al.*, 2017). The initial porosity is generated in the first phase, which is carried out using pyrolysis at

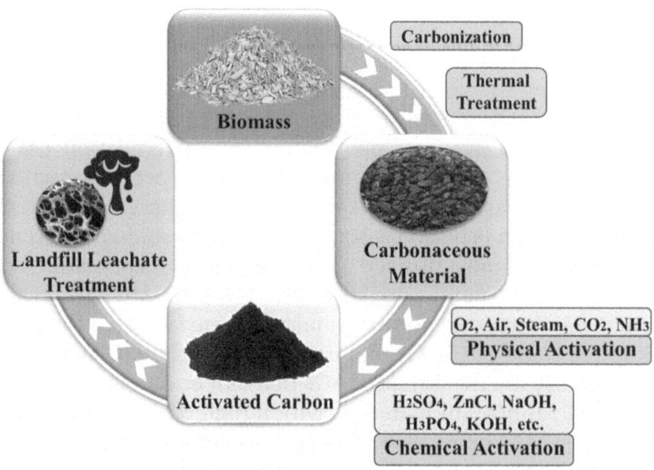

Figure 10.1 Schematic diagram of the activation techniques for AC production from biomass.

high temperatures and the second stage improves pore structure (Lam *et al.*, 2006). A pretreatment phase is also used before the processing. This step, which involves crushing, milling, and sifting afterward, is performed to obtain an acceptable size because it is one of the conditions for the ensuing process's success (Alslaibi *et al.*, 2013). Physical and chemical activation are two different types of activation processes. In the physical activation process, hot steam or CO_2 is used to activate the carbonized material (Matali *et al.*, 2013). In the activation of carbonized material chemically, it is combined with chemical activators before being heated in an inert environment. Compared to physical activation, lower temperatures are used by the chemical activation technique and it produces a higher surface area and improves porosity of the product (Allwar, 2012). Chemical activation techniques result in higher expenses for the activating agents and require more steps for cleaning (Shehzad *et al.*, 2015). Microwave chemical activation can improve adsorption effectiveness and increase surface area (Alslaibi *et al.*, 2013). Impregnation, which is often carried out using chemicals such as metal oxides (e.g., MgO, BaO, CuO, CeO_2, and TiO_2) or hydroxides, is an activation assist step for improving adsorption performance (Hidayu & Muda, 2016). Impregnation can be caused by alkaline hydroxides producing large surface areas (>2,500 m^2/g) (Tseng, 2006). The surface and physical properties of adsorbents can be identified using transmission electron microscopy and scanning electron microscopy. Waste utilization as a carbon source to produce ACs is another method for addressing both environmental and financial issues. Due to the lower moisture content of buried wastes, the produced adsorbent can be utilized as a valuable product for the treatment of leachate, which will also result in less formation of leachate and cheaper disposal costs (Shehzad *et al.*, 2015; Torretta *et al.*, 2016). The performance of leachate treatment using ACs has been the subject of numerous earlier research. For efficient treatment of leachate, adsorption is frequently utilized before or after the biological stage because biological processes cannot remove non-biodegradable materials (Torretta *et al.*, 2016). GAC has also been claimed to be able to remove 85% of non-biodegradable contaminants (Rodriguez *et al.*, 2004) and an 80% reduction in total carbon (Singh *et al.*, 2012). Additionally, using AC to adsorb the remaining 75% of the total iron and PO_4-P is a viable approach (Ching *et al.*, 2011). It can also remove ammonium of up to 40% and heavy metals of up to 90% (Kurniawan *et al.*, 2006). Adsorption can enhance sludge dewaterability in addition to increasing nitrification effectiveness and removing refractory chemicals (Aghamohammadi *et al.*, 2007). As mentioned earlier, both inorganic and organic contaminants can be effectively removed using ACs. Overall, AC can be utilized as an adsorbent for removing different contaminants from leachate.

10.5.2.2 Biochar

Biochar is produced as a by-product during the conversion of organic materials using different thermochemical techniques such as pyrolysis, gasification, and torrefaction. Pyrolysis can convert waste into biochar at a temperature of 300–900°C in the absence of oxygen (Kumar *et al.*, 2022a). Biochar can also be produced from different waste using gasification at a temperature >700°C using different gasifying agents such as steam, oxygen, and CO_2 (Mishra *et al.*, 2022c). Torrefaction can convert biomass into biochar at a mild temperature of 200–300°C in the absence of oxygen (Kumar *et al.*, 2022a). Waste can be converted into biochar using pyrolysis and other thermochemical techniques, which also yields valuable products such as gas, liquid, and solid hydrocarbons (Kumar *et al.*, 2021, 2022a; Mishra *et al.*, 2022c). As a result, biochar generated from waste can be utilized for generation of energy or adsorbents (Kumar *et al.*, 2020, 2022a). Figure 10.2 shows the biochar production from different feedstocks using various techniques and its application for landfill leachate treatment. Several research studies suggested that biochar generated from MSW might be utilized effectively to remove environmental pollutants (Singh *et al.*, 2022a, 2022b). Table 10.3 presents removal of various contaminants using biochar produced from different feedstocks. Arsenic and chromium were effectively removed from an aqueous solution using biochar treatment (Agrafioti *et al.*, 2014; Singh *et al.*, 2021b). Biochar could be employed directly or modified to provide a higher potential for organic contaminants and removal of heavy metals. The amount and types of functional groups in biochar may be manipulated by chemical changes, which could modify their porous structure and affect their adsorption ability (Kumar *et al.*, 2022a, 2022b). Additionally, chemical activation could be carried out at substantially lower temperatures with an improved catalytic oxidation potential because activation and carbonation were combined in a single phase (Kumar *et al.*, 2022a; Singh *et al.*, 2022b). Consequently, biochar activated with chemicals may offer an effective, easy, and affordable method for

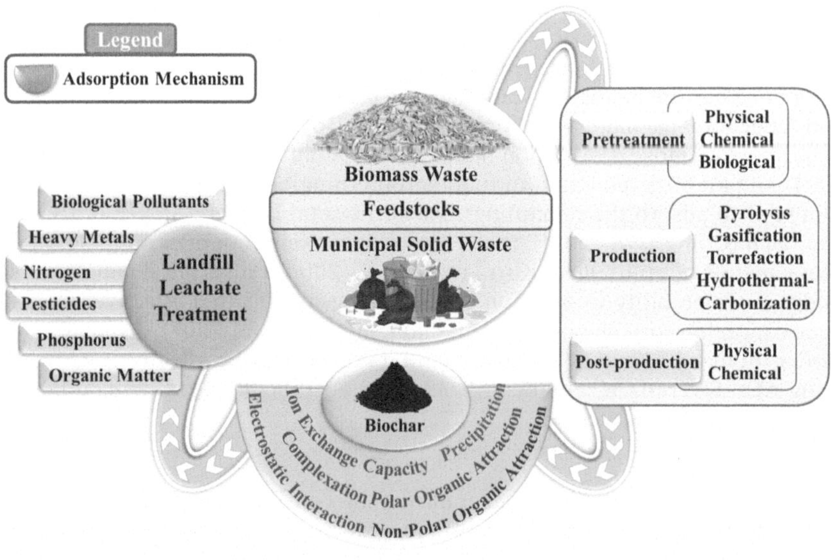

Figure 10.2 Biochar production from different feedstocks using various techniques and its application for landfill leachate treatment.

Table 10.3 Various contaminant removal using biochar produced from different feedstock.

Biochar Feedstock	Production Technique/ Temperature (°C)	Effect on Contaminant Removal	Reference
Pine-tree needles	Pyrolysis/500 and 700	Hg(II) adsorption of 0.35 g mg/min	Jeon *et al.* (2020)
Biomass and plastic waste	Pyrolysis/450–550	Arsenic removal of ~99.86 and 99.93% from biochar and plastic char, respectively	Singh *et al.* (2021b)
Bamboo, bagasse, and hickory chips	Pyrolysis/300, 450, and 600	Removal of SPY (89.6%) and SMX (83.3%)	Huang *et al.* (2020)
Ragweed	Pyrolysis/450	~139 and 358.7 mg/g removal of Cd^{2+} and Pb^{2+}	Lian *et al.* (2020)
Rice straw	Pyrolysis/500	Removal of Sr(II) and Ba(II)	Younis *et al.* (2020)
Pine-forestry	Pyrolysis/300	~97% removal of organic matter	Yazdani *et al.* (2019)
Banana pseudostem	Pyrolysis/300–600	~96.81% removal of nitrofuran antibiotic	Gurav *et al.* (2020)
Sewage sludge	Pyrolysis/300–700	~67–99% reduction of contaminant	Regkouzas and Diamadopoulos (2019)
Gingko (*Spiraea blumei*) leaf	Electric-furnace carbonization/800	~59.9 and 138.9 mg/g adsorption of Cu and Pb, respectively	Lee *et al.* (2019)
Loofah	Pyrolysis/700	~54.68 and 30.14 mg/g removal of Cu(II) and Cr(VI), respectively	Xiao *et al.* (2019)
Softwood pine	Pyrolysis/550–600	Ni(II) removal of ~73.5%	Sajjadi *et al.* (2019)
Enteromorpha	Hydrothermal liquefaction/250	~91 and 54% removal of Cu(II) and Pb(II), respectively	Yang *et al.* (2019)

Sulfapyridine (SPY); Sulfamethoxazole (SMX).

eliminating environmental pollutants (Kumar *et al.*, 2022a; Mishra *et al.*, 2021). Due to biochar's great effectiveness and inexpensive cost, chloride precursors such as NH_4Cl have been effectively employed to chemically transform biochar (Li *et al.*, 2015). Additionally, employing 2 M potassium hydroxide, Jin *et al.* (2014) modified biochar with a high capacity of adsorption. In addition to chemical activation, physical activation utilizing microwave energy and steam has both been shown to produce biochar (Li *et al.*, 2015). Experimental data modeling may reveal the effectiveness of various removal techniques using

MSW–Biochar (BC). Most investigations revealed that the pseudo-second-order kinetic model ascribed to the Langmuir isotherm and the chemisorption suggesting mono-layer sorption fit the heavy metal adsorption onto biochar (Kumar *et al.*, 2022a; Mishra *et al.*, 2022b). According to Agrafioti *et al.* (2014), electrostatic interaction between the metal cation and the negatively charged biochar surface may be the mechanism of biochar for the adsorption of heavy metals. Additionally, the high pore volume and large surface area speed up the mass transfer of heavy metals into the pores of biochar, increasing the potential for metal-binding sites (Kumar *et al.*, 2022a; Singh *et al.*, 2022b). As a result, biochar that has been chemically and physically modified has a higher adsorption potential than biochar without modification, perhaps because of the increased pore volume and surface area. Hydrogen bonding, electrostatic interactions, and van der Waals forces were also suggested as three ways for describing the rhodamine dyes and acridine orange's adsorption on biochar (Parshetti *et al.*, 2014). Batch desorption experiments for arsenic adsorption onto various biochar types created from various feedstocks, including different biomass and plastic waste, have been carried out (Kumar *et al.*, 2021; Singh *et al.*, 2021b). As a result, it appears that biochar created from waste has the ability to be utilized as an effective sorbent to remove pollutants from leachates. The adsorption of organic contaminants, that is, HAs and pesticides and HAs onto biochar in addition to heavy metals has also been documented in some literature. From the above-mentioned explanation, it can be confirmed that biochar can be utilized as an efficient adsorbent for leachate treatment.

10.5.2.3 Graphene oxide

The advantage of physical–chemical methods for GO synthesis is its applicability on a wider scale (Mishra *et al.*, 2022a). Additionally, functionalization, impregnation, deposition, and mixing are some of the technical benefits of turning graphene into colloidal hydrophilic dispersions (Mishra *et al.*, 2022a). Despite the synthesis process, the most unique feature of graphene oxidation is the multi-oxidant nature of the chemical reaction. In comparison to other graphene-based materials, it is easier to achieve surface changes in GO because of the hydrophilicity caused by functional groups with the content of oxygen at the GO edges (Stankovich *et al.*, 2006). The typical GO synthesis processes have been reported by Hummers and Offeman (1958), Staudenmaier (1898), and Brodie (1859), which involve chemically oxidizing and exfoliating pure graphite utilizing various oxidizing agents (Gascho *et al.*, 2019). Sujiono *et al.* (2020) synthesized GO from coconut shells using a modified Hummers method. First, high-carbon charcoal was produced from coconut shells at 600°C for 3 h. The carbonized sample was agitated for 3 h at 45°C with a sample-to-acid ratio of 1:3. The produced graphite powder was then dried in an oven at 110°C for 12 h after being rinsed with deionized water and NaOH until a pH of 6–7 was achieved. Materials created from GO have often been used for effective and efficient removal of heavy metals through adsorption. Table 10.4 presents the removal of different contaminants by various GO-based nanomaterials. Wang *et al.* (2018) have found that poly(γ-glutamic acid) (γ-PGA)–Fe_3O_4–GO–(*o*-multi-walled carbon nanotubes (MWCNTs)) may cause the adsorption of Ni(II), Cu(II), and

Table 10.4 Removal of different contaminants by various GO-based nanomaterials.

Adsorbent	Target Contaminant	Adsorption Capacity (mg/g)	Reference
Fe_2O_3 *x*% PANI@GO	Cr(VI)	143.54	Chinnathambi and Alahmadi (2021)
Al_2O_3/GO cellulose	F^-	5.34	Singh *et al.* (2021c)
Fe_2O_3–GO/CS	Cr(VI)	131.33	Shan *et al.*, (2021)
PV/GO/MnO_2	Co(II)	60.3	Huo *et al.* (2021)
Ash/GO/FeO_4	Pb(II)	47.16	Pelalak *et al.* (2021)
α-Fe_2O_3@rGO	Tetracycline	180.8	Zou *et al.* (2021)
GO–CMC	Oxytetracycline	102.05	Juengchareonpoon *et al.* (2021)
GO/g-C_3N_4–Fe_3O_4	Tetracycline	170	Sahoo *et al.* (2020)
Mn–PBA/GO	Ciprofloxacin	1,826.64	Khan *et al.* (2020)
CNT_5/LCys@GO/SA	Ciprofloxacin	200	Ma *et al.* (2020)
Magnetic GO/chitosan composite (MGOC)	I – Naphthylamine	142.7	Rebekah *et al.* (2021)
amino-poly(vinylamine) (PVAm)–GO–(*o*-(MWCNTs)–Fe_3O_4	Phenol	224.21	Zhou *et al.* (2020)
FFGOGA	Naphthalene	116.7	Arshad *et al.* (2020)
GO–EPI–L-Arg	Endotoxin	154.85(Eu/g)	Tapouk *et al.* (2020)
GO/MnO_x	*p*-Cresol	107.68	Zhang *et al.* (2020)

Cd(II). Graphene was produced electrochemically with the help of an ionic liquid. By using this technique, a magnetic Fe_3O_4–GO–(*o*-MWCNTs) hybrid nanocomposite with anionic polypeptide γ-PGA functionalization was generated through an easy one-pot synthesis (Obayomi *et al.*, 2022). The γ-PGA–Fe_3O_4–GO–(*o*-MWCNTs) aggregates showed potential absorbent material for removing different kinds of heavy metals, displaying high capabilities of adsorption. The highest adsorption removal capacities determined by the Langmuir model for Ni(II), Cu(II), and Cd(II) were 384.62, 574.71, and 625.00 mg/g, respectively, under ideal circumstances. A GO/carboxymethyl cellulose nanofibril (CMCNF) composite fiber (CF) was developed by Yu *et al.* (2020) as an adsorbent to remove Pb(II). Continuous production of cross-linked GO/CMCNF CF was achieved by using the coagulant Fe^{3+} ions during a conventional wet-spinning procedure. Due to several interactions, including electrostatic interactions and ionic bonding between Fe^{3+} and the carboxyl group on CMCNF, the CF has greater mechanical characteristics in comparison to pure GO fiber. Pb(II) could adsorb up to 101.4 mg/g of Pb onto GO/CMCNF–Fe^{3+} CF with effective adsorbent recovery. The adsorption properties are affected by different factors such as temperature, which affects the material surface binding characteristics and solutions thermodynamics; organic matters; or metal ions coexistence, which prevents the desirable molecules of adsorption from occurring on the solid surfaces and solid–liquid ratio, which impacts the numbers of functional groups and binding sites that are available. GO/manganese oxide composites were used by Zhang *et al.* (2020) to remove phenolic chemicals. For *p-tert*-butyl-phenol (*p*-TBP) and *p*-cresol, the highest adsorption capacities of GO/MnO_x composites were 135.41 and 107.68 mg/g, respectively. Even after five recycles, they claimed the composites of GO/MnO_x showed a high effectiveness of more than 90% for the removal of *p*-TBP. Additionally, they suggested that the presence of GO sheets boosted the MnO_x's oxidation capacity through an electron transfer interaction in addition to promoting the adsorption of phenolic compounds and by-products. Hu *et al.* (2015) have documented the adsorption of aniline and phenol from aqueous solutions onto GO/PPy composites. According to reports, aniline and phenol have sorption capabilities of 7.57 and 3.31 mmol/g on GO/PPy composites, respectively. The sorption of aniline and phenol was explained using the Langmuir model on the composites of GO/PPy, and π–π electron donor–acceptor interaction, Lewis' acid–base interaction, hydrophobic interaction, and ion exchange were primarily responsible. The thermodynamic properties of the temperature-dependent sorption isotherms showed that the adsorption of aniline and phenol on the composites of GO/PPy was a spontaneous and endothermic process. There was no discernible reduction in adsorption capacity as a result of the aniline and phenol being desorbed by ethanol and cycled for GO/PPy composites reuse.

10.5.2.4 Carbon nanotubes

Regarding the treatment of landfill leachate, another best option that has received considerable attention is CNTs. Chemical vapor deposition (CVD) is the primary method utilized to generate CNTs; however, laser ablation and arc-discharge synthesis have also been employed. CNTs are produced on the negative electrode via an electrical arc discharge between two carbon electrodes. A catalyst added to the electrode promotes the formation of single-walled nanotubes. Carbon vaporization and high temperatures, that is, 1,200°C are involved in laser ablation. Carbon-rich gases combine with catalysts during CVD procedure at temperatures of 600–1,200°C, resulting in the formation of CNTs on the catalyst surface (Bazargan & McKay, 2012). In landfill leachate studies, only single-walled carbon nanotubes (SWCNTs) seem to have been studied, and the findings show promise for high molecular weight organic adsorption (Lozano & Berge, 2012). MWCNTs are composed of graphene with numerous layers that are arranged in a concentric pattern around the smallest nanotube. SWCNTs are comprised of a single layer of graphene that normally includes tube-shaped carbon benzene rings. Chemical stability, high thermal and electrical properties, low mass, notable surface density, and higher diameter ratio are some unique characteristics of cylindrical CNTs (Kavosi *et al.*, 2018). Due to exceptional water-transport capabilities, exceptional chemical inertness, high mechanical strength, and greater specific surface area of MWCNTs and SWCNTs, they have received a lot of attention for removing pollutants (Ma *et al.*, 2017). However, there is much less research about their application, regardless of the possibility of landfill leachate treatment. As a result, given the efficacy of CNTs in the remediation of contaminated wastewater, this is considered as an important research subject that needs further study. It is also significant to highlight that while evaluating CNTs for environmental applications, any shortcomings must be taken into account. For instance, it has been observed that CNTs may be hazardous in a variety of cell types and animal studies (Zhao *et al.*, 2021). The evidence that is currently available, which is obviously limited, points to CNTs as being more problematic in comparison to other non-nanotube kinds of carbon. Exposure to CNTs can arise while working on their synthesis, utilizing them, or evaluating their characteristics. This emphasizes the requirement for a preventative strategy for controlling the exposure risk to these materials. The ecological impact and financial expense of CNTs have also been noted as disadvantages. Costs for MWCNTs have significantly lowered due to scaling up, using low-cost feedstock, increasing yields, reducing energy use, and producing less waste (De Volder *et al.*, 2013). The elimination of these contaminants frequently necessitates expensive thermal and chemical treatment and is a by-product of large-scale CNT generation processes. These processes have the potential to reduce and incorporate flaws into CNT sidewalls. MWCNTs are now offered in bulk for <$100/kg, which is 10 times more expensive in comparison to commercial carbon fibers (Clurman *et al.*, 2020). As a result, this shows how crucial it is to conduct additional research into CNTs, taking into account all of their positive qualities and whether or not the benefits can offset the considerable drawbacks mentioned earlier.

10.5.3 Other non-carbon-based materials

Magnetic particles are an intriguing new material that can be used to remediate landfill leachate. With the utilization of magnetic separation and particles, magnetic technologies have been utilized in order to remove contaminants (Rodriguez-Narvaez *et al.*, 2019). Nevertheless, despite the extremely optimistic findings achieved, only a few research studies have been carried out in regard to the treatment of leachate, even at a laboratory scale (Augusto *et al.*, 2019). Natural Fe_2O_3 nanoparticles were employed by Shadi *et al.* (2020) to remediate polluted landfill leachate. The color, NH_3-N, and COD were 2,851 Pt–Co, 449 mg/L, and 2,081 mg/L, respectively. The elimination of COD, NH_3-N, and color by adsorption on natural Fe_2O_3 was 75.9%, 43.8%, and 97%, respectively. Comparable outcomes have been achieved by several researchers using other magnetic nanoparticles alone or in combination with other methods of treatment for leachate (Kim & Ko, 2018). It is assumed that the lack of information on the prospective impacts on the ecology of these adsorbents after their usage in leachate treatment is an important gap demanding more in-depth investigations for handling any potential implications of large-scale implementation. This is because there are a large numbers of research studies currently being carried out on the utilization of magnetic nanomaterials for leachate treatment and uplifting findings have been confirmed so far for their implementation at the laboratory scale. Another significant topic for further study is the application of modified zeolites in leachate remediation. Due to the special qualities of zeolites, such as high selectivity, activity, and stability, the application of modified zeolites as adsorbents for the oxidation/reduction of some particular, simple chemicals has received growing interest (Bandala *et al.*, 2021). According to Vollprecht *et al.* (2019), processing clinoptilolite with NaOH eliminates Si while maintaining Al in the structure of the zeolite, increasing the amount of acidic adsorptive sites and the zeolite's capacity to adsorb NH_4^+ from landfill leachate for its recovery. Other research studies have discovered that physicochemically activated zeolites are particularly successful for the nutrient's adsorption (such as phosphate and ammonium) from landfill leachate when used with a zeolite dose of ~15 g/L and contact time of 12 h (Aziz *et al.*, 2020). However, there are only a small numbers of research studies that have evaluated the utilization of modified zeolites for removing the organic matter (TOC, BOD, or COD) or nutrients in landfill leachate. Such investigations have also evaluated the capacity of materials for removing particular contaminants, such as organic pollutants or heavy metals. This lack of knowledge leads to a significant information gap because modified zeolites have been shown to be able to act as catalysts for advanced oxidation processes in addition to being adsorbent, which allows for the potential of successive processes of degradation (Bandala *et al.*, 2021).

10.6 CONCLUSION

Municipal landfill leachate is a prevalent environmental load and a serious risk to surface and groundwater due to the presence of harmful substances such as nitrates, organic pollutants, and heavy metals. Diverse environmentally friendly strategies and techniques have been suggested and evaluated to remediate extremely contaminated leachates in order to comply with demanding quality standards for direct disposal into water bodies or sewer systems. The operating costs, plant flexibility, and investment as well as reliability, overall treatment performance, legal criteria for effluent levels of NH_4^+-N, COD, and heavy metals as well as leachate characteristics (such as landfill age, BOD/COD, and COD) play major roles in choosing the best treatment technology for landfill leachate. Physical, chemical, and biological treatments are the main types of landfill leachate treatment. The efficiency of the adsorption process for treating landfill leachate relies on environmentally friendly, cost-effective, and well-resistant pollutants. Due to high cost of commercial adsorbents, these are very occasionally used for leachate treatment. Numerous studies have noted that these adsorbents have been replaced with carbon-based adsorbents, which can be produced using different waste. Additionally, carbon-based adsorbents offer a powerful substitute to increase the effectiveness of the treatment. Several investigators are using the dispersion of nanoparticles (nanomaterial adsorbents) onto porous-supporting materials to reduce surface passivation, leaching, and aggregation of nanoparticles. Overall, it is clear that carbon-based adsorbents are a useful method for lowering the amounts of different leachate parameters, including heavy metals, ammonia nitrogen, color, and COD.

REFERENCES

Abdullah A. H., Kassim A., Zainal Z., Hussien M. Z., Kuang D., Ahmad F. and Wooi O. S. (2001). Preparation and characterization of activated carbon from gelam wood bark (*Melaleuca cajuputi*). *Malaysian Journal of Analytical Sciences*, **7**(1), 65–68.

Abuabdou S. M., Ahmad W., Aun N. C. and Bashir M. J. (2020). A review of anaerobic membrane bioreactors (AnMBR) for the treatment of highly contaminated landfill leachate and biogas production: effectiveness, limitations and future perspectives. *Journal of Cleaner Production*, **255**, 120215, https://doi.org/10.1016/j.jclepro.2020.120215

Aftab B. and Hur J. (2019). Unraveling complex removal behavior of landfill leachate upon the treatments of Fenton oxidation and MIEX® via two-dimensional correlation size exclusion chromatography (2D-CoSEC). *Journal of Hazardous Materials*, **362**, 36–44, https://doi.org/10.1016/j.jhazmat.2018.09.017

Aghamohammadi N., Bin Abdul Aziz H., Isa M. H. and Zinatizadeh A. A. (2007). Powdered activated carbon augmented activated sludge process for treatment of semi-aerobic landfill leachate using response surface methodology. *Bioresource Technology*, **98**(18), 3570–3578, https://doi.org/10.1016/j.biortech.2006.11.037

Agrafioti E., Kalderis D. and Diamadopoulos E. (2014). Arsenic and chromium removal from water using biochars derived from rice husk, organic solid wastes and sewage sludge. *Journal of Environmental Management*, **133**, 309–314, https://doi.org/10.1016/j.jenvman.2013.12.007

Allwar A. (2012). Characteristics of micro-and mesoporous structure and surface chemistry of activated carbons produced by oil palm shell. International Conference on Chemical, Ecology and Environmental Sciences Proceedings, Singapore, pp. 138–141.

Al-Qodah Z. and Shawabkah R. (2009). Production and characterization of granular activated carbon from activated sludge. *Brazilian Journal of Chemical Engineering*, **26**, 127–136, https://doi.org/10.1590/S0104-66322009000100012

Alslaibi T. M., Abustan I., Ahmad M. A. and Foul A. A. (2013). A review: production of activated carbon from agricultural byproducts via conventional and microwave heating. *Journal of Chemical Technology & Biotechnology*, **88**(7), 1183–1190, https://doi.org/10.1002/jctb.4028

Álvarez-Torrellas S., Martin-Martinez M., Gomes H. T., Ovejero G. and García J. (2017). Enhancement of *p*-nitrophenol adsorption capacity through N_2-thermal-based treatment of activated carbons. *Applied Surface Science*, **414**, 424–434, https://doi.org/10.1016/j.apsusc.2017.04.054

Arshad F., Selvaraj M., Banat F. and Haija M. A. (2020). Removal of metal ions and organics from real refinery wastewater using double-functionalized graphene oxide in alginate beads. *Journal of Water Process Engineering*, **38**, 101635, https://doi.org/10.1016/j.jwpe.2020.101635

Augusto P. A., Castelo-Grande T., Merchan L., Estevez A. M., Quintero X. and Barbosa D. (2019). Landfill leachate treatment by sorption in magnetic particles: preliminary study. *Science of the Total Environment*, **648**, 636–668, http://10.1016/j.scitotenv.2018.08.056

Aziz H. A., Noor A. F. M., Keat Y. W., Alazaiza M. Y. and Hamid A. A. (2020). Heat activated zeolite for the reduction of ammoniacal nitrogen, colour, and COD in landfill leachate. *International Journal of Environmental Research*, **14**(4), 463–478, https://doi.org/http://10.1007/s41742-020-00270-5

Babaei S., Sabour M. R. and Moftakhari Anasori Movahed S. (2021). Combined landfill leachate treatment methods: an overview. *Environmental Science and Pollution Research*, **28**(42), 59594–59607, https://doi.org/10.1007/s11356-021-16358-0

Babayemi J. O. and Dauda K. T. (2009). Evaluation of solid waste generation, categories and disposal options in developing countries: a case study of Nigeria. *Journal of Applied Sciences and Environmental Management*, **13**(3), 83–88, https://doi.org/10.4314/jasem.v13i3.55370

Bandala E. R., Sadek R., Gurgul J., Łątka K., Zimowska M., Valentin L., Rodriguez-Narvaez O. M. and Dzwigaj S. (2021). Assessment of the capability of Fe and Al modified BEA zeolites to promote advanced oxidation processes in aqueous phase. *Chemical Engineering Journal*, **409**, 127379, https://doi.org/10.1016/j.cej.2020.127379

Bazargan A. and McKay G. (2012). A review – synthesis of carbon nanotubes from plastic wastes. *Chemical Engineering Journal*, **195**, 377–391, https://doi.org/10.1016/j.cej.2012.03.077

Bhatt A. H., Karanjekar R. V., Altouqi S., Sattler M. L., Hossain M. S. and Chen V. P. (2017). Estimating landfill leachate BOD and COD based on rainfall, ambient temperature, and waste composition: exploration of a MARS statistical approach. *Environmental Technology & Innovation*, **8**, 1–16, https://doi.org/10.1016/j.eti.2017.03.003

Bolyard S. C., Reinhart D. R. and Richardson D. (2019). Conventional and Fourier transform infrared characterization of waste and leachate during municipal solid waste stabilization. *Chemosphere*, **227**, 34–42, https://doi.org/10.1016/j.chemosphere.2019.04.035

Brodie B. C. (1859). XIII. On the atomic weight of graphite. *Philosophical Transactions of the Royal Society of London*, **149**, 249–259, https://doi.org/10.1098/rstl.1859.0013

Chan L. S., Cheung W. H., Allen S. J. and McKay G. (2012). Error analysis of adsorption isotherm models for acid dyes onto bamboo derived activated carbon. *Chinese Journal of Chemical Engineering*, **20**(3), 535–542, https://doi.org/10.1016/S1004-9541(11)60216-4

Chen W., Liu X. Y., Qian C., Song X. N., Li W. W. and Yu H. Q. (2015). An UV–Vis spectroelectrochemical approach for rapid detection of phenazines and exploration of their redox characteristics. *Biosensors and Bioelectronics*, **64**, 25–29, https://doi.org/10.1016/j.bios.2014.08.032

Chen W., Luo Y., Ran G. and Li Q. (2019a). An investigation of refractory organics in membrane bioreactor effluent following the treatment of landfill leachate by the O_3/H_2O_2 and MW/PS processes. *Waste Management*, **97**, 1–9, https://doi.org/10.1016/j.wasman.2019.07.016

Chen W., Zhang A., Jiang G. and Li Q. (2019b). Transformation and degradation mechanism of landfill leachates in a combined process of SAARB and ozonation. *Waste Management*, **85**, 283–294, https://doi.org/10.1016/j.wasman.2018.12.038

Cheng S. Y., Show P. L., Juan J. C., Chang J. S., Lau B. F., Lai S. H., Ng E. P., Yian H. C. and Ling T. C. (2021). Landfill leachate wastewater treatment to facilitate resource recovery by a coagulation–flocculation process via hydrogen bond. *Chemosphere*, **262**, 127829, https://doi.org/10.1016/j.chemosphere.2020.127829

Ching S. L., Yusoff M. S., Aziz H. A. and Umar M. (2011). Influence of impregnation ratio on coffee ground activated carbon as landfill leachate adsorbent for removal of total iron and orthophosphate. *Desalination*, **279**(1–3), 225–234, https://doi.org/10.1016/j.desal.2011.06.011

Chinnathambi A. and Alahmadi T. A. (2021). Facile synthesis of Fe_3O_4 anchored polyaniline intercalated graphene oxide as an effective adsorbent for the removal of hexavalent chromium and phosphate ions. *Chemosphere*, **272**, 129851, https://doi.org/10.1016/j.chemosphere.2021.129851

Clurman A. M., Rodríguez-Narvaez O. M., Jayarathne A., De Silva G., Ranasinghe M. I., Goonetilleke A. and Bandala E. R. (2020). Influence of surface hydrophobicity/hydrophilicity of biochar on the removal of emerging contaminants. *Chemical Engineering Journal*, **402**, 126277, https://doi.org/10.1016/j.cej.2020.126277

Crini G., Lichtfouse E., Wilson L. D. and Morin-Crini N. (2019). Conventional and non-conventional adsorbents for wastewater treatment. *Environmental Chemistry Letters*, **17**(1), 195–213, https://doi.org/10.1007/s10311-018-0786-8

Cuhadaroglu D. and Uygun O. A. (2008). Production and characterization of activated carbon from a bituminous coal by chemical activation. *African Journal of Biotechnology*, **7**(20), 3703–3710.

Dąbrowski A. (2001). Adsorption – from theory to practice. *Advances in Colloid and Interface Science*, **93**(1–3), 135–224, https://doi.org/10.1016/S0001-8686(00)00082-8

Darweesh T. M. and Ahmed M. J. (2017). Adsorption of ciprofloxacin and norfloxacin from aqueous solution onto granular activated carbon in fixed bed column. *Ecotoxicology and Environmental Safety*, **138**, 139–145, https://doi.org/10.1016/j.ecoenv.2016.12.032

De Volder M. F., Tawfick S. H., Baughman R. H. and Hart A. J. (2013). Carbon nanotubes: present and future commercial applications. *Science*, **339**(6119), 535–539, https://doi.org/10.1126/science.1222453

Djeffal K., Bouranene S., Fievet P., Déon S. and Gheid A. (2021). Treatment of controlled discharge leachate by coagulation–flocculation: influence of operational conditions. *Separation Science and Technology*, **56**(1), 168–183, https://doi.org/10.1080/01496395.2019.1708114

Egun N. K. (2012). The waste to wealth concept; waste market operation in Delta state Nigeria. *Greener Journal of Social Sciences*, **2**(6), 206–212, https://doi.org/10.15580/GJSS.2012.6.110112191

Elmaadawy K., Liu B., Hu J., Hou H. and Yang J. (2020). Performance evaluation of microbial fuel cell for landfill leachate treatment: research updates and synergistic effects of hybrid systems. *Journal of Environmental Sciences*, **96**, 1–20, https://doi.org/10.1016/j.jes.2020.05.005

Foo K. Y. and Hameed B. H. (2009). An overview of landfill leachate treatment via activated carbon adsorption process. *Journal of Hazardous Materials*, **171**(1–3), 54–60, https://doi.org/10.1016/j.jhazmat.2009.06.038

Gascho J. L., Costa S. F., Recco A. A. and Pezzin S. H. (2019). Graphene oxide films obtained by vacuum filtration: X-ray diffraction evidence of crystalline reorganization. *Journal of Nanomaterials*, **2019**, 1–12, https://doi.org/10.1155/2019/5963148

Giraldo L. and Moreno-Piraján J. C. (2012). Synthesis of activated carbon mesoporous from coffee waste and its application in adsorption zinc and mercury ions from aqueous solution. *E-Journal of Chemistry*, **9**(2), 938–948, https://doi.org/10.1155/2012/120763

Gurav R., Bhatia S. K., Choi T. R., Park Y. L., Park J. Y., Han Y. H., Vyavahare G., Jadhav J., Song H. S., Yang P. and Yoon J. J. (2020). Treatment of furazolidone contaminated water using banana pseudostem biochar engineered with facile synthesized magnetic nanocomposites. *Bioresource Technology*, **297**, 122472, https://doi.org/10.1016/j.biortech.2019.122472

Hassan A. F. and Elhadidy H. (2017). Production of activated carbons from waste carpets and its application in methylene blue adsorption: kinetic and thermodynamic studies. *Journal of Environmental Chemical Engineering*, **5**(1), 955–963, https://doi.org/10.1016/j.jece.2017.01.003

He X. S., Xi B. D., Gao R. T., Zhang H., Dang Q. L., Li D. and Huang C. H. (2016). Insight into the composition and degradation potential of dissolved organic matter with different hydrophobicity in landfill leachates. *Chemosphere*, **144**, 75–80, https://doi.org/10.1016/j.chemosphere.2015.08.071

Hidayu A. R. and Muda N. (2016). Preparation and characterization of impregnated activated carbon from palm kernel shell and coconut shell for CO_2 capture. *Procedia Engineering*, **148**, 106–113, https://doi.org/10.1016/j.proeng.2016.06.463

Hong N., Cheng Q., Goonetilleke A., Bandala E. R. and Liu A. (2020). Assessing the effect of surface hydrophobicity/hydrophilicity on pollutant leaching potential of biochar in water treatment. *Journal of Industrial and Engineering Chemistry*, **89**, 222–232, https://doi.org/10.1016/j.jiec.2020.05.017

Hoornweg D. and Bhada-Tata P. (2012). What a waste: a global review of solid waste management.

Hu R., Dai S., Shao D., Alsaedi A., Ahmad B. and Wang X. (2015). Efficient removal of phenol and aniline from aqueous solutions using graphene oxide/polypyrrole composites. *Journal of Molecular Liquids*, **203**, 80–89, https://doi.org/10.1016/j.molliq.2014.12.046

Huang J., Zimmerman A. R., Chen H. and Gao B. (2020). Ball milled biochar effectively removes sulfamethoxazole and sulfapyridine antibiotics from water and wastewater. *Environmental Pollution*, **258**, 113809, https://doi.org/10.1016/j.envpol.2019.113809

Huguet A., Vacher L., Relexans S., Saubusse S., Froidefond J. M. and Parlanti E. (2009). Properties of fluorescent dissolved organic matter in the Gironde Estuary. *Organic Geochemistry*, **40**(6), 706–719, https://doi.org/10.1016/j.orggeochem.2009.03.002

Hummers Jr W. S. and Offeman R. E. (1958). Preparation of graphitic oxide. *Journal of the American Chemical Society*, **80**(6), 1339–1339, https://doi.org/10.1021/ja01539a017

Huo J., Yu G. and Wang J. (2021). Efficient removal of Co(II) and Sr(II) from aqueous solution using polyvinyl alcohol/graphene oxide/MnO_2 composite as a novel adsorbent. *Journal of Hazardous Materials*, **411**, 125117, https://doi.org/10.1016/j.jhazmat.2021.125117

Ioannidou O. and Zabaniotou A. (2007). Agricultural residues as precursors for activated carbon production – a review. *Renewable and Sustainable Energy Reviews*, **11**(9), 1966–2005, https://doi.org/10.1016/j.rser.2006.03.013

Ismail A. R. and Baek K. H. (2020). Lipase immobilization with support materials, preparation techniques, and applications: present and future aspects. *International Journal of Biological Macromolecules*, **163**, 1624–1639, https://doi.org/10.1016/j.ijbiomac.2020.09.021

Jeon C., Solis K. L., An H. R., Hong Y., Igalavithana A. D. and Ok Y. S. (2020). Sustainable removal of Hg(II) by sulfur-modified pine-needle biochar. *Journal of Hazardous Materials*, **388**, 122048, https://doi.org/10.1016/j.jhazmat.2020.122048

Jin H., Capareda S., Chang Z., Gao J., Xu Y. and Zhang J. (2014). Biochar pyrolytically produced from municipal solid wastes for aqueous As(V) removal: adsorption property and its improvement with KOH activation. *Bioresource Technology*, **169**, 622–629, https://doi.org/10.1016/j.biortech.2014.06.103

Juengchareonpoon K., Wanichpongpan P. and Boonamnuayvitaya V. (2021). Graphene oxide and carboxymethylcellulose film modified by citric acid for antibiotic removal. *Journal of Environmental Chemical Engineering*, **9**(1), 104637, https://doi.org/10.1016/j.jece.2020.104637

Kamaruddin M. A., Yusoff M. S., Rui L. M., Isa A. M., Zawawi M. H. and Alrozi R. (2017). An overview of municipal solid waste management and landfill leachate treatment: Malaysia and Asian perspectives. *Environmental Science and Pollution Research*, **24**(35), 26988–27020, https://doi.org/10.1007/s11356-017-0303-9

Kan Y., Yue Q., Li D., Wu Y. and Gao B. (2017). Preparation and characterization of activated carbons from waste tea by H_3PO_4 activation in different atmospheres for oxytetracycline removal. *Journal of the Taiwan Institute of Chemical Engineers*, **71**, 494–500, https://doi.org/10.1016/j.jtice.2016.12.012

Kavosi A., Hosseini Ghale Noei S., Madani S., Khalighfard S., Khodayari S., Khodayari H., Mirzaei M., Kalhori M. R., Yavarian M., Alizadeh A. M. and Falahati M. (2018). The toxicity and therapeutic effects of single- and multi-wall carbon nanotubes on mice breast cancer. *Scientific Reports*, **8**(1), 1–12, https://doi.org/10.1038/s41598-018-26790-x

Khan N. A., Najam T., Shah S. S. A., Hussain E., Ali H., Hussain S., Shaheen A., Ahmad K. and Ashfaq M. (2020). Development of Mn-PBA on GO sheets for adsorptive removal of ciprofloxacin from water: kinetics, isothermal, thermodynamic and mechanistic studies. *Materials Chemistry and Physics*, **245**, 122737, https://doi.org/10.1016/j.matchemphys.2020.122737

Kim D. G. and Ko S. O. (2018). Cu@ Fe_3O_4 core-shell nanoparticle-catalyzed oxidative degradation of the antibiotic oxytetracycline in pre-treated landfill leachate. *Chemosphere*, **191**, 639–650, https://doi.org/10.1016/j.chemosphere.2017.10.090

Kjeldsen P., Barlaz M. A., Rooker A. P., Baun A., Ledin A. and Christensen T. H. (2002). Present and long-term composition of MSW landfill leachate: a review. *Critical Reviews in Environmental Science and Technology*, **32**(4), 297–336, https://doi.org/10.1080/10643380290813462

Kumar A., Singh E., Mishra R. and Kumar S. (2020). Solid waste to energy: existing scenario in developing and developed countries. In: Handbook of Solid Waste Management: Sustainability through Circular Economy, C. Baskar, S. Ramakrishna, S. Baskar, R. Sharma, A. Chinnappan, R. Sehrawat (eds.), Springer Nature Singapore, Singapore, pp. 1–23.

Kumar S., Singh E., Mishra R., Kumar A. and Caucci S. (2021). Utilization of plastic wastes for sustainable environmental management: a review. *ChemSusChem*, **14**(19), 3985–4006, https://doi.org/10.1002/cssc.202101631

Kumar A., Singh E., Mishra R. and Kumar S. (2022a). Biochar as environmental armour and its diverse role towards protecting soil, water and air. *Science of the Total Environment*, **806**, 150444, https://doi.org/10.1016/j.scitotenv.2021.150444

Kumar A., Singh E., Mishra R., Lo S. L. and Kumar S. (2022b). A green approach towards sorption of CO_2 on waste derived biochar. *Environmental Research*, **214**, 113954, https://doi.org/10.1016/j.envres.2022.113954

Kurniawan T. A., Lo W. H. and Chan G. Y. (2006). Degradation of recalcitrant compounds from stabilized landfill leachate using a combination of ozone-GAC adsorption treatment. *Journal of Hazardous Materials*, **137**(1), 443–455, https://doi.org/10.1016/j.jhazmat.2006.02.020

Lam K. F., Yeung K. L. and McKay G. (2006). A rational approach in the design of selective mesoporous adsorbents. *Langmuir*, **22**(23), 9632–9641, https://doi.org/10.1021/la061410p

Lee M. E., Park J. H. and Chung J. W. (2019). Comparison of the lead and copper adsorption capacities of plant source materials and their biochars. *Journal of Environmental Management*, **236**, 118–124, https://doi.org/10.1016/j.jenvman.2019.01.100

Leton T. G. and Omotosho O. (2004). Landfill operations in the Niger Delta region of Nigeria. *Engineering Geology*, **73**(1–2), 171–177, https://doi.org/10.1016/j.enggeo.2003.12.006

Li G., Shen B., Li F., Tian L., Singh S. and Wang F. (2015). Elemental mercury removal using biochar pyrolyzed from municipal solid waste. *Fuel Processing Technology*, **133**, 43–50, https://doi.org/10.1016/j.fuproc.2014.12.042

Lian W., Yang L., Joseph S., Shi W., Bian R., Zheng J., Li L., Shan S. and Pan G. (2020). Utilization of biochar produced from invasive plant species to efficiently adsorb Cd(II) and Pb(II). *Bioresource Technology*, **317**, 124011, https://doi.org/10.1016/j.biortech.2020.124011

Lin J. H. and Wang S. B. (2017). An effective route to transform scrap tire carbons into highly-pure activated carbons with a high adsorption capacity of ethylene blue through thermal and chemical treatments. *Environmental Technology & Innovation*, **8**, 17–27, https://doi.org/10.1016/j.eti.2017.03.004

Liu Z., Wu W., Shi P., Guo J. and Cheng J. (2015). Characterization of dissolved organic matter in landfill leachate during the combined treatment process of air stripping, Fenton, SBR and coagulation. *Waste Management*, **41**, 111–118, https://doi.org/10.1016/j.wasman.2015.03.044

Lozano P. and Berge N. D. (2012). Single-walled carbon nanotube behavior in representative mature leachate. *Waste Management*, **32**(9), 1699–1711, https://doi.org/10.1016/j.wasman.2012.03.019

Luo Y., Li R., Sun X., Liu X. and Li D. (2019). The roles of phosphorus species formed in activated biochar from rice husk in the treatment of landfill leachate. *Bioresource Technology*, **288**, 121533, https://doi.org/10.1016/j.biortech.2019.121533

Luo H., Zeng Y., Cheng Y., He D. and Pan X. (2020). Recent advances in municipal landfill leachate: a review focusing on its characteristics, treatment, and toxicity assessment. *Science of the Total Environment*, **703**, 135468, https://doi.org/10.1016/j.scitotenv.2019.135468

Ma L., Dong X., Chen M., Zhu L., Wang C., Yang F. and Dong Y. (2017). Fabrication and water treatment application of carbon nanotubes (CNTs)-based composite membranes: a review. *Membranes*, **7**(1), 16, https://doi.org/10.3390/membranes7010016

Ma J., Jiang Z., Cao J. and Yu F. (2020). Enhanced adsorption for the removal of antibiotics by carbon nanotubes/graphene oxide/sodium alginate triple-network nanocomposite hydrogels in aqueous solutions. *Chemosphere*, **242**, 125188, https://doi.org/10.1016/j.chemosphere.2019.125188

Mandal P., Dubey B. K. and Gupta A. K. (2017). Review on landfill leachate treatment by electrochemical oxidation: drawbacks, challenges and future scope. *Waste Management*, **69**, 250–273, https://doi.org/10.1016/j.wasman.2017.08.034

Matali S., Khairuddin S. A., Sharifah A. S. A. K. and Hidayu A. R. (2013). Removal of selected gaseous effluent using activated carbon derived from oil palm waste: an overview. IEEE Symposium on Business, Engineering, and Industrial Applications, Indonesia.

McDougall G. J. (1991). The physical nature and manufacture of activated carbon. *Journal of the Southern African Institute of Mining and Metallurgy*, **91**(4), 109–120.

Mishra R., Singh E., Kumar A. and Kumar S. (2021). Artificial intelligence models for forecasting of municipal solid waste generation. In: Soft Computing Techniques in Solid Waste and Wastewater Management, R. R. Karri, G. Ravindran and M. H. Dehghani (eds.), Elsevier, Netherlands, pp. 289–304.

Mishra R., Kumar A., Singh E. and Kumar S. (2022a). Carbonaceous nanocomposites derived from waste material for wastewater treatment. In: Biorenewable Nanocomposite Materials, Vol. 2: Desalination and Wastewater Remediation, D. Pathania and L. Singh (eds.), American Chemical Society, Washington, DC, pp. 43–73.

Mishra R., Kumar A., Singh E., Kumar S., Tripathi V. K., Jha S. K. and Shukla S. K. (2022b). Current status of available techniques for removal of heavy metal contamination in the river ecosystem. In: Ecological Significance of River Ecosystems, S. Madhav, S. Kanhaiya, A. Srivastav, V. Singh, P. Singh (eds.), Elsevier, Netherlands, pp. 217–234.

Mishra R., Singh E., Kumar A., Ghosh A., Lo S. L. and Kumar S. (2022c). Co-gasification of solid waste and its impact on final product yields. Journal of Cleaner Production, **374**, 133989, https://doi.org/10.1016/j.jclepro.2022.133989

Mojiri A., Zhou J. L., Ratnaweera H., Ohashi A., Ozaki N., Kindaichi T. and Asakura H. (2021). Treatment of landfill leachate with different techniques: an overview. *Journal of Water Reuse and Desalination*, **11**(1), 66–96, https://doi.org/10.2166/wrd.2020.079

Mukherjee S., Mukhopadhyay S., Hashim M. A. and Sen Gupta B. (2015). Contemporary environmental issues of landfill leachate: assessment and remedies. *Critical Reviews in Environmental Science and Technology*, **45**(5), 472–590, https://doi.org/10.1080/10643389.2013.876524

Nadarajah K., Bandala E. R., Zhang Z., Mundree S. and Goonetilleke A. (2021). Removal of heavy metals from water using engineered hydrochar: kinetics and mechanistic approach. *Journal of Water Process Engineering*, **40**, 101929, https://doi.org/10.1016/j.jwpe.2021.101929

Nemchi F., Bestani B., Benderdouche N., Belhakem M. and Duclaux L. (2017). Enhancement of Ni^{2+} removal capacity of activated carbons obtained from Mediterranean *Ulva lactuca* and *Systoceira stricta* algal species. *Journal of Environmental Chemical Engineering*, **5**(3), 2337–2345, https://doi.org/10.1016/j.jece.2017.03.027

Obayomi K. S., Lau S. Y., Danquah M., Chiong T. and Takeo M. (2022). Advances in graphene oxide based nanobiocatalytic technology for wastewater treatment. *Environmental Nanotechnology, Monitoring & Management*, **17**, 100647, https://doi.org/10.1016/j.enmm.2022.100647

Oliveira L. F., Silva S. M. and Martinez C. B. (2014). Assessment of domestic landfill leachate toxicity to the Asian clam *Corbicula fluminea* via biomarkers. *Ecotoxicology and Environmental Safety*, **103**, 17–23, https://doi.org/10.1016/j.ecoenv.2014.01.034

Omran I. I., Al-Saati N. H., Al-Saati H. H., Hashim K. S. and Al-Saati Z. N. (2021). Sustainability assessment of wastewater treatment techniques in urban areas of Iraq using multi-criteria decision analysis (MCDA). *Water Practice and Technology*, **16**(2), 648–660, https://doi.org/10.2166/wpt.2021.013

Papastavrou C., Mantzavinos D. and Diamadopoulos E. (2009). A comparative treatment of stabilized landfill leachate: coagulation and activated carbon adsorption vs. electrochemical oxidation. *Environmental Technology*, **30**(14), 1547–1553, https://doi.org/10.1080/09593330903252240

Parshetti G. K., Chowdhury S. and Balasubramanian R. (2014). Hydrothermal conversion of urban food waste to chars for removal of textile dyes from contaminated waters. *Bioresource Technology*, **161**, 310–319, https://doi.org/10.1016/j.biortech.2014.03.087

Pelalak R., Heidari Z., Khatami S. M., Kurniawan T. A., Marjani A. and Shirazian S. (2021). Oak wood ash/GO/Fe_3O_4 adsorption efficiencies for cadmium and lead removal from aqueous solution: kinetics, equilibrium and thermodynamic evaluation. *Arabian Journal of Chemistry*, **14**(3), 102991, https://doi.org/10.1016/j.arabjc.2021.102991

Pi K. W., Li Z., Wan D. J. and Gao L. X. (2009). Pretreatment of municipal landfill leachate by a combined process. *Process Safety and Environmental Protection*, **87**(3), 191–196, https://doi.org/10.1016/j.psep.2009.01.002

Poblete R., Oller I., Maldonado M. I. and Cortes E. (2019). Improved landfill leachate quality using ozone, UV solar radiation, hydrogen peroxide, persulfate and adsorption processes. *Journal of Environmental Management*, **232**, 45–51, https://doi.org/10.1016/j.jenvman.2018.11.030

Rani A., Negi S., Hussain A. and Kumar S. (2020). Treatment of urban municipal landfill leachate utilizing garbage enzyme. *Bioresource Technology*, **297**, 122437, https://doi.org/10.1016/j.biortech.2019.122437

Rebekah A., Navadeepthy D., Bharath G., Viswanathan C. and Ponpandian N. (2021). Removal of 1-napthylamine using magnetic graphene and magnetic graphene oxide functionalized with chitosan. *Environmental Nanotechnology, Monitoring & Management*, **15**, 100450, https://doi.org/10.1016/j.enmm.2021.100450

Regkouzas P. and Diamadopoulos E. (2019). Adsorption of selected organic micro-pollutants on sewage sludge biochar. *Chemosphere*, **224**, 840–851, https://doi.org/10.1016/j.chemosphere.2019.02.165

Reis C. L. B., Sousa E. Y. A. D., Serpa J. D. F., Oliveira R. C. and Santos J. C. S. D. (2019). Design of immobilized enzyme biocatalysts: drawbacks and opportunities. *Química Nova*, **42**, 768–783.

Reshadi M. A. M., Bazargan A. and McKay G. (2020). A review of the application of adsorbents for landfill leachate treatment: focus on magnetic adsorption. *Science of the Total Environment*, **731**, 138863, https://doi.org/10.1016/j.scitotenv.2020.138863

Rodriguez J., Castrillon L., Maranon E., Sastre H. and Fernandez E. (2004). Removal of non-biodegradable organic matter from landfill leachates by adsorption. *Water Research*, **38**(14–15), 3297–3303, https://doi.org/10.1016/j.watres.2004.04.032

Rodriguez-Narvaez O. M., Peralta-Hernandez J. M., Goonetilleke A. and Bandala E. R. (2019). Biochar-supported nanomaterials for environmental applications. *Journal of Industrial and Engineering Chemistry*, **78**, 21–33, https://doi.org/10.1016/j.jiec.2019.06.008

Sabermahani F., Madadi Mahani N. and Pourgholamrezazadeh N. (2018). Adsorption of Pb(II) by activated carbon prepared from *Melia azedarach* fruit: equilibrium and thermodynamics. *Quarterly Journal of Iranian Chemical Communication*, **6**(2), 109–217, Serial No. (19), 148–159.

Sahoo S. K., Padhiari S., Biswal S. K., Panda B. B. and Hota G. (2020). Fe_3O_4 nanoparticles functionalized GO/g-C_3N_4 nanocomposite: an efficient magnetic nanoadsorbent for adsorptive removal of organic pollutants. *Materials Chemistry and Physics*, **244**, 122710, https://doi.org/10.1016/j.matchemphys.2020.122710

Sajjadi B., Broome J. W., Chen W. Y., Mattern D. L., Egiebor N. O., Hammer N. and Smith C. L. (2019). Urea functionalization of ultrasound-treated biochar: a feasible strategy for enhancing heavy metal adsorption capacity. *Ultrasonics Sonochemistry*, **51**, 20–30, https://doi.org/10.1016/j.ultsonch.2018.09.015

Shadi A. M. H., Kamaruddin M. A., Niza N. M., Emmanuel M. I., Hossain M. S. and Ismail N. (2020). Efficient treatment of raw leachate using magnetic ore iron oxide nanoparticles Fe_2O_3 as nanoadsorbents. *Journal of Water Process Engineering*, **38**, 101637, https://doi.org/10.1016/j.jwpe.2020.101637

Shadi A. M. H., Kamaruddin M. A., Niza N. M., Emmanuel M. I., Hossain M. S. and Ismail N. (2021). Electroflotation treatment of stabilized landfill leachate using titanium-based electrode. *International Journal of Environmental Science and Technology*, **18**(8), 2425–2440, https://doi.org/10.1007/s13762-020-03005-3

Shan H., Zeng C., Zhao C. and Zhan H. (2021). Iron oxides decorated graphene oxide/chitosan composite beads for enhanced Cr(VI) removal from aqueous solution. *International Journal of Biological Macromolecules*, **172**, 197–209, https://doi.org/10.1016/j.ijbiomac.2021.01.060

Shehzad A., Bashir M. J., Sethupathi S. and Lim J. W. (2015). An overview of heavily polluted landfill leachate treatment using food waste as an alternative and renewable source of activated carbon. *Process Safety and Environmental Protection*, **98**, 309–318, https://doi.org/10.1016/j.psep.2015.09.005

Shouliang H. U. O., Beidou X. I., Haichan Y. U., Liansheng H. E., Shilei F. A. N. and Hongliang L. I. U. (2008). Characteristics of dissolved organic matter (DOM) in leachate with different landfill ages. *Journal of Environmental Sciences*, **20**(4), 492–498, https://doi.org/10.1016/S1001-0742(08)62085-9

Silva T. F., Silva M. E. F., Cunha-Queda A. C., Fonseca A., Saraiva I., Boaventura R. A. and Vilar V. J. (2013). Sanitary landfill leachate treatment using combined solar photo-Fenton and biological oxidation processes at pre-industrial scale. *Chemical Engineering Journal*, **228**, 850–866, https://doi.org/10.1016/j.cej.2013.05.060

Singh S. K., Townsend T. G., Mazyck D. and Boyer T. H. (2012). Equilibrium and intra-particle diffusion of stabilized landfill leachate onto micro-and meso-porous activated carbon. *Water Research*, **46**(2), 491–499, https://doi.org/10.1016/j.watres.2011.11.007

Singh E., Kumar A., Mishra R. and Kumar S. (2021a). Olive-mill wastewater: a paradigm shift toward its sustainable management. Integrated and Hybrid Process Technology for Water and Wastewater Treatment, A. W. Mohammad and W. L. Ang (eds.), Elsevier, **2021**, 617–639. ISBN 9780128230312, https://doi.org/10.1016/B978-0-12-823031-2.00004-5

Singh E., Kumar A., Mishra R., You S., Singh L., Kumar S. and Kumar R. (2021b). Pyrolysis of waste biomass and plastics for production of biochar and its use for removal of heavy metals from aqueous solution. *Bioresource Technology*, **320**, 124278, https://doi.org/10.1016/j.biortech.2020.124278

Singh N., Kumari S., Goyal N. and Khan S. (2021c). Al_2O_3/GO cellulose based 3D-hydrogel for efficient fluoride removal from water. *Environmental Nanotechnology, Monitoring & Management*, **15**, 100444.

Singh E., Kumar A., Mishra R. and Kumar S. (2022a). Solid waste management during COVID-19 pandemic: recovery techniques and responses. *Chemosphere*, **288**, 132451, https://doi.org/10.1016/j.chemosphere.2021.132451

Singh E., Mishra R., Kumar A., Shukla S. K., Lo S. L. and Kumar S. (2022b). Circular economy-based environmental management using biochar: driving towards Sustainability. Process Safety and Environmental Protection, **163**, 585–600.

Skodras G., Diamantopoulou I., Zabaniotou A., Stavropoulos G. and Sakellaropoulos G. P. (2007). Enhanced mercury adsorption in activated carbons from biomass materials and waste tires. *Fuel Processing Technology*, **88**(8), 749–758, https://doi.org/10.1016/j.fuproc.2007.03.008

Stankovich S., Dikin D. A., Dommett G. H., Kohlhaas K. M., Zimney E. J., Stach E. A., Piner R. D., Nguyen S. T. and Ruoff R. S. (2006). Graphene-based composite materials. *Nature*, **442**(7100), 282–286, https://doi.org/10.1038/nature04969

Staudenmaier L. (1898). Verfahren zur Darstellung der Graphitsäure. *Berichte der Deutschen Chemischen Gesellschaft*, **31**, 1481–1487, https://doi.org/10.1002/cber.18980310237

Sun J., Zhang Z., Ji J., Dou M. and Wang F. (2017). Removal of Cr^{6+} from wastewater via adsorption with high-specific-surface-area nitrogen-doped hierarchical porous carbon derived from silkworm cocoon. *Applied Surface Science*, **405**, 372–379, http://dx.doi.org/10.1016/j.apsusc.2017.02.044

Sujiono E. H., Zabrian D., Dahlan M. Y., Amin B. D. and Agus J. (2020). Graphene oxide based coconut shell waste: synthesis by modified Hummers method and characterization. *Heliyon*, **6**(8), e04568, https://doi.org/10.1016/j.heliyon.2020.e04568

Tang C., Shu Y., Zhang R., Li X., Song J., Li B., Zhang Y. and Ou D. (2017). Comparison of the removal and adsorption mechanisms of cadmium and lead from aqueous solution by activated carbons prepared from *Typha angustifolia* and *Salix matsudana*. *RSC Advances*, **7**(26), 16092–16103, https://doi.org/10.1039/C6RA28035H

Tapouk F. A., Nabizadeh R., Nasseri S., Mesdaghinia A., Khorsandi H., Yousefi M., Alimohammadi M. and Khoobi M. (2020). Embedding of L-arginine into graphene oxide (GO) for endotoxin removal from water: modeling and optimization approach. *Colloids and Surfaces A: Physicochemical and Engineering Aspects*, **607**, 125491, https://doi.org/10.1016/j.colsurfa.2020.125491

Thue P. S., Lima E. C., Sieliechi J. M., Saucier C., Dias S. L., Vaghetti J. C., Rodembusch F. S. and Pavan F. A. (2017). Effects of first-row transition metals and impregnation ratios on the physicochemical properties of microwave-assisted activated carbons from wood biomass. *Journal of Colloid and Interface Science*, **486**, 163–175, https://doi.org/10.1016/j.jcis.2016.09.070

Tiwari R., Mishra R., Choubey A., Kumar S., Atabani A. E., Badruddin I. A. and Khan T. Y. (2023). Environmental and economic issues for renewable production of bio-jet fuel: a global prospective. *Fuel*, **332**, 125978, https://doi.org/10.1016/j.fuel.2022.125978

Tong H., Yin K., Giannis A., Ge L. and Wang J. Y. (2015). Influence of temperature on carbon and nitrogen dynamics during in situ aeration of aged waste in simulated landfill bioreactors. *Bioresource Technology*, **192**, 149–156, https://doi.org/10.1016/j.biortech.2015.05.049

Torretta V., Ferronato N., Katsoyiannis I. A., Tolkou A. K. and Airoldi M. (2016). Novel and conventional technologies for landfill leachates treatment: a review. *Sustainability*, **9**(1), 9, https://doi.org/10.3390/su9010009

Tseng R. L. (2006). Mesopore control of high surface area NaOH-activated carbon. *Journal of Colloid and Interface Science*, **303**(2), 494–502, https://doi.org/10.1016/j.jcis.2006.08.024

Ukundimana Z., Kobya M., Omwene P. I., Gengec E. and Can O. T. (2020). Anodic oxidation of effluents from stages of MBR-UF municipal landfill leachate treatment plant. *Environmental Engineering Science*, **37**(10), 702–714, https://doi.org/10.1089/ees.2020.0003

Van Thuan T., Quynh B. T. P., Nguyen T. D. and Bach L. G. (2017). Response surface methodology approach for optimization of Cu^{2+}, Ni^{2+} and Pb^{2+} adsorption using KOH-activated carbon from banana peel. *Surfaces and Interfaces*, **6**, 209–217, https://doi.org/10.1016/j.surfin.2016.10.007

Vollprecht D., Frühauf S., Stocker K. and Ellersdorfer M. (2019). Ammonium sorption from landfill leachates using natural and modified zeolites: pre-tests for a novel application of the ion exchanger loop stripping process. *Minerals*, **9**(8), 471, https://doi.org/10.3390/min9080471

Wang L., Li Y. J., Xiong Y., Tan W. B., Zhang L. Y., Li X., Wang X. S., Xu J. F., Li T. T., Wang J. S. and Cai M. X. (2017). Spectroscopic characterization of DOM and the nitrogen removal mechanism during wastewater reclamation plant. *PLoS ONE*, **12**(11), e0187355.

Wang L., Hu D., Kong X., Liu J., Li X., Zhou K., Zhao H. and Zhou C. (2018). Anionic polypeptide poly(γ-glutamic acid)-functionalized magnetic Fe_3O_4–GO–(*o*-MWCNTs) hybrid nanocomposite for high-efficiency removal of Cd(II), Cu(II) and Ni(II) heavy metal ions. *Chemical Engineering Journal*, **346**, 38–49, https://doi.org/10.1016/j.cej.2018.03.084

Wiszniowski J., Robert D., Surmacz-Gorska J., Miksch K. and Weber J. V. (2006). Landfill leachate treatment methods: a review. *Environmental Chemistry Letters*, **4**(1), 51–61, https://doi.org/10.1007/s10311-005-0016-z

Xiao F., Cheng J., Cao W., Yang C., Chen J. and Luo Z. (2019). Removal of heavy metals from aqueous solution using chitosan-combined magnetic biochars. *Journal of Colloid and Interface Science*, **540**, 579–584, https://doi.org/10.1016/j.jcis.2019.01.068

Xiaoli C., Shimaoka T., Qiang G. and Youcai Z. (2008). Characterization of humic and fulvic acids extracted from landfill by elemental composition, ^{13}C CP/MAS

NMR and TMAH-Py-GC/MS. *Waste Management*, **28**(5), 896–903, https://doi.org/10.1016/j.wasman.2007.02.004

Xiaoli C., Yongxia H., Guixiang L., Xin Z. and Youcai Z. (2013). Spectroscopic studies of the effect of aerobic conditions on the chemical characteristics of humic acid in landfill leachate and its implication for the environment. *Chemosphere*, **91**(7), 1058–1063, https://doi.org/10.1016/j.chemosphere.2013.01.052

Xue Q., Li J. S., Wang P., Liu L. and Li Z. Z. (2014). Removal of heavy metals from landfill leachate using municipal solid waste incineration fly ash as adsorbent. *Clean – Soil, Air, Water*, **42**(11), 1626–1631, https://doi.org/10.1002/clen.201300619

Yang W., Wang Z., Song S., Han J., Chen H., Wang X., Sun R. and Cheng J. (2019). Adsorption of copper(II) and lead(II) from seawater using hydrothermal biochar derived from *Enteromorpha*. *Marine Pollution Bulletin*, **149**, 110586, https://doi.org/10.1016/j.marpolbul.2019.110586

Yazdani M. R., Duimovich N., Tiraferri A., Laurell P., Borghei M., Zimmerman J. B. and Vahala R. (2019). Tailored mesoporous biochar sorbents from pinecone biomass for the adsorption of natural organic matter from lake water. *Journal of Molecular Liquids*, **291**, 111248, https://doi.org/10.1016/j.molliq.2019.111248

Younis S. A., El-Salamony R. A., Tsang Y. F. and Kim K. H. (2020). Use of rice straw-based biochar for batch sorption of barium/strontium from saline water: protection against scale formation in petroleum/desalination industries. *Journal of Cleaner Production*, **250**, 119442, https://doi.org/10.1016/j.jclepro.2019.119442

Yu H., Hong H. J., Kim S. M., Ko H. C. and Jeong H. S. (2020). Mechanically enhanced graphene oxide/carboxymethyl cellulose nanofibril composite fiber as a scalable adsorbent for heavy metal removal. *Carbohydrate Polymers*, **240**, 116348, https://doi.org/10.1016/j.carbpol.2020.116348

Yuan Z., He C., Shi Q., Xu C., Li Z., Wang C., Zhao H. and Ni J. (2017). Molecular insights into the transformation of dissolved organic matter in landfill leachate concentrate during biodegradation and coagulation processes using ESI FT-ICR MS. *Environmental Science & Technology*, **51**(14), 8110–8118, https://doi.org/10.1021/acs.est.7b02194

Zeng F., Liao X., Pan D. and Shi H. (2020). Adsorption of dissolved organic matter from landfill leachate using activated carbon prepared from sewage sludge and cabbage by $ZnCl_2$. *Environmental Science and Pollution Research*, **27**(5), 4891–4904, https://doi.org/10.1007/s11356-019-07233-0

Zhang B., Shan C., Hao Z., Liu J., Wu B. and Pan B. (2019). Transformation of dissolved organic matter during full-scale treatment of integrated chemical wastewater: molecular composition correlated with spectral indexes and acute toxicity. *Water Research*, **157**, 472–482, https://doi.org/10.1016/j.watres.2019.04.002

Zhang C., You S., Zhang J., Qi W., Su R. and He Z. (2020). An effective in-situ method for laccase immobilization: excellent activity, effective antibiotic removal rate and low potential ecological risk for degradation products. *Bioresource Technology*, **308**, 123271, https://doi.org/10.1016/j.biortech.2020.123271

Zhao R., Gupta A., Novak J. T. and Goldsmith C. D. (2017). Evolution of nitrogen species in landfill leachates under various stabilization states. *Waste Management*, **69**, 225–231, https://doi.org/10.1016/j.wasman.2017.07.041

Zhao J., Luo W., Xu Y., Ling J. and Deng L. (2021). Potential reproductive toxicity of multi-walled carbon nanotubes and their chronic exposure effects on the growth and development of *Xenopus tropicalis*. *Science of the Total Environment*, **766**, 142652, https://doi.org/10.1016/j.scitotenv.2020.142652

Zhou K., Zhang J., Xiao Y., Zhao Z., Zhang M., Wang L., Zhang X. and Zhou C. (2020). High-efficiency adsorption of and competition between phenol and hydroquinone in aqueous solution on highly cationic amino-poly(vinylamine)-functionalized

GO–(*o*-MWCNTs) magnetic nanohybrids. *Chemical Engineering Journal*, **389**, 124223, https://doi.org/10.1016/j.cej.2020.124223

Zou S. J., Chen Y. F., Zhang Y., Wang X. F., You N. and Fan H. T. (2021). A hybrid sorbent of α-iron oxide/reduced graphene oxide: studies for adsorptive removal of tetracycline antibiotics. *Journal of Alloys and Compounds*, **863**, 158475, https://doi.org/10.1016/j.jallcom.2020.158475

Zubair M., Daud M., McKay G., Shehzad F. and Al-Harthi M. A. (2017). Recent progress in layered double hydroxides (LDH)-containing hybrids as adsorbents for water remediation. *Applied Clay Science*, **143**, 279–292, https://doi.org/10.1016/j.clay.2017.04.002

doi: 10.2166/9781789063318_0243

Chapter 11

Landfill leachate-induced ultraviolet quenching substances

*Pubali Mandal[1] and Brajesh Kumar Dubey[2]**

[1]Department of Civil Engineering, Birla Institute of Technology and Science, Pilani 333031, India
[2]Department of Civil Engineering, Indian Institute of Technology, Kharagpur 721302, India
**Corresponding author: bkdubey@civil.iitkgp.ac.in*

ABSTRACT

Landfill leachate is a complex wastewater with a wide range of pollutants and requires proper treatment before discharge. Landfill leachate is either treated on-site by biological and membrane processes or treated transported to nearby wastewater treatment plant for co-treatment with municipal wastewater. Recalcitrant organics, especially for stabilized leachate will not be removed or degraded by conventional biological processes. Some of these pollutants may absorb ultraviolet (UV) lights and decrease the UV transmittance. The UV quenching substances (UVQS) can be broadly classified as hydrophobic substances (humic acid + fulvic acid) and hydrophilic substances. The UVQS may interfere with the UV disinfection process and make it less effective. It may also protect any co-pollutant from being degraded by UV. Therefore, understanding of the treatment methods capable of removing UVQS is necessary. The aim of this chapter is to present overview of different treatment technologies available to remove/degrade UVQS. The composition of solid waste responsible for the formation of UVQS, characteristics, chemical composition, distribution and problems of UVQS has also been discussed.

Keywords: Landfill leachate, UV quenching substances, wastewater treatment, recalcitrant organics.

11.1 INTRODUCTION

Global solid waste generation is booming due to the continuous growth of civilization, population, and economy. According to the report published by World Bank, projected global solid waste generation is 1.3–2.2 billion tons/year by 2025 (Hoornweg & Bhada-Tata, 2012). Although landfilling is the least favorable among the common waste management practices, its low-cost, easy operation, and maintenance make it adaptable. In terms of the global scenario, open dumping and landfilling are reported to cover 33% and 7.7% of solid waste management (Kaza *et al.*, 2018). The infiltrated precipitation and innate moisture content of the solid waste produce a complex liquid known as leachate through various reactions that take place in landfills during stabilization. The highly contaminated wastewater displays a wide range of pollutants, such as, biodegradable and recalcitrant organics, inorganics, heavy metals, xenobiotic organic compounds, and so on (Kjeldsen *et al.*, 2002; Mandal *et al.*, 2017). Landfill leachate may cause adverse impact on human health and the environment if not treated appropriately before discharge (Renou *et al.*, 2008).

A major component of leachate pollutants is dissolved organic matter. Whereas, biodegradable organics can be easily treated by biological treatment processes, recalcitrant fraction is resistant to microbial degradation. Physico-chemical process, electrochemical, bio-electrochemical, and advanced oxidation technologies are reported capable to deal with recalcitrant organics (Mandal *et al.*, 2017). Landfills can be categorized as young (<5 years), medium (5–10 years), and old (>10 years) depending on its age (Ahmed & Lan, 2012). With age, the generation of leachate decreases. However, the leachate constituents become more complex and recalcitrant. During the microbial degradation in landfills, low-molecular weight aliphatic compounds may be transformed into high-molecular weight aromatic humic substances due to polymerization and condensation processes (Bolyard & Reinhart, 2017). The aromatic components are hydrophobic in nature and consist of humic substances, that is, humic acid (HA) and fulvic acid (FA). The aliphatic components are mainly hydrophilic in nature (Iskander *et al.*, 2018).

The hydrophobic and hydrophilic substances present in landfill leachate strongly absorb ultraviolet (UV) light. The refractory nature makes it hard to remove by conventional biological processes adopted in wastewater treatment plants, thereby easily attains the downstream UV disinfection process and causes interference (Zhao *et al.*, 2013a). The hydrophobic and hydrophilic substances collectively named as UV quenching substances (UVQS) due to its UV quenching ability. The UVQS may also act as adsorption sites for co-pollutants, especially for pollutants targeted to be removed by UV disinfection (Xu *et al.*, 2016). Since the UVQS passed through the conventional treatment system, the organics may react and disinfectant by-products may be formed.

Landfill leachate is either treated on site or it is mixed at a certain percentage with municipal wastewater to treat in a municipal wastewater treatment plant. The disinfection of wastewater is conducted at the last stage to get rid of harmful pathogens, especially in case of an incident of pandemic outbreak. Therefore, understanding the treatment technologies efficiently in removing UVQS is of utmost importance and needs to be incorporated in treatment plants. The purpose of this chapter is to understand the sources, characteristics, distribution of landfill leachate, and treatment technologies with associated advantages and drawbacks to remove UVQS.

11.2 SOURCES OF UV QUENCHING SUBSTANCES IN LANDFILL LEACHATE

The components of UVQS in landfill leachate are hydrophobic and hydrophilic fractions. The degradation of organic solid waste generates hydrophilic aliphatic carbons. Humic substances are of high molecular weight and formed by humification process involving biological and chemical reactions (Castagnoli *et al.*, 1990). The process of humification involves polyphenols, lignin, and sugar amines (Jones & Bryan, 1998). Solid waste composition includes wood, plant leaves, vegetation, paper, and cardboard. Lignin is found very generously in these components of solid waste making them precursor material for UVQS generation in leachate. The characteristics of solid waste and degradation rate within landfills change the relationship or proportion of HA and FA. According to the report published by Kaza *et al.* (2018), global waste consists of majorly organic waste (44% food and green). Waste composition analysis also reveals generation of 17% of paper and cardboard, and 2% of wood. These components contribute to the formation of UVQS. The presence of proteins, carbohydrates, and fatty acids also assist in the humification process. Volatile fatty acids and biodegradable organic matters are usually of aliphatic structures and they contribute to the formation of HPI (hydrophilic fraction) (Bolyard & Reinhart, 2017).

Hydrophilic fraction has simple carbon containing structure making them vulnerable to microbial degradation. The initial high biochemical oxygen demand (BOD) and chemical oxygen demand (COD) generally found in young landfills has significant relation with the HPI. However, the contribution of HPI toward UV absorbance is less (Zhao *et al.*, 2013a). The HPI is readily biodegradable, which prevent them from entering the UV disinfection system as removed during the biological wastewater treatment process. Since, UV disinfection is usually the extreme downstream process, HPI does not hamper the wastewater disinfection process much in comparison with HA and FA.

11.3 PROPERTIES OF UV QUENCHING SUBSTANCES

Properties and chemical structures of humic substances change significantly depending on the percentage variation of the source materials, that is, solid waste composition. The humic substances make the wastewater yellowish or blackish in color. These are acidic heterogeneous macromolecules. The hydrophilic and hydrophobic fraction is the agglomeration of aliphatic and aromatic moieties, respectively, having covalent bonds. These substances generally contain phenolic, carboxyl, alkoxy groups and sometimes alanine, semiquinone, hydroquinone, sulfate esters, and phosphate ester groups (Jones & Bryan, 1998). The bonds responsible for comparatively larger molecular size of humic substances are van der Waals, hydrogen, π–π, and CH–π bonds (Piccolo, 2002). The landfill can be divided into three categories depending on its age, that is, young, medium, and old landfills (Kjeldsen *et al.*, 2002). The heterogeneous nature of the humic materials is mainly associated with characteristic changes over time. In the initial stages of landfill leachate generation, FA are formed and their concentration becomes predominant than HA (Kang *et al.*, 2002). However, HA concentration increases with time due to the humification and degradation process. Also, the humification rate has been observed initially higher and decreases over ages (Castagnoli *et al.*, 1990). The source materials for the generation of humic substances generally received substantial water, oxygen, and sunlight exposure. Due to this, humic substances become resistant to the conventional treatment processes (Jones & Bryan, 1998). The reversibility of redox sites and various redox states also make them unsuitable for microbial degradation (Ratasuk & Nanny, 2007). However, these compounds are susceptible to the advanced treatment methods.

11.4 CHEMICAL COMPOSITION

The major components of UVQS are FA, HA, and HPI. Concentrations of these components exhibit wide range due to variable factors, such as waste composition, landfill age, landfill operating conditions, and so on (Iskander *et al.*, 2018). The molecular weight of HA varies in the range of 50,000–100,000 g/mol, whereas, 500–2,000 g/mol of range is common for FA and HPI (Amaral *et al.*, 2015). In terms of elemental composition, HA consists of C, N, H, and O. Out of the mentioned elements, O content is reported to be lower in HA than FA and HPI (Kang *et al.*, 2002). Carbohydrates and phenolic groups of HA are higher but carboxylic acid functional groups are reported to be lower than the other two fractions (Christensen *et al.*, 1998). The aromatic properties of HA helps in precipitation at $pH < 2$ by charge neutralization. The FA fraction is usually soluble at all pH ranges (Jones & Bryan, 1998). In terms of order of aromaticity, the UVQS can be represented as HA > FA > HPI depending on the value of H to C ratio (Huo *et al.*, 2008).

11.5 PROBLEMS ASSOCIATED WITH UVQS

The properties of the UVQS are such that it decreased the transmittance of UV light due to which UV disinfection process suffers from poor performance. The UV transmittance and UV absorbance of any wastewater needs to be within $\geq$60–65% and <0.187/cm for effective disinfection (Iskander *et al.*, 2018). Among the hydrophobic and hydrophilic fractions, HA has been reported to exhibit more UV absorbance than FA and HPI fractions. HPI may contribute more towards the UV absorbance due to its comparatively higher concentration present in landfill leachate than HA and FA (Zhao *et al.*, 2013a). Metals entrapped within the humic substances may also contribute to UV absorbance properties.

The complex structure and continuous exposure with sunlight, oxygen, and so on make the humic substances recalcitrant. The presence of different aromatic compounds such as phenolic, aniline, benzoic, and so on makes the leachate a strong absorber of the UV light by π-π^* electron transitions in the range of near UV 270–280 nm (Chin *et al.*, 1994). UVQS substances are hard to remove by conventional biological treatment processes and therefore may interfere with the downstream UV disinfection system. The UVQS may also create difficulty in removal of co-pollutant by providing sorption site of HA surface (Xu *et al.*,

2016). Furthermore, the presence of carbonaceous and nitrogenous organic matter may increase the formation of disinfection-by-products if chlorination is done as disinfection process (Mitch & Sedlak, 2004).

11.6 DISTRIBUTION OF UVQS IN LANDFILL LEACHATE

Since the distribution of hydrophobic and hydrophilic fraction in landfill leachate directly affects the treatment process selection and efficiency, understanding the contributions of HA, FA, and HPI toward UVQS is necessary. The UV_{254} and $SUVA_{254}$ (specific UV absorbance) values are generally used to indicate the concentration of UVQS in leachate. $SUVA_{254}$ is calculated using UV_{254} and dissolved organic carbon (DOC). Furthermore, the ages of landfills are also reported to affect the formation and characteristics of UVQS (Zhao *et al.*, 2018). When the comparative characteristics evaluation of young and mature leachate was performed, mature leachate showed higher $SUVA_{254}$ value than young leachate. High $SUVA_{254}$ value indicates greater hydrophobic and aromatic content. It also suggests the presence of humic substances and high molecular weight compounds (>100 kDa). The low $SUVA_{254}$ obtained for young leachate indicates presence of non-humic and aliphatic compounds having low molecular weight (<100 kDa). The leachate $SUVA_{254}$ value increased with decreased biodegradability of organic matter. The details of the hydrophobic and hydrophilic fraction and parameters used as measures of UVQS is presented in Table 11.1. Bolyard and Reinhart (2017) collected 12 leachate samples from eight landfill sites and observed that bioavailable dissolved organic nitrogen (hydrophilic) contributed less toward UV_{254} absorbance (4.28 ± 3.12 1/cm) and COD (646 ± 556 mg/L) than recalcitrant dissolved organic nitrogen (hydrophobic, UV_{254}: 35 ± 33.9 1/cm, COD: $1{,}690 \pm 1{,}539$ mg/L). Castagnoli *et al.* (1990) studied the changes in HA and FA concentration (two components of humic substances) for a landfill from its initial days to 4 years. For both the HA and FA, the concentration was increased till 2 years after which it reduced. At the same time, FA/HA ratio showed increasing trend over the years and was always above one indicating the decomposition rate of solid waste within the landfill. HA and FA showed a greater contribution towards UV_{254} and $SUVA_{254}$ than HPI in a study conducted by Driskill *et al.* (2015). Four landfill sites were selected in this study and three landfills show $SUVA_{254}$ value in the increasing order of HA>FA>HPI except for one site where FA>HA>HPI. However, hydrophilic fraction (low molecular-weight fraction, <1 kDa) was predominant for the organic nitrogen.

In the study conducted by Zhao *et al.* (2018), four mature landfills and one laboratory simulated young landfill were selected for understanding the hydrophobic and hydrophilic fractions of dissolved organic matter. The mature leachate samples collected showed higher $SUVA_{254}$ value than young ones. The hydrophobic fraction contributed to higher $SUVA_{254}$ value than HPI due to the abundance of chromophores. Young leachate exhibited higher concentration of low molecular weight (<100 kDa) UVQS than mature leachate samples. The evaluation of comparison and characteristics of leachate humic substances and commercial humic acid depicted lower level of humification rate occurring

Table 11.1 Hydrophobic and hydrophilic fraction and parameters used to measure UVQS.

Type of Landfills	Location	No. of Landfills	Type of Leachate	Collection Point	UV_{254}, 1/cm	DOC, mg/L	$SUVA_{254}$, L/mg m	e_{280}, L/g cm	Hydrophobic and Hydrophilic Fractions		References
									UV_{254}, 1/cm	$SUVA_{254}$, L/mg m	
Municipal	Kentucky, USA				4.2–39.5	TOC: 205–1860.1	HA: 3.13–3.79, FA: 2.29–2.64, HPI: 1.45–1.87				Driskill *et al.* (2015)
Municipal	Kentucky, USA	1		Equalization tanks			HA: 2.18, FA: 2.6, HPI: 1.18		FA > HPI > HA	FA > HA > HPI	Driskill *et al.* (2015)
Municipal	Pennsylvania, USA	3					HA: 3.22, 4.23, 4.35, FA: 3.01, 2.69, 2.59, HPI: 1.23, 1.24, 1.22		HA > FA > HPI, FA > HA > HPI, FA > HPI > HA	HA > FA > HPI, HA > FA > HPI, HA > FA > HPI	Driskill *et al.* (2015)
Sanitary	Korea	1	Young	Leachate collection system		18,362		HA: 3, FA: 0.8			Kang *et al.* (2002)
Sanitary	Korea	1	Medium	Leachate collection system		1,630		HA: 11.4, FA: 7.3			Kang *et al.* (2002)
Unsanitary	Korea	1	Old	Slope of landfill		863		HA: 12.6, FA: 8.4			Kang *et al.* (2002)

Municipal	Pennsylvania, USA	1			15.3 ± 0.5	TOC: 734.26 ± 208.4	2.1				Zhao *et al.* (2012)
Municipal	New Hampshire, USA	1			19.8 ± 0.2	TOC: 722.38 ± 21.4	2.7				Zhao *et al.* (2012)
Municipal	Pennsylvania, USA				~14	TOC: ~730	HA: 2.77, FA: 1.55, HPI: 3.25				Zhao *et al.* (2013a)
Municipal	New Hampshire, USA				~12	TOC: ~740	HA: 2.65, FA: 2.31, HPI: 1.30				Zhao *et al.* (2013a)
Municipal	Pennsylvania, USA	3	Mature	Equalization tanks	14.2 ± 0.9, 5.0 ± 0.0, 20.6 ± 1.7	981 ± 26, 198 ± 1, 1,045 ± 54	1.45 ± 0.05, 2.50 ± 0.03, 1.97 ± 0.06		HPI > FA > HA, FA > HPI > HA, HPI > FA > HA	HA > FA > HPI, HA > FA = HPI, HA > HPI > FA	Zhao *et al.* (2018)
Municipal	New Hampshire, USA	1	Mature	Equalization tanks	13.7 ± 0. 8	516 ± 31	2.65 ± 0.01		HA > FA > HPI	HA > FA > HPI	Zhao *et al.* (2018)
Municipal	Laboratory simulated landfill	1	Young		1.22	805 ± 194	0.16 ± 0.01		HPI > FA > HA	FA > HA > HPI	Zhao *et al.* (2018)

within landfills in comparison with commercial HA where three differently aged leachate samples were evaluated (Kang *et al.*, 2002). The authors observed an increase in molecular size and aromatic compounds as landfill age increases. Zhao *et al.* (2012) selected two landfill sites located in USA. In both the sites, organic matter having size <1 kDa contributed to the major fractions of total organic carbon (TOC) indicating predominance of low molecular weight compounds. However, an increase in UV_{254} and $SUVA_{254}$ values were observed as the size fraction increases from 500 Da to 100 kDa for both the landfills. Zhao *et al.* (2013a) opted for three landfills to understand the characteristics of UVQS present in leachate. The $SUVA_{254}$ values were in the order of HA>FA>HPI for all the samples collected. The UV_{254} absorbance value was also higher for HPI fraction than HA and FA. Majority of the TOC is contributed by HPI fractions followed by FA and HA. However, the particle size of HA and FA was higher than HPI.

Overall, there are similarities in the findings obtained by UVQS-related studies. The characteristics of UVQS depend on landfill ages. As landfill age increases, concentration of hydrophobic fraction increases. As a result, high UV_{254} and $SUVA_{254}$ values were obtained. HPI fraction usually consists of lower molecular weight compounds of aliphatic nature and contributes more to TOC or DOC than HA and FA.

11.7 MEASUREMENT TECHNIQUES OF UV QUENCHING SUBSTANCES

The fractionation of UVQS for landfill leachate is generally performed resulting in three fractions, that is, HA, FA, and HPI. Several researchers used different analytical techniques for the determination of UVQS. Elemental analyzers can be used for this purpose to determine C, N, O, and H elements. The H/C value ≤ 1 indicates highly aromatic structure, whereas the high values delineate aliphatic structures. High O/C and N/C indicate prevalence of functional groups containing O and organic nitrogen (Christensen *et al.*, 1998; Kang *et al.*, 2002). The fractionation process of hydrophobic and hydrophilic UVQS based on their solubility characteristics and chemical nature is described here (Christensen *et al.*, 1998; Leenheer, 1981; Thurman & Malcolm, 1981). Of those, one popular method is described here. Initially, the raw leachate sample is filtered using either 1.5 or 0.45 μm filter paper. Once the suspended materials are eliminated, pH of the sample is brought down to <2 using hydrochloric or sulfuric acid to assist the agglomeration process of the HA. The precipitated HA is then separated from the supernatant using centrifuge or filtration techniques. The separated HA is then mixed in NaOH solution of 0.1 M strength. The supernatant contains FA and HPI. To separate FA fraction, a column containing resin is used. All FA components are adsorbed on the resin when the supernatant solution is passed through the column. HPI fraction does not get absorbed and thereby the filtrate needs to be analyzed for HPI fraction. The column is eluted with the solvent 0.1 M NaOH to recover the adsorbed FA fraction. These separated fractions are then analyzed for TOC or DOC, UV absorbance to quantify UVQS. The UV_{254} (UV absorbance at 254 wavelength) to DOC or TOC ratio is commonly known as $SUVA_{254}$, and $SUVA_{254}$ value

<2 L/mg m indicates aliphatic structure. Other than $SUVA_{254}$, UV_{253}/UV_{203}, E_{465}/E_{665} (ratio of absorbance values at 465 and 665 nm, and UV_{280}/DOC (e_{280}) ratios are also of great significance indicating aromaticity, molecular weight of organics, and humification (Chin *et al.*, 1994; Kang *et al.*, 2002; Weishaar *et al.*, 2003; Zhao *et al.*, 2013b). Gel filtration chromatography with different polyethylene glycol standards and high-pressure size-exclusion chromatography can also be employed to understand molecular weight and polydispersity index (weight-averaged/number-averaged molecular weight). In ideal scenarios, the HA component should have higher index followed by FA and HPI (Chin *et al.*, 1994; Wang *et al.*, 2016). Fourier transforms infrared spectroscopy is another method to understand the functional groups present and chemical composition (Kang *et al.*, 2002; Wang *et al.*, 2016). Excitation–emission matrix fluorescence spectroscopy can also be used to characterize the composition of dissolved organic matter by distinct peaks obtained at different excitation and emission wavelength. The humification index can be calculated from the excitation–emission matrix (Cui *et al.*, 2016). The characterization of organic matter can also be accomplished by analyzing oxidation states. Mean oxidation state can be calculated from the COD and TOC values. A negative value indicates higher concentration of humic substances and polycyclic aromatics (Wang *et al.*, 2016).

Aromaticity of natural organic matter can be measured directly using ^{13}C nuclear magnetic resonance. The obtained spectra depict different functional groups present and % aromaticity can also be calculated from peak area of the spectra (Weishaar *et al.*, 2003). ^{1}H nuclear magnetic resonance also helps in understanding different functional groups and % proton in different organic fractions (Kang *et al.*, 2002). Another technique to separate the hydrophobic and hydrophilic fractions is solid-phase extraction (Bolyard & Reinhart, 2017). In this method, the suspended material is separated similar to the above-mentioned method using 0.45 μm filter. Then the solution is directly passed through the XAD-8 column where hydrophobic fraction gets absorbed. The filtrate solution contains hydrophilic fraction and the hydrophobic fraction is re-dissolved in 0.1 M NaOH solution.

11.8 REMOVAL OF UVQS BY DIFFERENT TREATMENT TECHNOLOGIES

11.8.1 Biological process

Biological wastewater treatment process can be both aerobic and anaerobic. The selection and efficiency of biological treatment process depends on leachate composition. Biological processes show high efficiency for landfill leachate treatment when the biodegradability index is greater than 0.5 and volatile fatty acids concentration is high. However, as the landfill age increases, the recalcitrant nature of mature leachate tends to limit the efficiency of biological treatment (Renou *et al.*, 2008). Different treatment technologies adopted for removing UVQS and obtained results are presented in Table 11.2. Biological processes (sequencing batch reactor-activated sludge, fixed film bioreactor) are reported to be more efficient in removing TOC than UV_{254} absorbing substances. Humic substances are resistant to microbial degradation due to its recalcitrant nature. HA and FA are reported to exert majority of UV_{254}

Table 11.2 Performances of different treatment technologies for removing UVQS.

Biological Process

Landfills	Treatment Technologies	Treatment Conditions	Characteristics		References
			Before	After	
Raw leachate	SBR		$SUVA_{254}$, L/mg m: hydrophobic acid: ~1.8, hydrophobic neutral: ~1, hydrophilic fraction: ~0.2	$SUVA_{254}$: hydrophobic acid: ~2.7, hydrophobic neutral: ~1.3, hydrophilic fraction: ~0.6	Bu *et al.* (2010)
Raw leachate	SBR (biological with nitrification)	5.5 days HRT	$SUVA_{254}$: 2.18 for HA, 2.6 for FA, 1.18 for HPI	$SUVA_{254}$: 5.48 for HA, 3.39 for FA, 1.18 for HPI	Driskill *et al.* (2015)
Raw leachate	SBR-activated sludge system nitrification and denitrification	6–7 days HRT	$SUVA_{254}$: 3.22 for HA, 3.01 for FA, 1.23 for HPI	$SUVA_{254}$: 2.41 for HA, 3.15 for FA, 1.7 for HPI	Driskill *et al.* (2015)
Raw leachate	SBR-activated sludge system for ammonia and BOD removal without denitrification		$SUVA_{254}$: 4.23 for HA, 2.69 for FA, 1.24 for HPI	$SUVA_{254}$: 3.03 for HA, 3.63 for FA, 1.76 for HPI	Driskill *et al.* (2015)
Raw leachate	Two aerated tanks, Met-Pro fixed film bioreactors (ammonia and BOD removal without denitrification)		$SUVA_{254}$: 4.35 for HA, 2.59 for FA, 1.22 for HPI	$SUVA_{254}$: 5.07 for HA, 2.98 for FA, 1.65 for HPI	Driskill *et al.* (2015)
Raw leachate	Partial nitrification reactor and an anaerobic ammonium oxidation reactor		Aquatic humic substance, mg/L: 228, DOC: 288	Aquatic humic substance, mg/L: 91, DOC: 136	Liang *et al.* (2009)

Raw leachate	Two-stage Anaerobic Membrane Bioreactor	Thermophilic–mesophilic reactor, HRTs of 25 ± 5 days and 40 ± 5 days		HA: 28–59%, FA: 21–51%, HPI: 36–65%	Pathak *et al.* (2018)
Raw leachate	SBR activated sludge system nitrification and denitrification	6–7 days HRT	TOC: 734.26 ± 208.4, UV_{254}: 15.3±0.5	TOC: 418.47 ± 100.5, UV_{254}: 13.9±0.3	Zhao *et al.* (2012)
Raw leachate	Combination of activated sludge and powdered activated carbon system only nitrification	3.5 days HRT	TOC, mg/L: 722.38 ± 21.4, UV_{254}, 1/cm: 19.8±0.2	TOC, mg/L: 313.53 ± 27.6, UV_{254}, 1/cm: 9±0.3	Zhao *et al.* (2012)
Raw leachate	SBR activated sludge system nitrification and denitrification	6–7 days HRT	TOC: 734.26 ± 208 4, UV_{254}: 15.3 ± 0.5	TOC: 34.82%, UV_{254}: 2.51%	Zhao *et al.* (2013a)
Raw leachate	Combination of activated sludge and powdered activated carbon system only nitrification	3.5 days HRT	TOC, mg/L: 722.38 ± 21.4, UV_{254}, 1/cm: 19.8 ± 0.2	TOC: 55.78%, UV_{254}: 29.09%	Zhao *et al.* (2013a)
Raw leachate	Aeration	21–53 days	TOC: 205–1860.1, UV_{254}: 4.2–39.5	TOC: 36.49–42.25%, UV_{254}: −2.3–17.85%	Zhao *et al.* (2013a)
Raw leachate	Aerobic membrane bioreactor	18–44 h HRT	HA, mg/L: 52, FA: 156, HPI: 124	HA: 22, FA: 138, HPI: 75	Zolfaghari *et al.* (2017)

(*Continued*)

Table 11.2 Performances of different treatment technologies for removing UVQS.

Biological Process

Landfills	Treatment Technologies	Treatment Conditions	Characteristics		References
			Before	After	
Membrane Separation, Adsorption, and Ion exchange					
Raw landfill leachate	Lime precipitation–microfiltration–nanofiltration	Lime dose: 10.6 g/L, time: 0.66 h, permeate flow rate of microfiltration: 5 L/m^2h, nanofiltration feeding flow: feeding flow of 144 L/h	Humic substance, mg/L: 1840 ± 120	Humic substance, mg/L: 42 ± 4 (lime precipitation), 43 ± 5 (microfiltration), 86 ± 5 (nanofiltration)	Amaral *et al.* (2015)
Stabilized landfill leachate	Magnetic ion exchange (MIEX-Cl)	Mixing condition: 20 min, 40 mL/L	UV_{254}: 9.5, DOC: 570	UV_{254}: 57%, DOC: 34%	Boyer *et al.* (2011)
Raw leachate, SBR treated leachate	Powdered activated carbon	Dose: 10 g/L	$SUVA_{254}$: 1.3 for raw, 2.12 for SBR treated	$SUVA_{254}$: 0.121 for raw, 0.586 for SBR treated	Deng *et al.* (2018)
Anaerobic-aerobic pre-treated	Reverse osmosis		HA, mg/C L: 1076, FA: 2250, HPI: 960	HA, mg/C L: ~0, FA: 7, HPI: 15	Huo *et al.* (2009)
Raw leachate	Forward osmosis		TOC: 3075 ± 62, TOC: 67 ± 9	TOC: 98%	Iskander *et al.* (2017a)
Microbial fuel cell pre-treated	Granular activated carbon	Dose: 40 g/L, 6 h	UV_{254}: 10.29	UV_{254}: 88.8%	Iskander *et al.* (2017b)
Electrochemical process					
Activated sludge treated	Electron-beam radiation	30 kGy and pH 3	Aquatic humic substances: 68.6% of DOC	Aquatic humic substances: 52.9% of DOC	Bae *et al.* (1999)

Biologically pre-treated by aerated trickling bio-filter	Electrocoagulation	Cathode: stainless steel, current density: 10 for Al anode and 8 mA/cm² for Fe anode, time: 20 min	UV_{254}: 12.5 AU, TOC_{HA} mg/C L: 85, TOC_{FA}: 102, TOC_{HPI}: ~55	UV_{254}: 79–80%, TOC_{HA}: ~100%, TOC_{FA}: 57–60%, TOC_{HPI}: 37–46%	Dia *et al.* (2017)
Raw leachate	Microbial fuel cell	Anode: carbon brush, cathode: carbon cloth, HRT 40 days, continuous anolyte recirculation, and external resistance 10 Ω.	UV_{254}: ~15.9, $SUVA_{254}$: 0.89	UV_{254}: 25.4–49.5%, $SUVA_{254}$: 2.31	Iskander *et al.* (2017b)
Anaerobic, aerobic and coagulation pre-treated	Electrochemical process combined with UV irradiation	RuO_2/Ti electrode as anode and Ti cathode, light intensity: 1.36 mW/cm²		Humic and fulvic substances: 65% reduction in 30 min	Xiao *et al.* (2013)
Bio-treated leachate	High-voltage pulsed discharge combined with TiO_2	0.15 g TiO_2, 30 min, high-voltage pulsed discharge, 40 kV, 400 Hz, conductivity 4.05 ms/cm	UV_{254}: 0.062	UV_{254}: ~56%, Humic-like: 78.9%, fulvic-like: 85.3%, Visible humic like: 40.5%	Zhang *et al.* (2017)
Chemical Oxidation Treatment					
Raw leachate	Iron-coated granular activated carbon with H_2O_2	H_2O_2: 4 mL/L, pH: 6	HA: 150.9, FA: 262.8	HA: 84%, FA: 87%	Fan *et al.* (2007)
Biologically pre-treated leachate	Fenton	Fe^{2+}: 0.054, 0.036, H_2O_2: 0.088, molar H_2O_2/Fe^{2+}: 1.6, 2.4, pH: 4, contact time: 30 min	TOC: ~235, 510, UV_{254}: ~6.8, 14.3	TOC_{HA+FA}: 90, 96%, TOC_{HPI}: 52, 72%, $UV_{254\ HA+FA}$: 85, 97%, $UV_{254\ HPI}$: 90, 92%	Gupta *et al.* (2014)

(Continued)

Table 11.2 Performances of different treatment technologies for removing UVQS.

Biological Process

Landfills	Treatment Technologies	Treatment Conditions	Characteristics		References
			Before	After	
SBR-treated leachate	Fenton, ozonation	$[H_2O_2]$:[Fe(II)]: 3:1, pH: 6.0, O_3 flow rate: 639 mg/h, O_3: 639 mg/L	$SUVA_{254}$: 2.24	$SUVA_{254}$: 0.47, 1.36	Jung *et al.* (2017)
Raw leachate	Microwave-assisted persulfate oxidation	Sodium persulfate: 0.3 M	$SUVA_{254}$: 1.5	$SUVA_{254}$: 11.8	Kim and Ahn (2017)
Raw leachate	Fenton–microfiltration–nanofiltration	1.7 g H_2O_2/g COD, $FeSO_4 \cdot 7H_2O:H_2O_2$: 1:5.3, pH: 3.8, 115 rpm, 28 min	TOC: 850	TOC: 259, 257, 17	Moravia *et al.* (2013)
Raw leachate	$NaNO_2$-catalyzed wet co-oxidation	$NaNO_2$: 0.5 mM, time: 4 h, temperature: 150°C, oxygen pressure: 0.5 MPa	FA, mg/L: 200	TOC: 62.1%	Wang *et al.* (2014)
Reverse osmosis and nanofiltration concentrate	Ozonation	Time: 110 min	$SUVA_{254}$: 2.42, 2.67	$SUVA_{254}$: 1.13, 1.69	Wang *et al.* (2016)
Nanofiltration concentrate	O_3, O_3/H_2O_2, and O_3/UV	3.4 mg O_3/g COD, H_2O_2: 2.82 g/L, UV: 600 W	$SUVA_{254}$: 2.67	$SUVA_{254}$: 31.1% for O_3, 46.1% for O_3/H_2O_2, and 43.8% for O_3/UV	Wang *et al.* (2017)

Coagulation pre-treatment	O_3, O_3/H_2O_2, and O_3/UV	O_3: 1.2 g/L, UV: output intensity of 5.1 mW/cm^2, $H_2O_2:O_3$: 1 g/g	Distribution coefficient: 3.4 (the distribution range of molecular weights)	Distribution coefficient: 2.4 (O_3), 2.6 (O_3/UV), 2.2 (O3/H_2O_2)	Wu *et al.* (2004)
Reverse osmosis concentrate	Fenton	Fe^{2+}: 5 mM, H_2O_2: 240 mM, pH: 4.0, reaction time: 2 h, temperature: 70°C	HA: 40, FA: 43.3, HPI: 16.7% of DOC	Humic substances 91.5%	Wu *et al.* (2010a)
Reverse osmosis concentrate	Fenton	Fe^{2+}: 100 mM, H_2O_2: 240 mM, pH: 3.64	Humic substances: 83.3% of DOC	Humic substances 91.53%	Wu *et al.* (2010b)
SBR-treated leachate	Coagulation–Fenton oxidation		HA: ~100, FA: ~40, HPI: ~80	Coagulation: HA 71.1%, FA 37%, HPI 52.9%, Fenton: HA 52.1%, FA 32.6%, HPI 13.5%	Wu *et al.* (2011)
Pre-coagulated leachate membrane concentrate	Fenton	Fe^{2+}: 17.5 mmol/L, H_2O_2: 1 mol/L, pH: 2	TOC: 353 ± 25	TOC: 68.9%	Xu *et al.* (2017)

absorbance (Driskill *et al.*, 2015). Similar findings were observed by Zhao *et al.* (2013a) where TOC removal of 30%, and UV absorbance reduction of only 2% were obtained by 21-day aeration process for 30 years aged leachate sample. Sequencing batch reactor treatment of an active landfill exhibited almost same percentages of TOC and UV absorbance reduction. Hence, the biological process is less effective in removing refractory UV quenching substances. As a result, $SUVA_{254}$ (L/mg m) increased from 2.77 to 4.70 for HA, 1.92 to 3.25 for FA, and 1.55 to 2.05 for HPI. The application of activated sludge-sequencing batch reactor followed by powdered activated carbon improved the TOC removal percentage to 55.78% and UV absorbance to 29.09%. Not only for TOC, increased $SUVA_{254}$ removal efficiencies of 59.8%, 44.2%, and 53.1% were obtained for HA, FA, and HPI, respectively (Zhao *et al.*, 2013a). The increase in $SUVA_{254}$ indicates increase in aromaticity of the dissolved organic matter. Bu *et al.* (2010) observed significant increase in $SUVA_{254}$ after the sequencing batch reactor biological treatment due to the bio-degradation of simple molecules and formation of polycondensed humic substances. Another reason for the increase in aromatic structure after biological treatment could be the release of high molecular weight microbial by-products from the biomass. At the same time, polydispersity of dissolved organic matter decreased due to the consumption of readily available organics by microbes and adsorption of high molecular weight compounds into the sludge solids (Bu *et al.*, 2010). In an anaerobic ammonium oxidation reactor, aquatic humic substances were reduced from 228 to 91 mg/L and DOC reduction was from 288 to 136 mg/L. The contributions of FA and HA were 57% and 33% toward the total DOC removal indicating FA removal is preferred over HA by anammox process (Liang *et al.*, 2009). An anaerobic membrane bioreactor exhibited TOC removal of 50%. For separated fractions of UVQS removal, removal efficiencies of 39% for HA, 37% for FA, and 55% for HPI were obtained indicating significant removal of HA and hydrophilic fractions (Pathak *et al.*, 2018). Zhao *et al.* (2012) evaluated the performance of biological-based sequencing batch reactor for the treatment of landfill leachate and TOC removal efficiency of 43% was achieved. Membrane separation technique can be selected along with biological processes to lower down the concentration of UVQS. For example, nanofiltration technique can be used to separate the humic substances as molecular size of HA and FA is usually higher than the HPI. The use of nanofiltration before biological treatment helped in reducing UV absorbance value from 15.3 to 4.88/cm (Zhao *et al.*, 2012). Another study using membrane bioreactor showed up to 60% of HA removal due to adsorption by activated sludge (Zolfaghari *et al.*, 2017). Whereas only 11% of FA was removed due to its lesser interaction with the sludge. The treatment efficiency of biological processes in terms of removing UVQS is limited and suitable selection of other physico-chemical processes may be necessary.

11.8.2 Membrane separation, adsorption, and ion exchange

The removal of almost complete percentages of HA, FA, and HPI has been reported by Huo *et al.* (2009). The reverse osmosis process mainly contributes to the HA and FA removal due to their high molecular weight. Whereas biological process is reported to be responsible for the HPI fraction degradation. The forward osmosis process was able to remove approximately 98% of TOC from landfill leachate (Iskander *et al.*, 2017a). Magnetic ion-exchange resin (anion exchange-Cl) was used to treat stabilized landfill leachate by Boyer *et al.* (2011) and 57% of UV_{254} and 34% of DOC removal were achieved within 20 min. On contrary to the biological processes, UV_{254} removal percentages were higher than DOC for the resin. Adsorption technology has been reported to effectively abate all types of UVQS using activated carbon via chemisorption and physical adsorption (Deng *et al.*, 2018). Powdered activated carbon at a dose of 10 g/L was able to remove ≥92% of UV_{254} absorbance for all the three fractions isolated. However, inclination of adsorption toward chromophoric dissolved organic matter was observed (Deng *et al.*, 2018). Activated carbon in granular form was used by Iskander *et al.* (2017b) for the abatement of UVQS from microbial fuel cell pre-treated leachate. 89.1% of UV absorbance reduction was achieved with activated carbon dose of 40 g/L. Ateia *et al.* (2017) experienced preferential adsorption of synthetic HA solution in the molecular weight range of 0.5–2 kDa. For the FA, similar preferences were observed for molecular weight of more than 1–3 kDa. Besides the standalone physical processes, a series or combination of treatment technologies are also used for better removal performance of stabilized leachate (Amaral *et al.*, 2015). Chemical precipitation followed by microfiltration were able to remove 43 ± 5% of humic substances. Further treatment by nanofiltration improved the HA removal to 86 ± 5%. The TOC removal efficiency by lime precipitation and microfiltration were negligible (7% and 8%), but adaptation of nanofiltration technology significantly improved the TOC removal efficiency to 99%. Although the membrane technologies are quite effective in removing UVQS, the presence of HA may cause fouling. Reduction in the permeate flux by 18% was observed when 50 mg/L of HA was added in a pilot scale study (Šír *et al.*, 2012). In another study, it is reported that the extent of flux reduction not only is dependent on HA concentration, but also on the initial rate of flux when reverse osmosis and nanofiltration were used for landfill leachate treatment (Tang *et al.*, 2007). Among the three fractions of UVQS, HA and FA are mainly susceptible to the membrane separation processes due to their order of size (HA > FA > HPI). Almost all HA and FA can be removed by the membrane separation process with molecular weight cut-off of 1 kDa. The HPI fraction will be remaining due to their size <1 kDa (Zhao *et al.*, 2013a).

11.8.3 Electrochemical process

Electrocoagulation process showed excellent efficiency in removing ~100% HA contained in a pre-treated landfill leachate by aerated bio-filter system. Humic substances (HA + FA) contributed 90% of the TOC value. The FA and HPI fraction removal percentages vary from 57% to 60% and 37% to 46%, respectively. For all the fractions, ~90% of UV_{254} absorbance was reduced using either of aluminum or iron anode (Dia *et al.*, 2017). A hybrid electrochemical treatment, that is, high-voltage pulsed discharge combined with TiO_2 was employed by Zhang *et al.* (2017) to treat biologically pre-treated leachate. 33.6% improved reduction in aromatic compounds was observed evident from UV_{254} value when high-voltage pulsed discharge combined with TiO_2 was applied. Bae *et al.* (1999) attempted electron beam radiation after biological pre-treatment and aquatic humic substance content was reduced from 68.6% to 52.9% of the DOC. An electrocatalytic reduction followed by adsorption was adopted by Satyawali *et al.* (2007) to treat HA containing wastewater. The reduction of HA by graphite increased the hydrophobicity resulting in improved adsorption. Electrochemical process (RuO_2/Ti electrode as anode and Ti cathode) combined with UV irradiation also exhibited significant reduction of HA and FA (Xiao *et al.*, 2013). Microbial fuel cell has also been used to remove UVQS from landfill leachate (Iskander *et al.*, 2017b). Higher current density and hydraulic retention time increase the overall reduction in UVQS. Further post treatment by granular activated carbon improved the UV absorbance reduction from 50 to 89.1%. Highest decrease of UV absorbance after treatment by bio-electrochemical system was 48.2% for HA, 45.6% for FA, and 52.6% for HPI for a sample. $SUVA_{254}$ value was also increased after treatment indicating increase in aromaticity. More study on the applicability of bio-electrochemical system for landfill leachate treatment is needed.

11.8.4 Chemical oxidation

Technologies which use chemicals directly for removing UVQS from landfill leachate are described in this section. Fenton oxidation process was able to remove UV_{254} by 92% and 95% for two different leachate samples. The removal efficiency of UV_{254} (or humic substances) was greater than organic matter degradation by Fenton process probably due to conversion of organic matter into intermediates (Gupta *et al.*, 2014; Wu *et al.*, 2010a, 2010b). The Fenton process was tested better than the ozonation process as hydroxyl radical may remove both the hydrophilic and hydrophobic substances where ozonation removes only hydrophobic fraction (Jung *et al.*, 2017). $SUVA_{254}$ dropped from 2.24 to 0.47 L/mg m for Fenton and to 1.36 L/mg m for ozonation indicating better efficiency of the former technology. In another study, 50% of humic substances reduction was observed by Fenton treatment (Moravia *et al.*, 2013). Wu *et al.* (2011) adopted a series of technology for removing humic substances from landfill leachate and the three separated fractions show removal efficiencies of 71.1% for HA, 52.9% of FA, and 37% of HPI in the coagulation process and 52.1% for HA, 32.6% for FA, and 13.5% for HPI in the Fenton treatment process. Both the coagulation and Fenton process removes high molecular weight preferably (Wu *et al.*, 2011). The concentrate generated during the membrane treatment process may create problems of transferring pollutants from one media to another. The concentrates generated during the reverse osmosis or nanofiltration process can be treated by ozonation process (Wang *et al.*, 2016). Ozonation process reduced UVA_{254} absorbance values from 19.8 to 8.29/cm and $SUVA_{254}$ values from 2.42 to 1.13 L/mg m for reverse osmosis concentrate. Whereas, for nanofiltration concentrate, UVA_{254} absorbance values from 55.5 to 33.4/cm and $SUVA_{254}$ values from 2.67 to 1.69 L/mg m (Wang *et al.*, 2016). Among the separated fractions of dissolved organic matter, humic substances reduced significantly. However, the HPI fraction increased after treatment indicates recalcitrance of ozonation process for HPI removal or conversion of hydrophobic substances to HPI during the treatment process (Cataldo & Angelini, 2012; Wang *et al.*, 2016). Ozone in combination with other means of oxidation was also investigated for leachate treatment, for example, O_3/UV, O_3/H_2O_2, O_3/persulfate and so on, which helped in converting the higher molecular weight compounds into lower ones (Soubh & Mokhtarani, 2016; Wang *et al.*, 2017; Wu *et al.*, 2004). The major humic substances removal mechanism for the Fenton treatment is conversion to HPI (Xu *et al.*, 2017; Zhao *et al.*, 2013b). These smaller molecular weight compounds are more susceptible to treatment or degradation due to increased biodegradability. Similar phenomenon was also observed for catalytic wet oxidation treatment (Wang *et al.*, 2014). Several different types of Fenton processes were also attempted for the treatment of UVQS. Of those, iron-coated granular activated carbon with H_2O_2 process was able to achieve 84% of HA and 87% of FA removal (Fan *et al.*, 2007). Series of technologies (air stripping–Fenton–sequencing batch reactor–coagulation) are also adopted to treat landfill leachate. The combination of treatment technologies achieved COD removal efficiencies of 97.1% for HA, 95.8% for FA, and 71.7% for HPI (Liu *et al.*, 2015). Fenton treatment followed by biological process may increase the treatment efficiency by degrading intermediates formed by Fenton oxidation. Increase in $SUVA_{254}$ after treatment was observed when microwave-assisted persulfate oxidation was employed (Kim & Ahn, 2017). Among less adopted technologies, flotation (99% of HA removal) and bioflocculation (85% of HA removal) showed significant removal efficiency (Zouboulis *et al.*, 2003, 2004). Subcritical water catalytic oxidation exhibited 74.24% of TOC and 49.11% of HA removal efficiencies (Zhai *et al.*, 2016).

11.9 CONCLUSIONS AND PERSPECTIVES

The presence of UVQS reduced the UV transmittance during the disinfection of wastewater. Thus, understanding of the problems associated with HA, FA, and HPI and treatment remedies are of significant value. Based on the discussion so far, the following perspective and conclusions can be obtained.

- Landfill leachate consists of high concentration of UVQS and therefore on-site treatment and co-treatment with municipal wastewater may suffer decreased efficiency during UV disinfection.
- The characteristics of the UVQS changes with solid waste composition and landfill ages. Usually mature or stabilized leachate depicts higher fraction of hydrophobic substances.
- The HPI fraction is more biodegradable than humic substances and therefore easy to degrade by biological process. The humic substances are resistant to conventional biological processes but susceptible to advanced treatment methods.
- The contribution of humic substances towards UV absorbance is higher than hydrophilic if compared for equal concentrations. However, HPI contributes more towards COD or TOC. The SUVA values of the three fractions are usually obtained as HA>FA>HPI.

- The pore size of membrane controls the efficiency of UVQS removal. Though high efficiencies are obtained using nanofiltration and reverse osmosis, membrane clogging is a disadvantage of the process, especially in the presence of HA.
- The Fenton treatment, chemical oxidation, and electrochemical processes may be suitable for the combined application with biological processes to remove UVQS as these technologies are associated with high cost and chemical consumption.

The data related to UVQS in literature involving leachate characteristics and treatment is limited and therefore continuous documentation may be helpful in understanding the characteristic changes of UVQS. Further detailed exploration of the advanced technologies capable in removing UVQS is necessary which may be used for on-site treatment before discharging into wastewater stream.

REFERENCES

Ahmed F. N. and Lan C. Q. (2012). Treatment of landfill leachate using membrane bioreactors: a review. *Desalination*, **287**, 41–54, https://doi.org/10.1016/j.desal.2011.12.012

Amaral M. C. S., Pereira H. V., Nani E. and Lange L. C. (2015). Treatment of landfill leachate by hybrid precipitation/microfiltration/nanofiltration process. *Water Science and Technology*, **72**(2), 269–276, https://doi.org/10.2166/wst.2015.218

Ateia M., Apul O. G., Shimizu Y., Muflihah A., Yoshimura C. and Karanfil T. (2017). Elucidating adsorptive fractions of natural organic matter on carbon nanotubes. *Environmental Science and Technology*, **51**(12), 7101–7110, https://doi.org/10.1021/acs.est.7b01279

Bae B. U., Jung E. S., Kim Y. R. and Shin H. S. (1999). Treatment of landfill leachate using activated sludge process and electron-beam radiation. *Water Research*, **33**(11), 2669–2673, https://doi.org/10.1016/S0043-1354(98)00488-6

Bolyard S. C. and Reinhart D. R. (2017). Evaluation of leachate dissolved organic nitrogen discharge effect on wastewater effluent quality. *Waste Management*, **65**, 47–53, https://doi.org/10.1016/j.wasman.2017.03.025

Boyer T. H., Graf K. C., Comstock S. E. H. and Townsend T. G. (2011). Magnetic ion exchange treatment of stabilized landfill leachate. *Chemosphere*, **83**(9), 1220–1227, https://doi.org/10.1016/j.chemosphere.2011.03.040

Bu L., Wang K., Zhao Q. L., Wei L. L., Zhang J. and Yang J. C. (2010). Characterization of dissolved organic matter during landfill leachate treatment by sequencing batch reactor, aeration corrosive cell-Fenton, and granular activated carbon in series. *Journal of Hazardous Materials*, **179**(1–3), 1096–1105, https://doi.org/10.1016/j.jhazmat.2010.03.118

Castagnoli O., Musmeci L., Zavattiero E. and Chirico M. (1990). Humic substances and humification rate in a municipal refuse disposed of in a landfill. *Water, Air, and Soil Pollution*, **53**(1), 1–12.

Cataldo F. and Angelini G. (2012). Radiolysis and ozonolysis of a landfill leachate. *Journal of Radio Analytical and Nuclear Chemistry*, **293**(1), 141–148, https://doi.org/10.1007/s10967-012-1729-7

Chin Y. P., Alken G. and O'Loughlin E. (1994). Molecular weight, polydispersity, and spectroscopic properties of aquatic humic substances. *Environmental Science and Technology*, **28**(11), 1853–1858, https://doi.org/10.1021/es00060a015

Christensen J. B., Jensen D. L., GrØn C., Filip Z. and Christensen T. H. (1998). Characterization of the dissolved organic carbon in landfill leachate-polluted groundwater. *Water Research*, **32**(1), 125–135, https://doi.org/10.1016/S0043-1354(97)00202-9

Cui Y., Wu Q., Yang M. and Cui F. (2016). Three-dimensional excitation-emission matrix fluorescence spectroscopy and fractions of dissolved organic matter change in landfill leachate by biological treatment. *Environmental Science and Pollution Research*, **23**(1), 793–799, https://doi.org/10.1007/s11356-015-5226-8

Deng Y., Jung C., Zhao R., Torrens K. and Wu L. (2018). Adsorption of UV-quenching substances (UVQS) from landfill leachate with activated carbon. *Chemical Engineering Journal*, **350**, 739–746, https://doi.org/10.1016/j.cej.2018.04.056

Dia O., Drogui P., Buelna G., Dubé R. and Ihsen B. S. (2017). Electrocoagulation of bio-filtrated landfill leachate: fractionation of organic matter and influence of anode materials. *Chemosphere*, **168**, 1136–1141, https://doi.org/10.1016/j.chemosphere.2016.10.092

Driskill N. M., Novak J. T. and Goldsmith C. D. (2015). Hydrophobicity distribution of landfill leachates and evaluation of organic constituents that affect publicly owned treatment works. *Journal of Environmental Engineering*, **141**(4), 1–8, https://doi.org/10.1061/(ASCE)EE.1943-7870.0000904

Fan H. J., Chen I. W., Lee M. H. and Chiu T. (2007). Using FeGAC/H_2O_2 process for landfill leachate treatment. *Chemosphere*, **67**(8), 1647–1652, https://doi.org/10.1016/j.chemosphere.2006.09.075

Gupta A., Zhao R., Novak J. T. and Douglas Goldsmith C. (2014). Application of Fenton's reagent as a polishing step for removal of UV quenching organic constituents in biologically treated landfill leachates. *Chemosphere*, **105**, 82–86, https://doi.org/10.1016/j.chemosphere.2013.12.066

Hoornweg D. and Bhada-Tata P. (2012). What A Waste: A Global Review of Solid Waste Management. World Bank, Washington, DC, USA.

Huo S., Xi B., Yu H., He L., Fan S. and Liu H. (2008). Characteristics of dissolved organic matter (DOM) in leachate with different landfill ages. *Journal of Environmental Sciences*, **20**(4), 492–498, https://doi.org/10.1016/S1001-0742(08)62085-9

Huo S. L., Xi B. D., Yu H. C. and Liu H. L. (2009). Dissolved organic matter in leachate from different treatment processes. *Water and Environment Journal*, **23**(1), 15–22, https://doi.org/10.1111/j.1747-6593.2007.00104.x

Iskander S. M., Zou S., Brazil B., Novak J. T. and He Z. (2017a). Energy consumption by forward osmosis treatment of landfill leachate for water recovery. *Waste Management*, **63**, 284–291, https://doi.org/10.1016/j.wasman.2017.03.026

Iskander S. M., Novak J. T., Brazil B. and He Z. (2017b). Simultaneous energy generation and UV quencher removal from landfill leachate using a microbial fuel cell. *Environmental Science and Pollution Research*, **24**(33), 26040–26048, https://doi.org/10.1007/s11356-017-0231-8

Iskander S. M., Zhao R., Pathak A., Gupta A., Pruden A., Novak J. T. and He Z. (2018). A review of landfill leachate induced ultraviolet quenching substances: sources, characteristics, and treatment. *Water Research*, **145**, 297–311, https://doi.org/10.1016/j.watres.2018.08.035

Jones M. N. and Bryan N. D. (1998). Colloidal properties of humic substances. *Advances in Colloid and Interface Science*, **78**(1), 1–48, https://doi.org/10.1016/S0001-8686(98)00058-X

Jung C., Deng Y., Zhao R. and Torrens K. (2017). Chemical oxidation for mitigation of UV-quenching substances (UVQS) from municipal landfill leachate: Fenton process versus ozonation. *Water Research*, **108**, 260–270, https://doi.org/10.1016/j.watres.2016.11.005

Kang K. H., Shin H. S. and Park H. (2002). Characterization of humic substances present in landfill leachates with different landfill ages and its implications. *Water Research*, **36**(16), 4023–4032, https://doi.org/10.1016/S0043-1354(02)00114-8

Kaza S., Yao L., Bhada-Tata P. and Van Woerden F. (2018). What A Waste 2.0: A Global Snapshot of Solid Waste Management to 2050. World Bank Publications, Washington, DC, USA.

Kim Y. B. and Ahn J. H. (2017). Changes of absorption spectra, $SUVA_{254}$, and color in treating landfill leachate using microwave-assisted persulfate oxidation. *Korean Journal of Chemical Engineering*, **34**(7), 1980–1984, https://doi.org/10.1007/s11814-017-0104-3

Kjeldsen P., Barlaz M. A., Rooker A. P., Baun A., Ledin A. and Christensen T. H. (2002). Present and long-term composition of MSW landfill leachate: a review. *Critical Reviews in Environmental Science and Technology*, **32**(4), 297–336, https://doi.org/10.1080/10643380290813462

Leenheer J. A. (1981). Comprehensive approach to preparative isolation and fractionation of dissolved organic carbon from natural waters and wastewaters. *Environmental Science and Technology*, **15**(5), 578–587, https://doi.org/10.1021/es00087a010

Liang Z., Liu J. X. and Li J. (2009). Decomposition and mineralization of aquatic humic substances (AHS) in treating landfill leachate using the anammox process. *Chemosphere*, **74**(10), 1315–1320, https://doi.org/10.1016/j.chemosphere.2008.11.073

Liu Z. P., Wu W. H., Shi P., Guo J. S. and Cheng J. (2015). Characterization of dissolved organic matter in landfill leachate during the combined treatment process of air stripping, Fenton, SBR and coagulation. *Waste Management*, **41**, 111–118, https://doi.org/10.1016/j.wasman.2015.03.044

Mandal P., Dubey B. K. and Gupta A. K. (2017). Review on landfill leachate treatment by electrochemical oxidation: drawbacks, challenges and future scope. *Waste Management*, **69**, 250–273, https://doi.org/10.1016/j.wasman.2017.08.034

Mitch W. A. and Sedlak D. L. (2004). Characterization and fate of N-nitrosodimethylamine precursors in municipal wastewater treatment plants. *Environmental Science and Technology*, **38**(5), 1445–1454, https://doi.org/10.1021/es035025n

Moravia W. G., Amaral M. C. S. and Lange L. C. (2013). Evaluation of landfill leachate treatment by advanced oxidative process by Fenton's reagent combined with membrane separation system. *Waste Management*, **33**(1), 89–101, https://doi.org/10.1016/j.wasman.2012.08.009

Pathak A., Pruden A. and Novak J. T. (2018). Two-stage anaerobic membrane bioreactor (AnMBR) system to reduce UV absorbance in landfill leachates. *Bioresource Technology*, **251**, 135–142, https://doi.org/10.1016/j.biortech.2017.12.050

Piccolo A. (2002). The supramolecular structure of humic substances: a novel understanding of humus chemistry and implications in soil science. *Advances in Agronomy*, **75**, 57–134, https://doi.org/10.1016/S0065-2113(02)75003-7

Ratasuk N. and Nanny M. A. (2007). Characterization and quantification of reversible redox sites in humic substances. *Environmental Science and Technology*, **41**(22), 7844–7850, https://doi.org/10.1021/es071389u

Renou S., Givaudan J. G., Poulain S., Dirassouyan F. and Moulin P. (2008). Landfill leachate treatment: review and opportunity. *Journal of Hazardous Materials*, **150**(3), 468–493, https://doi.org/10.1016/j.jhazmat.2007.09.077

Satyawali Y., Wiele T. V. D., Saveyn H., Meeren P. V. D. and Verstraete W. (2007). Electrolytic reduction improves treatability of humic acids containing water streams. *Journal of Chemical Technology & Biotechnology*, **83**, 1163–1169, https://doi.org/10.1002/jctb.1715

Šír M., Podhola M., Patočka T., Honzajková Z., Kocurek P., Kubal M. and Kuraš M. (2012). The effect of humic acids on the reverse osmosis treatment of hazardous landfill leachate. *Journal of Hazardous Materials*, **207–208**, 86–90, https://doi.org/10.1016/j.jhazmat.2011.08.079

Soubh A. and Mokhtarani N. (2016). The post treatment of composting leachate with a combination of ozone and persulfate oxidation processes. *RSC Advances*, **6**(80), 76113–76122, https://doi.org/10.1039/C6RA09539A

Tang C. Y., Kwon Y. N. and Leckie J. O. (2007). Fouling of reverse osmosis and nanofiltration membranes by humic acid-effects of solution composition and hydrodynamic conditions. *Journal of Membrane Science*, **290**(1–2), 86–94, https://doi.org/10.1016/j.memsci.2006.12.017

Thurman E. M. and Malcolm R. L. (1981). Preparative isolation of aquatic humic substances. *Environmental Science & Technology*, **15**(4), 463–466, https://doi.org/10.1021/es00086a012

Wang P., Zeng G., Peng Y., Liu F., Zhang C., Huang B., Zhong Y., He Y. and Lai M. (2014). 2,4,6-Trichlorophenol-promoted catalytic wet oxidation of humic substances and stabilized landfill leachate. *Chemical Engineering Journal*, **247**, 216–222, https://doi.org/10.1016/j.cej.2014.03.014

Wang H., Wang Y. nan, Li X., Sun Y., Wu H. and Chen D. (2016). Removal of humic substances from reverse osmosis (RO) and nanofiltration (NF) concentrated leachate using continuously ozone generation-reaction treatment equipment. *Waste Management*, **56**, 271–279, https://doi.org/10.1016/j.wasman.2016.07.040

Wang H.-W., Li X.-Y., Hao Z.-P., Sun Y.-J., Wang Y.-N., Li W.-H. and Tsang Y. F. (2017). Transformation of dissolved organic matter in concentrated leachate from nanofiltration during ozone-based oxidation processes (O_3, O_3/H_2O_2 and O_3/UV). *Journal of Environmental Management*, **191**, 244–251, https://doi.org/10.1016/j.jenvman.2017.01.021

Weishaar J. L., Aiken G. R., Bergamaschi B. A., Fram M. S., Fujii R. and Mopper K. (2003). Evaluation of specific ultraviolet absorbance as an indicator of the chemical composition and reactivity of dissolved organic carbon. *Environmental Science and Technology*, **37**(20), 4702–4708, https://doi.org/10.1021/es030360x

Wu J. J., Wu C. C., Ma H. W. and Chang C. C. (2004). Treatment of landfill leachate by ozone-based advanced oxidation processes. *Chemosphere*, **54**(7), 997–1003, https://doi.org/10.1016/j.chemosphere.2003.10.006

Wu Y., Zhou S., Qin F., Peng H., Lai Y. and Lin Y. (2010a). Removal of humic substances from landfill leachate by Fenton oxidation and coagulation. *Process Safety and Environmental Protection*, **88**(4), 276–284, https://doi.org/10.1016/j.psep.2010.03.002

Wu Y., Zhou S., Qin F., Ye X. and Zheng K. (2010b). Modeling physical and oxidative removal properties of Fenton process for treatment of landfill leachate using response surface methodology (RSM). *Journal of Hazardous Materials*, **180**(1–3), 456–465, https://doi.org/10.1016/j.jhazmat.2010.04.052

Wu Y., Zhou S., Ye X., Chen D., Zheng K. and Qin F. (2011). Transformation of pollutants in landfill leachate treated by a combined sequence batch reactor, coagulation, Fenton oxidation and biological aerated filter technology. *Process Safety and Environmental Protection*, **89**(2), 112–120, https://doi.org/10.1016/j.psep.2010.10.005

Xiao S., Peng J., Song Y., Zhang D., Liu R. and Zeng P. (2013). Degradation of biologically treated landfill leachate by using electrochemical process combined with UV irradiation. *Separation and Purification Technology*, **117**, 24–29, https://doi.org/10.1016/j.seppur.2013.04.024

Xu J., Yu H. Q. and Sheng G. P. (2016). Kinetics and thermodynamics of interaction between sulfonamide antibiotics and humic acids: surface plasmon resonance and isothermal titration microcalorimetry analysis. *Journal of Hazardous Materials*, **302**, 262–266, https://doi.org/10.1016/j.jhazmat.2015.09.058

Xu J., Long Y., Shen D., Feng H. and Chen T. (2017). Optimization of Fenton treatment process for degradation of refractory organics in pre-coagulated leachate membrane concentrates. *Journal of Hazardous Materials*, **323**, 674–680, https://doi.org/10.1016/j.jhazmat.2016.10.031

Zhai Y., Zhu L., Zhu Y., Peng C., Wang T., Liu X., Li C. and Zeng G. (2016). Simultaneous total organic carbon and humic acid removals for landfill leachate using subcritical water catalytic oxidation based on response surface methodology. *Water, Air, and Soil Pollution*, **227**(8), 1–13, https://doi.org/10.1007/s11270-016-2977-x

Zhang C., Liu J., Yang X., Cao W. and Qian G. (2017). Degradation of refractory organics in biotreated landfill leachate using high voltage pulsed discharge combined with TiO_2. *Journal of Hazardous Materials*, **326**, 221–228, https://doi.org/10.1016/j.jhazmat.2016.12.034

Zhao R., Novak J. T. and Goldsmith C. D. (2012). Evaluation of on-site biological treatment for landfill leachates and its impact: a size distribution study. *Water Research*, **46**(12), 3837–3848, https://doi.org/10.1016/j.watres.2012.04.022

Zhao R., Gupta A., Novak J. T., Goldsmith C. D. and Driskill N. (2013a). Characterization and treatment of organic constituents in landfill leachates that influence the UV disinfection in the publicly owned treatment works (POTWs). *Journal of Hazardous Materials*, **258–259**, 1–9, https://doi.org/10.1016/j.jhazmat.2013.04.026

Zhao X., Wei X., Xia P., Liu H. and Qu J. (2013b). Removal and transformation characterization of refractory components from biologically treated landfill leachate by Fe^{2+}/NaClO and Fenton oxidation. *Separation and Purification Technology*, **116**, 107–113, https://doi.org/10.1016/j.seppur.2013.05.030

Zhao R., Jung C., Trzopek A., Torrens K. and Deng Y. (2018). Characterization of ultraviolet-quenching dissolved organic matter (DOM) in mature and young leachates before and after biological pre-treatment. *Environmental Science: Water Research and Technology*, **4**(5), 731–738, https://doi.org/10.1039/C7EW00544J

Zolfaghari M., Drogui P., Brar S. K., Buelna G. and Dubé R. (2017). Unwanted metals and hydrophobic contaminants in bioreactor effluents are associated with the presence of humic substances. *Environmental Chemistry Letters*, **15**(3), 489–494, https://doi.org/10.1007/s10311-016-0598-7

Zouboulis A. I., Jun W. and Katsoyiannis I. A. (2003). Removal of humic acids by flotation. *Colloids and Surfaces A: Physicochemical and Engineering Aspects*, **231**(1–3), 181–193, https://doi.org/10.1016/j.colsurfa.2003.09.004

Zouboulis A. I., Chai X. L. and Katsoyiannis I. A. (2004). The application of bioflocculant for the removal of humic acids from stabilized landfill leachates. *Journal of Environmental Management*, **70**(1), 35–41, https://doi.org/10.1016/j.jenvman.2003.10.003

doi: 10.2166/9781789063318_0267

Chapter 12

Microplastics in landfill leachate and its treatment

K. S. D. Premarathna[1], Sammani Ramanayaka[1,2], Thilakshani Atugoda[1], Madushika Sewwandi[1] and Meththika Vithanage[1,3,4]*

[1]Ecosphere Resilience Research Center, Faculty of Applied Sciences, University of Sri Jayewardenepura, Nugegoda, Sri Lanka
[2]Lancaster Environment Centre, Lancaster University, Lancaster LA1 4YQ, UK
[3]Sustainability Cluster, School of Engineering, University of Petroleum and Energy Studies, Dehradun, Uttarakhand 248007, India
[4]The UWA Institute of Agriculture, University of Western Australia, Perth, WA 6009, Australia
**Corresponding author: meththika@sjp.ac.lk*

ABSTRACT

Landfill leachate is a prominent source for microplastics into the environment. Various polymer types have been detected in leachate collected from landfills and among those polyethylene and polypropylene were the most prominently detected polymer types, with a size ranged from 20 to 5,000 μm. Fibers, foams, films, beads, and fragments were the mostly detected shapes, and a majority was transparent and yellow colored. The abundance of microplastics in leachate varied with the type of waste present in landfills. Microplastics present in leachate migrated to the surrounding soil, groundwater, and open waters located nearby landfills. The main migration pathways are direct leaching and usage of treated leachate as a soil conditioner, and usage of microplastic-contaminated sludge released from leachate-treatment facilities as fertilizers. To reduce the risks associated with microplastics released from landfill leachate, treatment of landfill leachate using proper treatment strategies is mandatory. The main techniques used to remove microplastics from landfill leachate are physical, chemical, biological, and land-based processes. After the treatment, abundance, size, and chemical composition of microplastics were altered considerably. However, the complexity and composition of leachate decreased the efficiency of treatment techniques substantially. This chapter provides a descriptive overview of the occurrence, abundance, and properties of microplastics extracted from leachate, the main migration pathways that reintroduce microplastics into the surrounding environment, the main treatment techniques, and challenges associated with the treatment strategies.

Keywords: Landfill leachate, microplastics, municipal solid waste, source segregation, wastewater treatment.

12.1 PLASTICS IN LANDFILLS AND LEACHATE

The increased usage of plastics has enhanced the annual production; for example, at present global plastic production is over 300 million tonnes (Plastics Europe, 2018). Concurrently, the release of plastic waste into the natural environment also increased rapidly. About 6.3 billion Mt of plastic waste was generated and more than 79% of plastic waste were released to landfill sites, 12% incinerated, and 9% recycled (Geyer *et al.*, 2017). After the disposal, plastics slowly undergo degradation via physical, chemical, and microbial weathering and form microplastics that are smaller than 5 mm (Hou *et al.*, 2021).

Open dumping and landfills are the most popular solid waste disposal methods practiced both in developed and developing countries around the world. Hence, annually open dumpsites receive a million of tonnes of municipal waste. For example, in 2018 the amount of municipal solid waste (MSW) that ended up in landfills was 146.1 million tons; more than 18% of it accounted for plastics (EPA, 2022). In 2017, more than 3/4 of plastics disposed by the citizens of the United State ended up in landfills, only 8% recycled (Heller *et al.*, 2020). Although landfills are considered as a waste management strategy, improper management and unregulated handling would increase the risk of contaminating the surrounding environment. In low- and middle-income countries, particularly in the Asian region, approximately 90% of landfills are non-engineered or open disposal facilities which is a common approach used as an option for solid waste management (Kumarathilaka *et al.*, 2017). Landfills are estimated to store up to 42% of global plastic waste production (Nizzetto *et al.*, 2016). Around 9.9–19.1% of MSW disposed in landfills in China accounted for rubber and plastics (Ding *et al.*, 2021). Plastics gathered in landfills degrade by continuous exposure to direct sunlight, mechanical forces, and microbial activities and produce microplastics (He *et al.*, 2019). Furthermore, plastics congregated inside open dump sites or landfills were subject to extreme conditions such as elevated temperatures, pH fluctuations, physical stress, and microbial degradation leading to the formation of microplastics. Therefore, MSW disposal sites serve as a huge reservoir of microplastics (Kawai *et al.*, 2012; Tupsakhare *et al.*, 2020).

Leachate is a liquid effluent that originates through the interaction between rainwater and degraded waste products from landfills (Wijekoon *et al.*, 2021). Landfills can generate a large volume of leachate especially after rainfall events which might take multiple pathways such as ground infiltration, surface runoff, evapotranspiration, or groundwater intrusion (Renou *et al.*, 2008). The quantity and the composition of leachate mainly depends upon the amount of water that infiltrates and percolates into landfills and the environmental degradation and fragmentation process taking place inside landfills (Upadhyay & Bajpai, 2021). It is considered as a severe contamination source that negatively influences the quality of surface and groundwater (Noerfitriyani *et al.*, 2018). Landfill leachate is an important source that carries pollutants present in landfills to surface water bodies, groundwater aquifers, and soil including microplastic particles. The main objective of this chapter is to discuss the occurrence, abundance, and characteristics of microplastics present in landfill leachate, the main migration pathways involved in the migration of microplastics from leachate to the surrounding soil and water, and current treatment techniques and challenges associated with those techniques.

12.2 OCCURRENCE AND ABUNDANCE OF MICROPLASTICS IN LANDFILL LEACHATE

Earlier studies have reported that polyethylene (PE) and polypropylene (PP) microplastics are the abundant polymer types in landfill leachate. The abundance of microplastics varied with the type of waste present in landfills.

12.2.1 Sources and formation of microplastics in landfills

Organic waste in landfill sites degrades fast with microbial activities under aerobic/anaerobic conditions producing leachate (He *et al.*, 2019). However, plastics which are mainly non-biodegradable will remain in landfills. Disposable diapers, carrier bags, films such as packaging sheets, container bottles, polyvinyl chloride (PVC) pipes, footwear, disposable cups, plastic utensils, household items such as shower curtains, and so on are considered to be the main plastic sources in landfills (Barnes *et al.*, 2009). Biochemical reactions result in temperature fluctuations (30–60°C), high salinity (ca. 3–41 mS/cm), and generation of toxic gases such as CH_4, CO, CO_2, and H_2S inside landfills that can create an environment to disintegrate and degrade (Sun *et al.*, 2021; Tupsakhare *et al.*, 2020). The maturity of the landfill governs the pH of the leachate; fresh landfills produce acidic leachate (acetogenic) and old landfills discharge leachate with high pH (methanogenic) (Kjeldsen *et al.*, 2002). The extreme environmental conditions enhance the fragmentation of plastic debris in landfill sites resulting in secondary microplastics including microbial degradation (Sun *et al.*, 2021). Furthermore, roadside tire particles which are a secondary microplastic due to waste transport and management can be ended up in landfills along with MSW (Figure 12.1) (Vijayan *et al.*, 2022). Nevertheless, compared to secondary microplastics, a small fraction of primary microplastics can reach landfills via used personal care products' packaging. Furthermore, discharged leachate acts as a media to transport microplastics originate from landfills (He *et al.*, 2019). Therefore, landfill leachate has been made a significant contribution to microplastic pollution in soils as well as freshwater environments (Alimi *et al.*, 2018; Sundt *et al.*, 2014).

Inside landfills, microplastic mainly forms via weathering of larger plastic particles through photo-degradation, thermal degradation, mechanical fragmentation, and biodegradation (Resmeriță *et al.*, 2018; Tu *et al.*, 2020). Physical degradation results in a change in the bulk structure whereas chemical degradation changes the molecular structure (Chamas *et al.*, 2020). Once plastics enter landfills, at the initial stage, they are subjected to aerobic biodegradation, but immediately shift into anaerobic conditions because of acids and gases generated due to organic solid waste degradation (Hou *et al.*, 2021). An anaerobic environment facilitates the survival of microorganisms that have the potential to degradation of plastic waste collected in landfills resulting in the formation of tiny plastic particles (Upadhyay & Bajpai, 2021). Microplastics

Figure 12.1 Plastic food packaging from household waste, face masks, tire particles, and plastic debris from MSW are the key sources of microplastics in MSW landfills.

undergo thermal degradation due to extreme temperature variations inside landfills; the presence of metals (iron, copper, and chromium) speeds up the thermal oxidation process (Hou *et al.*, 2021). Moreover, mechanical abrasion is caused due to anthropogenic activities and continuous exposure to sunlight breakdown of larger microplastics into smaller microplastic particles (Klein *et al.*, 2018). On the basis of the physical and chemical properties of plastics (size, structure, and crystallinity) and environmental conditions (pH, temperature, oxygen, moisture, and biofilms), time taken for the degradation varies (Chen *et al.*, 2020; Delaney 2013; Wang *et al.*, 2020a). The moisture content and biological enzymes present in landfill leachate predominantly weaken the polymer chain accelerating the fragmentation of plastic waste and thereby forming cracks and fractures of plastics, ultimately producing different shapes of microplastics in different sizes (Horton *et al.*, 2017; Shen *et al.*, 2022b). Furthermore, weathering of microplastics in landfill leachate would enrich the porosity of microplastics and form rough edges on the surfaces of microplastics. Consequently, the degradation and weathering process initiated in landfills plays a vital role in the formation of microplastics in different shapes and sizes and thereafter contaminating the leachate originated.

Accumulation of plastic debris in landfills can be extensively reduced by recycling household-generated plastic waste (Afroz *et al.*, 2017). Therefore, most of the large municipalities in the developed countries have their own plastic recycling plants (Al-Maaded *et al.*, 2012; Nkwachukwu *et al.*, 2013). However, a fraction of plastics still reach landfills such non-recyclable plastic food packaging which are being directly disposed to household organic waste (Jadhav *et al.*, 2021). Furthermore, authorities have introduced processes such as composting organic waste to manage the drawbacks caused by landfills (Vaverková *et al.*, 2020). Other than plastic recycling plants, they have introduced compost production plants and incineration plants to manage the organic waste in landfills. Unfortunately, non-recyclable plastics disposed along with the organic waste produce microplastics in the compost production process (Watteau *et al.*, 2018). Thus, MSW compost is also contaminated with microplastics and other toxicants such as potentially toxic elements, poly-aromatic hydrocarbons, pharmaceuticals, and personal care products (Ingrando *et al.*, 2022; Sahebdelfar *et al.*, 2022). Moreover, according to the World Health Organization's waste management guideline, healthcare waste is not allowed to dispose in landfill sites (World Health, 2005). However, thousands of single-use personal protection equipment such as face masks had been disposed to landfills during the COVID-19 pandemic. Face masks are a source of microplastics which could increase the generation of microplastics in landfills (Li *et al.*, 2022). Hence, it has become a serious concern for environmental engineers and researchers to find innovative technologies to remove microplastics from landfill leachate (Figure 12.2).

Figure 12.2 Photographs demonstrating open dump sites adjacent to water sources with leachate flowing directly to the surface water bodies (top) and origin of microplastics in leachate from plastics in the open dump site (bottom).

Table 12.1 Characteristics of microplastics found in landfill leachate.

Location of Study	Extraction Method	Abundance of Microplastics	Size of Microplastics (μm)	Polymer Type	Color of Microplastics	Shape of Microplastics	Reference
Landfills of Thailand (12 landfills)	Not mentioned	13.50–27.50 items/kg dry weight	330–5,000	PS, PP, PET	Not mentioned	Granules, films, fibers, irregular, spheres	Puthcharoen and Leungprasert (2019)
China	Filtration followed by wet peroxidation Identification through a micro-Raman spectrometer	235 40 ± 17.10 items/L and 11.40 ± 0.80 μg/L	<50 accounting for over 50%	PE, PP	Over 90% are transparent or yellowish; <10% were red, pink, purple, black, blue, and brown	Beads, fibers, films, fragments Fragments made up for the highest proportions	Sun *et al.* (2021)
Laogang, Shanghai, China	Filtration through sieves, wet peroxide digestion with 30% hydrogen peroxide	4–13	<500 500–1,000 1,000–5,000	Cellophane, PE, PP, PS	Not mentioned	Fibers, granules, fragments	Su *et al.* (2019)
Shanghai, China	Filtration followed by chemical digestion 10–30 mL of 30% H_2O_2 Identification through an FTIR microscope	291 ± 91 pieces/L	20–100 Approximately 90% of microplastics were ≤60	PP, PA rayon, PP, PES, PET	Not mentioned	Not mentioned	Xu *et al.* (2020)

Shanghai, Wuxi, Suzhou, and Changzhou, China	Filtration, wet peroxide digestion with 30% hydrogen peroxide and dual-density separation	0.42–24.58 items/L	1,000–5,000 and 100–1,000	PE and PP were the prominent; PVC, PS, ABS, PET, PU, EVA, PA, PES, EP, PF, PPC, PMMA, ALK, PMDS, PTFE in minor amounts	Not mentioned	Flakes, fragments, pellets, lines, foams	He *et al.* (2019)
Topinoja, Turku/Åbo, Finland	Hydrogen peroxide digestion after filtering the leachate samples	0.16 items/L	50–5,000	PE, PP	Not mentioned	Fragments, fibers, foils	Praagh *et al.* (2018)
Korvenmäki, Salo, Finland		1.10 items/L		PE, PP, PVC, PS, PET, PA, PU, PMMA			
Hollola, Lahti, Finland		1.97 items/L		PE, PS, PET, PU			
Gjerdrum, Ask, Norway		1.0 items/L		PE			
Southeast, Norway		1.40 items/L		PE, PET, PMMA			
Álfsnes, Iceland		4.51 items/L		PE, PP, PS, PU			
Tali, Helsinki	Filtration Stereomicroscope was used to identify and count the microlitter	0.002–0 017 items/L	<300, 100–300, and 100–20	Synthetic polymers	Not mentioned	Fibers, irregular particles	Kilponen (2016)
Jagodina, Serbia	Filtration, wet peroxide digestion, and density separation with $ZnCl_2$ Identification was carried out by a stereomicroscope	0.3–2.2 mg/L	Not mentioned	Not mentioned	Different colors	Irregular shape and rough edges Fragments were most abundant	Narevski *et al.* (2021)
Bijeljina, Bosnia, and Herzegovina		0.3–2.2 mg/L					
Novi Sad, Serbia		0.64–1.76 mg/L					

(*Continued*)

Table 12.1 Characteristics of microplastics found in landfill leachate. (*Continued*)

Location of Study	Extraction Method	Abundance of Microplastics	Size of Microplastics (μm)	Polymer Type	Color of Microplastics	Shape of Microplastics	Reference
Galuga landfill area, Indonesia	Fenton's reagent was used in the digestion process to extract micro- and mesoplastics from organic-rich samples Analyzed using FTIR and μ-FTIR spectroscopy	Highest: 57.04 ± 5.88 particles/m^3 Lowest: 5.19 ± 4.44 particles/m^3	<80 and 80–5,000	PE, PP, PS, PET, PES, cellophane	Not mentioned	Fiber, fragments, granules, foams	Nurhasanah *et al.* (2021)
South China	Density separation using $ZnCl_2$ solution with a density higher than 1.5 g/mL Visual analysis under a stereomicroscope (Nikon, Tokyo, Japan). Quantity, shapes, and colors of the suspected microplastics	3–25 items/L	85% in the range of 20–150	PU, PAT, PA, PEC, PP, PE, PS, PET, and others	Not mentioned	Not mentioned	Wan *et al.* (2022)
MSW landfill, Shanghai	Not mentioned	Not mentioned	Not mentioned Degraded from large plastics waste	PE, PP, PET	Not mentioned	Not mentioned	Yu *et al.* (2022)
Bushehr port, Iran	Leachate sample was filtered through a stack of stainless-steel sieves. Then H_2O_2 (30%, v/v) was used Density separation was done with NaI and vacuum filtration Micro-Raman analysis was applied to recognize the chemical composition	79.16 items/L	450–5,000	Nylon, PP, PC, PS	Black, white/transparent, blue, green, red, brown	Fragments, spherules, fibers, angular	Mohammadi *et al.* (2022)

Laogang leachate-treatment plant, Shanghai, China	Leachate sample was filtered and the residue on the membrane was digested with 20 mL of H_2O_2 (30%, v/v) for 3 days Detection was performed through a stereomicroscope Chemical characterization through FTIR and physical characterization by SEM	1.2 ± 0.57 items/L	<500 and 500–5,000	PE, PES, PP, PA, EPM, PVAC	Not mentioned	Fibers, fragments, particles, rods	Zhang *et al.* (2021)

Note: PU, polyurethane; PAT, polyacetal; PA, polyamide; PEC, polyethylene chlorinated; PP, polypropylene; PE, polyethylene; PS, polystyrene; PET, polyethylene terephthalate; PC, polycarbonate; PES, polyester; EPM, ethylene–propylene polymer; PVAC, polyvinyl acetate; PVC, polyvinyl chloride; ABS, acrylonitrile butadiene styrene; EVA, ethylene-vinyl acetate; EP, ethylene–propylene; PF, phenol–formaldehyde; PPC, polypropylene carbonate; PMMA, poly(methyl methacrylate); ALK, alkyne; PMDS, polydimethylsiloxane; PTFE, poly-tetra-fluoro-ethylene; SEM, scanning electron microscopy.

12.2.2 Abundance of microplastics in landfill leachate

Landfills and open dumps receive a massive amount of plastic wastes and has been estimated to be about 42% of 359 Mt of the global plastic waste generated in 2018 (Hahladakis *et al.*, 2018). Table 12.1 demonstrates the abundance and characteristics of microplastics from various landfill leachates in the world. PE and PP microplastics (the most common types of polymers in food packaging) have commonly reported in landfill leachate (Jadhav *et al.*, 2021). Selected MSW and industrial waste (IW) landfill sites in Nordic countries have been investigated for microplastics and the highest abundance of 4.51 particles/L of leachate was reported for the landfill 'Álfsnes' in Iceland (Praagh *et al.*, 2018). Interestingly, landfills with MSW and IW have demonstrated the elevated microplastic concentrations of >1.10 particles/L (Table 12.1). However, no microplastic particles have been found in non-treated leachate from landfills 'Kujala' in Finland and 'Fiflholt old cell' in Iceland (Praagh *et al.*, 2018).

12.3 CHARACTERISTICS OF MICROPLASTICS PRESENT IN LANDFILL LEACHATE

As the investigations of the presence of microplastics in landfill leachate are still in their early stage, limited information exists. The size distribution of microplastics and their polymer type was largely overlooked, whereas the discussion on other characteristics such as color, shape, surface morphology, and crystallinity were reported in a few studies. Thus, overall characterization does not cover and exists in the literature.

12.3.1 Size, color, and shape of microplastics

Most of the studies focused only on chemical characterization of microplastics whereas some focused their attention on both physical and chemical properties. Although the physical features such as size, shape, and color of microplastics were mainly discovered, it was unable to compare the abundance of the data because the extraction, separation, and identification techniques are different from each other (Table 12.1). It is understood that the landfill age crucially influences the size and shape of the microplastics because the weathering effect causes the formation of microplastics in different sizes and shapes (Su *et al.*, 2019). Therefore, a wide range of sizes of microplastics found in landfill leachate collected worldwide might be attributed to the degradation and fragmentation process undergoing in landfills. Finer particles <100 μm were observed in several studies (Su *et al.*, 2019; Wan *et al.*, 2022; Xu *et al.*, 2020). Approximately over 50% of microplastics found were <50 μm and 90% were <60 μm in the studies of Sun *et al.* (2021) and Xu *et al.* (2020), respectively. Detection of small particles is hard and therefore the abundance is always underestimated (Sun *et al.*, 2021). According to the existing literature, microplastics were found in spherical shapes (beads), slender shapes (fiber), lamellar shapes (film), and other irregular shapes (fragments). The most abundant shape is irregularly shaped fragments (Puthcharoen & Leungprasert, 2019; Sun *et al.*, 2021). Granules, foams, flakes, rods, and foils were other shapes of microplastics found in leachates (Praagh *et al.*, 2018; Zhang *et al.*, 2021).

Microplastics found in landfill leachate have not been largely categorized based on color to date. In most cases, enlarged images of the extracted microplastics were exhibited instead of recording the color of microplastics. However, transparent and yellow colors were dominant for most of the extracted microplastics (Sun *et al.*, 2021). Besides, there are pieces of evidence for the presence of microplastics in red, pink, purple, black, blue, and brown in landfill leachate (Mohammadi *et al.*, 2022; Sun *et al.*, 2021). The color can be faded and altered due to oxidative degradation and long-term existence in leachate. Exposure to different organic matters, sunlight, and different chemicals in leachate would cause the yellowing of microplastics. Accumulation of oxygen in the polymer chain further form yellowish microplastics in leachate. Identification of the shape was mainly carried out through visual inspection and stereomicroscopic analysis. The discussion on shapes can be varying from one study to another according to the bias with the recognition.

12.3.2 Polymer types

Microplastics in different polymer types were identified in landfill leachates (Table 12.1). Chemical characterization was largely performed in most studies by recording the presence of commonly utilized polymers such as polyamide (PA), PP, PE, polystyrene (PS), and polyethylene terephthalate (PET). The most prominently found polymer type was PE whereas PP was the second most abundant polymer type observed in landfill leachates (Praagh *et al.*, 2018; Sun *et al.*, 2021). The microplastics pool of some studies consisted of a wide range of polymers such as nylon, rayon, cellulose, polyurethane (PU), polyacetal, polyethylene chlorinated, polycarbonate, polyester, ethylene–propylene polymer, and polyvinyl acetate (Narevski *et al.*, 2021; Praagh *et al.*, 2018). Some of them are rarely found in other environmental vectors of microplastics: water, beach sand, soil, and compost. In the selected landfills, seven polymer types were identified with the order of PE > PU > PET > PS > PP > PVC > PA. Nevertheless, 'Laogang' landfill, in China has depicted fibrous microplastics as the highest available in leachate (Su *et al.*, 2019). Furthermore, oxidative degradation of PE microplastics was observed in micro-Fourier transform infrared (μ-FTIR) spectroscopy analysis with an increase in the landfill age (Su *et al.*, 2019). Sun *et al.* (2021) have reported 50% of 50–100 μm microplastics in leachate from a landfill in Suzhou, China.

The complexity of landfills with a wide variety of plastic waste causes the presence of microplastics belonging to a broad range of polymer types. For instance, in the study by Praagh *et al.* (2018), a minor amount of microplastics in landfill leachate were well matched with the chemical characteristics of PVC, acrylonitrile butadiene styrene, ethylene-vinyl acetate, ethylene–propylene, phenol–formaldehyde, polypropylene carbonate, poly(methyl methacrylate), alkyne, polydimethylsiloxane, poly-tetra-fluoro-ethylene. The existence of similar polymer types in both landfill leachate and plastic waste buried in landfills likely indicates that the landfill surrounding would be a potential source of microplastics (Narevski *et al.*, 2021). PE is the most frequently utilized type of plastic worldwide. Accordingly, the greater existence of PE microplastics in the landfill leachate is irrefragable. Chemical compositions of microplastics found in landfill leachate were rather different from their originals with the appearance of C–O and –OH peaks in their IR spectra. It would be due to the addition of oxygen to the polymer chain during the photo-oxidative weathering process (Fotopoulou & Karapanagioti, 2012; Sewwandi *et al.*, 2022). Thus, the naturally occurring aerobic biodegradation in the landfill leachate stimulates the activity of microbes ultimately enhancing the biological degradation of microplastics. Consequently, the chemical properties of the microplastics would be altered. Polymer properties such as crystallinity were decreased for PE microplastics and increased for PP microplastics in the landfill leachate with the age of a particular landfill exhibiting the effect of long-term exposure to degradation processes and changes taking place in the polymer chain (Yu *et al.*, 2022).

12.4 MIGRATION PATHWAYS OF MICROPLASTICS TO AQUATIC ECOSYSTEMS

In a typical landfill, the bottom is lined with an impervious liner to restrict the movement of leachate and then collected separately for treatment. The treatment systems can remove microplastics from the leachate with an efficiency of 70–100% (Poerio *et al.*, 2019; Praagh *et al.*, 2018). In some cases, smaller microplastics (nanoplastics) can permeate through the bottom liner reaching the underneath ground of the landfill which generally consists of a clay layer (Bozbey & Guler, 2006). Microplastics discharging from the leachate can ultimately alter the hydromechanical properties of the clay soil beneath (Bordoloi *et al.*, 2022). During the weathering process, the generated nanoplastics can enter the clay interlayer space reducing the thickness of the double-diffuse layer (Hou *et al.*, 2021). This happens as a result of surface hydroxyl ions on nanoplastics which can strongly bind to the clay surface restricting further development of the double-diffuse layer (Sridharan & Choudhury, 2008). The reduction of the double-diffuse layer increases the space in between the mineral layers enabling greater water flow along the void spaces, thus promoting the migration of microplastics through the clay soil (Bordoloi *et al.*, 2022).

Microplastics leaching into the soil can accumulate overtime and then percolate through the deeper soil layers reaching groundwater. Soil in vicinity to these landfills can contaminate with microplastics carried away by leachate seepage. However, there is limited evidence discovered by studies for groundwater and soil contamination beneath landfills. In south India, groundwater surrounding an open MSW dump site contained 2–80 items/L (Karuppasamy *et al.*, 2021). Higher microplastic concentration levels are closely associated with high-population density and improper waste disposal. Additionally, the level of contamination also relates to the soil composition. In the absence of a clay layer beneath the landfill, depending on the porosity there is higher possibility for percolation through the soil column (Gopinath *et al.*, 2020). A comprehensive study was undertaken by Wan *et al.* (2022) to analyze the extent of microplastic contamination in the surrounding environment of a landfill including refuse, underlying soil, leachate, and groundwater. The refuse had a higher particle concentration of 590–103,080 items/kg than the underlying soil with a particle concentration of 570–14,200 items/kg. Similarly, the particle concentration in leachate was 3–25 items/L and in groundwater the abundance was lower with 11–17 items/L. It is apparent that microplastics generated from the source can eventually transfer to the surrounding vicinity over time polluting the soil and groundwater (Wan *et al.*, 2022). The microplastic composition is diverse in the landfill refuse in terms of polymer and size composition; however, in the old refuse the smaller size fraction becomes higher than the fresh refuse due to weathering (He *et al.*, 2019). In groundwater, the majority of the microplastics are <150 μm which suggest that smaller particles are likely to migrate through soil than larger particles and accumulate in groundwater aquifers (O'Connor *et al.*, 2019).

In the river Cisadane, Indonesia, surface water samples collected in proximity to landfill areas had significantly a higher microplastic concentration than the samples collected from the rest of the locations at upstream and downstream. The average microplastic concentration in the river near the upstream and downstream landfills was 84.00 ± 12.11 and 89.33 ± 15.35 particles/m^3, respectively, which is four times the particle concentrations found in the rest of the samples (Sulistyowati *et al.*, 2022). A study estimated the daily average release of microplastics and mesoplastics into the Cianten river, Indonesia to be 80,640 ± 604.80 and 618,240 ± 1,905.45, respectively. The samples retrieved after the leachate drain showed microplastic concentrations three times and mesoplastic concentrations nine times the average concentration present in the rest of the samples (Nurhasanah *et al.*, 2021). These examples imply that leachate is a source to microplastics polluting surface freshwater environments.

Lakes are another source of inland freshwater bodies prone to microplastic pollution by terrestrial sources (Wu *et al.*, 2018). A case study in Vietnam provides evidence how the Trung Nghia lake became a microplastic hotspot from its receiving channel due to contamination from landfill leachate. The water and sediment samples retrieved from the input location of the channel to the lake had the highest microplastic concentrations of 3,840 items/m^3 and 14,500 items/kg, respectively (Tran-Nguyen *et al.*, 2022). In another study, in the Hoa Phu lake the leachate from the city landfill was assumed to be the major source of microplastic contamination. The higher proportion of fragmented microplastics in the sediment and water samples is associated with the weathering process of larger plastic debris in the landfill (Strady *et al.*, 2021).

There is serious threat to marine environments from landfills located in proximity to coastal areas. There have been earlier studies reported about landfill leachate reaching the coastal region causing critical adverse effects on the marine ecosystem from various hazardous contaminants and microplastics originating from leachate (Aghadadashi & Mehdinia, 2016). The urban runoffs from the Bushehr port, Persian Gulf had an average microplastic concentration of 1.86 items/L with fibers in the size range of 500–1,000 μm having the highest abundance (Hajiouni *et al.*, 2022). One of the major sources of these microplastics in the urban runoffs is assumed to be the landfill located adjacent to the Bushehr port proved by another study, with an average concentration of 79.16 items/L with the microplastic composition resembling somewhat similar to that of the runoff (Mohammadi *et al.*, 2022). Sea water samples collected near a landfill in northern France had 0.099–2.844 items/L of microplastic levels which was twice as greater as the levels found in a wastewater treatment effluent. The effect of microplastic pollution related to the abundance of microplastic ingested by mussels in the same location, indicates a significantly higher number than in the mussels that were collected from other locations (Kazour *et al.*, 2019).

Additionally, after treatment leachate has been considered to be used for irrigation purposes due to its high-nutrient content (Wijekoon *et al.*, 2021); for example, Aronsson *et al.* (2010) used pre-treated landfill leachate to irrigate mature poplar willow wood plants. However, the removal efficiency of microplastics is low; hence, once they are used as soil conditioners they act as a source of microplastics and reintroduce microplastics to the soil (Júnior *et al.*, 2016). Most of the time leachate treatment only changes the distribution and fate of microplastics. During the treatment process, some large microplastic particles remain trapped in sludge whereas the rest move to the final effluent and local sewage treatment plant (Shen *et al.*, 2022b). Sludge produced in wastewater-treatment plants (WWTPs) is used as a fertilizer in most countries. Through the application of sludge contaminated with microplastics to agricultural lands microplastics trapped in sludge re-enters the environment.

12.5 MITIGATION AND REMEDIATION OF MICROPLASTICS

Leachates from both active and closed landfills are contaminated with microplastics (He *et al.*, 2019). Therefore, treating leachate before it is released into the environment is one of the best solutions to reduce microplastic contamination. To minimize the environmental impact some landfill sites around the world have their own leachate-treatment plants (He *et al.*, 2019; Narevski *et al.*, 2021; Praagh *et al.*, 2018). Additionally, leachate is treated together with MSW in WWTPs. In both treatment facilities, the most used techniques are filtration process (ultrafiltration, nanofiltration, and reverse osmosis), flocculation process, and biological reaction process (Hou *et al.*, 2021).

12.5.1 Treatment methods of microplastics in landfill leachate

Physical, chemical, biological and land-based treatment techniques are used to treat leachate contaminated with microplastics (Figure 12.3). Traditional methods as well as advanced treatment techniques are used to treat microplastics present in landfill leachate (Silva *et al.*, 2021).

12.5.1.1 Physical and chemical treatments

Sand filtration, coagulation, and sedimentation are the most common traditional techniques used to treat wastewater (Shen *et al.*, 2020). Due to the close similarities between the physical properties of particulate matter and microplastics filtration techniques are effective for the remediation of

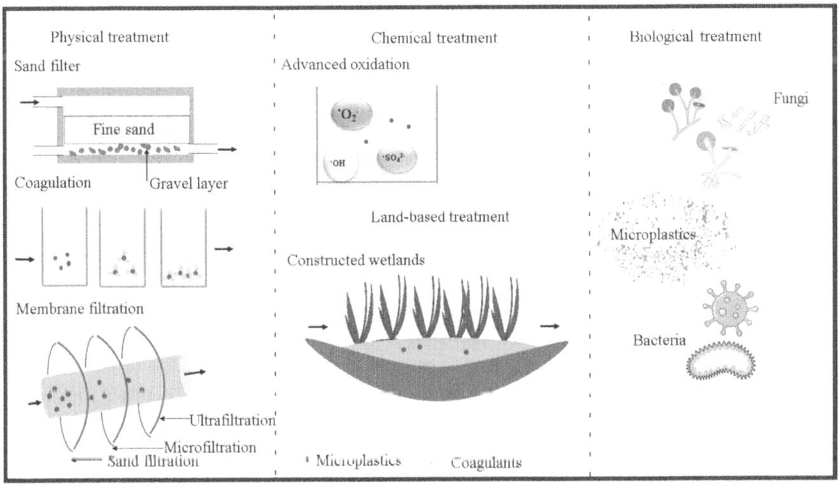

Figure 12.3 Key treatment methods used in microplastic treatment in leachate are physical and chemical, biological, and land-based treatment.

microplastic-contaminated leachate. In water samples collected from the outflow of three water-treatment plants (WTPs) which used different water-treatment techniques (WTP1 – coagulation/flocculation + sand filtration, WTP2 – coagulation/flocculation + sedimentation + sand filtration, WTP3 – coagulation/flocculation + flotation + sand filtration), WTP2 and WTP3 used an additional granular-activated carbon filter. A significant reduction in microplastic quantity was observed; the removal efficiencies were 70% (WTP1), 81% (WTP2), and 82% (WTP3) (Pivokonsky *et al.*, 2018). Flotation found to be a highly effective technique to remove microplastics due to their low density and high buoyancy (Da Costa *et al.*, 2016; Di & Wang, 2018).

Ultrafiltration, nanofiltration, and reverse osmosis are the main advanced physical treatment techniques applied in leachate-treatment facilities (Shen *et al.*, 2022b). Membrane filtration techniques are very effective for the removal of microplastic particles from leachate (Shen *et al.*, 2022b). The particle size of microplastics in landfill leachate is equal to the pore size of the physical membrane thereby, microplastics are removed efficiently and safely retained in small volumes (Shen *et al.*, 2020). Ziajahromi *et al.* (2017) observed a decrease in microplastic concentration after ultrafiltration and reverse osmosis. However, membrane fouling and trapping of large microplastic particles reduce the efficiency of the reverse osmosis process. Therefore, a system composed of membrane bioreactor, nanofiltrate, and reverse osmosis membrane could be much more effective, in which part of the microplastics could be biodegraded and captured in the membrane bioreactor, and the remaining microplastic particles are trapped by the nanofiltration and reverse osmosis units. The addition of a physical filtration unit containing biochar-mixed sand or zeolite also improved the microplastic removal efficiency; for example, a biochar filter showed more than 95% removal efficiency of microplastics with a particle size of 10 μm; its honeycomb structure captured, trapped, and entangled microplastics (Wang *et al.*, 2020b). The microplastic removal efficiency of an aluminosilicate filter media modified using cationic surfactant and rapid sand filters were >96 and 63%, respectively, and negatively charged PE and PA microplastics were bonded to the cationic surfactant headset through electrostatic interactions (Shen *et al.*, 2021). Usage of modified materials to remove microplastics lift the pressure generated on the subsequent membrane filtration system.

Advanced oxidation processes are mainly used for organic contaminants and ammonia from leachate; however, these can also be successfully used as alternative methods to remove microplastics from leachate (Shen *et al.*, 2022b). Oxidants (H_2O_2, O_3, and OCI^-) and free radicals ($\cdot OH$, $\cdot O_2^-$, and $\cdot SO_4^-$) generated during the oxidation process can erode microplastics which results in the breakage of long molecular chain and slow degradation of microplastics (Silva *et al.*, 2021). However, the high color intensity of leachate critically interferes with the degradation and oxidation process decreasing the efficiency. Electrocoagulation is an effective treatment process suitable for the removal of microplastics as well as heavy metals and suspended particles from landfill leachate. Shen *et al.* (2022a) used electrocoagulation to remove microplastics from sewage and the removal efficiency was higher than 91.7% for an Al anode. The removal rate of PE microbeads by electrocoagulation exceeded 90% (Perren *et al.*, 2018).

12.5.1.2 Biological treatments

Microorganisms have the ability to degrade plastic fragments using a range of active enzymes (Shen *et al.*, 2019). For instance, *Exiguobacterium* sp. YT2 present in mealworm intestine plays a major role in the degradation and mineralization of PS microplastics (Yang *et al.*, 2015), foam PS is their sole carbon source for *Zophobas atratus* that consume it at a 0.58 mg/day rate, and its intestinal microflora support degradation of PS (Yang *et al.*, 2020). Monomer styrene is used as the carbon source by many bacterial genera including *Flavobacterium*, *Rhodococcus*, *Nocardia*, and *Pseudomonas*; they degrade monomer styrene and produce precursors for the Krebs cycle (Danso *et al.*, 2019). *Zalerion maritimum*, a marine fungus, decreased the size and mass of PE microplastics (Paço *et al.*, 2017). Therefore, biotechnology can be used as a genetic engineering technology to treat environmental samples contaminated with microplastics (Puglisi *et al.*, 2019). Biodegradation can be used individually or combined with a physical and chemical process to remove microplastics available in landfill leachate. However, the complexity of leachate composition decreases the efficiency of the biodegradation process. Before using the biotechnological process for remediation four factors should be considered: the rate of microplastic biodegradation, the selectivity of microorganisms to microplastics, the effects of complex environmental conditions, and lastly, field-scale implementation and possible dangers.

The membrane bioreactor process was found to be a very effective biological treatment technology for microplastic removal. Also, this process omits most of the prominent drawbacks associated with traditional biological treatment technologies and results in a low amount of sludge (Sun *et al.*, 2019). Talvitie *et al.* (2017b) observed 99.9% removal of microplastics in wastewater. However, the membrane bioreactor process has been used less frequently; therefore, limited information is available on the removal of microplastics in landfill leachate.

12.5.1.3 Land-based treatments

Constructed wetlands are effectively used to degrade contaminants present in landfill leachate due to cost-effectiveness, simplicity, environmental friendliness, and high treatment efficiency (Kumarathilaka *et al.*, 2017). Plants, microorganisms, and medium contribute to remediating pollutants present in landfill leachate. When landfill leachate is released from a landfill site to the constructed wetlands insoluble pollutants are precipitated whereas soluble pollutants are directly absorbed by plant roots and degraded by microorganisms present in root biofilms. Aquatic vascular plants can effectively accumulate, stabilize, and capture microplastics and nanoplastics in the environment. Vascular plants have the ability to intercept microplastics floating in the water and they interact with floating plants; microplastics with rough surfaces are easily adsorbed by small plants (Taylor *et al.*, 2020). On the west coast of Italy, giant reed (*Arundo donax*) grown in the coastal areas assists in gathering microplastics released into the marine environment (Battisti *et al.*, 2020). Microplastics were detected in reed plants, and the quantity was high in reed clump sediments compared to the surrounding area (Yin *et al.*, 2021). Adsorption of microplastics by vascular plants depends on the electrostatic forces, particle size, and leaf morphology (Kalčíková, 2020). Negatively charged microplastics are attracted by cellulose components present in the plant cells, due to the roughness of the plant cellulose that improve adsorption by providing more binding sites. The complex leaf structure of vascular plants facilitates the capture of microplastics (Gutow *et al.*, 2016). On the basis of the aforementioned examples application of vascular plants to remove microplastics from landfill leachate is encouraged. The uprooting of plants in constructed wetlands from time to time completely removes the microplastics from landfill leachate. Furthermore, microplastics sink to sediment due to the formation of biofilms on microplastic surfaces. Microorganisms present in sediment biodegrade microplastics gradually.

12.5.2 Fate of microplastics in landfill leachate treatment

In the leachate-treatment facility, during the pre-treatment stage, larger microplastics were removed, and around 35–59% of microplastics could be retained due to the trapping in solid flocks, which are hard to remove, therefore, ended up in landfills (Blair *et al.*, 2019; Murphy *et al.*, 2016). During the primary treatment process, microplastics are further removed via skimming and settling processes, the removal percentage being 15–40% (Talvitie *et al.*, 2017a). The large microplastic particles are deposited on sludge, whereas tiny microplastics are attached to floating substances and grease. Microplastics

Table 12.2 Abundance of microplastics in landfill leachate, before and after the leachate treatment.

Country	Landfill	Location	Type of Waste	Leachate Treatment	Collection Point	Abundance of Microplastics (particles/L)	Microplastics in Untreated Leachate	Treated Leachate (Type or Amount of Microplastics)	Reference
Norway	Böler	Skedsmokorset	MSW, IW	SBR	Pump station	1.3	PE, PET, TR	TR	Praagh *et al.* (2018)
	Gjerdrum	Ask	Mixed	None	Well (with flow)	1	PE	–	
Finland	Topinoja	Turku/Åbo	MSW	None	Pump station	0.16	PE, PP, PBM	–	
	Korvenmäki	Salo	MSW, IW	None	Pump station	1.10	PE, PVC, PS, PET, PA, PU	–	
	Hollola	Lahti	MSW, IW	Artificial soil filtration	Leachate pond	1.97	PE, PS, PET, PU	PE, PS	
	Kujala	Lahti	MSW	None	Settlement pond	0	TR, PBM	–	
Iceland	Fiflholt new cell	Fiflholt	MSW, IW	Sand bed filtration	Bore hole	0.2	PE, PS, PET, PU	TR, PE	
	Fiflholt old cell	Fiflholt	MSW, IW	None	Outflow pipe	0	TR	–	
	Álfsnes	Álfsnes	MSW, IW	None	Leachate pond	4.51	PE, PP, PS, PU, PMB, TR	–	
China	–	Wuxi	MSW	–	Equalization basin Collection wells	0.96 0.42	PE, PP, PET, PS, PU	–	He *et al.* (2019)
	–	Changzhou	MSW	–	Collection wells	5.8	PE, PP, PET, PS, PU	–	

	–	Suzhou	MSW	BIOR and MF	Equalization basin	235.4 ± 17.1	PE, PP, PET, PS, PTFE	3.8 ± 0.5	Sun *et al.* (2021)
	Laogang	Shanghai	MSW	–	–	4–13	PP, PE, EPM, PET, PVC, PA, PS	–	Su *et al.* (2019)
Bosnia and Herzegovina	Brijesnica	Bijeljina	Sanitary landfill	Reverse osmosis by pump	Receiving basin	1.37 ± 0.65 mg/L	–	0.3–2.2 mg/L	Narevski *et al.* (2021)
Serbia	Gigoš	Jagodina	Sanitary landfill	Sedimentation and aeration	Aeration lagoon Sedimentation lagoon	2.12 ± 0.85 mg/L 2.16 ± 0.80 mg/L	–	2.12–2.16 mg/L	
	–	Novi Sad	Non-sanitary landfill	None	Lagoon Peripheral channel 1 Peripheral channel 2	0.64 ± 0.30 mg/L 1.40 ± 0.75 mg/L 1.76 ± 0.60 mg/L	–	0.64–1.76 mg/L	

IW, industrial waste; SBR: Sequencing Batch Biological Reactor, TR: tire particle.

trapped in sludge are separated by sedimentation (Hou *et al.*, 2021). After the pre-treatment and primary treatment processes microplastics larger than 1,000 μm were eliminated from leachate efficiently (Bayo *et al.*, 2020; Carr *et al.*, 2016). Furthermore, due to physical forces such as sand abrasion or water turbulence microplastics break down to more small particles (Enfrin *et al.*, 2019).

During the biological treatment processes, bacteria release specific enzymes which biodegrade microplastics. Moreover, microorganisms attracted to microplastics form biofilms on microplastics and alter their surface properties. Biofouling increases the deposition of microplastics in sludge. During the biodegradation long chains of plastic polymers were converted to monomers by enzymes released by microorganisms. For example, under aerobic conditions, microbes use microplastics as a substrate to produce CO_2 and H_2O, and under anaerobic conditions, microbes transform long carbon chains into acetic acid (Shen *et al.*, 2022b). *Flavobacterium*, *Rhodococcus*, *Nocardia*, and *Pseudomonas* degrade monomer styrene and produce precursors for the Krebs cycle (Danso *et al.*, 2019). Most of the byproducts formed during biological degradation are not harmful. Some microorganisms reduce the size of microplastics; for example, the marine fungi species *Z. maritimum* reduces the size and mass of microplastics (Paço *et al.*, 2017). During oxidation treatment, oxidants and free radicals break the long polymer chains of microplastics reducing their plasticity. Moreover, research studies have confirmed a substantial reduction in the number of microplastic particles after treating leachate (Table 12.2) (Narevski *et al.*, 2021; Sun *et al.*, 2021).

12.6 LIMITATIONS OF LANDFILL LEACHATE-TREATMENT TECHNIQUES

Conventional treatment methods used in WWTPs and leachate-treatment facilities may be suitable for the removal of large microplastic particles not for particles less than 10 μm (Xu *et al.*, 2020). Small microplastic particles eventually enter the environment via final effluents. Sand filters, gravel filters, and artificial soil filters are not suitable for microplastic removal because their ability to capture microplastics is very low (Sun *et al.*, 2019). Microplastics are lightweight materials mostly found floating in aqueous media such as leachate (Shen *et al.*, 2022b). Therefore, sedimentation and aeration will not be much suitable to remove microplastics successfully from leachate. For instance, landfill leachate from the 'Gigoš' landfill in Serbia has been treated with a similar method; however, no considerable reduction in the number of plastic particles has been observed in treated leachate (Narevski *et al.*, 2021). Furthermore, tire particles have been detected in treated leachate from the 'Böler' landfill in Norway and the 'Fiflholt new cell' landfill in Iceland (Praagh *et al.*, 2018).

Although reverse osmosis is effective in removing microplastics from leachate membranes designed for reverse osmosis is extremely sensitive to temperature, pH, and certain chemicals; therefore, the quality of influents should be strictly controlled. However, as leachate is a complex mixture of various contaminants, a pre-treatment process is required to improve the efficiency of treatment and reduce the risk of contamination of membrane filters. Furthermore, microorganisms carried on microplastic surface colonize on the surface of the filter increasing the risk of membrane fouling which decreases the efficiency of the filter making process expensive (Shen *et al.*, 2020, 2022b). The presence of a large number of microplastics in leachate could hinder the performance of leachate-treatment equipment. For example, a high quantity of microplastics trapped in the pores of the membrane reduces its efficiency (Enfrin *et al.*, 2019; Ma *et al.*, 2019). Although a number of studies have confirmed that plastics can be biodegraded by microorganisms, the rate of degradation process depends on the microbial species, polymer type, particle size, and initial biomass (Shen *et al.*, 2019). Furthermore, microplastic may degrade nanoplastics, which can pass through the filter due to their small size. Even though vascular plants have a high ability to adsorb microplastics plant epidermis and cell walls inhibit penetration of tissues subsequently reducing the adsorption of larger microplastic particles.

One of the main challenges related to microplastic removal techniques is the reintroduction of microplastics to the environment via sludge. Once the leachate passed through treatment processes concentration of microplastic in leachate was decreased and microplastic particles were trapped in sludge or filter cake. However, in most European countries usage of sewage sludge collected from WWTPs as fertilizers is popular; application of sludge contaminated with microplastics will reintroduce microplastics to agricultural lands. Unfortunately, the effectiveness of treatment technologies used in leachate-treatment facilities and WWTPs to remove microplastics has not been studied thoroughly, therefore, few records are available. Therefore, it is important to pay attention to improve leachate-treatment methodologies to remove microplastics as much as possible along with the nutrients and potentially toxic elements before releasing leachate into the environment.

12.7 CONCLUSIONS AND PERSPECTIVES

Landfills, especially open dump sites, generate leachate, which contains a large number of emerging contaminants including microplastics. Inside landfills, weathering of larger plastics via photo-degradation, thermal degradation, mechanical fragmentation, and biological degradation results in the formation of microplastics. The most abundant polymer type was PE, whereas PP was the second most prominent polymer type detected in landfill leachates. Irregular-shaped microplastics were most frequently found; other than that spherical-, slender-, and lamellar-shaped microplastics were also reported in leachate collected from landfills. Microplastics present in leachate transfer to the adjacent soil, groundwater, and open waters. The passage mainly occurred via direct leaching, usage of treated leachate as a soil conditioner, and usage of microplastic-contaminated sludge released from leachate-treatment facilities as fertilizers. Major leachate remediation practices used are physical and chemical processes, biological processes, and land-based processes.

Microplastic contamination can result in detrimental impacts on ecosystems, organisms, and humans. Therefore, expanding our understanding of microplastic pollution via landfill leachate and treatment strategies is mandatory.

- Data on the physical and chemical properties of microplastics in leachate are incomplete in many studies which makes it difficult to assess the environmental risks completely.
- The number of studies conducted for quantification and identification of microplastics in landfill leachate is limited, although they are very important to determine the risks associated with microplastics and to implement mitigatory measures.
- Microplastics in landfill leachate released from abandoned landfills have not been studied although they share the same risk as landfill leachate released from formal landfills.
- Microplastics in leachate are migrating to nearby ecosystems via various pathways; therefore, it is of paramount importance to understand the migratory mechanisms of microplastics from landfill leachate to the surrounding environment.
- Interactions between microplastics and other contaminants present in leachate are unknown; hence, laboratory studies under stimulated landfill environments are required to understand the microplastic interaction with other toxic chemicals and pathogens in landfill leachate.
- Records and research focus on microplastic removal using the current leachate-treatment process are inadequate; hence, it is necessary to strengthen research on relevant aspects in time, so as to provide a reliable reference basis for updating leachate-treatment processes.
- Development of novel, affordable, and highly efficient techniques to remove microplastics to be included in the operating procedures of landfill leachate-treatment facilities as well as in WWTPs.
- It is necessary to conduct a systematic study to explore the interactions that occur between microplastics and vascular plants in and around the landfills to understand the mechanisms of binding with plant surfaces.
- Lack of available information on the plastic additives in landfill leachate will give an indirect assessment of plastics products in leachate.

ACKNOWLEDGMENTS

The authors acknowledge financial support from the National Research Council (NRC 20-117), Sri Lanka and Research Council grant ASP/01/RE/SCI/2021/15 of University of Sri Jayewardenepura.

REFERENCES

Afroz R., Rahman A., Masud M. M. and Akhtar R. (2017). The knowledge, awareness, attitude and motivational analysis of plastic waste and household perspective in Malaysia. *Environmental Science and Pollution Research*, **24**, 2304–2315, https://doi.org/10.1007/s11356-016-7942-0

Aghadadashi V. and Mehdinia A. (2016). Occurrence, spatial deposition and footprint of polybrominated diphenyl ethers in surficial sediments of Bushehr Peninsula, the

Persian Gulf. *Marine Pollution Bulletin*, **112**, 211–217, https://doi.org/10.1016/j.marpolbul.2016.08.017

Alimi O. S., Farner Budarz J., Hernandez L. M. and Tufenkji N. (2018). Microplastics and nanoplastics in aquatic environments: aggregation, deposition, and enhanced contaminant transport. *Environmental Science & Technology*, **52**, 1704–1724, https://doi.org/10.1021/acs.est.7b05559

Al-Maaded M., Madi N. K., Kahraman R., Hodzic A. and Ozerkan N. G. (2012). An overview of solid waste management and plastic recycling in Qatar. *Journal of Polymers and the Environment*, **20**, 186–194, https://doi.org/10.1007/s10924-011-0332-2

Aronsson P., Dahlin T. and Dimitriou I. (2010). Treatment of landfill leachate by irrigation of willow coppice – plant response and treatment efficiency. *Environmental Pollution*, **158**, 795–804, https://doi.org/10.1016/j.envpol.2009.10.003

Barnes D. K., Galgani F., Thompson R. C. and Barlaz M. (2009). Accumulation and fragmentation of plastic debris in global environments. *Philosophical Transactions of the Royal Society B: Biological Sciences*, **364**, 1985–1998, https://doi.org/10.1098/rstb.2008.0205

Battisti C., Fanelli G., Filpa A. and Cerfolli F. (2020). Giant reed (*Arundo donax*) wrack as sink for plastic beach litter: first evidence and implication. *Marine Pollution Bulletin*, **155**, 111179, https://doi.org/10.1016/j.marpolbul.2020.111179

Bayo J., Olmos S. and López-Castellanos J. (2020). Microplastics in an urban wastewater treatment plant: the influence of physicochemical parameters and environmental factors. *Chemosphere*, **238**, 124593, https://doi.org/10.1016/j.chemosphere.2019.124593

Blair R. M., Waldron S. and Gauchotte-Lindsay C. (2019). Average daily flow of microplastics through a tertiary wastewater treatment plant over a ten-month period. *Water Research*, **163**, 114909, https://doi.org/10.1016/j.watres.2019.114909

Bordoloi S., Gupt C. B. and Sarmah A. K. (2022). Exploring the theoretical effects of landfill based microplastic accumulation on the hydro-mechanical properties of porous soil media. *Current Opinion in Environmental Science & Health*, **26**, 100332, https://doi.org/10.1016/j.coesh.2022.100332

Bozbey I. and Guler E. (2006). Laboratory and field testing for utilization of an excavated soil as landfill liner material. *Waste Management*, **26**, 1277–1286, https://doi.org/10.1016/j.wasman.2005.10.014

Carr S. A., Liu J. and Tesoro A. G. (2016). Transport and fate of microplastic particles in wastewater treatment plants. *Water Research*, **91**, 174–182, https://doi.org/10.1016/j.watres.2016.01.002

Chamas A., Moon H., Zheng J., Qiu Y., Tabassum T., Jang J. H., Abu-Omar M., Scott S. L. and Suh S. (2020). Degradation rates of plastics in the environment. *ACS Sustainable Chemistry & Engineering*, **8**, 3494–3511, https://doi.org/10.1021/acssuschemeng.9b06635

Chen Z., Zhao W., Xing R., Xie S., Yang X., Cui P., Lü J., Liao H., Yu Z., Wang S. and Zhou S. (2020). Enhanced in situ biodegradation of microplastics in sewage sludge using hyperthermophilic composting technology. *Journal of Hazardous Materials*, **384**, 121271, https://doi.org/10.1016/j.jhazmat.2019.121271

Da Costa J. P., Santos P. S. M., Duarte A. C. and Rocha-Santos T. (2016). (Nano)plastics in the environment – sources, fates and effects. *Science of the Total Environment*, **566–567**, 15–26, https://doi.org/10.1016/j.scitotenv.2016.05.041

Danso D., Chow J. and Streit W. R. (2019). Plastics: environmental and biotechnological perspectives on microbial degradation. *Applied and Environmental Microbiology*, **85**, e01095–, https://doi.org/10.1128/AEM.01095-19

Delaney P. (2013). How long it takes for some everyday items to decompose. Down2Earth Materials. 2022.

Di M. and Wang J. (2018). Microplastics in surface waters and sediments of the Three Gorges Reservoir, China. *Science of the Total Environment*, **616–617**, 1620–1627, https://doi.org/10.1016/j.scitotenv.2017.10.150

Ding Y., Zhao J., Liu J.-W., Zhou J., Cheng L., Zhao J., Shao Z., Iris Ç, Pan B., Li X. and Hu Z.-T. (2021). A review of China's municipal solid waste (MSW) and comparison with international regions: management and technologies in treatment and resource utilization. *Journal of Cleaner Production*, **293**, 126144, https://doi.org/10.1016/j.jclepro.2021.126144

Enfrin M., Dumée L. F. and Lee J. (2019). Nano/microplastics in water and wastewater treatment processes – origin, impact and potential solutions. *Water Research*, **161**, 621–638, https://doi.org/10.1016/j.watres.2019.06.049

EPA. (2022, July 31, 2022). National Overview: Facts and Figures on Materials, Wastes and Recycling. Retrieved 19.08.2022.

Fotopoulou K. N. and Karapanagioti H. K. (2012). Surface properties of beached plastic pellets. *Marine Environmental Research*, **81**, 70–77, https://doi.org/10.1016/j.marenvres.2012.08.010

Geyer R., Jambeck J. R. and Law K. L. (2017). Production, use, and fate of all plastics ever made. *Science Advances*, **3**, e1700782, https://doi.org/10.1126/sciadv.1700782

Gopinath K., Seshachalam S., Neelavannan K., Anburaj V., Rachel M., Ravi S., Bharath M. and Achyuthan H. (2020). Quantification of microplastic in red hills lake of Chennai city, Tamil Nadu, India. *Environmental Science and Pollution Research*, **27**, 33297–33306, https://doi.org/10.1007/s11356-020-09622-2

Gutow L., Eckerlebe A., Giménez L. and Saborowski R. (2016). Experimental evaluation of seaweeds as a vector for microplastics into marine food webs. *Environmental Science & Technology*, **50**, 915–923, https://doi.org/10.1021/acs.est.5b02431

Hahladakis J. N., Velis C. A., Weber R., Iacovidou E. and Purnell P. (2018). An overview of chemical additives present in plastics: migration, release, fate and environmental impact during their use, disposal and recycling. *Journal of Hazardous Materials*, **344**, 179–199, https://doi.org/10.1016/j.jhazmat.2017.10.014

Hajiouni S., Mohammadi A., Ramavandi B., Arfaeinia H., De-La-Torre G. E., Tekle-Röttering A. and Dobaradaran S. (2022). Occurrence of microplastics and phthalate esters in urban runoff: a focus on the Persian Gulf coastline. *Science of the Total Environment*, **806**, 150559, https://doi.org/10.1016/j.scitotenv.2021.150559

He P., Chen L., Shao L., Zhang H. and Lü F. (2019). Municipal solid waste (MSW) landfill: a source of microplastics? –evidence of microplastics in landfill leachate. *Water Research*, **159**, 38–45, https://doi.org/10.1016/j.watres.2019.04.060

Heller M. C., Mazor M. H. and Keoleian G. A. (2020). Plastics in the US: toward a material flow characterization of production, markets and end of life. *Environmental Research Letters*, **15**, 094034, https://doi.org/10.1088/1748-9326/ab9e1e

Horton A. A., Walton A., Spurgeon D. J., Lahive E. and Svendsen C. (2017). Microplastics in freshwater and terrestrial environments: evaluating the current understanding to identify the knowledge gaps and future research priorities. *Science of the Total Environment*, **586**, 127–141, https://doi.org/10.1016/j.scitotenv.2017.01.190

Hou L., Kumar D., Yoo C. G., Gitsov I. and Majumder E. L. W. (2021). Conversion and removal strategies for microplastics in wastewater treatment plants and landfills. *Chemical Engineering Journal*, **406**, 126715, https://doi.org/10.1016/j.cej.2020.126715

Ingrando I., Rivoira L., Castiglioni M., Tumiatti V., Lenzi F., Pagliano A. and Bruzzoniti M. C. (2022). Microwave-assisted extraction and gas chromatographic determination of thirty priority micropollutants in biowaste fraction derived from municipal solid waste for material recovery in the circular-economy approach. *Talanta*, **241**, 123268, https://doi.org/10.1016/j.talanta.2022.123268

Jadhav E. B., Sankhla M. S., Bhat R. A. and Bhagat D. S. (2021). Microplastics from food packaging: an overview of human consumption, health threats, and alternative solutions. *Environmental Nanotechnology, Monitoring & Management*, **16**, 100608, https://doi.org/10.1016/j.enmm.2021.100608

Júnior F., Gondim F., Pereira M., Braga B., Filho R. and Barbosa F. E. L. (2016). Sanitary landfill leachate as a source of nutrients on the initial growth of sunflower plants. *Revista Brasileira de Engenharia Agrícola e Ambiental*, **20**, 746–750, https://doi.org/10.1590/1807-1929/agriambi.v20n8p746-750

Kalčíková G. (2020). Aquatic vascular plants – a forgotten piece of nature in microplastic research. *Environmental Pollution*, **262**, 114354, https://doi.org/10.1016/j.envpol.2020.114354

Karuppasamy M. B., Natesan U., Vaikunth R., Praveen Kumar R., Ruthra R. and Srinivasalu S. (2021). Spatial distribution of microplastic concentration around landfill sites and its potential risk on groundwater. *Chemosphere*, **277**, 130263, https://doi.org/10.1016/j.chemosphere.2021.130263

Kawai M., Purwanti I. F., Nagao N., Slamet A., Hermana J. and Toda T. (2012). Seasonal variation in chemical properties and degradability by anaerobic digestion of landfill leachate at Benowo in Surabaya, Indonesia. *Journal of Environmental Management*, **110**, 267–275, https://doi.org/10.1016/j.jenvman.2012.06.022

Kazour M., Terki S., Rabhi K., Jemaa S., Khalaf G. and Amara R. (2019). Sources of microplastics pollution in the marine environment: importance of wastewater treatment plant and coastal landfill. *Marine Pollution Bulletin*, **146**, 608–618, https://doi.org/10.1016/j.marpolbul.2019.06.066

Kilponen J. (2016). Microplastics and harmful substances in urban runoffs and landfill leachates: possible emission sources to marine environment.

Kjeldsen P., Barlaz M. A., Rooker A. P., Baun A., Ledin A. and Christensen T. H. (2002). Present and long-term composition of MSW landfill leachate: a review. *Critical Reviews in Environmental Science and Technology*, **32**, 297–336, https://doi.org/10.1080/10643380290813462

Klein S., Dimzon I. K., Eubeler J. and Knepper T. P. (2018). Analysis, occurrence, and degradation of microplastics in the aqueous environment. In: Freshwater Microplastics: Emerging Environmental Contaminants?, M. Wagner and S. Lambert (eds.), Springer International Publishing, Cham, pp. 51–67, https://doi.org/10.1007/978-3-319-61615-5_3

Kumarathilaka P., Wijesekara H., Bolan N., Kunhikrishnan A. and Vithanage M. (2017). Phytoremediation of landfill leachates. In: Phytoremediation: Management of Environmental Contaminants, A. A. Ansari, S. S. Gill, R. Gill, G. R. Lanza, and N. Lee (eds.), Springer, Switzerland, pp. 439–467, https://doi.org/10.1007/978-3-319-52381-1_17

Li A. S. H., Sathishkumar P., Selahuddeen M. L., Mahmood W. M. A. W., Abidin M. H. Z., Wahab R. A., Huri M. A. M. and Abdullah F. (2022). Adverse environmental effects of disposable face masks due to the excess usage. *Environmental Pollution*, **308**, 119674, https://doi.org/10.1016/j.envpol.2022.119674

Ma B., Xue W., Hu C., Liu H., Qu J. and Li L. (2019). Characteristics of microplastic removal via coagulation and ultrafiltration during drinking water treatment. *Chemical Engineering Journal*, **359**, 159–167, https://doi.org/10.1016/j.cej.2018.11.155

Mohammadi A., Malakootian M., Dobaradaran S., Hashemi M. and Jaafarzadeh N. (2022). Occurrence, seasonal distribution, and ecological risk assessment of microplastics and phthalate esters in leachates of a landfill site located near the marine environment: Bushehr port, Iran as a case. *Science of the Total Environment*, **842**, 156838, https://doi.org/10.1016/j.scitotenv.2022.156838

Murphy F., Ewins C., Carbonnier F. and Quinn B. (2016). Wastewater treatment works (WwTW) as a source of microplastics in the aquatic environment. *Environmental Science & Technology*, **50**, 5800–5808, https://doi.org/10.1021/acs.est.5b05416

Narevski A. C., Novaković M. I., Petrović M. Z., Mihajlović I. J., Maoduš N. B. and Vujić G. V. (2021). Occurrence of bisphenol A and microplastics in landfill leachate: lessons from South East Europe. *Environmental Science and Pollution Research*, **28**, 42196–42203, https://doi.org/10.1007/s11356-021-13705-z

Nizzetto L., Bussi G., Futter M. N., Butterfield D. and Whitehead P. G. (2016). A theoretical assessment of microplastic transport in river catchments and their retention by soils and river sediments. *Environmental Science: Processes & Impacts*, **18**, 1050–1059, https://doi.org/10.1039/C6EM00206D

Nkwachukwu O. I., Chima C. H., Ikenna A. O. and Albert L. (2013). Focus on potential environmental issues on plastic world towards a sustainable plastic recycling in developing countries. *International Journal of Industrial Chemistry*, **4**, 1–13, https://doi.org/10.1186/2228-5547-4-34

Noerfitriyani E., Hartono D. M., Moersidik S. S. and Gusniani I. (2018). Impact of leachate discharge from Cipayung landfill on water quality of Pesanggrahan River, Indonesia. *IOP Conference Series: Earth and Environmental Science*, **120**, 012002, https://doi.org/10.1088/1755-1315/120/1/012002

Nurhasanah Cordova M. R. and Riani E. (2021). Micro- and mesoplastics release from the Indonesian municipal solid waste landfill leachate to the aquatic environment: case study in Galuga landfill area, Indonesia. *Marine Pollution Bulletin*, **163**, 111986, https://doi.org/10.1016/j.marpolbul.2021.111986

O'connor D., Pan S., Shen Z., Song Y., Jin Y., Wu W.-M. and Hou D. (2019). Microplastics undergo accelerated vertical migration in sand soil due to small size and wet–dry cycles. *Environmental Pollution*, **249**, 527–534, https://doi.org/10.1016/j.envpol.2019.03.092

Paço A., Duarte K., Da Costa J. P., Santos P. S. M., Pereira R., Pereira M. E., Freitas A. C., Duarte A. C. and Rocha-Santos T. A. P. (2017). Biodegradation of polyethylene microplastics by the marine fungus *Zalerion maritimum*. *Science of the Total Environment*, **586**, 10–15, https://doi.org/10.1016/j.scitotenv.2017.02.017

Perren W., Wojtasik A. and Cai Q. (2018). Removal of microbeads from wastewater using electrocoagulation. *ACS Omega*, **3**, 3357–3364, https://doi.org/10.1021/acsomega.7b02037

Pivokonsky M., Cermakova L., Novotna K., Peer P., Cajthaml T. and Janda V. (2018). Occurrence of microplastics in raw and treated drinking water. *Science of the Total Environment*, **643**, 1644–1651, https://doi.org/10.1016/j.scitotenv.2018.08.102

Plastics Europe. (2018). European Association of Plastics Recycling and Recovery Organisations (EPRO), Plastics – the Facts 2017. https://search.issuelab.org/resource/plastics-the-facts-2017-an-analysis-of-european-plastics-production-demand-and-waste-data.html, 2022.

Poerio T., Piacentini E. and Mazzei R. (2019). Membrane processes for microplastic removal. *Molecules*, **24**, 4148, https://doi.org/10.3390/molecules24224148

Praagh M. V., Hartman C. and Brandmyr E. (2018). Microplastics in landfill leachates in the Nordic countries, Nordisk Ministerråd.

Puglisi E., Romaniello F., Galletti S., Boccaleri E., Frache A. and Cocconcelli P. S. (2019). Selective bacterial colonization processes on polyethylene waste samples in an abandoned landfill site. *Scientific Reports*, **9**, 14138, https://doi.org/10.1038/s41598-019-50740-w

Puthcharoen A. and Leungprasert S. (2019). Determination of microplastics in soil and leachate from the landfills. *Thai Environmental Engineering Journal*, **33**, 39–46.

Renou S., Givaudan J., Poulain S., Dirassouyan F. and Moulin P. (2008). Landfill leachate treatment: review and opportunity. *Journal of Hazardous Materials*, **150**, 468–493, https://doi.org/10.1016/j.jhazmat.2007.09.077

Resmeriță A.-M., Coroaba A., Darie R., Doroftei F., Spiridon I., Simionescu B. C. and Navard P. (2018). Erosion as a possible mechanism for the decrease of size of plastic pieces floating in oceans. *Marine Pollution Bulletin*, **127**, 387–395, https://doi.org/10.1016/j.marpolbul.2017.12.025

Sahebdelfar N., Khorasani R. and Astaraei A. (2022). Effect of some additives on heavy metals behavior and phytoavailability in municipal solid waste compost-amended soil. *International Journal of Environmental Science and Technology*, **19**, 307–318, https://doi.org/10.1007/s13762-021-03146-z

Sewwandi M., Hettithanthri O., Egodage S. M., Amarathunga A. A. D. and Vithanage M. (2022). Unprecedented marine microplastic contamination from the *Xpress Pearl* container vessel disaster. *Science of the Total Environment*, **828**, 154374, https://doi.org/10.1016/j.scitotenv.2022.154374

Shen M., Zeng G., Zhang Y., Wen X., Song B. and Tang W. (2019). Can biotechnology strategies effectively manage environmental (micro)plastics? *Science of the Total Environment*, **697**, 134200, https://doi.org/10.1016/j.scitotenv.2019.134200

Shen M., Song B., Zhu Y., Zeng G., Zhang Y., Yang Y., Wen X., Chen M. and Yi H. (2020). Removal of microplastics via drinking water treatment: current knowledge and future directions. *Chemosphere*, **251**, 126612, https://doi.org/10.1016/j.chemosphere.2020.126612

Shen M., Hu T., Huang W., Song B., Zeng G. and Zhang Y. (2021). Removal of microplastics from wastewater with aluminosilicate filter media and their surfactant-modified products: performance, mechanism and utilization. *Chemical Engineering Journal*, **421**, 129918, https://doi.org/10.1016/j.cej.2021.129918

Shen M., Zhang Y., Almatrafi E., Hu T., Zhou C., Song B., Zeng Z. and Zeng G. (2022a). Efficient removal of microplastics from wastewater by an electrocoagulation process. *Chemical Engineering Journal*, **428**, 131161, https://doi.org/10.1016/j.cej.2021.131161

Shen M., Xiong W., Song B., Zhou C., Almatrafi E., Zeng G. and Zhang Y. (2022b). Microplastics in landfill and leachate: occurrence, environmental behavior and removal strategies. *Chemosphere*, **305**, 135325, https://doi.org/10.1016/j.chemosphere.2022.135325

Silva A. L. P., Prata J. C., Duarte A. C., Soares A. M. V. M., Barceló D. and Rocha-Santos T. (2021). Microplastics in landfill leachates: the need for reconnaissance studies and remediation technologies. *Case Studies in Chemical and Environmental Engineering*, **3**, 100072, https://doi.org/10.1016/j.cscee.2020.100072

Sridharan A. and Choudhury D. (2008). Computation of hydraulic conductivity of montmorillonitic clays by diffuse double layer theory. *International Journal of Geotechnical Engineering*, **2**, 1–10, https://doi.org/10.3328/IJGE.2008.02.01.1-10

Strady E., Dang T. H., Dao T. D., Dinh H. N., Do T. T. D., Duong T. N., Duong T. T., Hoang D. A., Kieu-Le T. C., Le T. P. Q., Mai H., Trinh D. M., Nguyen Q. H., Tran-Nguyen Q. A., Tran Q. V., Truong T. N. S., Chu V. H. and Vo V. C. (2021). Baseline assessment of microplastic concentrations in marine and freshwater environments of a developing Southeast Asian country, Viet Nam. *Marine Pollution Bulletin*, **162**, 111870, https://doi.org/10.1016/j.marpolbul.2020.111870

Su Y., Zhang Z., Wu D., Zhan L., Shi H. and Xie B. (2019). Occurrence of microplastics in landfill systems and their fate with landfill age. *Water Research*, **164**, 114968, https://doi.org/10.1016/j.watres.2019.114968

Sulistyowati L., Nurhasanah Riani E. and Cordova M. R. (2022). The occurrence and abundance of microplastics in surface water of the midstream and downstream of the Cisadane River, Indonesia. *Chemosphere*, **291**, 133071, https://doi.org/10.1016/j.chemosphere.2021.133071

Sun J., Dai X., Wang Q., Van Loosdrecht M. C. M. and Ni B. J. (2019). Microplastics in wastewater treatment plants: detection, occurrence and removal. *Water Research*, **152**, 21–37, https://doi.org/10.1016/j.watres.2018.12.050

Sun J., Zhu Z.-R., Li W.-H., Yan X., Wang L.-K., Zhang L., Jin J., Dai X. and Ni B.-J. (2021). Revisiting microplastics in landfill leachate: unnoticed tiny microplastics and their fate in treatment works. *Water Research*, **190**, 116784, https://doi.org/10.1016/j.watres.2020.116784

Sundt P., Schulze P.-E. and Syversen F. (2014). Sources of microplastic-pollution to the marine environment. *Mepex for the Norwegian Environment Agency*, **86**, 20.

Talvitie J., Mikola A., Setälä O., Heinonen M. and Koistinen A. (2017a). How well is microlitter purified from wastewater? – a detailed study on the stepwise removal of microlitter in a tertiary level wastewater treatment plant. *Water Research*, **109**, 164–172, https://doi.org/10.1016/j.watres.2016.11.046

Talvitie J., Mikola A., Koistinen A. and Setälä O. (2017b). Solutions to microplastic pollution – removal of microplastics from wastewater effluent with advanced wastewater treatment technologies. *Water Research*, **123**, 401–407, https://doi.org/10.1016/j.watres.2017.07.005

Taylor S. E., Pearce C. I., Sanguinet K. A., Hu D., Chrisler W. B., Kim Y.-M., Wang Z. and Flury M. (2020). Polystyrene nano- and microplastic accumulation at *Arabidopsis* and wheat root cap cells, but no evidence for uptake into roots. *Environmental Science: Nano*, **7**, 1942–1953, https://doi.org/10.1039/D0EN00309C

Tran-Nguyen Q. A., Vu T. B. H., Nguyen Q. T., Nguyen H. N. Y., Le T. M., Vo V. M. and Trinh-Dang M. (2022). Urban drainage channels as microplastics pollution hotspots in developing areas: a case study in Da Nang, Vietnam. *Marine Pollution Bulletin*, **175**, 113323, https://doi.org/10.1016/j.marpolbul.2022.113323

Tu C., Chen T., Zhou Q., Liu Y., Wei J., Waniek J. J. and Luo Y. (2020). Biofilm formation and its influences on the properties of microplastics as affected by exposure time and depth in the seawater. *Science of the Total Environment*, **734**, 139237, https://doi.org/10.1016/j.scitotenv.2020.139237

Tupsakhare S., Moutushi T., Castaldi M. J., Barlaz M. A., Luettich S. and Benson C. H. (2020). The impact of pressure, moisture and temperature on pyrolysis of municipal solid waste under simulated landfill conditions and relevance to the field data from elevated temperature landfill. *Science of the Total Environment*, **723**, 138031, https://doi.org/10.1016/j.scitotenv.2020.138031

Upadhyay K. and Bajpai S. (2021). Microplastics in landfills: a comprehensive review on occurrence, characteristics and pathways to the aquatic environment. *Nature Environment & Pollution Technology*, **20**, 1935–1945, https://doi.org/10.46488/NEPT.2021.v20i05.009

Vaverková M. D., Adamcová D., Winkler J., Koda E., Petrželová L. and Maxianová A. (2020). Alternative method of composting on a reclaimed municipal waste landfill in accordance with the circular economy: benefits and risks. *Science of the Total Environment*, **723**, 137971, https://doi.org/10.1016/j.scitotenv.2020.137971

Vijayan A., Österlund H., Magnusson K., Marsalek J. and Viklander M. (2022). Microplastics (MPs) in urban roadside snowbanks: quantities, size fractions and dynamics of release. *Science of the Total Environment*, **851**, 158306, https://doi.org/10.1016/j.scitotenv.2022.158306

Wan Y., Chen X., Liu Q., Hu H., Wu C. and Xue Q. (2022). Informal landfill contributes to the pollution of microplastics in the surrounding environment. *Environmental Pollution*, **293**, 118586, https://doi.org/10.1016/j.envpol.2021.118586

Wang C., Xian Z., Jin X., Liang S., Chen Z., Pan B., Wu B., Ok Y. S. and Gu C. (2020a). Photo-aging of polyvinyl chloride microplastic in the presence of natural organic acids. *Water Research*, **183**, 116082, https://doi.org/10.1016/j.watres.2020.116082

Wang Z., Sedighi M. and Lea-Langton A. (2020b). Filtration of microplastic spheres by biochar: removal efficiency and immobilisation mechanisms. *Water Research*, **184**, 116165, https://doi.org/10.1016/j.watres.2020.116165

Watteau F., Dignac M.-F., Bouchard A., Revallier A. and Houot S. (2018). Microplastic detection in soil amended with municipal solid waste composts as revealed by transmission electronic microscopy and pyrolysis/GC/MS. *Frontiers in Sustainable Food Systems*, **2**, 81, https://doi.org/10.3389/fsufs.2018.00081

Wijekoon P., Koliyabandara P., Cooray A., Lam S. S., Athapattu B. and Vithanage M. (2021). Progress and prospects in mitigation of landfill leachate pollution: risk, pollution potential, treatment and challenges. *Journal of Hazardous Materials*, **421**, 126627, https://doi.org/10.1016/j.jhazmat.2021.126627

World Health Organization. (2005). Management of solid health-care waste at primary health-care centres: A decision-making guide.

Wu C., Zhang K. and Xiong X. (2018). Microplastic pollution in inland waters focusing on Asia. In: Freshwater Microplastics, M. Wagner and S. Lambert (eds.), Springer, Cham, pp. 85–99, https://doi.org/10.1007/978-3-319-61615-5_5

Xu Z., Sui Q., Li A., Sun M., Zhang L., Lyu S. and Zhao W. (2020). How to detect small microplastics (20–100 μm) in freshwater, municipal wastewaters and landfill leachates? A trial from sampling to identification. *Science of the Total Environment*, **733**, 139218, https://doi.org/10.1016/j.scitotenv.2020.139218

Yang Y., Yang J., Wu W.-M., Zhao J., Song Y., Gao L., Yang R. and Jiang L. (2015). Biodegradation and mineralization of polystyrene by plastic-eating mealworms: part 2. Role of gut microorganisms. *Environmental Science & Technology*, **49**, 12087–12093, https://doi.org/10.1021/acs.est.5b02663

Yang Y., Wang J. and Xia M. (2020). Biodegradation and mineralization of polystyrene by plastic-eating superworms *Zophobas atratus*. *Science of the Total Environment*, **708**, 135233, https://doi.org/10.1016/j.scitotenv.2019.135233

Yin L., Wen X., Huang D., Zeng G., Deng R., Liu R., Zhou Z., Tao J., Xiao R. and Pan H. (2021). Microplastics retention by reeds in freshwater environment. *Science of the Total Environment*, **790**, 148200, https://doi.org/10.1016/j.scitotenv.2021.148200

Yu F., Wu Z., Wang J., Li Y., Chu R., Pei Y. and Ma J. (2022). Effect of landfill age on the physical and chemical characteristics of waste plastics/microplastics in a waste landfill sites. *Environmental Pollution*, **306**, 119366, https://doi.org/10.1016/j.envpol.2022.119366

Zhang Z., Su Y., Zhu J., Shi J., Huang H. and Xie B. (2021). Distribution and removal characteristics of microplastics in different processes of the leachate treatment system. *Waste Management*, **120**, 240–247, https://doi.org/10.1016/j.wasman.2020.11.025

Ziajahromi S., Neale P. A., Rintoul L. and Leusch F. D. L. (2017). Wastewater treatment plants as a pathway for microplastics: development of a new approach to sample wastewater-based microplastics. *Water Research*, **112**, 93–99, https://doi.org/10.1016/j.watres.2017.01.042

doi: 10.2166/9781789063318_0297

Chapter 13

Separation treatment of landfill leachate

Brahim Arhoun, Juan Manuel Paz-Garcia, Jose Miguel Rodriguez-Maroto and Maria Villen-Guzman*

Chemical Engineering Department, Faculty of Sciences, University of Malaga, 29071 Malaga, Spain
**Corresponding author: arhoun.b@uma.es*

ABSTRACT

Landfilling municipal waste is the most common way to manage waste in the world. An important problem associated with landfills is the production of leachates. Landfill leachate is heterogeneous wastewater and often possesses potential pollution with high concentrations of organic and inorganic contaminants. The generated leachate must be appropriately collected and treated before being discharged into the environment. Several technologies have been used for leachate treatment which can be classified as biological, chemical, and physical processes. Among physical processes, membrane separation including microfiltration, ultrafiltration, nanofiltration, and reverse osmosis has been used for landfill leachate treatment. However, single-membrane processes are not very effective unless they are combined with biological and physicochemical treatment processes. This is because membrane separation presents two main drawbacks: membrane concentrates (MCs) and fouling. Here, we briefly review the main processes currently used for MC treatment. Finally, the fouling membrane is also discussed.

Keywords: Landfill leachate, microfiltration, ultrafiltration, nanofiltration and reverse osmosis, membrane concentrates, fouling.

13.1 INTRODUCTION

In recent decades, the increase in the world population and economic development has generated an increase in urban solid waste. These residues constitute one of the most severe environmental problems of modern societies, particularly the most advanced and industrialized ones. This problem continues to worsen due to the relationship between income levels, quality of life, and the volume of waste generated.

The world generates 2.01 billion tonnes of municipal solid waste annually. High-income countries generate about 34% of the world's waste. Worldwide, daily waste generated per person averages to 0.74 kg but ranges widely from 0.11 to 4.54 kg. Global waste is expected to grow up to 3.40 billion tonnes by 2050 (World Bank, 2022). Adequate waste disposal or treatment depends on the income of each country. Figure 13.1 shows the global treatment and disposal of waste.

The main waste management strategies are recycling, incineration, and landfilling. The latter remains the cheapest option; up to 90% of the municipal solid waste collected worldwide is managed by the landfill method. Moreover, in developing countries it is estimated to be 70% of the total waste (Sabour & Amiri, 2017). Hence, critical factors for sustainable landfilling are landfill liners, the thickness of the soil cover, leachate collection, landfill gas recovery, and flaring facilities (Nanda & Berruti, 2021).

The EU Landfill Directive (1999/31/EC) indicates the requirements with which landfill sites must comply, including location, conditioning, management, control, closure, and aftercare of landfills. Also, it indicates types of waste to be landfilled. The generated leachate collected from a landfill must be treated to comply with the appropriate standards required for discharge (Council of the European Union, 1999).

Landfilling is not a sustainable method; as a consequence of the deposit of municipal waste in landfills, a series of sanitary and environmental problems are generated, such as greenhouse gas emissions, unpleasant odor, high toxicity, and discharges of liquid effluents called leachate, which are caused by waste moisture and percolation of rainwater through the waste. The main drawback of leachates is that they are high in contamination, which can cause a significant threat to groundwater and surface water resources. As an example of the high toxicity of leachate, the leakage of a small volume of leachate into aquatic systems may contaminate a large volume of surface and groundwater (Keyikoglu *et al.*,

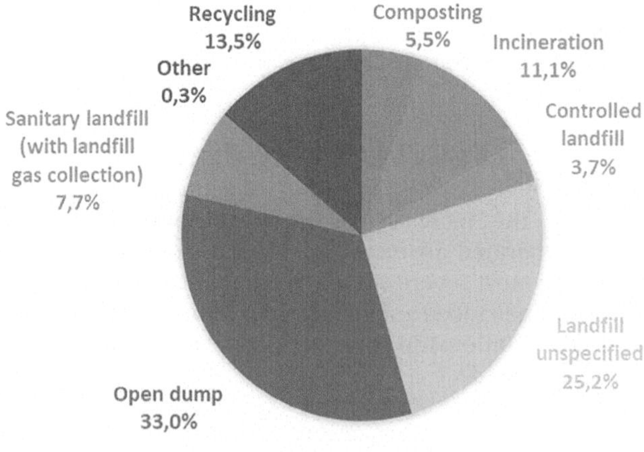

Figure 13.1 Global treatment and disposal of waste (World Bank, 2022).

2021). Landfill leachate exhibits acute and chronic toxicity, which is classified as hazardous. Leachate needs to be treated to meet strict quality standards before direct discharge into natural water bodies (Mishra, 2017).

Conventional landfill leachate treatments include biological and physicochemical processes. However, these treatments are not enough to reach the required level of pollutant removal (Luo *et al.*, 2019; Peng, 2017). The combination of two or more technologies has been used with the aim of removal of organic content and, thus, obtaining a stream within discharge standards (Luo *et al.*, 2019). Membrane separation includes microfiltration (MF), ultrafiltration (UF), nanofiltration (NF), and reverse osmosis (RO), which has also been experimented to treat leachate from sanitary landfills. Membrane separation processes have been used as a main step in a landfill leachate treatment chain or as a pre-treatment or post-treatment step to achieve purification (Abbas *et al.*, 2009). However, single membrane processes are ineffective for leachate treatment. In addition, the drawback of membrane processes is the production of a membrane concentrate (MC) which is even more hazardous and needs an appropriate treatment (Keyikoglu *et al.*, 2021). Combined membrane processes with physicochemical and biological methods improve the leachate treatment.

This chapter describes the treatment of landfill leachate by membrane separation. First, an introduction to the characteristics of landfill leachate and an overview of the currently employed processes are presented. Second, it focuses on the treatment of landfill leachate by MF, UF, NF, and RO, and their combination with biological and physical–chemical processes to improve the efficiency of membrane separation for landfill leachate treatment. However, membrane fouling and MCs are the significant limitations of applying membrane technologies. Finally, a brief state-of-the-art of the characteristics of MCs, treatment processes, and also membrane fouling is presented.

13.2 LANDFILL LEACHATE

Landfill leachate results from complex physical, chemical, and biological transformations within landfill sites which act as an ecological reactor. Leachate generation and its composition depends on different parameters, including the type of waste, local weather, and leachate age. To select the adequate process for leachate treatment, data about the composition and rate of production of leachate are a critical requirement in leachate management.

13.2.1 Generation of landfill leachate

The decomposition of landfilled waste causes landfill leachate generation through a series of physical–chemical and biological processes. The landfill conditions, such as construction, geometry, weather, temperature, moisture, pH, amount of biodegradable matter, and hydrogeological parameters, influence the generation of landfill gases and leachate (Nanda & Berruti, 2021). During the decomposition of waste, leachate is formed due to intrinsic moisture contained in waste and by excess rainwater percolating through the waste layers, which dissolves the different components. A landfill can produce leachate even 50 years after its closure (Shafiq & Yusof, 2015). The leachate generated has an intense black/brown color, an unpleasant odor, and high toxicity, and is also composed of a high load of refractory organic matter, heavy metals, ammonia nitrogen, inorganic salts, and chlorinated organic, as well as some humus components (Anqi *et al.*, 2020). A contact with the environment can cause a significant risk to the natural ecosystem, public health, and investable pollution of groundwater.

The quantity of leachates depends on rainwater percolation through wastes, biochemical processes, the moisture content in wastes, and its degree of compaction into the waste layers in a landfill. The leachate production is generally higher when the waste is less compacted because compaction reduces the percolation rate (Abbas *et al.*, 2009).

13.2.2 Composition of landfill leachate

Many factors affect landfill leachate composition, including waste type and composition, precipitation, landfilling technology, seasonal weather variation, age, and hydrological conditions (Mojiri *et al.*, 2020). In particular, the composition of landfill leachates varies greatly depending on the age of the landfill. When water penetrates waste in a landfill, multiple contaminants are transferred to leachate. The characteristics and composition of the leachate change significantly depending on its age, therefore, may provide information to select a suitable treatment process. Therefore, landfill leachate is usually classified as young, intermediate, and mature leachate (Abbas *et al.*, 2009; Mojiri *et al.*, 2020). Landfill leachate is generally characterized by its high chemical oxygen demand (COD) and biological oxygen demand (BOD). Usually, it consists of undesirable substances such as organic and inorganic contaminants. Table 13.1 presents the typical composition of landfill leachates.

Table 13.1 Landfill leachate classification vs age.

Landfill Leachate Characteristics	Young	Intermediate	Mature
Age (years)	<5	5–10	>10
pH	<6.5	6.5–7.5	>7.5
COD (mg/L)	>10,000	5,000–1,000	<5,000
BOD_5/COD	0.5–1	0.1–0.5	>0.1
TOC/COD	<0.3	0.3–0.5	>0.5
NH_3-N (mg/L)	<400	NA	>400
Heavy metals (mg/L)	>2 (medium to low)	<2 (low)	<2 (low)
Organic compounds	VFA (80%)	VFA (5–30%) + HFA	HFA (80%)
Biodegradability	High	Medium	low

Source: Abbas *et al.* (2009) and Mojiri *et al.* (2020).
VFA, volatile fatty acids; HFA, humic and fulvic acids.

Table 13.2 Type of each pollutant of each group.

Pollutants	Specification
Dissolved organic matter	Determined by COD or TOC, volatile fatty acids, fulvic-like and humic-like compounds
Heavy metals	Cd, Cr, Cu, Pb, Ni, and Zn
Inorganic macrocomponents	Ca^{2+}, Mg^{2+}, Na^{+}, K^{+}, NH^{4+}, iron Fe^{2+}, Mn^{2+}, Cl^{-}, SO_4^{2-}, and HCO_3^{-}
XOCs	Variety of aromatic hydrocarbons, phenols, chlorinated aliphatics, pesticides, and plasticizers

Source: Kjeldsen *et al.* (2002).

Landfill leachate generally contains various pollutants, including dissolved organic matter, heavy metals, inorganic macro components, and xenobiotic organic compounds (XOCs). Table 13.2 summarizes the type of each pollutant of each group. Other compounds, such as borate, sulfide, arsenate, selenate, barium, lithium, mercury, and cobalt may be found in low concentrations in leachate (Kjeldsen *et al.*, 2002). Of about 1,000 organic compounds, more than 200 have been detected in groundwater due to migration of leachate into the groundwater (Schwarzbauer *et al.*, 2002).

13.3 LANDFILL LEACHATE TREATMENT

The composition, volumetric flow rate, and strict regulation as well as economics are critical factors in leachate management. Landfill leachate must be appropriately treated and managed, maximizing recovery and minimizing waste disposal (Topal & Atasoy, 2022). Conventionally, landfill leachate has been treated through leachate transfer (with domestic sewage, recycling), biological processes (aerobic and anaerobic stabilization lagoons, natural filters, and activated sludge), and physicochemical processes (adsorption, flotation, chemical precipitation, coagulation/flocculation, air stripping, ion exchange, membrane filtration, and advanced oxidation treatments) (Keyikoglu *et al.*, 2021). The most used processes are classified according to their efficiency concerning the age of the leachate (see Table 8.1 in Chapter 8).

The combination of landfills with domestic sewage in the municipal sewage treatment plant entails easy and low-operating costs. The recirculation of leachate increases the moisture content in a controlled reactor and provides the distribution of nutrients and enzymes between methanogens and solid/liquid. Given that leachate contains organic inhibitory compounds with low biodegradability and heavy metals, it can lead to the inhibition of methanogenesis in both options of leachate transfer (Renou *et al.*, 2008).

Biological treatments are usually applied to remove ammonia, COD and BOD_5 from young leachate whose organic fraction is mainly composed of volatile fatty acids (Renou *et al.*, 2008). However, single-biological processes are not very effective at treating low biodegradable compounds namely humic and fulvic acids in stabilized leachate (Trebouet *et al.*, 2001).

Physical/chemical treatment of landfill leachate is also used for treatment lines (pre-treatment or final purification) or for treatment of specific pollutants (ammonia stripping) (Anqi *et al.*, 2020). The physical and chemical treatments are limited by a large amount of sludge produced and the need to consume a large amount of coagulant (Verma & Kumar, 2017).

Due to the large variability of characteristics and complex nature of landfill leachate, the leachate treatment requires the combination of technologies to fulfill the legal standards for discharging.

An integration of membrane processes such as membrane MF, UF, RO, and NF as possible processes for the treatment of landfill leachate has also been increasingly explored. A membrane process has the advantages of stable effluent and a high rejection rate for refractory organic matter (Anqi *et al.*, 2020).

Given the strict regulation of the application of treatment of landfill leachate, the use of membranes as a single process or combined with biological and physicochemical processes could play an important role in the development of treatments for achieving the required limit for discharge of the leachate.

13.4 MEMBRANE SEPARATION PROCESSES

Membrane separation technology has gained considerable attention in wastewater, including landfill leachate treatment. A membrane is a thin film or barrier with selective permeability. Membrane separation involves the selective filtration of influent through different-sized pores (Warsinger *et al.*, 2018). According to the pore size of membrane materials, the leading membrane processes can be categorized into four groups: MF, UF, NF, and RO. The driving force for material transport through this membrane is a pressure difference. These processes are called pressure-driven membrane processes. The range of operating pressures and limit particle size for the four pressure-driven membrane processes are given in Table 13.3. The wastewater that goes through the membrane is called permeate, and that which remains on the other side of the membrane containing the solute is referred to as the retentate (or concentrate). When landfill leachate passes through the membrane only the particles smaller than the membrane's pore can penetrate the membrane.

Membrane filtration technologies offer great promise for removal of pollutants. Membrane filtration has various advantages including higher

Table 13.3 Typical range of application of pressure-driven membrane separation processes.

Process	Typical Operating Pressure Range (bar)	Limit Particle Size Range (nm) or (MW)
MF	0.2–5	100–10,000
UF	1–10	1–100 (10^2–10^6 Da)
NF	5–10	0.5–5 (10^2–10^3 Da)
RO	10–150	(10^1–10^2 Da)

Source: Berk (2009).

separation efficiency, conservation of energy, lack of phase change, eco-friendly, and scaled-up simplicity (Zhu *et al.*, 2014). However, MCs and membrane fouling are considered the main drawbacks of the process.

Membrane filtration is generally considered an efficient and economical way of separating components from wastewater. It requires less energy compared with other separation technologies, such as distillation or evaporation, because it relies on pressure differences to push the feed solution through the membrane rather than heat. Several factors can impact the energy efficiency of membrane filtration technology, including the type of membrane used, the concentration of the feed solution, and the occurrence of membrane fouling.

According to the study by Pearce (2008), the average energy requirement for treating pre-treated wastewater using an RO system is 1.2–1.5 kWh/m^3. Iskander *et al.* (2017) found that the energy consumption for forward osmosis treatment of landfill leachate was 0.276 $\pm$ 0.033 kWh/m^3, with a recirculation rate of 110 mL/min. The energy consumption was lower for leachate with lower concentrations of contaminants. Zhou *et al.* (2015a, 2015b) and Singh *et al.* (2020) reported that the energy consumption for pressurized MF/UF, cross-flow MF, NF–NF, and NF–RO was in the range of 0.1–0.2, 2.5–3.0, 2.54–4.2, and 2.9–5.858 kWh/m^3, respectively.

The cost of a leachate treatment process can vary widely depending on the specific treatment method and the size and complexity of the system. For example, the cost of using a combination of lime application and NF for landfill leachate treatment was evaluated by Almeida *et al.* (2020). The authors considered that the landfill would operate for 25 years, and after its closure, the leachate treatment plant would continue its activities for an additional 15 years, for a total of 40 years. The total cost to treat 1 m^3 of leachate was estimated at \$10.54 (for 25 years) and \$11.33 (for 40 years).

In another study, Santos *et al.* (2019) examined the effectiveness of three combinations of advanced oxidation processes (AOPs) with Fenton's reagent, MF, and NF for the treatment of landfill leachate. They found that these treatments resulted in high removal efficiencies for COD (94–96%) and color (96–99%), based on an average leachate flow of 300 m^3/day. The cost per cubic meter of the treated leachate was in the range of \$29.32–69.36. In comparison, Calabrò *et al.* (2018) estimated that the on-site treatment cost per cubic meter using RO membranes was in the range of \$15–40.

Overall, the energy efficiency and cost of membrane filtration technology can vary widely depending on the specific application and operating conditions. Proper design and operation of a membrane filtration system can maximize energy efficiency and minimize operating costs.

13.4.1 Membrane separation in landfill leachate treatment

The complexity of treatment of landfill leachate is related to the variations in the composition and volume rate of the leachate. It is crucial to highlight that no universal treatment applies to all leachate, given the high complexity and heterogeneity of each storage site. To comply with local regulatory constraints, the treatment of leachate often requires a combination of processes to adapt to the variations in composition and amount of the leachate (Renou *et al.*, 2008).

In any case, pressure-driven membrane processes including MF, UF, NF, and RO are considered one of the most promising and efficient methods for landfill leachate treatment (Li *et al.*, 2010, 2007; Renou *et al.*, 2008). The membranes can achieve high yields of contaminant removal from the leachate; they are generally applied as combined processes with biological and physical–chemical treatments, achieving thorough and optimal treatment of the leachate. The integrated membrane technologies achieve a pollutant removal efficiency of higher than 95% with a large volume of treated leachate and a low capital cost investment (Keyikoglu *et al.*, 2021). Renou *et al.* (2008) reported that the use of membrane technologies in leachate treatment plants has shown to be an indispensable means of achieving purification. This technology can be used either as a main step in landfill leachate treatment plants or as a single post-treatment step.

Figure 13.2 summarizes some combination of membrane processes with biological and physical chemical treatments as the pre-treatment stage (UF,

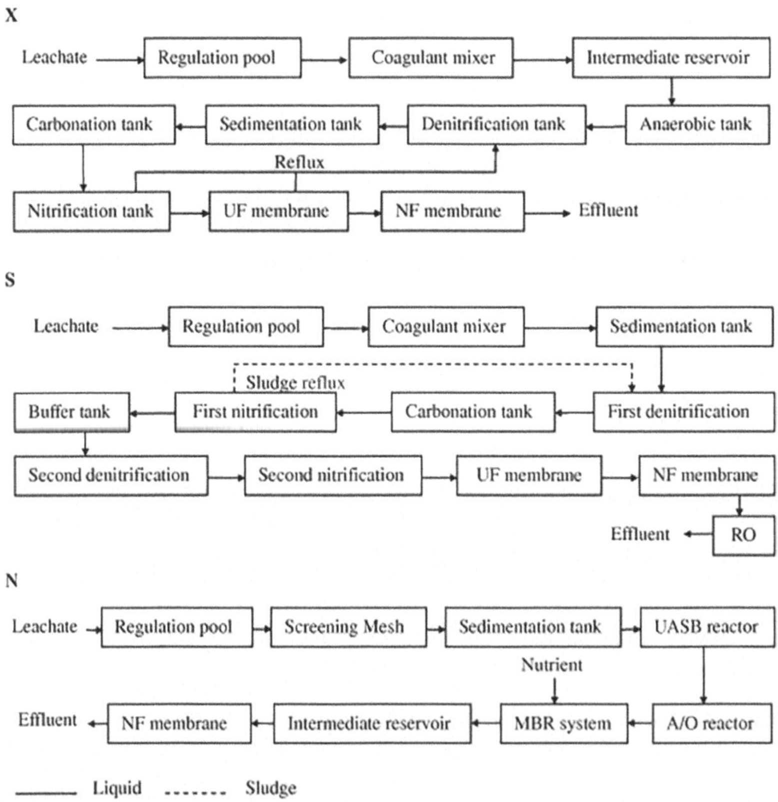

Figure 13.2 Process flow diagram of leachate treatment with combination of membrane processes with biological and physical–chemical processes in three landfill leachate treatment plants. (Reprinted from Zhang *et al.* (2013), with permission from Elsevier.)

NF) or as the polishing stage (NF, RO) using in three full-scale plants in China. All three leachate plants incorporated physical/chemical and biological processes as the first step for the removal of ammonia, COD, and BOD and UF and/or NF membrane as a second step. RO was also used in landfill leachate treatment plants (Zhang *et al.*, 2013).

Such hybrid processes must be designed individually considering the following aspects (Rautenbach & Mellis, 1994):

- effluent quality
- flexibility concerning flow rate and concentration of the wastewater
- residues
- required chemicals
- energy consumption
- investment costs
- required area
- limit of discharging.

After a brief description of membrane processes in landfill leachate treatment, this section presents the processes employing MF, UF, NF, and RO.

13.4.1.1 Microfiltration

MF is classified as a low-pressure cross-flow membrane process. The relatively large pore size of the membrane limits the removal of suspended solids or particles. Dissolved organic and inorganic impurities could not be effectively treated (Warsinger *et al.*, 2018).

As a single process, no significant retention rate was achieved. Piatkiewicz *et al.* (2001) reported using MF as a pre-filtration stage that COD reduction was between 25 and 35%. MF is not applied alone as their effectiveness is poor for limit standard requirements.

A landfill leachate-collected sedimentation tank was treated by the MF process after an activated sludge treatment, allowing the elimination of COD and BOD_5 up to 43 and 63%, respectively. It was found that MF was insufficient for inclusion in the regulatory standards for effluent discharge (Pertile *et al.*, 2018).

MF is generally used as a pre-treatment for other membrane processes (UF, NF, or RO) or to assist chemical treatments (Abbas *et al.*, 2009). The use of MF has been considered a good pre-treatment method for RO processes since early reports (Chakravorty & Layson, 1997; Ebrahim *et al.*, 1997). This is mainly due to its ability to retain microorganisms and suspended solids, avoiding fouling of RO membranes (Anis *et al.*, 2019).

More recently, combinations between ozonation process (O_3) and submerged MF demonstrate the great capacity as fouling reducing pre-treatments for RO in the treatment of landfill leachate. Figure 13.3 shows the combination of ozonation and MF scheme to condition landfill leachate avoiding fouling of RO membranes and increasing leachate treatment efficiency (Gripa *et al.*, 2021).

The use of a microfiltration–powdered activated carbon (MF–PAC) hybrid system met the COD discharge limit value when 8 g PAC/L concentration was used compared with the NF unit. According to Ince *et al.* (2010), hybrid Fenton–MF process had the best removal efficiency of color and COD with 47 and 67%,

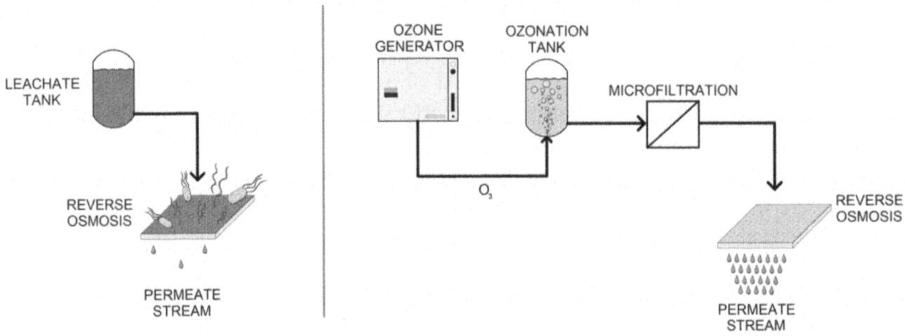

Figure 13.3 Pre-treatment of landfill leachate for RO treatment. (Reprinted from Gripa *et al.* (2021), with permission from Elsevier.)

respectively. Moravia *et al.* (2013) reported that combining the Fenton process with MF improves the effluent quality and eliminates the sedimentation stage used for sludge separation, drastically reducing the treatment time.

13.4.1.2 Ultrafiltration

UF is a membrane separation technique between MF and NF, with pressure as the driving force. UF is commonly operated mainly as sieving for separation, concentration, and fractionation. UF contains a wide separation range compared with MF and allows us to remove macromolecular substances, pathogens, particles, and colloids from landfill leachate (Warsinger *et al.*, 2018). However, the removal efficiency greatly depends on the type of membrane material and the range of pore size.

As single-step UF membranes have been successfully studied in lab scale and ensured that scaling up was tested using industrial modules at bench scale. Higher permeate fluxes can be obtained using mineral membrane, which in addition can be used under higher pressures and membrane regeneration is easier (Tabet *et al.*, 2002). Therefore, UF has a lower treatment cost. However, it is characterized by low rejection of organics with a low molecular weight (MW) (Kulikowska *et al.*, 2019).

In a treatment of landfill leachate using air stripping followed by a combined process of coagulation/UF, the removal efficiency of COD, color, and NH_4-N was 84.2, 75, and 22.5%, respectively. The BOD/COD ratio of effluent increased from 0.049 to 0.311 (Pi *et al.*, 2009). The implementation of the pre-treatment with lime may allow reducing the costs of a UF unit by about 50% in terms of investment and from 5 to 30% for operating costs (Renou *et al.*, 2008).

In terms of leachate treatment, similar to MF membranes, it is also usually used as a pre-treatment process for NF or RO. In general, UF can be used as a pre-treatment for RO or carried out after biological treatment (Anqi *et al.*, 2020). UF can be used to remove the larger MW components of leachate that tend to foul NF and RO membranes. UF membranes integrated with membrane bioreactors (MBRs) have also been successfully used in full-scale plants as pre-treatment for NF or RO.

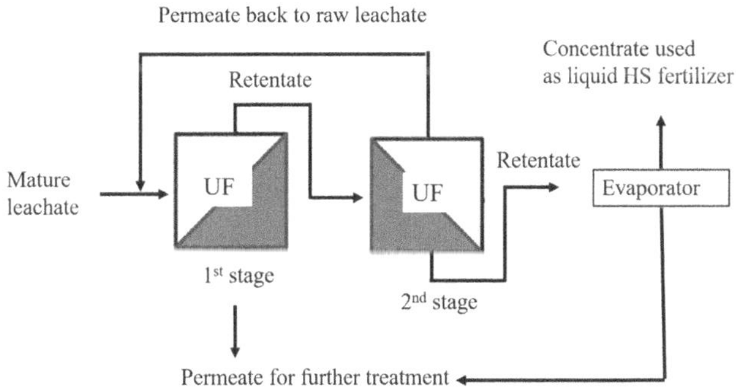

Figure 13.4 Recycling concept for mature leachate. (Reprinted from Xu *et al.* (2006), with permission from Elsevier.)

It was found that UF was effective as a pre-treatment process for RO; UF membranes with pore diameters greater than RO membranes remove particles that can foul RO membranes. In addition, given that UF works under lower operating pressures, it allows for easier control of fouling compared with RO that requires much higher operating pressures (Syzdek & Ahlert, 1984). UF was also used as a post-treatment for conventional biological processes and 50% of COD were separated successfully (Bohdziewicz *et al.*, 2001). In another study, old leachate was recycled using two UF stages and evaporation combined processes (Figure 13.4) (Xu *et al.*, 2006).

Therefore, to increase the efficiency of leachate treatment, integrated systems, two systems, adsorption–UF and coagulation/UF, are used as pre-treatment for treating stabilized landfill leachate using NF and RO. Coagulation provided better efficiencies for removing organic compounds (COD and total organic carbon (TOC) decreased by 65 and 86%, respectively) compared with adsorption. In such systems, the role of activated carbon is to adsorb low-MW organic compounds that cannot be separated by UF alone. Both pre-treatment systems showed reversible fouling because flux was recovered very close to the initial flux of the RO/NF membranes (Dolar *et al.*, 2016).

In another study, the combination of adsorption with fine-UF membrane has been proposed obtaining high rejection of contaminants. The role of adding the adsorption before membrane filtration is to reduce membrane fouling (Kulikowska *et al.*, 2019).

13.4.1.3 Nanofiltration

NF can eliminate ions contributing to osmotic pressure; thus, it requires lower operating pressures than those used in RO (Mojiri *et al.*, 2020). NF is able to remove recalcitrant organic compounds and heavy metals from landfill leachate. In NF, the operating pressure is 5–15 bar and allows materials dissolved in water to be separated into monovalent and bivalent ions (Top *et al.*, 2011). The advantage of using an NF membrane is that it requires lower operating pressures, has higher fluxes compared to RO membranes, and has better retention compared to UF membranes (Li *et al.*, 2010). Moreover, NF presents high retention of charged organics, multivalent ions, and heavy metals, requires relatively low investment, and has low operational and maintenance costs (Linde *et al.*, 1995). However, NF cannot remove most contaminants from landfill leachate. This can be considered a disadvantage because applying NF in leachate treatment requires additional processes for removing nitrogen compounds and monovalent salts.

NF separation occurs in two stages: neutral species are rejected according to their size, and the inorganic ions are separated by electrostatic interaction between the ions and the membrane (Yacubowicz & Yacubowicz, 2005). However, the type of membrane used significantly affects the separation of organic compounds and the operational parameters (Trebouet *et al.*, 2001). Table 13.4 summarizes a comparison of the efficiency of two types of membranes for leachate treatment.

NF as a single process does not play a significant role in landfill leachate treatment. In a laboratory-scale unit, COD removal was 52–66% (Marttinen *et al.*, 2002). A pilot study shows that the NF process efficiently treats the refractory organic matter as COD removal was 70–80% (Trebouet *et al.*, 1999, 2001). Although they used stabilized and low-strength leachate, the NF permeate did not meet the strict legal standards for nitrogen compounds (see Table 13.4). The nitrogen concentration is given in total Kjeldahl nitrogen (TKN). The modification of pH of the stabilized leachate and coagulation as pre-treatments did not enhance the permeation flux and the retention rate of the membrane but influenced the performance of the membrane (Trebouet *et al.*, 2001).

Table 13.4 Stabilized leachate and permeates characteristics, DP = 2 MPa, $U = 3$ m/s and constant concentration.

Parameters	Raw Leachate	MPT-20 Membrane		MPT-31 Membrane		Discharge Limit of Leachate, (Saint-Nazaire, France)
		Permeate	*R* (%)	Permeate	*R* (%)	
pH	7.5	8.3	–	8.4	–	5.5 ≪ 8.5
SS (mg/L)	130	0	–	–	–	30
COD (mg/L)	500	130	74	100	80	120
BOD_5 (mg/L)	7.1	1	85	0.11	98	30
TKN (mg/L)	540	420	22	380	30	30
$N\text{-}NH_4^+$ (mg/L)	430	380	12	340	21	–
Na^+ (mg/L)	520	440	15	435	16	–
Ca^{2+} (mg/L)	140	90	36	61	56	–
Cl^- (mg/L)	700	600	14	620	11	–
Fe (mg/L)	10	≈0	>99	≈0	>99	10
Ni (mg/L)	0.15	≈0	>99	≈0	>99	–

Source: Reprinted from Trebouet *et al.* (2001), with permission from Elsevier.

NF has been used to treat filtered landfill leachate to remove inorganic compounds such as heavy metals, which are multivalent cations and can be rejected by the NF membrane. For example, the rejection of Cd, Zn, Pb, and Cr was found to be higher than 70%. In contrast, the monovalent cations pass through the membrane, while the retention of potassium and sodium was <10% (Linde *et al.*, 1995). Because NF as a single-treatment step presents some limitations, combined processes could produce a purified permeate in compliance with legislative discharge standards.

NF process offers a versatile approach to meet multiple water quality objectives, such as control of organic, inorganic, and microbial contaminants when used as post-treatment. The combination of NF with different processes can lead to more complex but, at the same time, more cost-efficient processes (Meier *et al.*, 2021).

The quality of the leachate treated using MBRs did not comply with the standard limits for discharging and needed an additional process for polish leachate treatment (Keyikoglu *et al.*, 2021). Combining NF as a polishing step for MBR effluents allows achieving a higher efficiency in removing the organic and inorganic compounds from leachate. NF process seems to be an economical and viable solution for minimizing environmental risks (Li *et al.*, 2010).

Campagna *et al.* (2013) investigated the NF of an MBR effluent and characterized the MW distribution of the leachate's organic carbon and nitrogen fractions. According to the results, the NF stage can remove 55% of the COD after the MBR stage but cannot effectively remove the fraction smaller than 500 Da.

In another study, landfill leachate from a sanitary landfill site in Malaysia was filtered through an NF membrane in order to determine its rejection capability toward pollutants such as COD, conductivity, nitrate, ammonia nitrogen, and heavy metals such as Pb, Cd, Cu, Zn, and Fe. The NF membrane used was an HL membrane. The overall rejections of total suspended solids, heavy metals, conductivity, and COD were more than 85%, except for nitrate and ammonia nitrogen being much lower at 45.4 and 20.5%, respectively. NF can be considered an alternative for advanced filtration, especially within a hybrid treatment system combining conventional biological–physical processes and membrane filtration (Mohammad *et al.*, 2004).

An integration of air stripping, MBR, and NF processes in a piloting landfill leachate treatment is evaluated. After pre-treatment by air stripping, 65% of $N\text{-}NH_3$ was removed without pH adjustment. After pre-treatment, the effluent was treated using MBR, obtaining 44% COD removal, and part of $N\text{-}NH_3$ was converted to nitrite and nitrate. In the last step, NF improves the removal of organic compounds, high toxicity, and nitrite and nitrate that are generated in the MBR. The integration of air stripping, MBR, and NF process obtained an excellent efficiency in removing COD, ammonia, color, and toxicity of 88, 95, 100, and 100%, respectively (Amaral *et al.*, 2015).

Regarding the removal of organic pollutants using the Fenton process, the results show that the removal efficiencies were 62.5% for COD and 71.0% for TOC. Nonetheless, the NF process was effective as the final polishing stage of the effluent, achieving an average removal of 98.4 and 98.1% for COD and

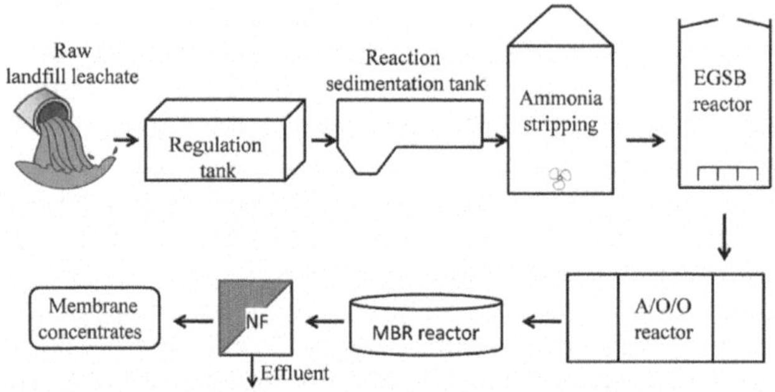

Figure 13.5 Schematic diagram of leachate treatment process. (Reprinted from Hou *et al.* (2017, with permission from Elsevier.)

TOC, respectively. The combination of Fenton–MF with an NF system also allows for improved removal efficiency of pollutants, besides reducing the costs of treatment (Moravia *et al.*, 2013).

Moreover, combining Fenton process with membrane processes can provide a wide range of advantages. Santos *et al.* (2019) evaluated the performance of different routes involving the integration of Fenton's reagents and MF and NF membrane processes for the treatment of landfill leachate. High removal efficiencies of COD (94–96%) and color (96–99%) were obtained.

NF membrane combined with other processes was widely used in the full scale for landfill leachate, an example, a leachate treatment plant in China. The detailed flow diagram for the raw landfill leachate treatment process is shown in Figure 13.5. The process consists of mixing leachate uniformly in regulation tanks, and pH was adjusted in the reaction sedimentation tank. After treatment with an ammonia stripping tower and the subsequent expanded granular sludge bed process to remove COD, the treated leachate entered a combined processing area, including the anaerobic/oxic/oxic reactor, for denitrification. Finally, MBR and NF membranes were used to comply with the effluent discharge standards (Hou *et al.*, 2017).

Applying NF for landfill leachate treatment requires effective pre-treatment for heavily contaminated wastewater. Keyikoglu *et al.* (2021) provide an overview of the combination of NF with a biological and physical–chemical process in landfill leachate treatment. Table 13.5 summarizes several combinations in full-scale plants for landfill leachate treatment.

13.4.1.4 Reverse osmosis

Among the procedures for landfill leachate treatment, RO is one of the most promising and effective techniques. RO can concentrate low-MW dissolved and suspended solids to purify wastewater (Abbas *et al.*, 2009). RO has the ability to work at high fluxes and to operate over wide temperature and pH ranges, with a 98–99% rejection rate for organic and inorganic contaminants. In addition, RO can be used to remove heavy metals, suspended/colloidal materials, and dissolved solids from landfill leachate (Kurniawan *et al.*, 2006). This membrane

Table 13.5 Combinations of biological and physical chemical processes with NF in full-scale plants for landfill leachate treatment.

Option No.	Treatment Processes						Location of Landfill	Reference
1	S	UASB	A/O	MBR	UF	NF	Zhejiang (China)	He *et al.* (2015)
2	S	AS	EGSB	AN-O-O	MBR	NF	Shenzhen (China)	Hou *et al.* (2017)
3	S	UASB	A/O	MBR-UF	NF		Zhejiang (China)	Zhang *et al.* (2013)
4	UASB	MBR	NF				Zhejiang and Qingdao (China)	Qiao *et al.* (2018)
5	UASB	A1-A2-O	MBR	NF			Shenzhen and Beijing (China)	Li *et al.* (2016)
6	A2-O	MBR-UF	NF				Istanbul (Turkey)	Top *et al.* (2011)
7	A1O	MBR	NF				Chengdu (China)	Chen *et al.* (2019)
8	AS	AN	O-A2	S	NF		Shenzhen and Beijing (China)	Li *et al.* (2016)
9	A2-A1	AN	MBR	UF	NF		Beijing (China)	Soomro *et al.* (2020)
10	AN	DN-N	MBR	NF			Jiangsu (China)	Cui *et al.* (2018)
11	DN-N	MBR	UF	NF			Qingdao (China)	Wang *et al.* (2016)
12	AN	MBR	NF				Wuhan (China)	Xiong *et al.* (2014)
13	C-F	AN	DF-N	UF	NF		Zhejiang (China)	Zhang *et al.* (2013)
15	MBR	NF					Changzhou (China)	Xu *et al.* (2017a)

UASB, up-flow anaerobic sludge blanket; AN, anaerobic oxidation; CF, coagulation–flocculation, S, sedimentation; DN-N, denitrification–nitrification; A1-A2-O, anoxic–aerobic–oxic; EGSB, subsequent expanded granular sludge bed process; AS, air stripping.

separation process is now the most widely used in many countries in western Europe, North America, and Asia, due to its good ability to retain both organic and inorganic contaminants.

In general, biological and physical–chemical treatments have been widely used for leachate treatment. However, they are not good end-of-pipe solutions to comply with the strict legislation of discharge. RO is one approach being implemented as a complement for other treatments or, in some cases, is the only step for leachate treatment and has proved to be an efficient purification step to comply with limit standards (Labiadh *et al.*, 2016; Zhou *et al.*, 2016).

An integrated membrane process MBR + RO has been proposed for the treatment of leachate. This way, the MBR process enhances the removal efficiency of biodegradable organic matters and nitrification rate, and RO aims to remove physically the remaining inorganic nitrogenous ions and non-biodegradable matters. An MBR + RO system was found to be more economical, and reduces operating cost from $5/ton to $3/ton for conventional treatment (Ahn *et al.*, 2002).

Moreover, an MBR + NF/RO process can effectively separate organic and inorganic pollutants that could not be degraded during biological treatment and produces effluent with stable quality. The global removal efficiency of the MBR + NF/RO process was >97% for both COD and BOD_5 (Keyikoglu *et al.*, 2021). In addition, Li *et al.* (2007) used an integrated biological treatment process to treat leachate. The process combines an up-flow anaerobic sludge blanket, a sequencing batch reactor, a continuous MF, and RO process to treat leachate. The combined process could achieve removal efficiencies of 99.5, 99.8, and 99.8% for total nitrogen, COD, and BOD_5, respectively. Also, the effluent's final quality met the water reuse standards.

In a pilot-scale study, membrane filtration (UF + RO), struvite (magnesium ammonium phosphate (MAP)) precipitation, and ammonia stripping alternatives were studied on biologically pre-treated landfill leachate. The results indicated that the system, including the up-flow anaerobic sludge blanket reactor and membrane reactors (UF + RO), could be proposed as an appropriate treatment alternative for young landfill leachates. This system provided high removals of COD, color, and conductivity (>98–99%) (Ozturk *et al.*, 2003).

Combinations between the ozonation process (O_3) and submerged MF were evaluated as fouling-reducing pre-treatments for RO in the treatment of leachate. The results demonstrate that a first ozonation step followed by the MF step represents the best route, reducing the leachate's modified fouling index over 170 times. The integration of O_3 + MF + RO results in the removal rate of 99.09 and 99.6%, respectively, for COD and $N\text{-}NH_3$ (Gripa *et al.*, 2021).

Several studies were performed to evaluate the efficiency of RO systems combined with conventional biological and physicochemical processes on the purification of landfill leachate at full scale. Among these combinations RO and MBR processes have been applied for the removal of organic compounds, heavy metals, and inorganic from leachate. Table 13.6 summarizes multiple processes combined with RO for leachate treatment.

RO was also employed for the treatment of leachate from the Spillepeng (Sweden) and the Wijster landfills (Netherlands). Both plants consist of two stages: the first stage is equipped with tubular modules and the second with spiral wound modules. Adjusting the pH to a value below 6.5 makes possible to reduce fouling (Linde *et al.*, 1995).

The reduction of COD and $N\text{-}NH_4^+$ was found to be 98% with the initial concentrations of 335 and 140 mg/L, respectively. In addition, the application of RO for the treatment of stabilized leachate from Lipowa landfill (Poland)

Table 13.6 Combination of biological and physical chemical processes with RO for landfill leachate treatment.

Option No.	Treatment Processes					Location of Landfill	Reference
1	RO					Bohemia (Czech)	Hendrych *et al.* (2019)
	RO					Kolenfeld (Germany)	Li *et al.* (2009)
	RO1	RO2				Wijster (Holland) Spillepeng (Sweden)	Linde *et al.* (1995)
2	C-F	2 stage (DN-N)	UF	NF	RO	Zhejiang (China)	Zhang *et al.* (2013)
3	UASB	SBR	CMF	RO		Guangzhou (China)	Wu *et al.* (2010)
4	A2-O	MBR	UF	RO		Qingdao (China)	Wang *et al.* (2016)
5	A2-O/ A2-O	MBR-UF	RO			Xiangtan (China)	Zhou *et al.* (2016)
6	SF	RO1	RO2			Erzurum (Turkey)	Hunce *et al.* (2012)
7	AS	RO1	RO2			Tunisia	Labiadh *et al.* (2016)
8	AS	Tertiary MBR	DT RO			Shanghai (China)	Xingxing *et al.* (2015)
9	AN	A-O1	MBR	RO		Wuhan (China)	Song *et al.* (2019)
10	C-F	MBR	RO			Sfax (Tunisia)	Kallel *et al.* (2017)
11	SF	CF	RO			Warminsko-Mazurskie (Poland)	Talalaj (2015)

UASB, up-flow anaerobic sludge blanket; SBR, sequencing batch reactor; CMF, continuous micro-filtration; DT, disk tube; AN, anaerobic oxidation; C-F, coagulation–flocculation; SF, sand filter; CF, cartridge filter; DN-N, denitrification–nitrification; A1–A2–O, anoxic–aerobic–oxic; AS, air stripping.

was also explored. With the initial concentration of 1,780 mg/L, 97% of COD was removed (Kurniawan *et al.*, 2006).

The stabilized leachate from Kolenfeld (Germany) was treated by two systems. First, leachate was pre-treated through an activated sludge process, followed by coagulation/flocculation, sedimentation, and RO. The maximum removal of COD and N-NH_4^+ with initial concentrations of 3,100 and 1,000 mg/L was found to be 99.7 and 99.9%, respectively. In the second system, raw leachate was fed directly into the RO system. The maximum removal of COD and N-NH_4^+ was found to be 99.5 and 98.9%, respectively. The use of direct RO membrane filtration with thin open channel spiral wound modules is able to achieve satisfactory results in terms of water quality, process stability, and membrane flux (Li *et al.*, 2009).

The treatment of intermediate leachate collected from Siedliska landfill (Poland) was carried out using RO. Before separation into the RO membrane modules, suspended particles (>50 μm) were removed by a sand filter and for suspended particles (>10 μm) a cartridge filter was employed. Then, to make inorganic salts more soluble, the leachate's pH value was adjusted between 6.0 and 6.5 using H_2SO_4 before treatment process. Furthermore, to calculate the effectiveness of removing organic and inorganic pollutants from landfill leachate, a long-term RO was used. Investigations were carried out during 1 year of RO exploitation. Some parameters were analyzed. The results showed that the average removal of COD, conductivity, N-NH_4^+, total inorganic nitrogen, CN^-, Fe, and Cl^- attained 97, 97.2, 98.7, 99, 93, 97.6, and 98%, respectively. Despite these results, the permeate quality did not meet the Polish discharge standards for S^- (Talalaj, 2015).

An integrated two-stage disk tube-reverse osmosis (DTRO) membrane system was also used in leachate treatment. The separating size of an integrated DTRO membrane is <0.1 nm (approximately the size of many ions), which makes DTRO the most sophisticated membrane separation technology available. Global removals for each stage were 92.4–99.2, 46.2–95.8, 84.8–97.9, and 88–95.5% for COD, NH_4^+, NO_x, and total dissolved solids (TDS), respectively (Cingolani *et al.*, 2018).

In a recent study, landfill leachate was first filtered through a sand filter, cloth bag filter, and another cartridge filter. Then, the filtered leachate was pumped to DTRO. The DTRO system achieved >83% water recovery rate, reduced the electrical conductivity of effluent to 0.15–0.22 mS/cm, and reduced carbonaceous and nitrogenous pollutants to a level suitable for discharge. Furthermore, a DTRO system effectively removes salts (e.g., Cl^- and SO_4^{2-}) and refractory dissolved organic matter (DOM) (e.g., humic- and fulvic-like substances). A DTRO system can effectively remove a large number of macromolecular dissolved organic compounds with carbon number >23, and highly unsaturated compounds with double-bond equivalent >12. Additionally, >80% of the molecules assigned to the dissolved OM were removed; even CHONS compounds with complex molecular structures were wholly removed (Wu *et al.*, 2021). Figure 13.6 shows an integrated two-stage DTRO membrane system for leachate treatment.

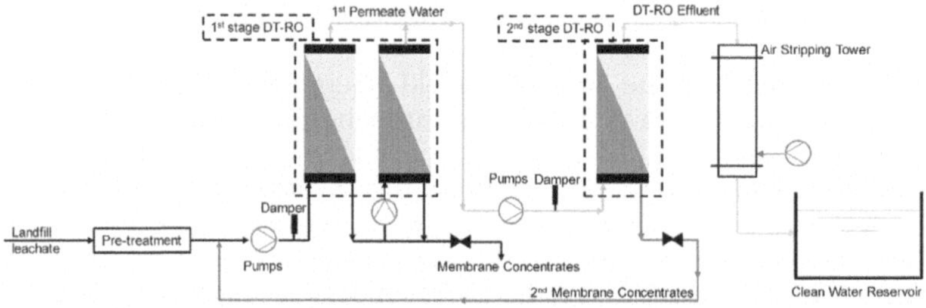

Figure 13.6 Integration of two-stage DTRO membrane system for leachate treatment. (Reprinted from Wu *et al.* (2021), reproduced with permission from Elsevier.)

Table 13.7 Features of NF and RO membranes in landfill leachate applications.

Parameter	NF	RO
COD retention (%)	90–99	92–99
AOX retention (%)	85–93	90–96
NO_3-N retention (%)	10–20	83–93
NH_4-N retention (%)	10–20	86–94
Cl^- retention (%)	5–10	98–99.6
SO_4^{2-} retention (%)	94–98	99–99.9
Heavy metal retention (%)	85–96	88–97
Permeate flux (L/m² h)	20–80	10–20
Water recovery (%)	76–85	70–80
Transmembrane pressure (bar)	10–30	40–65

Source: adapted from as cited in Meier *et al.* (2021).

RO is highly efficient for treating leachate with rejection coefficients on an average >0.98. NF compared with RO can be operated at lower pressures; offers higher fluxes; rejects organics and multivalent ions selectively; and needs lower investment, operation, and maintenance costs (Ramaswami *et al.*, 2018). Due to these advantages, some studies in the recent past have preferred NF over RO and have explored its ability for the treatment of landfill leachates (Košutić *et al.*, 2015; Mohammad *et al.*, 2004). However, it should be considered that monovalent ions can permeate an NF membrane, whereas organic solutes and other electrolytes are mostly rejected.

A comparative study of the treatment of stabilized leachate from the Hallenberg landfill (Germany) was carried out using RO and NF. At a permeate flux of 15 L/(m² h), RO gave a higher removal efficiency of COD (99.2%) and NH_3-N (99.9%) when compared with NF (COD: 95.88%; NH_3-N: 57.61%) (Peters, 1998).

In another study, NF–RO and RO–NF systems were used and compared for the treatment of an old leachate. In this case, both NF and RO stages in NF–RO can be operated at lower pressures compared with NF and RO stages in RO–NF. For similar operating conditions, individual stages in NF–RO provided higher water fluxes than those in RO–NF, showing NF–RO to be more energy efficient (Ramaswami *et al.*, 2018). Table 13.7 summarizes a comparison of features of NF and RO membranes in landfill leachate applications.

Membrane separation processes such as NF and RO are widely used as a polishing step for landfill leachate treatment. However, two factors can be identified as significant limitations for the implementation of membrane separation processes: (1) the need for further treatment of the MCs and (2) membrane fouling.

13.4.2 MC landfill leachate removal

13.4.2.1 Composition of MCs

Membrane separation processes do not destroy the pollutants but merely concentrate them into smaller volumes of wastewater called MCs (Safarpour *et al.*, 2017). MC results in a secondary pollution problem that must be appropriately treated before being discharged into the environment. For example, during NF and RO membrane processes, as much as 13–30% of the total incoming leachate resulted in a concentrated leachate (Wang *et al.*, 2016). Other studies reported approximately 25–50% of untreated leachate (Hendrych *et al.*, 2019; Kallel *et al.*, 2017). MC composition depends on various factors, such as initial leachate composition and treatment process (Zhang *et al.*, 2018).

MC is a dark-colored solution and heavily loaded with refractory pollutants and inorganics salts, which are highly toxic and carcinogenic, and their treatment is an urgent issue (Hou *et al.*, 2017). Membrane usually concentrates five main groups of pollutants (Keyikoglu *et al.*, 2021):

(i) major pollutants: COD, BOD_5, BOD_5/COD, TOC, TN, and NH_3-N
(ii) ions of inorganic salts: Cl^-, NO_3^-, and SO_4^{2-}
(iii) heavy metals: As, Cd, Cr, Co, Cu, Ni, Pb, and Zn
(iv) humic substances: humic and fulvic-like compounds
(v) XOCs: aromatic compounds, halogenated hydrocarbons, and endocrine-disrupting chemicals.

The main characteristics of the MCs are summarized in Table 13.8.

As observed from Table 13.8, MCs are a type of low-biodegradable wastewater because of high COD and low BOD_5 concentrations resulting in a low BOD_5/COD ratio (<0.10).

Table 13.8 Main characteristics of the MCs of landfill leachate treatment processes.

Parameter	NF	RO
EC (mS/cm)	10–100	10–100
COD (mg/L)	1,100–6,800	400–49,500
BOD_5 (mg/L)	3–290	95–4,800
BOD_5/COD ratio	<0.10	<0.10
NH_3-N	1–490	240–3,000
NO_3-N	90–1,045	20–170
Cl^-	2,500–6,500	2,500–6,500
SO_4^{2-}	400–2,000	400–2,000
TDS	8,000–67,000	8,000–67,000
As	0.060–3.7	0.1–1.5
Cd	0.001–3	0.004–1.2
Co	0.01–7.7	0.01–0.2
Cr	0.06–26.9	0.32–2.2
Cu	0.01–23.2	0.02–3.7
Ni	0.04–26.6	0.18–1.02
Pb	0.06–57	0.05–22.79
Zn	0.079–6,656.3	0.008–2,662.5

Source: Keyikoglu *et al.* (2021).

13.4.2.2 Treatment of MCs

A high concentration of salts and refractory pollutants in MCs significantly reduces biodegradability, making it untreatable by biological treatment processes. Therefore, physicochemical methods such as evaporation (Yang *et al.*, 2018), recirculation (Sun *et al.*, 2011), solidification/stabilization (Hunce *et al.*, 2012), adsorption (Wang *et al.*, 2013), membrane distillation (Xingxing *et al.*, 2015), coagulation (Long *et al.*, 2017), chemical coagulation, and oxidation (Ye *et al.*, 2017) are used. However, physicochemical methods could not degrade the pollutants and only separate/concentrate the pollutants from MCs (Keyikoglu *et al.*, 2021).

On the contrary, AOPs such as ultraviolet (UV) radiation-based processes ($UV/O_3/H_2O_2$), Fenton-based processes (H_2O_2/Fe^{2+}), chemical oxidation (ozonation or H_2O_2 oxidation), photo-Fenton processes, and electrochemical processes, are known for their effectiveness in the degradation of refractory pollutants. Table 13.9 summarizes some examples of advanced processes for treating MCs.

Several studies have implemented the combination of processes of physical, chemical, and AOPs to improve the efficiency of MCs, including photo-electrocatalytic oxidation (Zhou *et al.*, 2015b), microwave-Fenton (Zhang *et al.*, 2018), O_3/Fenton (Huang *et al.*, 2019), coagulation–ozonation (Chen *et al.*, 2019), and combination of electrocoagulation and electro-oxidation (Soomro *et al.*, 2020).

Another perspective in the management of MCs could be the recovery of some by-products such as phosphate (Kumar *et al.*, 2007), minerals (Mohammadesmaeili *et al.*, 2010), potassium, chlorine (Li *et al.*, 2015), and humic substances (Xu *et al.*, 2017b).

13.4.3 Membrane fouling

One of the major limiting factors of membrane separation processes is fouling or biofouling of the membrane, induced by deposits of inorganic, organic, and microbiological substances on both the membrane surface and inside the membrane pores, leading to membrane's efficiency deterioration (Gkotsis *et al.*, 2017; Wiszniowski *et al.*, 2006). Moreover, membrane fouling leads to a decrease in permeate flux, causing the flux to become independent of the pressure across the membrane, which can affect the economic efficiency of the membrane plant.

The fouling index of landfill leachate has been reported to range between 3.5 and 5.4, which indicates a high tendency to fouling (Trebouet *et al.*, 1999). In addition, membrane fouling increased at low pH is explained by electrostatic effects and also depends on membrane material (Trebouet *et al.*, 2001).

Membrane fouling requires extensive pre-treatment or chemical cleaning of the membranes and results in a short lifetime of the membranes and decreases process productivity which involves the increase of the operational costs (Renou *et al.*, 2008). Given that leachate contains a variety of contaminants deposited on the membrane surface, membrane cleaning is a complex issue, which should be addressed for the selection of the most economical and effective cleaning solution (Peng, 2017).

Table 13.9 Examples of advanced processes for treating MCs.

Process	MC Origin	MC Characteristics	Experimental Conditions	Treatment Efficiency	Reference
O_3/H_2O_2	RO	pH = 7.1, COD = 1,380 mg/L	pH = 9, H_2O_2 = 4 g/L, gas flow rate = 0.5 L/min, O_3 conc. = 10 mg/min, t = 180 min	COD = 45% Color = 89%	Amaral-Silva *et al.* (2016)
Fenton	NF	pH = 7.8–8.2, COD = 3,300 mg/L, TOC = 1,080 mg/L	pH = 3, H_2O_2 = 45 mM, Fe^{2+} = 15 mM, H_2O_2/Fe^{2+} = 3, t = 40 min	COD = 78.9% TOC = 70.2%	Teng *et al.* (2020)
Photo-Fenton	NF	pH = 7.52 COD = 1,280 mg/L, BOD_5 = 121 mg/L	pH = 3, $FeSO_4{\cdot}7H_2O$ = 7.2 mM, H_2O_2 = 400 mM, reaction time = 3 h	COD = 93%	Zhao *et al.* (2020)
Electro-Fenton	NF	pH = 7.95 COD = 5,250 mg/L, BOD_5/COD = 0.035, Cl^- = 7,285 mg/L	pH = 2.9, cathode = Fe, H_2O_2/COD = 1.42, current = 2.27 A	COD = 60.8%	Varank *et al.* (2020)
Electro-oxidation	RO	pH = 5.16 COD = 3,778.56 mg/L, NH_3-N = 791.52 mg/L, BOD_5 = 201.06 mg/L	pH = 5.16, anode = BDD, cathode = stainless steel, j = 50 mA/cm^2, electrolysis time = 6 h	COD = 87.5% NH_3-N = 74.06	Zhou *et al.* (2016)

To sum up, the appropriate selection of feeding characteristics, the hydrodynamic conditions of the system, and cleaning techniques can minimize fouling. Thus, knowledge of fouling mechanisms is necessary to adequately select methods for mitigating it (Moravia *et al.*, 2013).

13.5 CONCLUSIONS

Currently, membrane separation processes are widely used as a pre-treatment or polishing step combined with biological and physical–chemical processes for leachate treatment and have been successfully implanted in multiple full-scale plants. NF and RO can eliminate refractory organic matter, total nitrogen, inorganic salts, and heavy metals. NF and RO have proved to be an efficient purification step achieving limit water reuse standards.

Nevertheless, NF and RO present some economic and technical problems, such as membrane fouling, replacement, and MCs. For membrane fouling, it requires extensive pre-treatment or chemical cleaning. A large volume of concentrate remaining at the end of the treatment process has to be appropriately treated before it can be discharged into the environment.

REFERENCES

Abbas A. A., Jingsong G., Ping L. Z., Ya P. Y. and Al-Rekabi W. S. (2009). Review on landfill leachate treatments. *American Journal of Applied Sciences*, **6**, 672–684, https://doi.org/10.3844/ajassp.2009.672.684

Ahn W.-Y., Kang M.-S., Yim S.-K. and Choi K.-H. (2002). Advanced landfill leachate treatment using an integrated membrane process. *Desalination*, **149**, 109–114, https://doi.org/10.1016/S0011-9164(02)00740-3

Almeida R. de, Campos J. and Oroski F. de A. (2020). Techno-economic evaluation of landfill leachate treatment by hybrid lime application and nanofiltration process. *Detritus*, **170**, 170–181, https://doi.org/10.31025/2611-4135/2020.13897

Amaral M. C. S., Moravia W. G., Lange L. C., Roberto M. M. Z., Magalhães N. C. and dos Santos T. L. (2015). Nanofiltration as post-treatment of MBR treating landfill leachate. *Desalination and Water Treatment*, **53**, 1482–1491, https://doi.org/10.1080/19443994.2014.943061

Amaral-Silva N., Martins R. C., Castro-Silva S. and Quinta-Ferreira R. M. (2016). Ozonation and perozonation on the biodegradability improvement of a landfill leachate. *Journal of Environmental Chemical Engineering*, **4**, 527–533, https://doi.org/10.1016/j.jece.2015.12.002

Anis S. F., Hashaikeh R. and Hilal N. (2019). Reverse osmosis pretreatment technologies and future trends: a comprehensive review. *Desalination*, **452**, 159–195, https://doi.org/10.1016/j.desal.2018.11.006

Anqi T., Zhang Z., Suhua H. and Xia L. (2020). Review on landfill leachate treatment methods. *IOP Conference Series: Earth and Environmental Science*, **565**, 012038, https://doi.org/10.10.1088/1755-1315/565/1/012038

Berk Z. (2009). Chapter 10 – Membrane processes. In: Food Process Engineering and Technology, Food Science and Technology, Z. Berk (ed.), Academic Press, San Diego, pp. 233–257, https://doi.org/10.1016/B978-0-12-373660-4.00010-7

Bohdziewicz J., Bodzek M. and Górska J. (2001). Application of pressure-driven membrane techniques to biological treatment of landfill leachate. *Process Biochemistry*, **36**, 641–646, https://doi.org/10.1016/S0032-9592(00)00259-4

Calabrò P. S., Gentili E., Meoni C., Orsi S. and Komilis D. (2018). Effect of the recirculation of a reverse osmosis concentrate on leachate generation: a case study in an Italian landfill. *Waste Management*, **76**, 643–651, https://doi.org/10.1016/j.wasman.2018.03.007

Campagna M., Çakmakcı M., Büşra Yaman F. and Özkaya B. (2013). Molecular weight distribution of a full-scale landfill leachate treatment by membrane bioreactor and nanofiltration membrane. *Waste Management*, **33**, 866–870, https://doi.org/10.1016/j.wasman.2012.12.010

Chakravorty B. and Layson A. (1997). Ideal feed pretreatment for reverse osmosis by continuous microfiltration. *Desalination*, **110**, 143–149, https://doi.org/10.1016/S0011-9164(97)00093-3

Chen W., Gu Z., Wen P. and Li Q. (2019). Degradation of refractory organic contaminants in membrane concentrates from landfill leachate by a combined coagulation–ozonation process. *Chemosphere*, **217**, 411–422, https://doi.org/10.1016/j.chemosphere.2018.11.002

Cingolani D., Fatone F., Frison N., Spinelli M. and Eusebi A. L. (2018). Pilot-scale multistage reverse osmosis (DT-RO) for water recovery from landfill leachate. *Waste Management*, **76**, 566–574, https://doi.org/10.1016/j.wasman.2018.03.014

Council of the European Union (1999). Directiva 1999/31/CE del Consejo, de 26 de abril de 1999, relativa al vertido de residuos.

Cui Y.-H., Xue W.-J., Yang S.-Q., Tu J.-L., Guo X.-L. and Liu Z.-Q. (2018). Electrochemical/peroxydisulfate/Fe^{3+} treatment of landfill leachate nanofiltration concentrate after ultrafiltration. *Chemical Engineering Journal*, **353**, 208–217, https://doi.org/10.1016/j.cej.2018.07.101

Dolar D., Košutić K. and Strmecky T. (2016). Hybrid processes for treatment of landfill leachate: coagulation/UF/NF–RO and adsorption/UF/NF–RO. *Separation and Purification Technology*, **168**, 39–46, https://doi.org/10.1016/j.seppur.2016.05.016

Ebrahim S., Bou-Hamed S., Abdel-Jawad M. and Burney N. (1997). Microfiltration system as a pretreatment for RO units: technical and economic assessment. *Desalination*, **109**, 165–175, https://doi.org/10.1016/S0011-9164(97)00062-3

Gkotsis P. K., Batsari E. L., Peleka E. N., Tolkou A. K. and Zouboulis A. I. (2017). Fouling control in a lab-scale MBR system: comparison of several commercially applied coagulants. *Journal of Environmental Management*, **203**, 838–846, https://doi.org/10.1016/j.jenvman.2016.03.003

Gripa E., Campos J. C. and da Fonseca F. V. (2021). Combination of ozonation and microfiltration to condition landfill leachate for reverse osmosis treatment. *Journal of Water Process Engineering*, **43**, 102264, https://doi.org/10.1016/j.jwpe.2021.102264

He R., Tian B.-H., Zhang Q.-Q. and Zhang H.-T. (2015). Effect of Fenton oxidation on biodegradability, biotoxicity and dissolved organic matter distribution of concentrated landfill leachate derived from a membrane process. *Waste Management*, **38**, 232–239, https://doi.org/10.1016/j.wasman.2015.01.006

Hendrych J., Hejralová R., Kroužek J., Špaček P. and Sobek J. (2019). Stabilisation/solidification of landfill leachate concentrate and its residue obtained by partial evaporation. *Waste Management*, **95**, 560–568, https://doi.org/10.1016/j.wasman.2019.06.046

Hou C., Lu G., Zhao L., Yin P. and Zhu L. (2017). Estrogenicity assessment of membrane concentrates from landfill leachate treated by the UV–Fenton process using a human breast carcinoma cell line. *Chemosphere*, **180**, 192–200, https://doi.org/10.1016/j.chemosphere.2017.04.033

Huang Z., Gu Z., Wang Y. and Zhang A. (2019). Improved oxidation of refractory organics in concentrated leachate by a Fe^{2+}-enhanced O_3/H_2O_2 process. *Environmental Science and Pollution Research*, **26**, 35797–35806, https://doi.org/10.1007/s11356-019-06592-y

Hunce S. Y., Akgul D., Demir G. and Mertoglu B. (2012). Solidification/stabilization of landfill leachate concentrate using different aggregate materials. *Waste Management*, **32**, 1394–1400, https://doi.org/10.1016/j.wasman.2012.03.010

Ince M., Senturk E., Onkal Engin G. and Keskinler B. (2010). Further treatment of landfill leachate by nanofiltration and microfiltration–PAC hybrid process. *Desalination*, **255**, 52–60, https://doi.org/10.1016/j.desal.2010.01.017

Iskander S. M., Zou S., Brazil B., Novak J. T. and He Z. (2017). Energy consumption by forward osmosis treatment of landfill leachate for water recovery. *Waste Management*, **63**, 284–291, https://doi.org/10.1016/j.wasman.2017.03.026

Kallel A., Ellouze M. and Trabelsi I. (2017). Co-management of landfill leachate concentrate with brick waste by solidification/stabilization treatment. *Arabian Journal of Geosciences*, **10**, 81, https://doi.org/10.1007/s12517-017-2871-x

Keyikoglu R., Karatas O., Rezania H., Kobya M., Vatanpour V. and Khataee A. (2021). A review on treatment of membrane concentrates generated from landfill leachate treatment processes. *Separation and Purification Technology*, **259**, 118182, https://doi.org/10.1016/j.seppur.2020.118182

Kjeldsen P., Barlaz M. A., Rooker A. P., Baun A., Ledin A. and Christensen T. H. (2002). Present and long-term composition of MSW landfill leachate: a review. *Critical Reviews in Environmental Science and Technology*, **32**, 297–336, https://doi.org/10.1080/10643380290813462

Košutić K., Dolar D. and Strmecky T. (2015). Treatment of landfill leachate by membrane processes of nanofiltration and reverse osmosis. *Desalination and Water Treatment*, **55**, 2680–2689, https://doi.org/10.1080/19443994.2014.939863

Kulikowska D., Zielińska M. and Konopka K. (2019). Treatment of stabilized landfill leachate in an integrated adsorption–fine-ultrafiltration system. *International Journal of Environmental Science and Technology*, **16**, 423–430, https://doi.org/10.1007/s13762-018-1685-z

Kumar M., Badruzzaman M., Adham S. and Oppenheimer J. (2007). Beneficial phosphate recovery from reverse osmosis (RO) concentrate of an integrated membrane system using polymeric ligand exchanger (PLE). *Water Research*, **41**, 2211–2219, https://doi.org/10.1016/j.watres.2007.01.042

Kurniawan T. A., Lo W. and Chan G. Y. (2006). Physico-chemical treatments for removal of recalcitrant contaminants from landfill leachate. *Journal of Hazardous Materials*, **129**, 80–100, https://doi.org/10.1016/j.jhazmat.2005.08.010

Labiadh L., Fernandes A., Ciríaco L., Pacheco M. J., Gadri A., Ammar S. and Lopes A. (2016). Electrochemical treatment of concentrate from reverse osmosis of sanitary landfill leachate. *Journal of Environmental Management*, **181**, 515–521, https://doi.org/10.1016/j.jenvman.2016.06.069

Li Z., Zhou S. and Qiu J. (2007). Combined treatment of landfill leachate by biological and membrane filtration technology. *Environmental Engineering Science*, **24**, 1245–1256, https://doi.org/10.1089/ees.2006.0169

Li F., Wichmann K. and Heine W. (2009). Treatment of the methanogenic landfill leachate with thin open channel reverse osmosis membrane modules. *Waste Management*, **29**, 960–964, https://doi.org/10.1016/j.wasman.2008.06.012

Li W., Zhou Q. and Hua T. (2010). Removal of organic matter from landfill leachate by advanced oxidation processes: a review. *International Journal of Chemical Engineering*, **2010**, e270532–10, https://doi.org/10.1155/2010/270532

Li X., Zhu W., Wu Y., Wang C., Zheng J., Xu K. and Li J. (2015). Recovery of potassium from landfill leachate concentrates using a combination of cation-exchange membrane electrolysis and magnesium potassium phosphate crystallization. *Separation and Purification Technology*, **144**, 1–7, https://doi.org/10.1016/j.seppur.2015.01.035

Li J., Zhao L., Qin L., Tian X., Wang A., Zhou Y., Meng L. and Chen Y. (2016). Removal of refractory organics in nanofiltration concentrates of municipal solid waste leachate treatment plants by combined Fenton oxidative-coagulation with photo-Fenton processes. *Chemosphere*, **146**, 442–449, https://doi.org/10.1016/j.chemosphere.2015.12.069

Linde K., Jönsson A. and Wimmerstedt R. (1995). Treatment of three types of landfill leachate with reverse osmosis. *Desalination*, **101**, 21–30, https://doi.org/10.1016/0011-9164(95)00004-L

Long Y., Xu J., Shen D., Du Y. and Feng H. (2017). Effective removal of contaminants in landfill leachate membrane concentrates by coagulation. *Chemosphere*, **167**, 512–519, https://doi.org/10.1016/j.chemosphere.2016.10.016

Luo H., Cheng Y., He D. and Yang E.-H. (2019). Review of leaching behavior of municipal solid waste incineration (MSWI) ash. *Science of the Total Environment*, **668**, 90–103, https://doi.org/10.1016/j.scitotenv.2019.03.004

Marttinen S. K., Kettunen R. H., Sormunen K. M., Soimasuo R. M. and Rintala J. A. (2002). Screening of physical–chemical methods for removal of organic material, nitrogen and toxicity from low strength landfill leachates. *Chemosphere*, **46**, 851–858, https://doi.org/10.1016/S0045-6535(01)00150-3

Meier J., Remmen K., Wintgens T. and Melin T. (2021). Nanofiltration in landfill leachate treatment. In: Nanofiltration, A. I. Schäefer and A. G. Fane (eds), John Wiley & Sons, Ltd, Germany, pp. 663–690, https://doi.org/10.1002/9783527824984.ch15

Mishra V. (2017). Modeling of batch Sorber system: kinetic, mechanistic, and thermodynamic modeling. *Applied Water Science*, **7**, 3173–3180, https://doi.org/10.1007/s13201-016-0463-7

Mohammad A. W., Hilal N. and Pei L. Y. (2004). Treatment of landfill leachate wastewater by nanofiltration membrane. *International Journal of Green Energy*, **1**, 251–263, https://doi.org/10.1081/GE-120038756

Mohammadesmaeili F., Badr M. K., Abbaszadegan M. and Fox P. (2010). Mineral recovery from inland reverse osmosis concentrate using isothermal evaporation. *Water Research*, **44**, 6021–6030, https://doi.org/10.1016/j.watres.2010.07.070

Mojiri A., Zhou J. L., Ratnaweera H., Ohashi A., Ozaki N., Kindaichi T. and Asakura H. (2020). Treatment of landfill leachate with different techniques: an overview. *Journal of Water Reuse and Desalination*, **11**, 66–96, https://doi.org/10.2166/wrd.2020.079

Moravia W. G., Amaral M. C. S. and Lange L. C. (2013). Evaluation of landfill leachate treatment by advanced oxidative process by Fenton's reagent combined with membrane separation system. *Waste Management*, **33**, 89–101, https://doi.org/10.1016/j.wasman.2012.08.009

Nanda S. and Berruti F. (2021). Municipal solid waste management and landfilling technologies: a review. *Environmental Chemistry Letters*, **19**, 1433–1456, https://doi.org/10.1007/s10311-020-01100-y

Ozturk I., Altinbas M., Koyuncu I., Arikan O. and Gomec-Yangin C. (2003). Advanced physico-chemical treatment experiences on young municipal landfill leachates. *Waste Management*, **23**, 441–446, https://doi.org/10.1016/S0956-053X(03)00061-8

Pearce G. K. (2008). UF/MF pre-treatment to RO in seawater and wastewater reuse applications: a comparison of energy costs. *Desalination*, **222**, 66–73, https://doi.org/10.1016/j.desal.2007.05.029

Peng Y. (2017). Perspectives on technology for landfill leachate treatment. *Arabian Journal of Chemistry*, **10**, S2567–S2574, https://doi.org/10.1016/j.arabjc.2013.09.031

Pertile C., Zanini M., Baldasso C., Andrade M. Z. and Tessaro I. C. (2018). Evaluation of membrane microfiltration fouling in landfill leachate treatment. *Matéria (Rio de Janeiro)*, **23**, https://doi.org/10.1590/s1517-707620170001.0297

Peters T. A. (1998). Purification of landfill leachate with membrane filtration. *Filtration & Separation*, **35**, 33–36, https://doi.org/10.1016/S0015-1882(97)83113-8

Pi K. W., Li Z., Wan D. J. and Gao L. X. (2009). Pretreatment of municipal landfill leachate by a combined process. *Process Safety and Environmental Protection*, **87**, 191–196, https://doi.org/10.1016/j.psep.2009.01.002

Piatkiewicz W., Biemacka E. and Suchecka T. (2001). A Polish study: treating landfill leachate with membranes. *Filtration & Separation*, **38**, 22–26, https://doi.org/10.1016/S0015-1882(01)80377-3

Qiao M., Zhao X. and Wei X. (2018). Characterization and treatment of landfill leachate membrane concentrate by Fe^{2+}/NaClO combined with advanced oxidation processes. *Scientific Reports*, **8**, 12525, https://doi.org/10.1038/s41598-018-30917-5

Ramaswami S., Behrendt J. and Otterpohl R. (2018). Comparison of NF–RO and RO–NF for the treatment of mature landfill leachates: a guide for landfill operators. *Membranes*, **8**, 17, https://doi.org/10.3390/membranes8020017

Rautenbach R. and Mellis R. (1994). Waste water treatment by a combination of bioreactor and nanofiltration. *Desalination*, **95**, 171–188, https://doi.org/10.1016/0011-9164(94)00012-3

Renou S., Givaudan J. G., Poulain S., Dirassouyan F. and Moulin P. (2008). Landfill leachate treatment: review and opportunity. *Journal of Hazardous Materials*, **150**, 468–493, https://doi.org/10.1016/j.jhazmat.2007.09.077

Sabour M. R. and Amiri A. (2017). Comparative study of ANN and RSM for simultaneous optimization of multiple targets in Fenton treatment of landfill leachate. *Waste Management*, **65**, 54–62, https://doi.org/10.1016/j.wasman.2017.03.048

Safarpour M., Vatanpour V., Khataee A., Zarrabi H., Gholami P. and Yekavalangi M. E. (2017). High flux and fouling resistant reverse osmosis membrane modified with plasma treated natural zeolite. *Desalination*, **411**, 89–100, https://doi.org/10.1016/j.desal.2017.02.012

Santos A. V., Andrade L. H. d., Amaral M. C. S. and Lange L. C. (2019). Integration of membrane separation and Fenton processes for sanitary landfill leachate treatment. *Environmental Technology*, **40**, 2897–2905, https://doi.org/10.1080/09593330.2018.1458337

Schwarzbauer J., Heim S., Brinker S. and Littke R. (2002). Occurrence and alteration of organic contaminants in seepage and leakage water from a waste deposit landfill. *Water Research*, **36**, 2275–2287, https://doi.org/10.1016/S0043-1354(01)00452-3

Shafiq M. D. and Yusof M. B. M. (2015). Engineering Towards a Sustainable Future. Penerbit USM, Malaysia.

Singh V., Phuleria H. C. and Chandel M. K. (2020). Estimation of energy recovery potential of sewage sludge in India: waste to watt approach. *Journal of Cleaner Production*, **276**, 122538, https://doi.org/10.1016/j.jclepro.2020.122538

Song J., Li W., Li Y., Mosa A., Wang H. and Jin Y. (2019). Treatment of landfill leachate RO concentration by iron–carbon micro-electrolysis (ICME) coupled with H_2O_2 with emphasis on convex optimization method. *Environmental Pollutants and Bioavailability*, **31**, 49–55, https://doi.org/10.1080/09542299.2018.1552086

Soomro G. S., Qu C., Ren N., Meng S., Li X., Liang D., Zhang S. and Li Y. (2020). Efficient removal of refractory organics in landfill leachate concentrates by electrocoagulation in tandem with simultaneous electro-oxidation and in-situ peroxone. *Environmental Research*, **183**, 109249, https://doi.org/10.1016/j.envres.2020.109249

Sun Y., Sun X. and Zhao Y. (2011). Comparison of semi-aerobic and anaerobic degradation of refuse with recirculation after leachate treatment by aged refuse bioreactor. *Waste Management*, **31**, 1202–1209, https://doi.org/10.1016/j.wasman.2011.01.022

Syzdek A. C. and Ahlert R. C. (1984). Separation of landfill leachate with polymeric ultrafiltration membranes. *Journal of Hazardous Materials*, **9**, 209–220, https://doi.org/10.1016/0304-3894(84)80018-7

Tabet K., Moulin P., Vilomet J. D., Amberto A. and Charbit F. (2002). Purification of landfill leachate with membrane processes: preliminary studies for an industrial plant. *Separation Science and Technology*, **37**, 1041–1063, https://doi.org/10.1081/SS-120002240

Talalaj I. A. (2015). Removal of organic and inorganic compounds from landfill leachate using reverse osmosis. *International Journal of Environmental Science and Technology*, **12**, 2791–2800, https://doi.org/10.1007/s13762-014-0661-5

Teng C., Zhou K., Zhang Z., Peng C. and Chen W. (2020). Elucidating the structural variation of membrane concentrated landfill leachate during Fenton oxidation process using spectroscopic analyses. *Environmental Pollution*, **256**, 113467, https://doi.org/10.1016/j.envpol.2019.113467

Top S., Sekman E., Hoşver S. and Bilgili M. S. (2011). Characterization and electrocaogulative treatment of nanofiltration concentrate of a full-scale landfill leachate treatment plant. *Desalination*, **268**, 158–162, https://doi.org/10.1016/j.desal.2010.10.012

Topal A. D. and Atasoy A. D. (2022). Reverse osmosis treatment system for landfill leachate: operation conditions, advantages and challenges. *Environmental Research & Technology*, **5**, 119–127, https://doi.org/10.35208/ert.1027553

Trebouet D., Schlumpf J. P., Jaouen P., Maleriat J. P. and Quemeneur F. (1999). Effect of operating conditions on the nanofiltration of landfill leachates: pilot-scale studies. *Environmental Technology*, **20**, 587–596, https://doi.org/10.1080/09593332008616853

Trebouet D., Schlumpf J. P., Jaouen P. and Quemeneur F. (2001). Stabilized landfill leachate treatment by combined physicochemical–nanofiltration processes. *Water Research*, **35**, 2935–2942, https://doi.org/10.1016/S0043-1354(01)00005-7

Varank G., Yazici Guvenc S., Dincer K. and Demir A. (2020). Concentrated leachate treatment by electro-Fenton and electro-persulfate processes using central composite design. *International Journal of Environmental Research*, **14**, 439–461, https://doi.org/10.1007/s41742-020-00269-y

Verma M. and Kumar R. N. (2017). Coagulation and electrocoagulation for co-treatment of stabilized landfill leachate and municipal wastewater. *Journal of Water Reuse and Desalination*, **8**, 234–243, https://doi.org/10.2166/wrd.2017.102

Wang D. M., Liu D., Gong Z. J., Tao L. X. and Liu Q. M. (2013). Treatment of landfill leachate reverse osmosis concentrate by Fenton oxidation–coagulation–adsorption. *Science and Technology Engineering*, **13**, 5423–5426.

Wang H., Wang Y., Li X., Sun Y., Wu H. and Chen D. (2016). Removal of humic substances from reverse osmosis (RO) and nanofiltration (NF) concentrated leachate using continuously ozone generation–reaction treatment equipment. *Waste Management*, **56**, 271–279, https://doi.org/10.1016/j.wasman.2016.07.040

Warsinger D. M., Chakraborty S., Tow E. W., Plumlee M. H., Bellona C., Loutatidou S., Karimi L., Mikelonis A. M., Achilli A., Ghassemi A., Padhye L. P., Snyder S. A., Curcio S., Vecitis C. D., Arafat H. A. and Lienhard J. H. (2018). A review of polymeric membranes and processes for potable water reuse. *Progress in Polymer Science*, **81**, 209–237, https://doi.org/10.1016/j.progpolymsci.2018.01.004

Wiszniowski J., Robert D., Surmacz-Gorska J., Miksch K. and Weber J. V. (2006). Landfill leachate treatment methods: a review. *Environmental Chemistry Letters*, **4**, 51–61, https://doi.org/10.1007/s10311-005-0016-z

World Bank (2022). Trends in Solid Waste Management [WWW Document]. URL: https://datatopics.worldbank.org/what-awaste/trends_in_solid_waste_management.html (accessed 10.6.22)

Wu Y., Zhou S., Qin F., Ye X. and Zheng K. (2010). Modeling physical and oxidative removal properties of Fenton process for treatment of landfill leachate using response surface methodology (RSM). *Journal of Hazardous Materials*, **180**, 456–465, https://doi.org/10.1016/j.jhazmat.2010.04.052

Wu C., Chen W., Gu Z. and Li Q. (2021). A review of the characteristics of Fenton and ozonation systems in landfill leachate treatment. *Science of the Total Environment*, **762**, 143131, 10.1016/j.scitotenv.2020.143131

Xingxing Q., Chaojie Z. and Ying Z. (2015). Treatment of Landfill Leachate RO Concentrate by VMD. Presented at the 2015 AASRI International Conference on Circuits and Systems (CAS 2015), Atlantis Press, pp. 13–17, https://doi.org/10.2991/cas-15.2015.4

Xiong C., Li G., Zhang Z., Xia Z., Li J. and Ye H. (2014). Technique for advanced electrochemical oxidation treatment of nanofiltration concentrate of landfill leachate. *Wuhan University Journal of Natural Sciences*, **19**, 355–360, https://doi.org/10.1007/s11859-014-1025-1

Xu Y.-D., Yue D.-B., Zhu Y. and Nie Y.-F. (2006). Fractionation of dissolved organic matter in mature landfill leachate and its recycling by ultrafiltration and evaporation combined processes. *Chemosphere*, **64**, 903–911, https://doi.org/10.1016/j.chemosphere.2006.01.039

Xu J., Long Y., Shen D., Feng H. and Chen T. (2017a). Optimization of Fenton treatment process for degradation of refractory organics in pre-coagulated leachate membrane concentrates. *Journal of Hazardous Materials*, **323**, 674–680, https://doi.org/10.1016/j.jhazmat.2016.10.031

Xu Y., Chen C., Li X., Lin J., Liao Y. and Jin Z. (2017b). Recovery of humic substances from leachate nanofiltration concentrate by a two-stage process of tight ultrafiltration membrane. *Journal of Cleaner Production*, **161**, 84–94, https://doi.org/10.1016/j.jclepro.2017.05.095

Yacubowicz H. and Yacubowicz J. (2005). Nanofiltration: properties and uses. *Filtration & Separation*, **42**, 16–21, https://doi.org/10.1016/S0015-1882(05)70617-0

Yang B.-Q., Yang J.-M., Yang H., Liu Y.-M., Li X.-K., Wang Q.-Z. and Pan X.-J. (2018). Co-bioevaporation treatment of concentrated landfill leachate with addition of food waste. *Biochemical Engineering Journal*, **130**, 76–82, https://doi.org/10.1016/j.bej.2017.11.010

Ye Z.-L., Hong Y., Pan S., Huang Z., Chen S. and Wang W. (2017). Full-scale treatment of landfill leachate by using the mechanical vapor recompression combined with coagulation pretreatment. *Waste Management*, **66**, 88–96, https://doi.org/10.1016/j.wasman.2017.04.026

Zhang Q. Q., Tian B. H., Zhang X., Ghulam A., Fang C.-R. and He R. (2013). Investigation on characteristics of leachate and concentrated leachate in three landfill leachate treatment plants. *Waste Management*, **33**, 2277–2286, https://doi.org/10.1016/j.wasman.2013.07.021

Zhang A., Gu Z., Chen W., Li Q. and Jiang G. (2018). Removal of refractory organic pollutants in reverse-osmosis concentrated leachate by microwave–Fenton process. *Environmental Science and Pollution Research*, **25**, 28907–28916, https://doi.org/10.1007/s11356-018-2900-7

Zhao J., Ouyang F., Yang Y. and Tang W. (2020). Degradation of recalcitrant organics in nanofiltration concentrate from biologically pretreated landfill leachate by ultraviolet–Fenton method. *Separation and Purification Technology*, **235**, 116076, https://doi.org/10.1016/j.seppur.2019.116076

Zhou D., Zhu L., Fu Y., Zhu M. and Xue L. (2015a). Development of lower cost seawater desalination processes using nanofiltration technologies – a review. *Desalination*, **376**, 109–116, https://doi.org/10.1016/j.desal.2015.08.020

Zhou X., Zheng Y., Zhou J. and Zhou S. (2015b). Degradation kinetics of photoelectrocatalysis on landfill leachate using codoped TiO_2/Ti photoelectrodes. *Journal of Nanomaterials*, **2015**, e810579–11, 10.1155/2015/810579

Zhou B., Yu Z., Wei Q., Long H., Xie Y. and Wang Y. (2016). Electrochemical oxidation of biological pretreated and membrane separated landfill leachate concentrates on boron doped diamond anode. *Applied Surface Science*, **377**, 406–415, https://doi.org/10.1016/j.apsusc.2016.03.045

Zhu W.-P., Sun S.-P., Gao J., Fu F.-J. and Chung T.-S. (2014). Dual-layer polybenzimidazole/polyethersulfone (PBI/PES) nanofiltration (NF) hollow fiber membranes for heavy metals removal from wastewater. *Journal of Membrane Science*, **456**, 117–127, https://doi.org/10.1016/j.memsci.2014.01.001

doi: 10.2166/9781789063318_0327

Chapter 14

Modeling and optimization of hybrid leachate treatment processes and scale-up of the process

Fabiano Luiz Naves[1], Cristiane Medina Finzi Quintão[1], Mateus de Souza Amaral[2], Renata Carolina Zanetti Lofrano[1], Alexandre Boscaro França[1], José Izaquiel Santos da Silva[3], Clarissa Nascimento de Oliveira[3] and Edilailsa Januário de Melo[4]*

[1]Department in Chemical Engineering, Federal University of Sao João Del Rei, Ouro Branco, Research Group on Waste Treatment and Management Processes, Ouro Branco, MG, Brazil
[2]Laboratory of Technological and Environmental Chemistry, Department of Chemistry, Institute of Exact and Biological Sciences, Federal University of Ouro Preto, Ouro Preto, MG, Brazil
[3]Institute of Science and Technology (ICT), Federal University of the Jequitinhonha and Mucuri Valleys (UFVJM), Diamantina, MG, Brazil
[4]Faculty of Chemical Engineering (FEQ), UNICAMP, Campinas, SP, Brazil
**Corresponding author: fabianonavesengenheiro@ufsj.edu.br*

ABSTRACT

Optimized models of landfill leachate processes that could lead to the production of fuel oils from the pyrolysis of plastics and cultivation of microalgae can be highly efficient and have sustainable activities. Studies using advanced oxidation processes in combination with biological processes showed to be efficient in treating leachate. The complexity of this type of waste, as well as the continuous modification of its properties due to local seasonality, prevents kinetic and thermodynamic studies to predict the steady state at the final disposal. Environmental management systems using Lean Six Sigma accompanied by process modeling and optimization tools, whether statistical or computational, generate reliable models. These models can reproduce reductions in contaminants by confirming the initial conditions of each leachate enabling the scale-up of the process. This chapter provides data survey, types of treatment, and models linked to optimization tools in wastewater decontamination processes from municipal solid waste. Also, this chapter includes the synthesis and proposal for treatment and management of the classification of garbage until the generation of leachate, aiming to contribute to future research in the area of leachate treatment.

Keywords: Leachate, NBI, modeling, optimization.

14.1 INTRODUCTION

The goal to achieve the steady state of a process may be more important than the removal of organic load from the leachate treatment. This indicates that the disposal parameters can be properly controlled over time (Naves, 2019). A modeling and optimization process must be contextualized according to Figure 14.1.

The optimization of the leachate treatment process must be modeled robustly, considering all or most of the process variables in a range of levels that can represent their respective changes in different seasonal periods. There is a significant increase in the recalcitrant organic load with time in landfills. This increase, as already demonstrated in the literature, completely changes the kinetics of the treatment. On certain seasonal periods, leachate can have very specific characteristics even in very short time intervals, alternating in properties from high-organic loads in dry seasons and low-organic loads in periods of heavy rainfall. It is underestimated that in high-organic loads, there is a need for higher concentrations of oxidants to remove organic matter. The low-organic load of the leachate dilution caused by greater rainfall requires a lower concentration of oxidants. There are currently no models that can predict these changes.

Hamid *et al.* (2020) used an electrocoagulation process to remove NH_3-N from leachate. To evaluate the levels and factors (zeolite dosage, current density, electrolysis time, and pH), a response surface was used to create a model of removal of NH_3-N. All factors were significant and consequently can change the response when the levels are adjusted under different conditions. The model was convex in terms of its convexity, and a point of maximum removal of NH_3-N was easily detected. The optimization using desirability algorithm determined the optimum levels of each of the factors: content of zeolite 105 g,

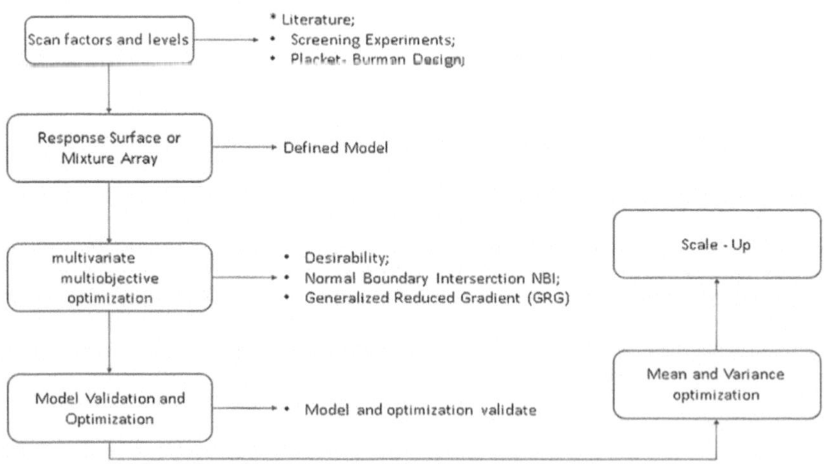

Figure 14.1 Flowchart for process modeling and process optimization.

current density 600 A/m^2, duration of the electrolytic process 60 min, and pH 8.2. A removal of 72.51% of NH_3-N was obtained under these conditions.

Wang *et al.* (2016) used a statistical model based on the Box–Behnken arrangement, to evaluate chemical oxygen demand (COD) removal and the biological oxygen demand (BOD)/COD ratio during treatment using a micro-electrolysis Fenton process. The factors studied were Fe–C dosage, H_2O_2 concentration, and initial pH; the latter showing no significance to the model for both responses. The 'optimization' of the process was carried out through the peaks related to the response surface, which could lead to erroneous results. The bi-objective problem to reduce COD and increase the BOD/COD ratio is not simply trivial, and leads to trade-off responses. These responses can be spelled out as an inverse relationship of increased COD and decreased BOD/COD ratio. A pair of responses was stated in an optimization problem using the normal boundary intersection (NBI) algorithm; it would be possible to build a Pareto frontier based on the simultaneous optimization of both responses. From the frontier, it would be possible to choose setups that represent association of different levels that lead to strategic results of COD reduction or increase in BOD/COD.

Aziz *et al.* (2011b) studied a leachate treatment process using a reactor with activated carbon for 5.5 h and reported a removal of 64.1% of COD, 71.2% of color, 81.4% of NH_3-N, and 33% of total dissolved solid. The factors monitored for the construction of a quadratic model from a response surface were aeration rate and contact time. The responses were optimized individually, but it was not possible to determine which levels of each factor led to the best responses.

In Gopikumar *et al.* (2021), Arduino-computer-server-sensor sequence was used to control the properties of the landfill leachate treatment, using pre-treatment with ultrasound and later with an upflow anaerobic sludge blanket (UASB) reactor. Turbidity, dissolved oxygen, and liquid level sensors were used. On the contrary, leachate entry patterns were not considered, and may vary over different seasonal periods, which consequently would not have been previously programmed in the dynamic control system.

Mahtab *et al.* (2020), using a response surface, modeled the Fenton process in the treatment of landfill leachate, with fixed entry conditions, obtaining 61% of COD removal in 36 min. Under practically the same experimental conditions as Mahtab *et al.* (2020), but with an even shorter reaction time and a much higher initial organic load of leachate of 17,988 mg O_2/L, Zhang *et al.* (2005) studied the removal of 87% COD through the Fenton process; however, the difference between COD removal techniques is completely unclear in both studies. Venkatesh Reddy *et al.* (2020) studied the behavior of factors such as pH, COD, BOD, and solids influenced by volumetric precipitation. In that study, it is possible to note that the pH undergoes a slight increase with an increase in rainfall, characterized by a decrease in volatile acids. The other variables analyzed during the period of volumetric precipitation were characterized by a behavior with a high Pearson correlation. This correlation demonstrates that based on the determination of one of the parameters, the others can be estimated and, controlled, during the treatment process.

Some researchers have used techniques of artificial neural networks (ANN), aiming precisely to understand the relationships between the inputs and outputs of the process (Bunsan *et al.*, 2013). One of the great difficulties of the application of ANN is that they must be trained to simulate such correlations between the independent variables (Jalili Ghazi Zade & Noori, 2008). The use of models that can directly predict the leachate treatment process in an experimental way may be more viable, because it is optimized in a multivariate and multiobjective way (Pinto *et al.*, 2019).

Applied models can represent the treatment process but the need for a more robust modeling is justified when different properties of analyzed responses have to be assessed. The use of algorithms to optimize the models becomes essential to model several setups referring to different initial leachate conditions.

14.2 OPTIMIZATION OF LEACHATE TREATMENT MODELS

The high complexity of the composition of real effluents such as landfill leachate requires robust models to predict possible changes in the environment, which would enable the adaptability of the treatment process considering the levels and factors from the desired responses. Most industrial controls are programmed to obtain stability from a defined set point. Several studies in the literature showed that optimization is carried out only for a possibly stationary regime, where the initial leachate properties do not vary. The properties of real effluents are constantly changing. Even if the treatment plant is operating in a satisfactory way, it can lead to undesirable values of the desired responses at any moment.

Initial parameters of the leachate can be characterized in optimization problems by using a multivariate context and the use of algorithms that can predict, according to the leachate's properties, the chemical and physical conditions related to a certain type of treatment that may lead responses within the ranges established by regulatory bodies.

One of the most robust methods for multivariate and multiobjective analysis in environmental processes is the NBI algorithm (Naves, 2019), developed by Das and Dennis (1998). Some authors have applied this tool for modeling environmental processes (Braga *et al.*, 2020). Paixão *et al.* (2019) studied the optimization of a black eriochrome T dye treatment process. This type of dye has some similarities with the landfill leachate. It uses advanced oxidation treatment based on molecular ozone; the color removal is high when compared to the removal of COD. This is based on the characteristic of the steric effect and recalcitrance of the molecules. This study showed a possible improvement in the removal of COD to 90% when both Pareto boundary and NBI algorithm were used.

This method favors the construction of the Pareto frontier from two points considered by Nadir and Utopia (Das & Dennis, 1998). Once the variation of properties in a landfill leachate within a range with their respective deviations is known, a boundary can be built and that can represent these properties under sufficient experimental conditions to remove the environmental parameters.

Researchers have commonly selected some variables of importance for study, as well as the initial conditions. Bakraouy *et al.* (2017) using a coagulation and flocculation process using ferric chloride together with a coagulant studied the removal of the percentage of phenol, color, turbidity, COD, and abs 254. The initial organic loads of the leachate in terms of COD and BOD were 11,520 and 6,710 mg/L, respectively, yielding a BOD/COD ratio of 0.58. This ratio would provide a biological treatment to remove organic matter at the time of collection to carry out the experiments. At the time of operation of the landfill, there will be a considerable reduction in the BOD/COD ratio, which would directly influence the efficiency of a possible biological treatment. The current efficiency for removing up to 96% of COD, by coagulation and flocculation, could be substantially reduced.

Tawakkoly *et al.* (2019) also used a coagulation process to verify COD removal. The coagulant was the Hispanic Salvia extract. The COD reduction for this study was on average 39.76% with an initial COD of 62,500 mg/L and BOD of 22,500 mg/L. Although COD removal is less when compared to Bakraouy *et al.* (2017), the BOD/COD ratio is much lower with an average value of 0.36. Mohajeri *et al.* (2010) using an electro Fenton process in the treatment of leachate with an initial COD of 2,950 mg/L, obtained a COD removal of 94.07% after 43 min. On the basis of the COD value, it would be possible to propose a biological treatment, which would considerably decrease the cost of application on larger scales (Oloibiri *et al.*, 2017).

COD and BOD as initial factors, and the levels of each of the factors studied, pH, coagulant and flocculant concentration, Fe^{2+}/H_2O_2 ratio, ozone concentration, treatment time among others could be determined for various dilutions other than the leachate. These dilutions could temporarily represent the quality of the leachate. Multiobjective problem could be described according to equation (14.1) for the optimization problem:

$$\begin{aligned} &\text{Min COD, BOD} \\ &\text{st, } \frac{\text{BOD}}{\text{COD}} > 0.33 \end{aligned} \tag{14.1}$$

The above equation referring to the optimization of the COD and BOD removal process, aims to establish the disposal parameters.

Leachate dilutions can be simulated under various leachate conditions throughout the landfill operation. The solutions will be linked to the disposal parameters of each country linked to regulatory bodies. Paixão *et al.* (2019) used an NBI algorithm to predict, from the Pareto boundary, the volume of a reactor for the treatment of black dye eriochrome. It is always possible to predict a boundary for a bi-objective problem or a Pareto surface for a number of responses greater than two, which usually occurs in an environmental problem at real scales. The Pareto surface draws the best curve that represents the set of points referring to the levels of optimized responses within a pre-specified pattern.

14.3 PROPOSED OPTIMIZATION USING NBI

This chapter proposes a further application of the technological frontier in subsequent publications. Table 14.1 presents the data referring to the studies already published on the theme, which contains the data on the characterization of the leachate.

Table 14.1 Studies that optimized leachate treatment processes.

Reference	Studied Parameters				
	COD (mg/L)	BOD (mg/L)	BOD/ COD	$N\text{-}NH_3$ (mg/L)	pH
Braga *et al.* (2020)	860	116	0.13		8.3
Gomes *et al.* (2019)	5,581	190	0.033	1,771	8.1
Moreno-Garcia *et al.* (2019)				0.74	
Amirian *et al.* (2017)	4,083.5	1,960	0.49		
Aziz *et al.* (2011a)	400–2,860	50–1,024	0.09–0.4	400–1,039	6.93–8.31
Yilmaz *et al.* (2010)	38,200	22,000	0.58		7.25
Amiri and Sabour (2014)	11,260	2,930	0.26		7.21
Bashir *et al.* (2010)	2,615	100	0.038	2,010	8.7
Wang *et al.* (2016)	4,980	548	0.11	1,850	7.52
Farrokhi *et al.* (2015)	19,840	180	0.009	1,750	6.8
Amr *et al.* (2013)	1,780	107	0.06	780	8.6
Liu *et al.* (2020)	1,200	350	0.29	125	5.5
Zhao *et al.* (2014)	900			1,381	8.4
Saranga *et al.* (2020)	20,000				7.5
Adlan *et al.* (2011)	2,610			1,975	8.13
Aziz *et al.* (2011b)	1,655	373	0.218	600	7.87
Yeh *et al.* (2018)	233	29.7	0.127	325	7.4
Azadi *et al.* (2018)	678	130	0.19	15	8.7
Zhou *et al.* (2017)	4,378	29.1	0.01		7.8
Hilles and Abu Amr (2016)	20,000	1,821	0.09	2,487	8.42
Shaylinda *et al.* (2018)	770	109	0.14	500	8.1
Lessoued *et al.* (2017)	4,219	1,100	0.26		8.7
Cheng *et al.* (2020)	1,020			720	8.5
Bakraouy *et al.* (2017)	11,520	6,710	0.58		8.4
Jegan Durai *et al.* (2020)	2,990	135	0.045		8.8
Tripathy *et al.* (2019)	2,240	80	0.036	2,293	6.61
Huda *et al.* (2017)	7,230	540	0.075		8.43
Mohd-Salleh *et al.* (2018)	2,954	166.89	0.058	920.4	8.76
Sabour and Amiri (2017)	11,250	2,810	0.25		7.17
Azmi *et al.* (2015)	2,740	193.2	0.07	1,113.2	8.3
Kabuk *et al.* (2014)	9,800			3,710	8.05

The survey showed that the initial leachate parameters studied by several researchers are indeed different; it may have a direct relationship with the time of operation of the landfill. Figures 14.2 and 14.3 present the catalogued data in graphic terms.

Figures 14.2 and 14.3 present a trend that may predict the considerable variation for the initial leachate parameters studied by researchers. The levels of the factors presented can initially be used to predict intervals for each of the factors used in the treatment of leachate. Figures 14.4–14.6 present these intervals as well as existing outliers in some of the factors.

Despite the pH, BOD, and COD parameters having an outlier, it is possible to predict the range values of each parameter very accurately and are shown in Table 14.2.

The listed parameters are the factors with respective input levels and response times which must be controlled after treatment. An experimental model composed of a response surface could be used, and later optimized. A considerable range of levels of the associated factors can generate a robust model, applicable in any landfill. The construction of a single-objective function could optimize both responses. In this case, NBI algorithm could be used, which would generate a point cloud that could be better represented by a trend line as shown in Figure 14.7.

Each point at the frontier presents an optimal point, corresponding to a certain experimental condition regarding the levels of the input factors. For example, a given point has an initial leachate value of pH, COD, BOD, BOD/COD, and N-NH_3, which optimizes the output responses.

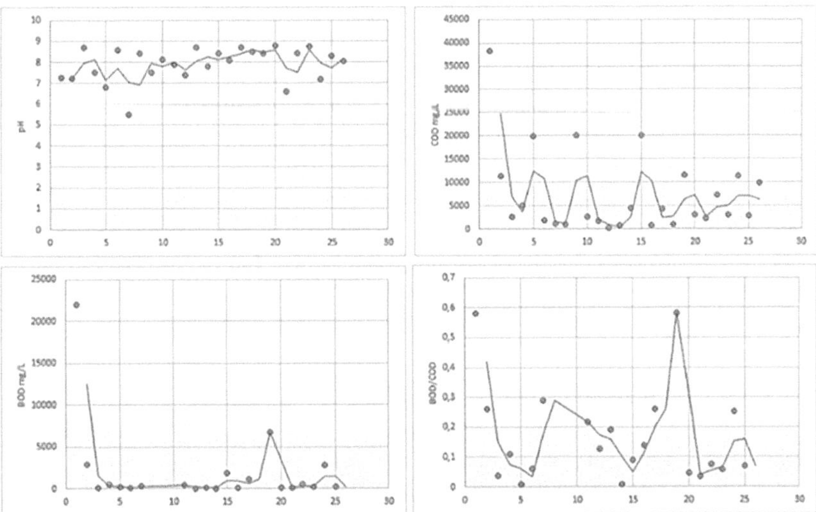

Figure 14.2 Behavior of leachate parameters most studied by researchers (pH, hydrogenionic potential; BOD, biochemical oxygen demand; COD, chemical oxygen demand; BOD/COD, recalcitrance ration).

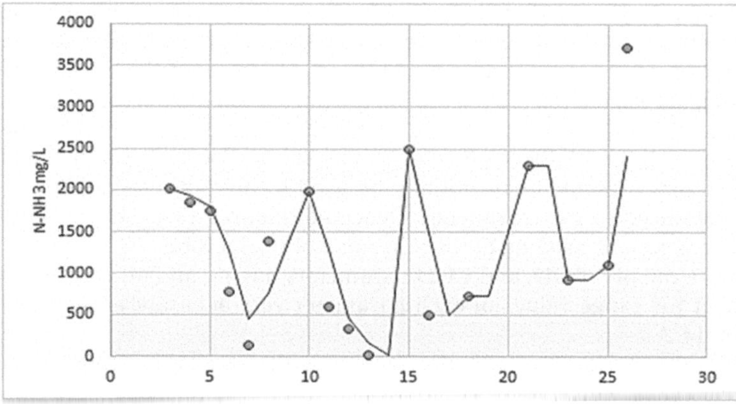

Figure 14.3 Oscillation of ammoniacal nitrogen concentration from historical data in landfill leachate (N-NH_3, ammonia nitrogen concentration).

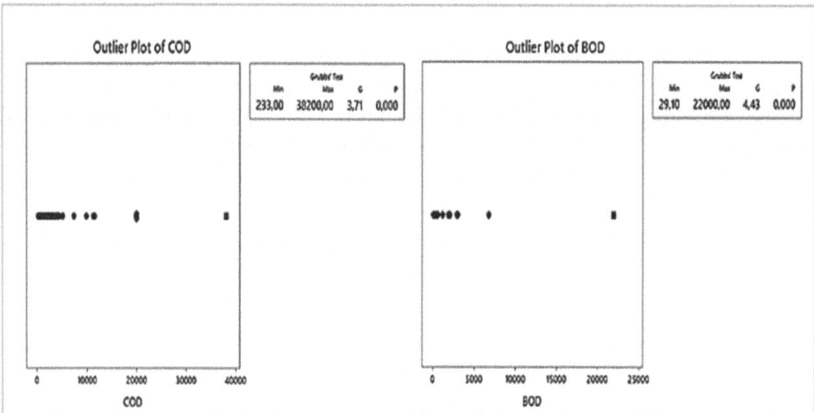

Figure 14.4 Presentation of intervals and outlier verification at the levels of factors established by selected researchers BOD and COD data (BOD, biochemical oxygen demand; COD, chemical oxygen demand; G, test of grubber; P, P-value).

14.4 IMPLEMENTATION OF NBI IN ENVIRONMENTAL CONTROLS

The NBI generally represents a function that simultaneously optimizes both desired responses. Thus, any point on the frontier corresponds to a different process setup within an optimal region. The manual adjustment of input parameters can lead to inaccurate adjustments. For example, Amiri and Sabour (2014) used response surface models to study COD removal and iron and organic compounds present in sludge. The Fenton process analyzed the factors: iron peroxide ratio, pH, and Fe^{2+} dosage. The leachate was characterized with a

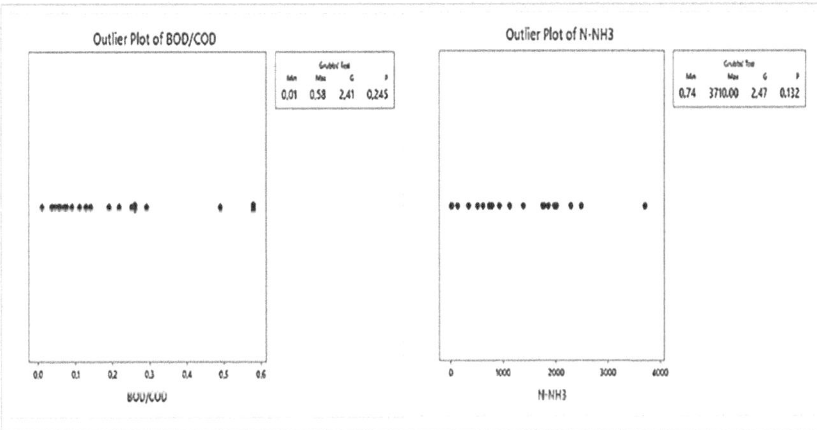

Figure 14.5 Presentation of intervals and outlier verification at the levels of factors established by selected researchers BOD/COD and $N\text{-}NH_3$ data (BOD, biochemical oxygen demand; COD, chemical oxygen demand; G, test of grubber; P, P-value).

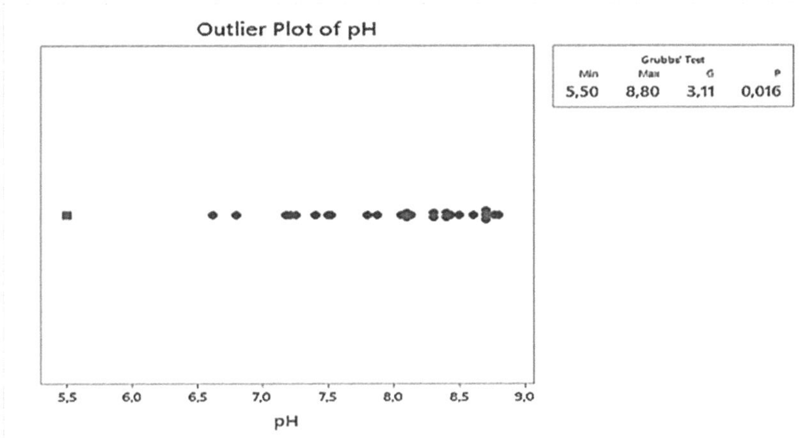

Figure 14.6 Presentation of intervals and outlier verification at the levels of pH established by selected researchers (G, grubber test; Min, minimo; Max, maximum; P, P-value).

COD of 11,260 mg/L, pH 7.21, BOD of 2,930 mg/L, and turbidity of 125 NTU. Both COD and BOD directly interfere in the effect of the treatment and final disposal of the leachate; this demonstrates the optimization from pairs of responses.

The use of NBI from the models created by the response surface optimizes both responses and builds a single model that would represent the overall process. Each point from the Pareto frontier determines the optimal levels of

Table 14.2 Limits of each parameter.

Parameters	Minimum	Medium	Maximum
pH	6.61	7.7	8.8
BOD (mg/L)	29.10	3,369.6	6,710
COD (mg/L)	233	10,036.5	19,840
N-NH_3 (mg/L)	0.74	1,855.4	3,710
BOD/COD	0.01	0.3	0.58

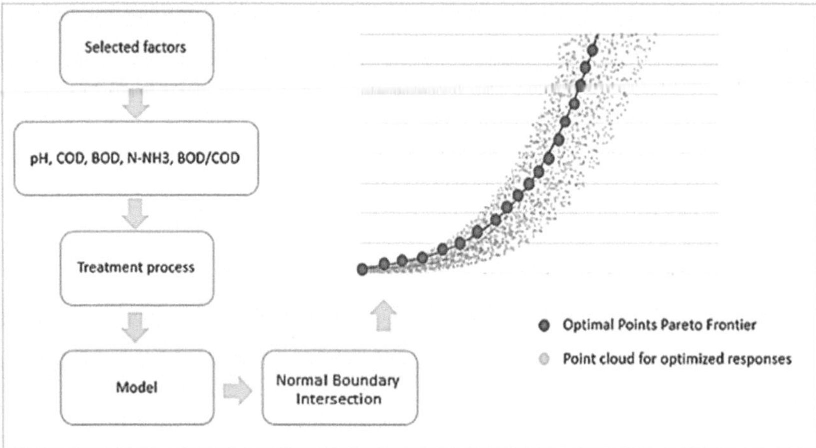

Figure 14.7 Presentation of the proposal for multivariate and multiobjective optimization in the treatment of leachate using normal boundary intersection.

factors, such as pH, $[H_2O_2]/[Fe^{2+}]$ ratio, and Fe^{2+} dosage. This frontier would represent only the treatment process for initial leachate conditions as presented in this study. Changes in the leachate's properties over time could reflect on different levels to reach the same responses. However, if the researcher could have also used COD and BOD values from the studied landfills, the change in COD and BOD could be predicted by adjusting the pH parameters, $[H_2O_2]/[Fe^{2+}]$ ratio, and Fe^{2+} dosage. This could supply and maintain the desired response.

The incorporation of the optimized function through the NBI in programmable logic controllers leads to several set points, levels of the input factors in the leachate treatment process, which generally does not occur with this type of specific set point instrumentation. As the nature of the leachate is extremely dynamic, this controller could act under the optimal points of the Pareto frontier, and, the natural variations of the leachate would require changes in the process parameters such as pH, $[H_2O_2]/[Fe^{2+}]$ ratio, and Fe^{2+} dosage.

The proposal presented in this study considers a leachate treatment that adopts the following characteristics:

- Treatment process – ozonation;
- Reactor type – rotating packed bed (RPB);
- RPB entry factors – pH, reaction medium, ozone flow, reactor rotation frequency;
- Leachate entry factors (controlled by dilution) – COD, BOD, pH;
- Modeling tool – surface response central composite design;
- Responses analyzed after treatment – COD, BOD, TOC, N-NH_3;
- Optimization tool – NBI algorithm.

The proposed ozonation process was chosen due to a better color removal capacity and increased BOD/COD ratio (Braga *et al.*, 2020). An amount of leachate can be introduced to the RPB reactor, which has the ability to increase the diffusion of ozone to the medium, increases the formation of hydroxyl radicals; minimize reaction time and operating costs. Controllable factors of the model linked to the reactor, and at the same time factors of the leachate can be controlled within certain levels of dilutions and pH correction using standard solutions of sulfuric acid and sodium hydroxide. It would take 90 experiments to validate a second-order model with 10 central points and only a single block. This model, after validation, could be applied to any leachate with any input characteristic.

The modeling and optimization using the NBI algorithm would result in a solution to the optimization problem so that the variance determined from the standard error calculated by the ordinary least square was minimized and the removal of COD, BOD, TOC, and N-NH_3 was maximized (Naves, 2019). The application of the NBI algorithm leads to the construction of the Pareto frontier (Figure 11), a control parameter in the treatment of leachate.

There are two possibilities: to use regression and determine a polynomial function or use the point cloud itself in controllers. The proposal would be schematically similar to Figure 14.8.

Gupta and Paulraj (2017) evaluated the properties of the landfill leachate as well as their interactions. The COD and BOD parameters presented a Pearson correlation of 81.7%. As for this proposal, the initial treatment of leachate must be at a BOD/COD ratio above 0.33 so that the subsequent biological process of cultivating microalgae is not impaired. A colorimetric sensor for COD identification can be used to predict other proposed parameters. A simplified form of NBI controller block diagram is shown in Figure 14.9.

The great versatility of the NBI control system, as shown in Figure 14.9, implements process models within countless boundary conditions of the input factors, built by response surface statistical arrangements. It allows multiobjective optimization to the detriment of numerous set points associated with the Pareto boundary or surface. The boundary conditions of the factors associated with the entry of the model relevant to the leachate can be applied at any time of operation of the landfill.

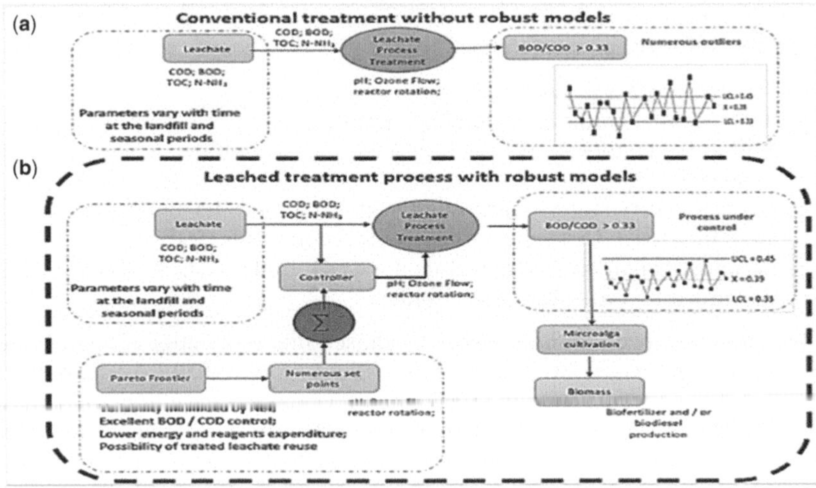

Figure 14.8 Proposed leachate treatment using robust models with minimization of associated variance and improvement of the proposed advanced oxidation treatment (pH, hydrogenionic potential; BOD, biochemical oxygen demand; COD, chemical oxygen demand; BOD/COD, recalcitrance ration; TOC, organic carbon total).

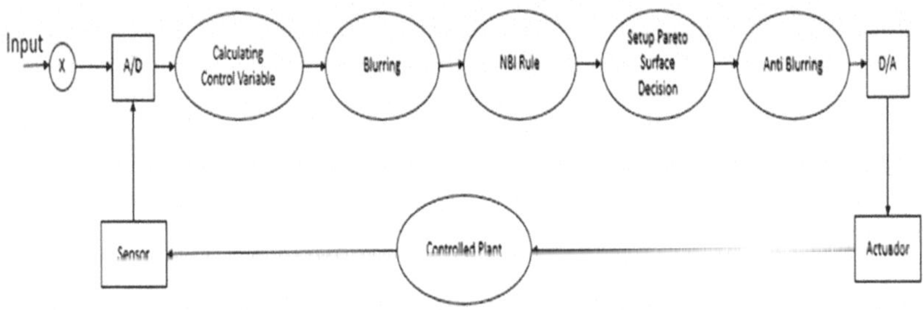

Figure 14.9 Block diagram of NBI controller.

14.5 SUSTAINABLE IMPLICATIONS IN LEACHATE TREATMENT

The application of modeling and optimization techniques to treat landfill leachate, a worldwide concern, as proposed in this chapter, leads to more effective results in reducing organic and inorganic load, in a more sustainable way. In the general context of the process shown in Figure 14.1, the destination of plastic waste for energy generation with bio oil production can lead to considerable reductions in CO_2 released into the atmosphere (Joshi & Seay, 2020). These reductions in CO_2 in the atmosphere can be considerably improved by applying models built from pyrolytic processes (Zhang *et al.*, 2020).

The nutrients nitrogen and phosphorus are difficult to remove from the leachate by advanced oxidation processes. In contrast, the organic load can be easily removed by advanced oxidation processes, increasing the biodegradability of the leachate. The control of the leachate's biodegradability can be applied to the cultivation of microalgae, which may be able to use nutrients that are difficult to remove using advanced oxidation processes. The bio oil generated from microalgae may, from optimized models, be used in machines, which could reduce the use of fossil fuels (Bhuana *et al.*, 2020). The reduction of CO_2 released into the atmosphere would be considerably reduced by adding value to the waste.

Ordinary diesel from fossil fuels generates 96,440 kg of CO_2/MJ of energy from the fuel. Fuels generated from pyrolysis of recyclable plastics generate an average of 20% less carbon dioxide than those released into the atmosphere (Joshi & Seay, 2019).

14.6 CONCLUSION

This chapter proposes the use of hybrid treatment processes to yield better results when they are associated with the construction of models and optimization using management tools. The characteristic of landfill leachate is quite complex and at the same time unstable. The need for models that predict these changes as input factors and, at the same time, the responses to be evaluated, can lead to better process generalizations with possible application in any type of landfill. These models can be subsequently optimized using the NBI algorithm. The transfer function can be used in controllers that might maintain the leachate treatment process, even under unstable conditions, within specification limits defined by the environmental agencies. The addition of value to the waste by applying management tools for the generation of biodiesel and bio oil is supported by a sustainability and social aspect.

ACKNOWLEDGMENTS

The authors acknowledge the Fundação de Amparo a Pesquisa de Minas Gerais FAPEMIG and the Federal University of São João del Rei – UFSJ and Group on Waste Treatment and Management Processes.

REFERENCES

Adlan M. N., Palaniandy P. and Aziz H. A. (2011). Optimization of coagulation and dissolved air flotation (DAF) treatment of semi-aerobic landfill leachate using response surface methodology (RSM). *Desalination*, **277**, 74–82, https://doi.org/10.1016/j.desal.2011.04.006

Amiri A. and Sabour M. R. (2014). Multi-response optimization of Fenton process for applicability assessment in landfill leachate treatment. *Waste Management*, **34**, 2528–2536, https://doi.org/10.1016/j.wasman.2014.08.010

Amirian P., Bazrafshan E. and Payandeh A. (2017). Optimisation of chemical oxygen demand removal from landfill leachate by sonocatalytic degradation in the presence of cupric oxide nanoparticles. *Waste Management & Research*, **35**, 636–646, https://doi.org/10.1177/0734242X17704715

Amr S. S. A., Aziz H. A., Adlan M. N. and Bashir M. J. K. (2013). Optimization of semi-aerobic stabilized leachate treatment using ozone/Fenton's reagent in the advanced oxidation process. *Journal of Environmental Science and Health, Part A: Toxic/Hazardous Substances and Environmental Engineering*, **48**, 720–729, https://doi.org/10.1080/10934529.2013.744611

Azadi S., Karimi-Jashni A. and Javadpour S. (2018). Modeling and optimization of photocatalytic treatment of landfill leachate using tungsten-doped TiO_2 nano-photocatalysts: application of artificial neural network and genetic algorithm. *Process Safety and Environmental Protection*, **117**, 267–277, https://doi.org/10.1016/j.psep.2018.03.038

Aziz S. Q., Aziz H. A. and Yusoff M. S. (2011a). Optimum process parameters for the treatment of landfill leachate using powdered activated carbon augmented sequencing batch reactor (SBR) technology. *Separation Science and Technology*, **46**, 2348–2359, https://doi.org/10.1080/01496395.2011.595753

Aziz S. Q., Aziz H. A., Yusoff M. S. and Bashir M. J. K. (2011b). Landfill leachate treatment using powdered activated carbon augmented sequencing batch reactor (SBR) process: optimization by response surface methodology. *Journal of Hazardous Materials*, **189**, 404–413, https://doi.org/10.1016/j.jhazmat.2011.02.052

Azmi N. B., Bashir M. J. K., Sethupathi S., Wei L. J. and Aun N. C. (2015). Stabilized landfill leachate treatment by sugarcane bagasse derived activated carbon for removal of color, COD and NH_3-N – optimization of preparation conditions by RSM. *Journal of Environmental Chemical Engineering*, **3**, 1287–1294, https://doi.org/10.1016/j.jece.2014.12.002

Bakraouy H., Souabi S., Digua K., Dkhissi O., Sabar M. and Fadil M. (2017). Optimization of the treatment of an anaerobic pretreated landfill leachate by a coagulation–flocculation process using experimental design methodology. *Process Safety and Environmental Protection*, **109**, 621–630, https://doi.org/10.1016/j.psep.2017.04.017

Bashir M. J. K., Aziz H. A., Yusoff M. S., Aziz S. Q. and Mohajeri S. (2010). Stabilized sanitary landfill leachate treatment using anionic resin: treatment optimization by response surface methodology. *Journal of Hazardous Materials*, **182**, 115–122, https://doi.org/10.1016/j.jhazmat.2010.06.005

Bhuana D. S., Qadariyah L., Panjaitan R. and Mahfud M. (2020). Optimization of biodiesel production from *Chlorella* sp. through in-situ microwave-assisted acid-catalyzed trans-esterification. *IOP Conference Series: Materials Science and Engineering*, **732**, 012004, https://doi.org/10.1088/1757-899X/732/1/012004

Braga W. L. M., de Melo D. H. A., de Morais D., Samanamud G. R. L., França A. B., Finzi Quintão C. M., Loures C. C. A., de Urzedo A. P. F. M., Naves L. L. R., de Freitas Gomes J. H. and Naves F. L. (2020). Optimization of the treatment of sanitary landfill by the ozonization catalysed by modified nanovermiculite in a rotating packed bed. *Journal of Cleaner Production*, **249**, 119395, https://doi.org/10.1016/j.jclepro.2019.119395

Bunsan S., Chen W. Y., Chen H. W., Chuang Y. H. and Grisdanurak N. (2013). Modeling the dioxin emission of a municipal solid waste incinerator using neural networks. *Chemosphere*, **92**, 258–264, https://doi.org/10.1016/j.chemosphere.2013.01.083

Cheng S. Y., Show P. L., Juan J. C., Ling T. C., Lau B. F., Lai S. H. and Ng E. P. (2020). Sustainable landfill leachate treatment: optimize use of guar gum as natural coagulant and floc characterization. *Environmental Research*, **188**, 109737, https://doi.org/10.1016/j.envres.2020.109737

Das I. and Dennis J. E. (1998). Normal-boundary intersection: a new method for generating the Pareto surface in nonlinear multicriteria optimization problems. *SIAM Journal on Optimization*, **8**, 631–657, https://doi.org/10.1137/S1052623496307510

Farrokhi M., Dindarloo K. and Jamali H. A. (2015). Optimization of Fenton oxidation process for biodegradability enhancement of a mature landfill leachate using response surface methodology. *Research Journal of Environmental Toxicology*, **9**, 274–289, https://doi.org/10.3923/rjet.2015.274.289

Gomes A. I., Santos S. G. S., Silva T. F. C. V., Boaventura R. A. R. and Vilar V. J. P. (2019). Treatment train for mature landfill leachates: optimization studies. *Science of the Total Environment*, **673**, 470–479, https://doi.org/10.1016/j.scitotenv.2019.04.027

Gopikumar S., Raja S., Robinson Y. H., Shanmuganathan V., Chang H. and Rho S. (2021). A method of landfill leachate management using internet of things for sustainable smart city development. *Sustainable Cities and Society*, **66**, 102521, https://doi.org/10.1016/j.scs.2020.102521

Gupta A. and Paulraj R. (2017). Leachate composition and toxicity assessment: an integrated approach correlating physicochemical parameters and toxicity of leachates from MSW landfill in Delhi. *Environmental Technology*, **38**, 1599–1605, https://doi.org/10.1080/09593330.2016.1238515

Hamid M. A. A., Aziz H. A., Yusoff M. S. and Rezan S. A. (2020). Optimization and analysis of zeolite augmented electrocoagulation process in the reduction of high-strength ammonia in saline landfill leachate. *Water (Switzerland)*, **12**, 247, https://doi.org/10.3390/w12010247

Hilles A. H. and Abu Amr S. S. (2016). Factorial design and optimization of leachate treatment using persulfate oxidation. *Global NEST Journal*, **18**, 842–854, https://doi.org/10.30955/gnj.002035

Huda N., Raman A. A. A., Bello M. M. and Ramesh S. (2017). Electrocoagulation treatment of raw landfill leachate using iron-based electrodes: effects of process parameters and optimization. *Journal of Environmental Management*, **204**, 75–81, https://doi.org/10.1016/j.jenvman.2017.08.028

Jalili Ghazi Zade M. and Noori R. (2008). Prediction of municipal solid waste generation by use of artificial neural network: a case study of Mashhad. *International Journal of Environmental Research*, **2**, 13–22.

Jegan Durai N., Gopalakrishna G. V. T., Padmanaban V. C. and Selvaraju N. (2020). Oxidative removal of stabilized landfill leachate by Fenton's process: process modeling, optimization & analysis of degraded products. *RSC Advances*, **10**, 3916–3925, https://doi.org/10.1039/c9ra09415f

Joshi C. A. and Seay J. R. (2019). Total generation and combustion emissions of plastic derived fuels: a trash to tank approach. *Environmental Progress & Sustainable Energy*, **39**, 1–10, https://doi.org/10.1002/ep.13151

Joshi C. A. and Seay J. R. (2020). Total generation and combustion emissions of plastic derived fuels: a trash to tank approach. *Environmental Progress & Sustainable Energy*, **39**, 1–10, https://doi.org/10.1002/ep.13151

Kabuk H. A., Ilhan F., Avsar Y., Kurt U., Apaydin O. and Gonullu M. T. (2014). Investigation of leachate treatment with electrocoagulation and optimization by response surface methodology. *Clean – Soil, Air, Water*, **42**, 571–577, https://doi.org/10.1002/clen.201300086

Lessoued R., Souahi F. and Pelaez L. C. (2017). Modelization and statistical optimization of coagulation–flocculation treatment of an old leachate. *Water Environment Research*, **89**, 2136–2141, https://doi.org/10.2175/106143017x14839994523703

Liu Z., Pan L., Hu F. and Hu Y. (2020). Advanced landfill leachate biochemical effluent treatment using Fe-Mn/AC activates $O_3/Na_2S_2O_8$ process: process optimization, wastewater quality analysis, and activator characterization. *Environmental Science and Pollution Research*, **27**, 15337–15349, https://doi.org/10.1007/s11356-020-08046-2

Mahtab M. S., Islam D. T. and Farooqi I. H. (2020). Optimization of the process variables for landfill leachate treatment using Fenton based advanced oxidation technique. *Engineering Science and Technology, An International Journal*, **24**, 428–435, https://doi.org/10.1016/j.jestch.2020.08.013

Mohajeri S., Aziz H. A., Isa M. H., Zahed M. A. and Adlan M. N. (2010). Statistical optimization of process parameters for landfill leachate treatment using electro-Fenton technique. *Journal of Hazardous Materials*, **176**, 749–758, https://doi.org/10.1016/j.jhazmat.2009.11.099

Mohd-Salleh S. N. A., Mohd-Zin N. S., Othman N., Mohd-Amdan N. S. and Mohd-Shahli F. (2018). Dosage and pH optimization on stabilized landfill leachate via coagulation–flocculation process. *MATEC Web of Conferences*, **250**, 1–11, https://doi.org/10.1051/matecconf/201825006007

Moreno-Garcia L., Gariépy Y., Bourdeau N., Barnabé S. and Raghavan G. S. V. (2019). Optimization of the proportions of four wastewaters in a blend for the cultivation of microalgae using a mixture design. *Bioresource Technology*, **283**, 168–173, https://doi.org/10.1016/j.biortech.2019.03.067

Naves F. L. (2019). Normal boundary intersection applied to controllers in environmental controls. In: Biosensors for Environmental Monitoring, T. Rinken (ed.), p. 204, https://doi.org/10.5772/intechopen.83662

Oloibiri V., Chys M., De Wandel S., Demeestere K. and Van Hulle S. W. H. (2017). Removal of organic matter and ammonium from landfill leachate through different scenarios: operational cost evaluation in a full-scale case study of a Flemish landfill. *Journal of Environmental Management*, **203**, 774–781, https://doi.org/10.1016/j.jenvman.2016.09.055

Paixão K., Abreu E., Lamas Samanamud G. R., Boscaro França A., Almeida Loures C. C., Prado Baston E., Lima Rezende Naves L., Canellas Bosch J. and Naves F. L. (2019). Normal boundary intersection applied in the scale-up for the treatment process of eriochrome black T through the $UV/TiO_2/O_3$ system. *Journal of Environmental Chemical Engineering*, **7**, 102801, https://doi.org/10.1016/j.jece.2018.11.045

Pinto M. B., Samanamud G. R. L., Baston E. P., França A. B., Naves L. L. R., Loures C. C. A. and Naves F. L. (2019). Multivariate and multiobjective optimization of tannery industry effluent treatment using *Musa* sp. flower extract in the coagulation and flocculation process. *Journal of Cleaner Production*, **219**, 655–666, https://doi.org/10.1016/j.jclepro.2019.02.060

Sabour M. R. and Amiri A. (2017). Comparative study of ANN and RSM for simultaneous optimization of multiple targets in Fenton treatment of landfill leachate. *Waste Management*, **65**, 54–62, https://doi.org/10.1016/j.wasman.2017.03.048

Saranga V. K., Kumar P. K., Verma K., Bhagawan D., Himabindu V. and Narasu M. L. (2020). Effect of biohythane production from distillery spent wash with addition of landfill leachate and sewage wastewater. *Applied Biochemistry and Biotechnology*, **190**, 30–43, https://doi.org/10.1007/s12010-019-03087-x

Shaylinda M. Z. N., Hamidi A. A., Mohd N. A., Ariffin A., Irvan D., Hazreek Z. A. M. and Nizam Z. M. (2018). Optimization of composite coagulant made from polyferric chloride and tapioca starch in landfill leachate treatment. *Journal of Physics: Conference Series*, **995**, 012019, https://doi.org/10.1088/1742-6596/995/1/012019

Tawakkoly B., Alizadehdakhel A. and Dorosti F. (2019). Evaluation of COD and turbidity removal from compost leachate wastewater using *Salvia hispanica* as a natural coagulant. *Industrial Crops and Products*, **137**, 323–331, https://doi.org/10.1016/j.indcrop.2019.05.038

Tripathy B. K., Ramesh G., Debnath A. and Kumar M. (2019). Mature landfill leachate treatment using sonolytic-persulfate/hydrogen peroxide oxidation: optimization of process parameters. *Ultrasonics Sonochemistry*, **54**, 210–219, https://doi.org/10.1016/j.ultsonch.2019.01.036

Venkatesh Reddy C., Shekhar Rao D. and Kalamdhad A. S. (2020). Statistical modelling and assessment of landfill leachate emission from fresh municipal solid waste: a laboratory-scale anaerobic landfill simulation reactor study. *Waste Management & Research*, **38**, 1161–1175, https://doi.org/10.1177/0734242X20954280

Wang L., Yang Q., Wang D., Li X., Zeng G., Li Z., Deng Y., Liu J. and Yi K. (2016). Advanced landfill leachate treatment using iron-carbon microelectrolysis-Fenton process: process optimization and column experiments. *Journal of Hazardous Materials*, **318**, 460–467, https://doi.org/10.1016/j.jhazmat.2016.07.033

Yeh C. J., Lo S. L., Kuo J. and Chou Y. C. (2018). Optimization of landfill leachate treatment by microwave oxidation using the Taguchi method. *International Journal of Environmental Science and Technology*, **15**, 2075–2086, https://doi.org/10.1007/s13762-017-1567-9

Yilmaz T., Aygün A., Berktay A. and Nas B. (2010). Removal of COD and colour from young municipal landfill leachate by Fenton process. *Environmental Technology*, **31**, 1635–1640, https://doi.org/10.1080/09593330.2010.494692

Zhang H., Heung J. C. and Huang C. P. (2005). Optimization of Fenton process for the treatment of landfill leachate. *Journal of Hazardous Materials*, **125**, 166–174, https://doi.org/10.1016/j.jhazmat.2005.05.025

Zhang W., Zhang J., Ding Y., He Q., Lu K. and Chen H. (2020). Pyrolysis kinetics and reaction mechanism of expandable polystyrene by multiple kinetics methods. *Journal of Cleaner Production*, **285**, 125042, https://doi.org/10.1016/j.jclepro.2020.125042

Zhao X., Zhou Y., Huang S., Qiu D., Schideman L., Chai X. and Zhao Y. (2014). Characterization of microalgae–bacteria consortium cultured in landfill leachate for carbon fixation and lipid production. *Bioresource Technology*, **156**, 322–328, https://doi.org/10.1016/j.biortech.2013.12.112

Zhou X. Zhou S. and Feng X. (2017). Optimization of the photoelectrocatalytic oxidation of landfill leachate using copper and nitrate co-doped TiO_2 (Ti) by response surface methodology. *PLoS ONE*, **2**, 1–18.

doi: 10.2166/9781789063318_0345

Chapter 15

Energy and resource recovery from landfill leachate

*Rickwinder Singh, Apoorva Upadhyay, Prakhar Talwar, Nikita Verma and Vivekanand Vivekanand**

Centre for Energy and Environment, Malaviya National Institute of Technology, Jaipur 302017, Rajasthan, India

**Corresponding author: vivekanand.cee@mnit.ac.in*

ABSTRACT

Landfill leachate contains a significant proportion of pollutants, organic matter, and heavy metals which negatively affect the surrounding ecosystem, groundwater, and soil. Leachate has the potential to generate energy and produce other valuable products such as biogas, bio-hydrogen, and microalgae for biofuel production. However, there is a need to explore various processes such as anaerobic digestion (AD), microbial processes for fuel cells, cultivation of microalgae, and extraction of heavy metals. On the basis of this, these processes have been explored to examine various valuable products that can be recovered from leachate and challenges related to these processes are also discussed. AD and microbial fuel cells are processes that tend to produce energy sources such as biomethane and electricity, respectively. Similarly, the cultivation of microalgae in leachate is the main source to capture carbon dioxide gas to a great extent. These processes aim to recover energy and other chemical products further can be developed in the future socially and economically.

Keywords: Leachate, resource recovery, emission mitigation, bioenergy

15.1 INTRODUCTION

Municipal solid waste (MSW) generation, which typically amounts to 2.2 billion metric tonnes per year, is growing continuously around the world as a result of fast industrialization and uncontrolled increase in population. Most nations favor landfills for the disposal of MSW due to the economic benefits and simplicity of operation (Wijekoon *et al.*, 2022). However, a huge amount of leachate is produced at the bottom of landfills attributable to the physicochemical and biological changes that occur in MSW. Furthermore, these changes along with rainwater and waste moisture result in leachate formation which is a serious issue nowadays (Bakonyi *et al.*, 2019). Hazardous substances found in leachate include heavy metals (Al, As, Cd, Cr, Cu, Fe, Hg, Mn, Ni, Pb and Zn) (Martínez-Ruiz *et al.*, 2022), inorganic macro-pollutants (Ca^{2+}, Mg^{2+}, Na^{+}, K^{+}, NH^{4+}, Fe^{2+}, Cl^{-} and SO_4^{2-}), halogenated compounds, phenols, organic compounds, pesticides, and other micropollutants. These hazardous compounds get mixed with groundwater, polluting it and near-surface water bodies (Bakonyi *et al.*, 2019), because engineered landfills are preferred to avert infiltration of leachate to groundwater (Martínez-Ruiz *et al.*, 2022). But usually underdeveloped and developing countries have opted for non-engineered landfills and open dumps, which have a severe effect on living beings (Wijekoon *et al.*, 2022). As shown in Figure 15.1, 33% of total MSW (which is maximum) disposed into open dumps and only 11% to engineered landfills, which signifies that more awareness is required to avoid disposal of MSW in open areas. In addition, the composition of leachate varies depending on where it is produced. For example, leachate produced in Africa would be different from leachate produced in Asian or American countries due to their disparate waste production, cultural practices, weather patterns, geographic differences, and regulations (Martínez-Ruiz *et al.*, 2022). Leachate quality is also affected by its properties such as chemical oxygen demand (COD), biological oxygen demand (BOD), suspended solids (SSs), pH, and ammonia content. However, environmental factors including the amount of rainfall, infiltration, and surface run-off have a significant impact on leachate quantity (Abuabdou *et al.*, 2020). The composition of leachate depends on MSW type, age of landfill, and site hydrology. Generally, high concentration of ammonia-nitrogen (1,000–3,000 mg/L), COD (10,000–70,000 mg/L) (Gu *et al.*, 2019), and low BOD_5/COD ratio (<0.1) are found in leachate. Among all pollutants, ammonia-nitrogen is one of the important pollutants as it causes nitrification and depletes dissolved oxygen resulting in toxic effects on living organisms and also increasing chlorine demand for sterilization treatment (Kurniawan *et al.*, 2021). However, recovery of these pollutants (organic matter and ammonia-nitrogen) will be valuable from an economic point of view as

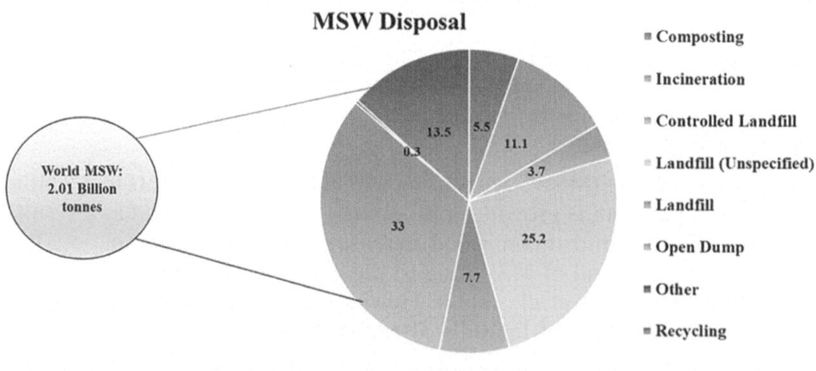

Figure 15.1 Global production of MSW and its sources.

bioenergy (biomethane, biohydrogen, and microbial fuel cells (MFCs)) can be obtained from organic matter, fertilizers recovered from humid substance (Gu *et al.*, 2019), and ammonia recovered from ammonia-nitrogen, which is important due to its industrial and commercial applications (Zico *et al.*, 2021). Therefore, treatment of leachate is required before its final disposal. For landfill leachate treatment, many biological and physicochemical approaches are preferred. Physicochemical techniques such as coagulation/flocculation, adsorption, chemical oxidation, chemical precipitation, air stripping, and filtration are used. Similar to this, biological methods such as aerobic and anaerobic treatment or some methods to improve fundamental processes of these methods play a significant role in recovering organic matter and nitrogenous contaminants (Wijekoon *et al.*, 2022). Common methods used for treatment are air stripping, nitrification–denitrification, and struvite precipitation. Volatile compounds from leachate get transferred to air due to intensive contact between leachate and air during air stripping methods. Nitrosomonas bacteria are used to convert ammonium ions into nitrite to nitrate and clostridium or pseudonymous bacteria to convert that nitrate to nitrogen gas during nitrification–denitrification. Struvite precipitation is used for recovery of phosphorous. But these techniques have a severe impact on environment and loss of valuable resources such as the release of nitrogen gas into atmosphere, volatile compounds mixed with air, wastage of ammonia, and so on. To remove ammonia and hydrogen sulfide, membrane-based treatment is an efficient technique. When an ammonia-rich nutrient solution is passed through a microporous hydrophobic membrane, the resulting products are absorbed by an absorbing solution (such as sulfuric acid), and valuable products are then recovered, such as sulfate ammonium (Zico *et al.*, 2021). Another efficient method for removing refractory contaminants from leachate is adsorption (Kurniawan *et al.*, 2021). It is a simple operational, cost-effective, and less energy consumption method in which via electrostatic interactions, the dispersive force between π-electrons of absorbent and absorbates or hydrophobic interactions, targeted pollutants get relocated from liquid surface to solid surface. Due to its vast surface area, reactivity, and inexpensive, activated carbon is considered a superb absorbent material (Kurniawan *et al.* 2021). Above-discussed methods possess some challenges either in terms of cost or environmental impacts such as additional additives, poor removal performance, and secondary emissions. Therefore, other methods such as anaerobic digestion (AD), MFCs, and treatment with microalgae or macroalgae are used to overcome these challenges and recover the energy from the leachate (Saleem *et al.*, 2022). AD is an environmentally friendly method for producing cleaner energy by biological conversion of organic matter. MSW leachate proved as a favorable for co-digestion in AD and provides an optimistic condition by improving buffering capacity due to high ammonium nitrogen concentration and enhancing the stable supply of nutrients for anaerobic bacteria due to the presence of trace metals. MFC is another effective approach toward COD and ammonium removal from leachate with the efficiency of ~97 and 98%, respectively, with an algae cathode. MFC not just helps in removing the pollutant but also generates energy with the benefits of higher energy conversion. Secondary pollutants have some setbacks such as large internal resistance, low output power, and no storage for electrical energy (the energy has to be utilized as soon as it is generated, otherwise it would be wasted) (Feng *et al.*, 2020).

15.2 CHARACTERIZATION OF LANDFILL LEACHATE

Leachate from landfills is characterized using a number of variables, including pH, ammonia, heavy metals, COD, total organic carbon (TOC), BOD, and SSs. BOD to COD ratio and COD to TOC ratio are popular indicators of the degradability of organic compounds and the conditions of oxidized carbon. Several variables impact the quality of landfill leachate including trash type, operational parameters, weather, groundwater study, and landfill maturity (Dabaghian *et al.*, 2018). Changes in waste content and water content, as well as temperature and rainfall throughout the year, have a major impact on the toxicity and appearance of landfill leachate (Costa *et al.*, 2019). Due to high temperatures in summers, a drop in dissolved organic carbon and COD is observed, whereas winters result in an increase in pH, total nitrogen, and electrical conductivity of leachate. By contrast, there is little to no seasonal fluctuation in oxidation–reduction potential or amounts of metals and total or volatile SSs (Zhao *et al.*, 2013). Higher values in summer for the ratio of absorbance at 254 nm to total organic carbon concentration (specific ultraviolet absorbance at 254 nm (SUVA254)) suggest greater aromaticity in landfill leachate than values in winter (Yang *et al.*, 2019). Due to the increased effectiveness of biodegradation in landfills when temperatures are higher, aromatic structures make up a larger percentage of the dissolved organic matter (DOM) found in leachate throughout the summer. The composition and characteristics of landfill leachate are significantly influenced by the age of the landfill. Leachate can be divided into three kinds based on the age of the waste (Miao *et al.*, 2019). Hydrophilic organic matter with low molecular weight, high biodegradability index (BOD/COD), and low pH are the major constituents of fresh landfill leachate. In contrast, leachate from abandoned landfills is predominately made up of humic acid (HA) and fulvic acid (FA) with high molecular weights, which causes a high pH and a low biodegradability index. Notably, when pH increases with age, the concentration of heavy metals decreases as the metals become less soluble. DOM can be split into discrete fractions according on various criteria, and these fractions respond differently to various therapies. DOM fractions are categorized as chromophoric DOM (CDOM) and fluorescent DOM (FDOM) based on their fluorescence emittance (Helms *et al.*, 2008). The high UV and visible light absorption of CDOM contributes to the overall DOM's dark hue in solution (Lozinski *et al.*, 2019). However, FDOM only comprises a fraction of the overall DOM. A combination of XAD-8 resin and cation exchange resin is used to extract the HA, FA, and hydrophilic fractions from landfill leachate DOM because of their different solubilities in various solvents (Zhang *et al.*, 2013). In addition, landfill leachate can be separated into four parts depending on hydrophobicity by employing the XAD-8 resin method: hydrophilic component, hydrophobic base, hydrophobic neutral, and hydrophobic acid (He *et al.*, 2016). Table 15.1 depicts the overview of techniques used to analyze different characteristics of leachate.

Table 15.1 Techniques used for the characterization of leachate.

Techniques	Components	Relevance	Reference
Basic characteristics	BOD	Quantity of organic substance capable of being reduced	Babuponnusami and Muthukumar (2014)
	COD	Quantity of all biodegradable materials	Wang *et al.* (2015)
	BOD_5/COD	Biodegradability indicator	Liu *et al.* (2020)
	TOC	Quantity of all carbon-containing chemical compounds	Zhang *et al.* (2020)
	Ammonium, TKN, TON	Variables that reveal the nitrogen content.	Santos *et al.* (2018)
Procedures based on molecular weight	Cut-offs for membranes and suspended materials	As a measure of sedimentation efficiency, the rate at which membranes of varying molecular weights cross one another is used.	Oloibiri *et al.* (2017)
Structured analytic methods	Analysis of the elements, FT-ICR MS, ^{1}H, ^{15}N, and ^{13}C NMR	Different H-type, N-type, and C-type functional groups can be determined using elemental analysis and the H/C, O/C, and N/C ratios to accurately identify elements and their combinations.	Xiaoli *et al.* (2008)
Strategies based on spectroscopy	UV-Vis spectroscopy	Aromaticity and hydrophobicity.	Yan *et al.* (2018)
	FTIR spectroscopy	Characteristic peaks can be used to identify various functional categories.	Wang *et al.* (2017)
	Fluorescence spectroscopy	Independent groups of fluorescent components.	Ishii & Boyer (2012)
	2D-COS spectroscopy	To determine the exact ordering and relative orientations of spectrum shifts caused by an external perturbation.	Teng *et al.* (2020)

The aromatic moieties present in the hydrophobic fraction of DOM have been shown to contribute to generation of disinfection by-products as chlorination takes place (Korshin *et al.*, 2007). Biodegradable and non-biodegradable DOM fragments can be distinguished. The non-biodegradable fractions are composed of mostly humic substances formed by the condensation and polymerization of microbial breakdown by-products (Iskander *et al.*, 2018). On the contrary, biodegradable organic matter generally arises from the early stages of degradation of organic solid waste. On the contrary, biodegradable organic matter derives mostly from the first stages of decay of organic solid waste. This portion is made up of molecules that are all just linked together by carbon atoms, and it can be easily filtered by biological treatment methods. Emerging organic contaminants (EOCs) such as organic contaminants, endocannabinoid chemical products, medicinal and personal care products, antibiotic resistance genes (ARGs) and disinfection by-products, which have high prevalence in landfill leachate and adverse impacts on the environment and human health have garnered a lot of attention in recent years. Different types of microplastics in the range of 0.42–24.58 items/L were found in landfill leachate (He *et al.*, 2019). In addition, phenolic chemicals, such as bisphenol A and 2,4-di-test-butyl phenol, were commonly found at extremely high levels (>10 mg/L) in landfill leachate (Aziz *et al.*, 2018). Antibiotics, non-steroidal anti-inflammatory drugs, stimulants, and beta-blockers are just some of the 172 pharmaceutical and personal care products found in landfill leachate around the world in the last two decades (Yu *et al.*, 2020). Furthermore, landfill leachate is regarded as an antibiotic resistance reservoir due to the presence of significant amount of ARGs. As landfill leachate is a major source of EOCs, it must be properly treated.

15.3 RESOURCE AND ENERGY RECOVERY TECHNIQUES FROM LANDFILL LEACHATE

15.3.1 Microbial fuel cells

An MFC is a type of bio-electrochemical system that removes different types of substrates from leachate and generates electricity from the combinations of biological and electrochemical technologies (Feng *et al.*, 2020). In MFCs, anaerobic reaction of organic substance (e^- donor) occurs at anode compartment by variety of microorganisms; electrons and protons produced from this process will transfer to the cathode chamber either indirectly via mediators or directly (electroactive bacteria) as depicted in Figure 15.2. Compared to an aerobic-activated sludge process, production is 2.4–24.6 times less in MFCs. Due to the high organic content of MFCs and the presence of inorganic metals, landfill leachate offers a possible fuel source and is found to have strong electrical conductivity, which could boost power output (Elmaadawy *et al.*, 2020). Another important factor is the material of electrode used in MFCs, as open circuit voltage (OCV) depends on it. Carbon cloth is the most common material used for anode through which maximum OCV is achieved and for the same leachate condition, a pyrrhotite cathode MFC achieves more OCV compared to graphite cathode (Cheng *et al.*, 2022).

15.3.2 Nutrition removal

15.3.2.1 Carbon removal

Exoelectrogen bacteria are normally used for microbial metabolism serving as electron donors at the anode and generating electrons by oxidizing organic and inorganic matter to anode electrode exogenously serving as electron acceptors as depicted in equation (15.1). Protons infiltrate from the separator to the cathode chamber, where they interact with terminal electron acceptors,

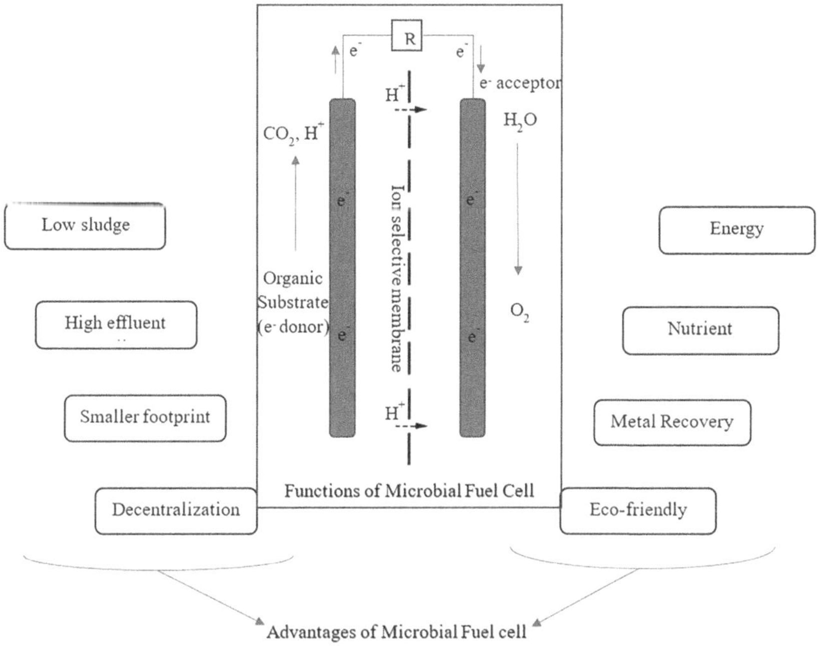

Figure 15.2 Function and advantages of microbial fuel cell.

while electrons move from the anode to the cathode electrode along the electric circuit as shown in equation (15.2):

$$CH_3COO^- + 4H_2O \rightarrow 2HCO_3^- + 9H^+ + 8e^- \tag{15.1}$$

$$O_2 + 4H^+ + 4e^- \rightarrow 2H_2O \tag{15.2}$$

15.3.2.2 Nitrogen removal

Landfill leachate consists of N_2 in several forms such as ammonium ion, NH_3, nitrite, N_2, and nitrate. A conventional method via bacteria (aerobic autotrophic bacteria) converts aerobically ammonia nitrogen (NH_4-N) to nitrite (NO_2^--N) and then to nitrate nitrogen (NO_3^--N) as depicted in equations (15.3) and (15.4):

$$2NH_4^+ + 3O_2 \rightarrow 2NO_2^- + 2H_2O + 4H^+ \tag{15.3}$$

$$2NO_2^- + O_2 \rightarrow 2NO_3^- \tag{15.4}$$

Furthermore, the denitrification process takes place in which, the reduction of nitrate to nitrite, nitrous oxide, nitric oxide, and ultimately to N_2 by heterotrophic bacteria occurs using organic carbon in the absence of dissolved oxygen (equation (15.5)). Other methods can be carried out either at the anode or cathode electrode during the bio-electrochemical denitrification process.

Heterotrophic nitrification occurs in the anode chamber under anaerobic circumstances as nitrate from nitrification (aerobic) is transferred from the cathode to the anode through an anion exchange membrane (equation (15.5)):

$$5C_6H_6O + 28NO_3^- \rightarrow 14N_2 + 30CO_2 + 28OH^- + H_2O \quad (15.5)$$

15.3.2.3 Phosphorous removal

Electrochemical precipitation as struvite or magnesium ammonium phosphate hexahydrate ($MgNH_4PO_4{\cdot}6H_2O$) in the cathode chamber is used for removal of phosphorous in MFCs. Alkaline pH (8–10) is the optimal operating conditions of MFCs; this pH is induced from oxidation–reduction reaction that took place near the cathode. The agglomeration of struvite crystals is limited by insufficient levels of Ca^{2+} and Mg^{2+} cations. To increase the struvite formation, pH of the cathode is increased using an external power supply (Elmaadawy *et al.*, 2020).

15.4 AD OF LEACHATE FOR BIOGAS PRODUCTION

Landfilling is an affordable and quickly adopted technique for managing solid waste; however, it also has environmental drawbacks including managing leachate (Yarimtepe & Oz, 2015). Landfill leachate, which is wastewater made up primarily of organic matter that results from trash percolation, can be a useful resource for the production of electricity (Oz & Yarimtepe, 2014). Various treatment techniques have been used to treat leachate and gain substantial benefits including low energy usage, production of biogas, and others (Yarimtepe & Oz, 2015).

Rapid exploitation of conventional energy resources has intended to uncover the potential of renewable energy sources. Leachate tends to lessen greenhouse gas emissions from landfills while still having the potential to provide energy in the form of gas through biological processes including AD and dark fermentation to produce methane and hydrogen, respectively. The degradation of organic compounds by anaerobic microorganisms under dark fermentation in leachate for biogas production depends on the characteristics of waste landfill leachate components, their fermentation, reactor design, and co-fermentation of organic biomass in leachate. The available technologies include chemical oxidation, reverse osmosis, coagulation and flocculation, adsorption, and so on (Bakonyi *et al.*, 2019; Yang *et al.*, 2022). Leachate has been reportedly used as a suitable substrate in AD with notable biomethane potential (Ma *et al.*, 2018). Figure 15.3 depicts the steps involved in anaerobic degradation of organic matter. Various green treatment schemes are available to ensure sustainability as the emphasis has been laid on reducing environmental footprint and feasible resource recovery. Microalgae-based treatment is one such technique that helps to utilize the nutrients from leachate which leads to further purification (Ali *et al.*, 2021). To enhance biogas production, the age of leachate also plays a vital role. The new leachate has a high amount of contamination with active acidogenic reactions, whereas old leachate is more stabilized. As COD determines the

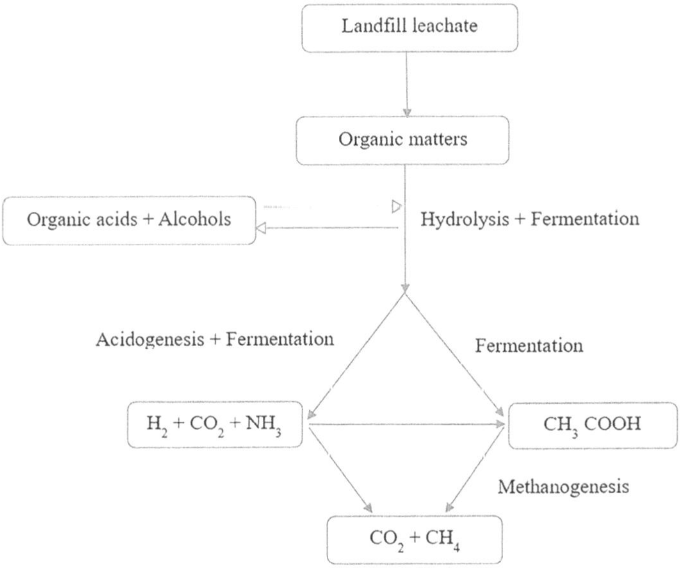

Figure 15.3 Anaerobic degradation of organic matters.

organic strength of leachate, new leachate has a COD value extending up to even hundreds of thousands (mg/L) determined by waste composition. Studies conducted on different blends and recirculation ratios to improve methane yield showed that biogas production can be improved by recirculating old leachate with new blended leachate (Aromolaran & Sartaj, 2021). Volatile fatty acids (VFAs) are viewed as promising ingredients for manufacturing sustainable biofuels through an anaerobic route, in addition to the generation of biogas from leachate by AD. Figure 15.4 shows the schematic flow of obtaining end products through processes involved in single- and two-stage AD. With the growing concerns about energy crisis and environmental pollution, using cheap and readily available feedstock is a promising alternative. Research has been carried out to assess the impact of parameters such as pH, the effect of

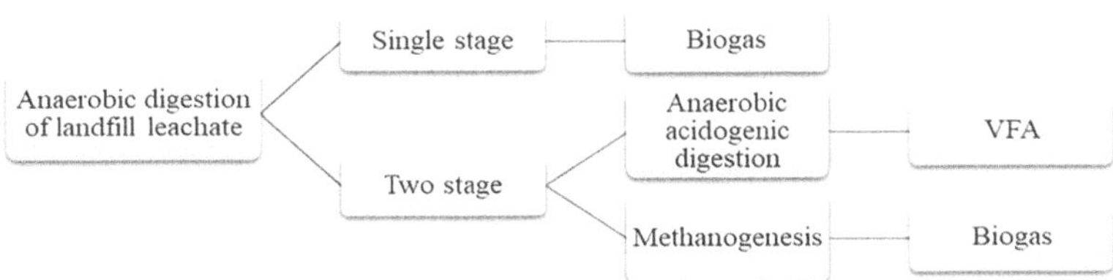

Figure 15.4 Schematic diagram of two-stage AD process.

loading rate on COD, and so on (Begum *et al.*, 2018). To enhance biogas yield, various pre-treatment methods before the anaerobic process are also used, which enhances the degradation of complex organic matter, thereby ensuring higher methane content. Ultrasound pre-treatment is one such method that has been analyzed for its effect on parameters such as turbidity, soluble COD, SS concentration, and so on. Ultrasound pre-treatment performed at different ultra-sonication durations increases the soluble COD because of high cavitation energy. The organic matter solubilization was determined by calculating the ratio of soluble COD to total COD. For various energy inputs, the total COD was essentially steady, but the ratio increased.

According to reports, particulate COD is converted into soluble COD when low-frequency ultrasound is applied, boosting the solubilization by 2–35%. Due to the application of ultrasound energy, the disintegration of SSs was observed, marking an increase in turbidity, approximately two times higher. Although disintegration of suspended particles reduces particle size and increases soluble COD removal with the reduction in concentration of SSs, it ameliorates AD, which is indicated by an enhanced biogas yield. With the above-mentioned effects, conclusion can be drawn that ultrasound can enhance the production of biogas with a higher methane content (Oz & Yarimtepe, 2014). An anaerobic membrane bioreactor (AnMBR), which functions in the absence of oxygen and utilizes a membrane for solid–liquid separation, is another type of treatment procedure. It combines the advantages of anaerobic processes and membrane technology. One of the benefits of AnMBR is biogas production. Further ensuring overall performance of AnMBR in terms of economic and commercial viability is an efficient methane recovery. Biogas production rate depends on the retention of methanogenic bacteria by controlling solid and hydraulic retention time. The composition of methane from AnMBR is ~80%, directly proportional to the organic loading rate. The recovery of methane from AnMBR poses a challenging issue due to the solubility of methane in the effluent, which is significantly affected by operational time. Compared to a conventional biogas plant, which has a methane content of ~70%, AnMBR has a methane content in the range of 70–90% with nitrogen (0–15%) and carbon dioxide (3–15%). The biogas yield is increased by the thermophilic AnMBR's quick reaction time and high loading rate. Polluted biogas can occasionally be produced by landfill leachate that contains high levels of organic contaminants, particularly ammonia. Another critical issue is the varying composition of refractory compounds in landfill leachate, making sustainability of overall process uncertain (Abuabdou *et al.*, 2020). Studies also reported that air injection in aerobic bioreactors could increase the pH of leachate by quickly decomposing acid products. By enhancing leachate quality, it speeds up waste degradation and creates a window for biogas to electricity conversion. Hybrid bioreactor systems that apply both aerobic and anaerobic approaches have also been proposed in studies. The aeration affects the composition of biogas as high methane content was observed from hybrid bioreactors. The temporary aeration also increased alkalinity (Bonu *et al.*, 2022; Xu *et al.*, 2015)(Figure 15.5 shows the parameters affecting AnMBR process efficiency in terms of end product efficiency (Table 15.2).

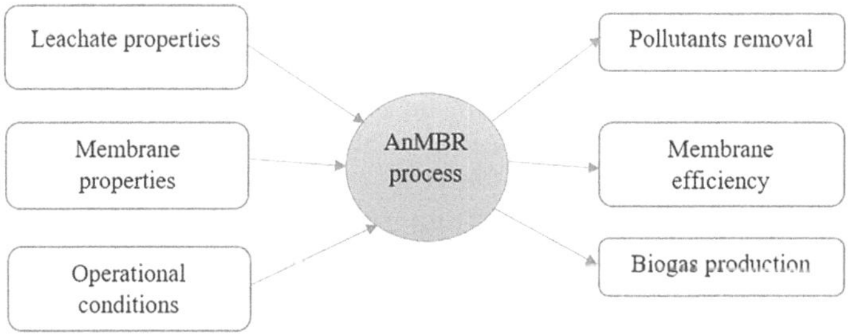

Figure 15.5 Factors affecting the performance of AnMBR process (*source*: Bonu *et al.*, 2022).

Table 15.2 Studies on leachate treatment by AD.

Reference	Sample Collection	Method	Result
Yarimtepe and Oz (2015)	Leachate treatment system, Istanbul Environmental Management in Industry & Trade	AD in integration with low-frequency ultrasound	A considerable rise in soluble COD content was observed, that further enhanced the biogas and methane yield as compared to control reactors.
Aromolaran and Sartaj (2021)	Local landfill site, Moose Creek, ON, Canada	AD through recirculation of blended leachate	Enhanced biogas production through recirculating leachate blended with old leachate. In comparison with control reactor recirculating only young leachate, biogas production increased between 77.2 and 193.2% when a blend of 1N2O (1 new and 2 old leachate parts) is used.
Begum *et al.* (2018)	Municipal landfill dumpsite, Hyderabad	Single- and two-stage AD process to study the influence of pH and initial organic load (IOL) on VFAs	A 21% rise in COD removal in both two-stage AD. CH_4 yields in single- and two-stage were AD between 0.21 and 0.34 L CH_4/(g COD removed) and 0.2 and 0.32 L CH_4/(g COD removed), respectively. VFA yield with varying pH and IOL was in the range of 0.26–0.36 g VFA/(g COD removed).
Abuabdou *et al.* (2020)	Local landfill dumpsite	Use an AnMBR	Biogas production from AnMBR contains a methane content of 80%.
Xu *et al.* (2015)	Shenzhen University Town, China	AD with hybrid bioreactors (both aerobic and anaerobic ways)	High methane content (>60%) was observed from aerated hybrid bioreactors.

15.5 CHALLENGES AND FUTURE PERSPECTIVE OF ENERGY RECOVERY FROM LANDFILL LEACHATE

Most energy production is dependent on fossil fuels such as petroleum, coal, and natural gas. However, usage of renewable energy has been implemented for replacing these conventional fuels. In contrast, bioenergy is eco-friendly and involved or generated by converting waste into energy and other valuable products. Currently, the high potential of waste and leachate is the main concern of research in which various valuable nutrients are recovered from biodegradable waste by AD. Although AD is a promising and sustainable approach to recovering energy in the form of biomethane from leachate, this area of research is still underdeveloped. Additional research on energy and nutrient recovery from leachate is required to be carried out on a lab scale and large scale. There are several challenges such as the concentration of leachate, trace elements, C/N ratio, and pH which are the major barriers to implementing AD for leachate treatment. Also, mono-digestion of leachate is difficult due to concerns related to stability of AD, inhibitors, and economic perspective. For this problem, co-digestion of different substrates with leachate may help utilize waste and produce energy (Liu *et al.*, 2022). On leachate, research should be conducted to extract nitrogen, potassium, and phosphorus using different methods such as chemical crystallization and cultivation of microalgae. The economic and technical perspectives should be considered for further research before implementing this whole process for leachate treatment. On the contrary, there are some studies on MFCs that have been investigated by several researchers. In MFCs, leachate is supplied in cells and nutrients are utilized for directly generating electricity. However, some barriers also exist for using it in MFCs including need of pre-treatment, inhibitors, instability, high internal resistances, fouling of membrane, and low growth rate of microbes. These challenges need to be addressed during the use of leachate in MFCs (Do *et al.*, 2018; Feng *et al.*, 2020). Additionally, the economics of leachate usage for energy and resource recovery will be a barrier, but it will be beneficial if energy recovery, reduced environmental degradation, and utilization of leachate are considered. Still, there are numerous parameters that will hinder the processes used for energy and nutrient recovery from leachate (Gu *et al.*, 2019). Also, researchers can work on an integrated system of MSW and leachate for proper utilization of waste for generating energy. This integrated system will mitigate emission of leachate and create new opportunities for young professionals and local people.

15.6 CONCLUSION

As reported, many pollutants, including dissolved organic and inorganic pollutants, heavy metals, and other NH_3-producing chemicals, are present in landfill leachate and have an adverse effect on the ecosystem and living organisms. There are numerous techniques such as AD, MFCs, microalgae production, and resource recovery methods which tend to utilize leachate for valuable assets. The performance of AD processes can be increased when different types of membranes are used. Additionally, carbonation and ammonia may be used for removal of calcium from leachate and recovery of NH_3 simultaneously. On the contrary, MFCs are a better approach to nutrient removal from leachate as they increase the removal efficiency as well as generate electricity. However, it is reported that the performance of MFCs depends on quality, age, and source of leachate. There is a recommendation for an integrated system that should be applied for recovery from both MSW and leachate. This proposed integrated system has numerous advantages such as proper utilization of waste (including leachate), and environmental and economic benefits as compared to traditional processes of leachate. There are various ways to use landfill leachate for industrial and commercial purposes that can provide leachate management that is sustainable. It also reported that energy and resource recovery from leachate will support landfill management because utilization of leachate for generating energy will provide economic benefits to involved governments and industries. However, it will require an initial investment and other expenses and the feasibility of processes that will be used.

REFERENCES

Abuabdou S. M. A., Ahmad W., Aun N. C. and Bashir M. J. K. (2020). A review of anaerobic membrane bioreactors (AnMBR) for the treatment of highly contaminated landfill leachate and biogas production: effectiveness, limitations and future perspectives. *Journal of Cleaner Production*, **255**, 120215, https://doi.org/10.1016/j.jclepro.2020.120215

Ali S., Paul Peter A., Chew K. W., Munawaroh H. S. H. and Show P. L. (2021). Resource recovery from industrial effluents through the cultivation of microalgae: a review. *Bioresource Technology*, **337**, 125461, https://doi.org/10.1016/j.biortech.2021.125461

Aromolaran A. and Sartaj M. (2021). Enhancing biogas production from municipal solid waste through recirculation of blended leachate in simulated bioreactor landfills. *Biomass Conversion and Biorefinery*, **13**(4), 19–26, https://doi.org/10.1007/s13399-020-01260-5

Aziz A., Agamuthu P. and Fauziah S. H. (2018). Removal of bisphenol A and 2,4-di-*tert*-butylphenol from landfill leachate using plant-based coagulant. *Waste Management and Research*, **36**(10), 975–984, https://doi.org/10.1177/0734242X18790360

Babuponnusami A. and Muthukumar K. (2014). A review on Fenton and improvements to the Fenton process for wastewater treatment. *Journal of Environmental Chemical Engineering*, **2**(1), 557–572, https://doi.org/10.1016/J.JECE.2013.10.011

Bakonyi P., Dharmaraja J., Shobana S., Koók L., Rózsenberszki T., Nemestóthy N., Banu J. R., Bélafi-Bakó K. and Kumar G. (2019). Leachate valorization in anaerobic biosystems: towards the realization of waste-to-energy concept via biohydrogen, biogas and bioelectrochemical processes. *International Journal of Hydrogen Energy*, **44**, 17278–17296, https://doi.org/10.1016/j.ijhydene.2019.02.028

Begum S., Anupoju G. R., Sridhar S., Bhargava S. K., Jegatheesan V. and Eshtiaghi N. (2018). Evaluation of single and two stage anaerobic digestion of landfill leachate: effect of pH and initial organic loading rate on volatile fatty acid (VFA) and biogas production. *Bioresource Technology*, **251**, 364–373, https://doi.org/10.1016/j.biortech.2017.12.069

Bonu R., Anand N. and Palani S. G. (2022). Impact of thermal pre-treatment on anaerobic co-digestion of sewage sludge and landfill leachate. *Materials Today: Proceedings*, **72**, 6–10, https://doi.org/10.1016/j.matpr.2022.06.130

Cheng P., Shan R., Yuan H. R., Tan X., Chen Y. and Wu J. (2022). Synchronous biodegradation and bio-electricity generation in a microbial fuel cell with aged and fresh leachate from the identical subtropical area. *Journal of Environmental Management*, **316**, 115017, https://doi.org/10.1016/j.jenvman.2022.115017

Costa A. M., Alfaia R. G., de S. M. and Campos J. C. (2019). Landfill leachate treatment in Brazil – an overview. *Journal of Environmental Management*, **232**, 110–116, https://doi.org/10.1016/J.JENVMAN.2018.11.006

Dabaghian Z., Peyravi M., Jahanshahi M. and Rad A. S. (2018). Potential of advanced nano-structured membranes for landfill leachate treatment: a review. *ChemBioEng Reviews*, **5**(2), 119–138, https://doi.org/10.1002/CBEN.201600020

Do M. H., Ngo H. H., Guo W. S., Liu Y., Chang S. W., Nguyen D. D., Nghiem L. D. and Ni B. J. (2018). Challenges in the application of microbial fuel cells to wastewater treatment and energy production: a mini review. *Science of the Total Environment*, **639**, 910–920, https://doi.org/10.1016/j.scitotenv.2018.05.136

Elmaadawy K., Liu B., Hu J., Hou H. and Yang J. (2020). Performance evaluation of microbial fuel cell for landfill leachate treatment: research updates and synergistic effects of hybrid systems. *Journal of Environmental Sciences*, **96**, 1–20, https://doi.org/10.1016/j.jes.2020.05.005

Feng Q., Xu L., Xu Y., Liu C., Lu Y., Wang H., Wu T., Wang R., Chen Y. and Cheng Y. (2020). Treatment of aged landfill leachate by a self-sustained microbial fuel cell-microbial electrolysis cell system. *International Journal of Electrochemical Science*, **15**, 1022–1033, https://doi.org/10.20964/2020.01.19

Gu N., Liu J., Ye J., Chang N. and Li Y. Y. (2019). Bioenergy, ammonia and humic substances recovery from municipal solid waste leachate: a review and process integration. *Bioresource Technology*, **293**, 122159, https://doi.org/10.1016/j.biortech.2019.122159

He X. S., Xi B. D., Gao R. T., Zhang H., Dang Q. L., Li D. and Huang C. H. (2016). Insight into the composition and degradation potential of dissolved organic matter with different hydrophobicity in landfill leachates. *Chemosphere*, **144**, 75–80, https://doi.org/10.1016/J.CHEMOSPHERE.2015.08.071

He P., Chen L., Shao L., Zhang H. and Lü F. (2019). Municipal solid waste (MSW) landfill: a source of microplastics? –evidence of microplastics in landfill leachate. *Water Research*, **159**, 38–45, https://doi.org/10.1016/J.WATRES.2019.04.060

Helms J. R., Stubbins A., Ritchie J. D., Minor E. C., Kieber D. J. and Mopper K. (2008). Absorption spectral slopes and slope ratios as indicators of molecular weight, source, and photobleaching of chromophoric dissolved organic matter. *Limnology and Oceanography*, **53**(3), 955–969, https://doi.org/10.4319/lo.2008.53.3.0955

Ishii S. K. L. and Boyer T. H. (2012). Behavior of reoccurring parafac components in fluorescent dissolved organic matter in natural and engineered systems: a critical review. *Environmental Science and Technology*, **46**(4), 2006–2017, https://doi.org/10.1021/ES2043504/SUPPL_FILE/ES2043504_SI_001.PDF

Iskander S. M., Zhao R., Pathak A., Gupta A., Pruden A., Novak J. T. and He Z. (2018). A review of landfill leachate induced ultraviolet quenching substances: sources, characteristics, and treatment. *Water Research*, **145**, 297–311, https://doi.org/10.1016/J.WATRES.2018.08.035

Korshin G. V., Benjamin M. M., Chang H. S. and Gallard H. (2007). Examination of NOM chlorination reactions by conventional and stop-flow differential absorbance spectroscopy. *Environmental Science and Technology*, **41**(8), 2776–2781, https://doi.org/10.1021/ES062268H/SUPPL_FILE/ES062268HSI20070130_103411.PDF

Kurniawan T. A., Singh D., Xue W., Avtar R., Othman M. H. D., Hwang G. H., Setiadi T., Albadarin A. B. and Shirazian S. (2021). Resource recovery toward sustainability through nutrient removal from landfill leachate. *Journal of Environmental Management*, **287**, 112265, https://doi.org/10.1016/j.jenvman.2021.112265

Liu X., Novak J. T. and He Z. (2020). Synergistically coupling membrane electrochemical reactor with Fenton process to enhance landfill leachate treatment. *Chemosphere*, **247**, 125954, https://doi.org/10.1016/J.CHEMOSPHERE.2020.125954

Liu Y., Lv Y., Chen H., Zou L., Li Y. Y. and Liu J. (2022). High-efficiency anaerobic co-digestion of food waste and mature leachate using expanded granular sludge blanket reactor. *Bioresource Technology*, **362**, 127847, https://doi.org/10.1016/j.biortech.2022.127847

Lozinski D., Bolyard S. C., Reinhart D. R. and Motlagh A. M. (2019). Treatment of leachate organic matter through sunlight driven processes. *Waste Management*, **94**, 18–26, https://doi.org/10.1016/J.WASMAN.2019.05.015

Ma H., Guo Y., Qin Y. and Li Y. Y. (2018). Nutrient recovery technologies integrated with energy recovery by waste biomass anaerobic digestion. *Bioresource Technology*, **269**, 520–531, https://doi.org/10.1016/j.biortech.2018.08.114

Martínez-Ruiz M., Molina-Vázquez A., Santiesteban-Romero B., Reyes-Pardo H., Villaseñor-Zepeda K. R., Meléndez-Sánchez E. R., Araújo R. G., Sosa-Hernández J. E., Bilal M., Iqbal H. M. N. and Parra-Saldivar R. (2022). Micro-algae assisted green bioremediation of water pollutants rich leachate and source products recovery. *Environmental Pollution*, **306**, 119422, https://doi.org/10.1016/j.envpol.2022.119422

Miao L., Yang G., Tao T. and Peng Y. (2019). Recent advances in nitrogen removal from landfill leachate using biological treatments – a review. *Journal of Environmental Management*, **235**, 178–185, https://doi.org/10.1016/J.JENVMAN.2019.01.057

Oloibiri V., Michael C., Stijn D., Kristof D. and Stijn W. (2017). Removal of organic matter and ammonium from landfill leachate through different scenarios: Operational cost evaluation in a full-scale case study of a Flemish landfill. *Journal of Environmental Management*, **203**, 774–781, https://doi.org/10.1016/j.jenvman.2016.09.055

Oz N. A. and Yarimtepe C. C. (2014). Ultrasound assisted biogas production from landfill leachate. *Waste Management*, **34**, 1165–1170, https://doi.org/10.1016/j.wasman.2014.03.003

Saleem S., Zeshan Iftikhar R., Zafar M. I. and Sohail N. F. (2022). Growth kinetics of microalgae cultivated in different dilutions of fresh leachate for sustainable nutrient recovery and carbon fixation. *Biochemical Engineering Journal*, **178**, 108299, https://doi.org/10.1016/j.bej.2021.108299

Santos A. V., Andrade L. H., Amaral M. C. S. and Lange L. C. (2018). Integration of membrane separation and Fenton processes for sanitary landfill leachate treatment. *Environmental Technology*, **40**(22), 2897–2905, https://doi.org/10.1080/09593330.2018.1458337

Teng C., Zhou K., Zhang Z., Peng C. and Chen W. (2020). Elucidating the structural variation of membrane concentrated landfill leachate during Fenton oxidation process using spectroscopic analyses. *Environmental Pollution*, **256**, 113467, https://doi.org/10.1016/J.ENVPOL.2019.113467

Wang Z., Guo Y., Liu Z., Feng X., Chen Y. and Tao T. (2015). Catechin as a new improving agent for a photo-Fenton-like system at near-neutral pH for the removal of inderal. *Photochemical and Photobiological Sciences*, **14**(2), 473–480, https://doi.org/10.1039/C4PP00371C

Wang L., Li Y. J., Xiong Y., Tan W. B., Zhang L. Y., Li X., Wang X. S., Xu J., Li T. T. and Wang J. S. (2017). Spectroscopic characterization of DOM and the nitrogen removal mechanism during wastewater reclamation plant. *PLoS ONE*, **12**(11), e0187355.

Wijekoon P., Koliyabandara P. A., Cooray A. T., Lam S. S., Athapattu B. C. L. and Vithanage M. (2022). Progress and prospects in mitigation of landfill leachate pollution: risk, pollution potential, treatment and challenges. *Journal of Hazardous Materials*, **421**, 126627, https://doi.org/10.1016/j.jhazmat.2021.126627

Xiaoli C., Shimaoka T., Qiang G. and Youcai Z. (2008). Characterization of humic and fulvic acids extracted from landfill by elemental composition, ^{13}C CP/MAS NMR and TMAH-Py-GC/MS. *Waste Management*, **28**(5), 896–903, https://doi.org/10.1016/j.wasman.2007.02.004

Xu Q., Tian Y., Wang S. and Ko J. H. (2015). A comparative study of leachate quality and biogas generation in simulated anaerobic and hybrid bioreactors. *Waste Management*, **41**, 94–100, https://doi.org/10.1016/j.wasman.2015.03.023

Yan M., Roccaro P., Fabbricino M. and Korshin G. V. (2018). Comparison of the effects of chloramine and chlorine on the aromaticity of dissolved organic matter and yields of disinfection by-products. *Chemosphere*, **191**, 477–484, https://doi.org/10.1016/J.CHEMOSPHERE.2017.10.063

Yang X., Meng L. and Meng F. (2019). Combination of self-organizing map and parallel factor analysis to characterize the evolution of fluorescent dissolved organic matter in a full-scale landfill leachate treatment plant. *Science of the Total Environment*, **654**, 1187–1195, https://doi.org/10.1016/J.SCITOTENV.2018.11.135

Yang S., Luo J., Li Y. Y. and Liu J. (2022). Contradictory effects of calcium on biological and membrane treatment of municipal solid waste leachate: a review and process optimization via biogas recirculation. *Journal of Cleaner Production*, **338**, 130568, https://doi.org/10.1016/j.jclepro.2022.130568

Yarimtepe C. C. and Oz N. A. (2015). Enhanced biogas production from landfill leachate by low frequency ultrasound. Sustainable Development, G. Passerini (ed.), WIT Transactions on The Built Environment, WIT press, UK, Vol. **168**, pp. 225–234, https://doi.org/10.2495/sd150201

Yu X., Sui Q., Lyu S., Zhao W., Liu J., Cai Z., Yu G. and Barcelo D. (2020). Municipal solid waste landfills: an underestimated source of pharmaceutical and personal care products in the water environment. *Environmental Science and Technology*, **54**(16), 9757–9768, https://doi.org/10.1021/ACS.EST.0C00565/SUPPL_FILE/ES0C00565_SI_001.PDF

Zhang Q. Q., Tian B. H., Zhang X., Ghulam A., Fang C. R. and He R. (2013). Investigation on characteristics of leachate and concentrated leachate in three landfill leachate treatment plants. *Waste Management*, **33**(11), 2277–2286, https://doi.org/10.1016/J.WASMAN.2013.07.021

Zhang Z., Teng C., Zhou K., Peng C. and Chen W. (2020). Degradation characteristics of dissolved organic matter in nanofiltration concentrated landfill leachate during electrocatalytic oxidation. *Chemosphere*, **255**, 127055, https://doi.org/10.1016/J.CHEMOSPHERE.2020.127055

Zhao J., Lu X. Q., Luo J. H., Liu J. Y., Xu Y. F., Zhao A. H., Liu F., Tai J., Qian G. R. and Peng B. (2013). Characterization of fresh leachate from a refuse transfer station under different seasons. *International Biodeterioration & Biodegradation*, **85**, 631–637, https://doi.org/10.1016/J.IBIOD.2013.05.012

Zico M. M., Ricci B. C., Reis B. G., Magalhães N. C. and Amaral M. C. S. (2021). Sustainable ammonia resource recovery from landfill leachate by solar-driven modified direct contact membrane distillation. *Separation and Purification Technology*, **264**, 118356, https://doi.org/10.1016/j.seppur.2021.118356

doi: 10.2166/9781789063318_0361

Chapter 16

Efficient and economical landfill leachate management with phytoremediation

Anjali Bhagwat[1,2]* *and Chandra Shekhar Prasad Ojha*[2]

[1]*Hydrological Investigation Division, National Institute of Hydrology, Roorkee, Uttarakhand, India*
[2]*Department of Civil Engineering, Indian Institute of Technology, Roorkee, Uttarakhand, India*
**Corresponding author: anjali.nihr@gov.in*

ABSTRACT

Landfill and plant species have a symbiotic relation; nutrient-rich leachate supports plant growth, whereas plants absorb heavy metals from landfills and accumulate them in different parts, reducing leaching of heavy metals. Thus, the current study aims to understand how nutrient-rich leachate obtained from the Gazipur landfill supports the growth of Indian marigold (*Tagetes erecta* L.) by comparing the growth observed in stem height, leaf width, leaf length, root length, and chlorophyll content when treated with raw leachate, inorganic fertilizers and under controlled conditions. Furthermore, marigold was used for remediating landfill soil under three treatment conditions. It was observed that the order of growth in marigold follows control > leachate > inorganic fertilizers. However, the chlorophyll content follows the order leachate > inorganic fertilizer > control; thus, plants are more productive when treatments are applied. When the number of flowers is compared, maximum number of flowers is produced in the presence of leachate. When total metal removal is considered, maximum remediation observed is of the order leachate > control > inorganic fertilizers; this can be attributed to the root length which is the least in the case of inorganic fertilizers. Furthermore, it was also observed that heavy metal accumulation is least in marigold flowers, thus application of leachate makes the use of marigold flowers safer in comparison with other treatments. Finally, the economic potential of the three treatments is evaluated with refund values of 5.702, 5.165, and 5.09 for marigolds grown in leachate, with fertilizers and under control conditions, respectively. The chapter will be a value-add for policymakers in ensuring sustainable development by minimizing the production of chemical, toxic, and hazardous waste.

Keywords: Landfill leachate, marigold, phytoremediation, growth and economy.

16.1 INTRODUCTION

Leachates are results of various waste decomposing processes taking place in a landfill. They are highly enriched in various organic and inorganic compounds such as chloride, sulfate, calcium, magnesium, pesticides, polychlorinated biphenyls, dioxins, humic/fulvic acids, and metals such as arsenic, lead, chromium, cadmium, nickel, and so on (Bhagwat *et al.*, 2021). Besides being rich in toxic chemicals, they can also be rich in organic matters such as nitrogen, phosphorous, potassium (NPK), carbon, and trace metals used as biofertilizers (Romero *et al.*, 2013; Sanadi *et al.*, 2019). Thus, landfill leachate supports the growth of various plant species such as sugarcane, hemp, sunflower, marigold, and so on (Abrile *et al.*, 2021; Bhagwat *et al.*, 2023; Kalousek *et al.*, 2020a; Kalousek *et al.*, 2020b; Shaarani *et al.*, 2019). Landfills and plant species have a symbiotic relation; the nutrient-rich leachate supports the growth and development of plant species, whereas the plants grown absorb heavy metals from landfills and accumulate them in different parts and help in reducing the leaching of heavy metals in the leachate.

Thus, several studies have focused on remediating heavy metals from contaminated soil using different plant species. Abbas *et al.* (2019) tested the removal of heavy metals from landfill leachate with aquatic plants such as water hyacinth and water lettuce over 15 days and found that these plants are good candidates for removing Zn, Fe, and Pb. Furthermore, Awan *et al.* (2020) observed that marigold is the best choice for the remediation of Zn-contaminated water. Madanan *et al.* (2021) used marigold for removing heavy metals from landfill leachate and found that marigold is a good candidate for phytoextraction of Zn and Cd but not for Pb. Barasarathi *et al.* (2022) used four ornamental plant species, *Cordyline fruticosa*, *Duranta variegated*, *Tradescantia spathacea*, and *Chlorophytum comosum* for phytoremediation of leachate-contaminated soil in Malaysia and found that *C. fruticosa* has the highest potential toward heavy metal uptake. Even though studies have shown phytoremediation as an efficient method for soil remediation, its field applicability still needs to be evaluated. Włóka *et al.* (2019) pointed out that the cost of growing plants for phytoremediation is many times higher than the income generated by selling the biomass; thus, it becomes important to assess the economy and efficiency of the phytoremediation process.

As marigolds showed the potential for removing more than one heavy metal under different experimental conditions, the Indian marigold (*Tagetes erecta* L.) is considered the best choice for remediating the leachate produced from the Gazipur landfill site. Additionally, it has a short life cycle (50–60 days), a rapid and well-grown root system, and can quickly grow in laterite soil (Goswami & Das, 2017). Marigolds can quickly grow on nutrient-deprived soils and show quick heavy metal uptake in their initial growth phases (Choudhury *et al.*, 2016). Furthermore, marigold flowers are considered aesthetic and used for worship and other religious purposes in India, as natural coloring agents and hedges in gardens and farms for attracting pollinators, increasing their market demand and economic value.

The current study was thus undertaken with the following objectives:

(i) To assess the phytoremediation potential of Indian marigolds.
(ii) To evaluate the growth potential of Indian marigolds when treated with landfill leachate, fertilizers, and under natural (control) conditions.
(iii) Comparative analysis of the economic potential of marigold when treated with landfill leachate, fertilizers, and under natural (control) condition

The study will be particularly useful for policymakers in ensuring sustainable development goal 6: 'Clean water and sanitation.' By eliminating dumping, minimizing the production of chemical, toxic, and hazardous waste, reduced proportion of untreated wastewater, and increased safe recycling and reuse can improve water quality by 2030.

16.2 MATERIALS AND METHODS

16.2.1 Experimental setup

Indian marigolds (*T. erecta* L.) were selected for experiments. Two to three plantlets of marigold were first allowed to acclimatize into pots of dimensions $10 \times 10 \times 12$ cm^3. Plants were grown for 14 days in hydroponic culture, under controlled conditions of the cultivation room. The average temperature was 25°C at day/22°C at night; relative humidity was 55–65%. The photoperiod was 14 h a day and night with average sunshine hours of 12 h/day.

The pot experiments were performed at the National Institute of Hydrology, Roorkee, India. The sapling of 5–10 cm in length was sown in undistributed evenly mixed alluvial soil collected from the campus. All the debris and impurities were carefully removed before filling the pots with soil. An 8 kg pot was filled with soil, and plant sapling was sown in the marigold growing period from 18 November 2020 to February 15, 2021 (i.e., ~84 days). The soil was treated with different concentrations of landfill leachate every week (Table 16.1). The control samples were treated with distilled water (DW) only. There are three treatments with three replicates each, that is, a total of nine pot treatments (Table 16.1). Daily growth parameters such as shoot height, leafs

Table 16.1 Detailed description of the three treatment procedures namely: (a) raw leachate, (b) inorganic fertilizers, and (c) control conditions.

Treatment	No. of Replicates	Details of Treatments	Analysis
100% raw leachate	3	100% raw leachate	Chlorophyll quantification, root shoot, and leaf length and diameter
100% inorganic fertilizers	3	Inorganic fertilizers (NPK) + 100% DW	Chlorophyll quantification, root shoot, and leaf length and diameter
Control	3	100% DW (control)	Chlorophyll quantification, root shoot, and leaf length and diameter

height and width, stem diameter, and the number of buds and flowers were observed and recorded.

16.2.2 Determination of growth parameters

For evaluating the growth in the plant, stem height, leaf length, leaf width, root length, and the number of flowers were calculated. The total height was measured from the base of the stem to the top of the apical bud and the total growth in the stem was calculated using the below equation:

$$\text{Stem height} = \text{Total Height of the plant} - \text{Transplanted Height} \tag{16.1}$$

For the measurement of leaf length and leaf width, the leaves of the plants from the experimental variants were measured with a vernier caliper. The number of flowers is calculated by counting the number of flowers the marigold bears every 5 days considering the life of the flower to be 5 days. The root length of the plants for each treatment is observed at the end of the growth period, that is, 184 days.

16.2.3 Analysis of chlorophyll content

The leaves were homogenized and extracted with acetone. The content of chlorophylls a, b, and c in the extract was subsequently determined using a spectrophotometer (Hach, Mfg., USA) by measuring the absorbance of the extracts at wavelengths 663, 645, and 440 nm and calculating according to Sang *et al.* (2010). From each experimental variant, one mixed sample was considered for analysis. The results presented are the average values of three independent repeats of the experiment:

(a) $C_a = 11.85(\text{absorbance } 664) - 1.54(\text{absorbance } 647) - 0.08(\text{absorbance } 630)$
(b) $C_b = 21.03(\text{absorbance } 647) - 5.43(\text{absorbance } 664) - 2.66(\text{absorbance } 630)$
(c) $C_c = 24.52(\text{absorbance } 630) - 7.60(\text{absorbance } 647) - 1.67(\text{absorbance } 664)$

where C_a, C_b, and C_c are the concentrations of chlorophylls a, b, and c, respectively, mg/L, and absorbance 664, 647, and 630 are the corrected optical densities (with a 1 cm light path) at the respective wavelengths.

After determining the concentration of pigment in the extract, calculate the amount of pigment per unit volume for the three types of chlorophylls given by:

$$\text{Chlorophyll } a/b/c\,(\text{mg/m}^3) = \frac{(C_a/C_b/C_c) \times \text{Extract volume (L)}}{\text{Volume of sample (m}^3)} \tag{16.2}$$

16.2.4 Statistical analysis

All the heavy metal analysis was carried out in triplicates. Standard deviation is used for calculating the precision of data. One-way analysis of variance is used for analyzing the data with 95% confidence levels. Fisher's least-significant difference post-hoc test is used for finding the variance in and between the groups. Minitab 17 is used for the analysis of data.

16.3 PHYTOREMEDIATION

After the end of 14 weeks, the plants were harvested and stored in aluminum foils. The harvested plants were sorted into stems, leaves, and flowers and gently washed with tap water and air-dried, and the dry weight of biomass was determined. The samples are now considered for heavy metal analysis using complete acid digestion.

16.3.1 Heavy metal digestion

The heavy metal digestion in various parts of plants is carried out using the procedure suggested by Zehra *et al.* (2020). A weight of 10 g of the sample (stem, leave, and flower) is taken and grounded finely (2 mm) with the help of a pestle and mortar. The sample was then transferred to a 100 mL beaker and subjected to acid digestion. In acid digestion, 10 mL HNO_3 and DW of 1:1 is added and digested at a temperature of 95°C until the solution is reduced to half of its volume using a hot plate. Again, 5 mL HNO_3 is added and heated until the volume is reduced to half. Then, 5 mL HCl is carefully added, and if white fumes appear, then again, 5 mL HCl is added and digested to half the volume. This makes the solution aqua regia (HNO_3 + HCl) and helps in the fast digestion of the plant materials. After the volume is reduced to half, add 3 mL H_2O_2 (30%) and check for effervescence; this would indicate the presence of undigested plant material. This step is subsequently repeated until there is no visible effervescence. Then, 2 mL DW is added, and the sample is digested to half of its volume. Finally, after the digestion the sample is cooled and filtered through Whatman no. 42 filter paper; the final volume is makeup to 50 mL with DW.

16.4 ECONOMIC EVALUATION OF PHYTOREMEDIATION WITH MARIGOLD

The economic evolution of phytoremediation was based on the percentage refund factor (R) as proposed by Włóka *et al.* (2019). With the economic refund, a balance between annual costs and annual income is evaluated. The refund is calculated using the below equation:

$$\text{Refund}\,(\%) = \frac{\text{Predicted Income}}{\text{Estimated Cost}} \tag{16.3}$$

The predicted income includes the cost of soil remediation and the cost of selling the flowers/biomass, whereas the estimated cost of phytoremediation includes the cost of hiring manpower, the cost of transplantation of the marigold plant, and the cost of fertilizers.

16.5 RESULTS AND DISCUSSION

16.5.1 Growth parameters

For evaluating the efficiency of leachate in supporting the growth and development of the marigold plant, a comparative study is undertaken where the growth parameters, namely shoot height, shoot girth, leaf length, leaf width, root length, chlorophyll content, and the number of flowers are compared among the plants treated with leachate, inorganic fertilizers, and the control, that is, where only tap water is used for growing the plants. The results of this comparative study are shown in Figure 16.1. In the initial 45 days (before the harvest), the rate of increase in plant height is fastest when the marigold is treated with fertilizer, followed by the leachate, and least in the case of tap water. However, after 45 days, the maximum shoot height follows the order: control > leachate > fertilizer as shown in Table 16.2. In the case of growth in shoot girth, leaf length, and leaf width, during the first 45 days before the flowering period, the growth pattern follows the order: leachate > fertilizer > control. However, after 45 days the order of growth becomes control > leachate > fertilizer. Thus, it is observed that the use of inorganic fertilizers accelerates the growth of the marigold in the initial period. However, the plants treated with fertilizers are dwarf on all parameters in comparison with their natural growth. For marigold plants treated with landfill leachate, although the plants are shorter in comparison with natural growth. However, the difference is not as significant as in the case of the growth observed with the inorganic fertilizer treatment.

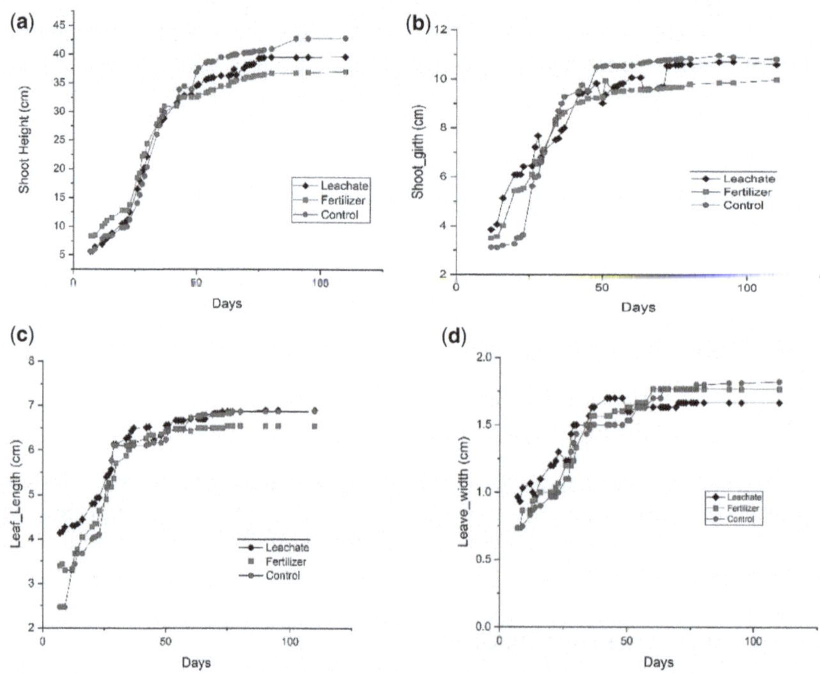

Figure 16.1 Temporal changes in plant growth parameters: (a) shoot height, (b) shoot girth, (c) leaf length, and (d) leaf width.

Table 16.2 Growth parameters in marigold at the end of the experiment when treated with leachate, fertilizers, and control.

Parameters	Leachate	Inorganic Fertilizer	Control
Shoot height	39.67	37.07	42.90
Shoot girth	8.68	8.34	8.74
Leaf length	6.9	6.53	6.88
Leaf width	1.67	1.77	1.82
Root length	8.34	5.3	9
Chlorophyll a	18.54	16.21	12.02
Chlorophyll b	34.57	24.52	12.57
Chlorophyll c	19.21	12.57	11.24
Number of flowers	345	324	314

After completion of the harvest period, the marigold plants were taken out and their root lengths were measured to analyze the root development under different treatment conditions. The root length in the case of control is observed as 9 cm, in the case of treatment with landfill leachate 8.34 cm, and when treated with inorganic fertilizers 5.3 cm. This is because the plants are obtaining the essential nutrients through the application of inorganic fertilizers which hinders root development. In the case of landfill leachate, the essential nutrients nitrogen, phosphorous, and potassium are available but their quantity and forms are not always optimal for needs of plants, which impacts the development of roots of marigolds. It can be concluded that the application of fertilizers hinders the growth of roots and thus the plants do not hold the soil strongly to protect themselves against the strong winds and movement of stray animals which are very commonly found at landfills. Furthermore, shorter root length hinders the uptake of heavy metals from the landfill soil which is an essential objective of phytoremediation.

To understand the health of the marigold plant and its response to environmental factors, three types namely chlorophylls a, b, and c were calculated for the two sets of treatment and control; the values so obtained are given in Table 16.2. Chlorophyll a which is primary for photosynthesis is found to be 18.54, 16.21, and 12.02 μmol/m^2 when treated with landfill leachate, inorganic fertilizers and under control conditions, respectively. Thus, the photosynthesis potential of marigolds increases when treated with leachate. Landfill leachates are loaded with heavy metals and thus various trace metals responsible for the creation of chlorophyll formation achieve their optimal value and enhance photosynthesis (Andresen *et al.*, 2018).

On comparing chlorophyll b content in treatment with landfill leachate, inorganic fertilizers, and under control conditions, the values are 34.57, 24.52, and 12.57 μmol/m^2, respectively. The increased chlorophyll b reflects that the addition of landfill leachate and fertilizers improves marigold efficiency to absorb a broader spectrum of sunlight and is more adaptive to grow in shade (Khaleghi *et al.*, 2012). In addition to chlorophylls a and b, the non-essential photosynthesis chlorophyll c, which is mostly found in algae (Ferroni *et al.*, 2020), is also compared among the three treatments and the values are 19.21, 12.57, and 11.24 μmol/m^2 for the treatment with landfill leachate, fertilizer and as a control, respectively. The presence of high chlorophyll c represents the growth in algae in the flowering pot which is representative of over-watering. Thus, the water requirement for the growth of marigold is the least with landfill leachate in comparison with both inorganic fertilizers and natural conditions.

Marigold holds economic value because of their flowers. Marigold flowers hold aesthetic value, are used for worshipping at religious places, and are popular as raw material for manufacturing natural yellow dyes. Therefore, the total number of marigold flowers produced with each treatment in 105 days of the growing period is calculated as given in Table 16.2. The total number of flowers harvested during the growth period when treated with leachate, inorganic fertilizers, and control conditions are 345, 321, and 311, respectively.

16.5.2 Phytoremediation

To understand the effect of the treatment method on phytoremediation, landfill soil from the Gazipur landfill site with chemical characteristics given in Table 16.3 is subjected to phytoremediation with Indian marigold. Thus, the first pot is treated with 100 mL of leachate every week, to the second pot

Table 16.3 Physico-chemical characteristics of landfill soil and leachate of Gazipur landfill site, New Delhi, India.

Sl. No.	Parameter	Soil	Leachate
1	%Sand	60	–
2	%Silt	22	–
3	%Clay	18	–
4	Porosity	50.184	–
5	EC	1,600 μS	35,000 μS
6	pH	7.75	8.43
7	Ca^{2+}	211.63 ± 77.18 mg/kg	2,985.45 mg/L
8	Mg^{2+}	91.13 ± 23.27 mg/kg	932.275 mg/L
9	K^+	164.75 ± 88.29 mg/kg	419.917 mg/L
10	Na^+	225.70 ± 97.15 mg/kg	226.905 mg/L
11	HCO_3^+	944.64 ± 408.09 mg/kg	11,924 mg/L
12	%TOC	0.97 ± 0.12 mg/kg	1,277.6 mg/L
13	As	0.006921 mg/kg	0.181035 mg/L
14	Cd	0.757 mg/kg	0.642 mg/L
15	Cr	0.632 mg/kg	1.300 mg/L
16	Pb	0.856 mg/kg	4.659 mg/L
17	Ni	0.038 mg/kg	0.517 mg/L
18	Zn	0.448 mg/kg	0.650 mg/L

100 mL of inorganic fertilizer is added every week, and the third pot was under control conditions. At the end of the harvest period, heavy metal accumulation in different parts of the marigold, that is, roots, stem, leaves, and flowers were evaluated and the results are given in Table 16.4.

The result shows that the chromium absorption both in the areal and root is maximum when treated with leachate followed by marigold accumulation under control conditions and the one treated with fertilizers has lower developed roots and shoots leading to lower accumulation of chromium. The cumulative lead accumulation of 103.27, 26.18, and 50.56 mg/kg is reported when treated with landfill leachate, fertilizers, and under the control conditions, respectively. The cumulative accumulation of cadmium and nickel is 37.6 mg/kg and 63.99 μg/kg when treated with landfill leachate, 8.532 mg/kg and 27.673 μg/kg when treated with fertilizers, and 13.64 mg/kg and 46.46 μg/kg under the control conditions. Thus, it can be concluded that marigold is an effective phytoremediator in the presence of landfill leachate enriched in heavy metals in comparison with situations where the marigold is treated with fertilizers in the landfill for supporting its growth.

Furthermore, when the heavy metal absorption under leachate treatment is studied in different parts of the plant, the order of accumulation for Cr follows: roots > stem > leaves > flowers; Pb accumulation: leaves > roots > stem > flowers; Cd gets accumulated in roots > leaves > stem > flowers (Bhagwat *et al.*, 2023). Thus, it can be observed that when marigold is treated

Table 16.4 Heavy metal accumulation in stem, leaves, roots, and flowers of a marigold plant under treatment: (a) leachate, (b) inorganic fertilizers; and (c) control.

Treatment Method	Cr	Pb	Cd	Ni
	Heavy metal accumulation in marigold stem (mg/kg)			
Leachate	62 (9.81%)	15.98 (1.87%)	3.80 (0.50%)	6.62 (0.17%)
Inorganic fertilizer	26.42 (4.18%)	7.5 (0.88%)	2.75 (0.36%)	4.10 (0.11%)
Control	15.71 (2.49%)	10.7 (1.25%)	2.68 (0.35%)	6.84 (0.18%)
	Heavy metal accumulation in marigold leaves (mg/kg)			
Leachate	12.79 (2.02%)	64.95 (7.59%)	12.79 (1.69%)	29.6 (0.78%)
Inorganic fertilizer	0.27 (0.04%)	1.58 (0.18%)	0.27 (0.04%)	0.963 (0.03%)
Control	0.44(0.07%)	0	0.44 (0.06%)	2.99 (0.08%)
	Heavy metal accumulation in marigold roots (mg/kg)			
Leachate	126.78 (20.06%)	20.58 (2.4%)	20.56 (2.72%)	26.09 (0.69%)
Inorganic fertilizer	25.43 (4.02%)	2.03 (0.24%)	2.232 (0.29%)	7.47 (0.20%)
Control	38.21 (6.05%)	4.27 (0.50%)	2.33 (0.31%)	7.23 (0.19%)
	Heavy metal accumulation in marigold flowers (mg/kg)			
Leachate	1.25 (0.08%)	1.76 (0.21%)	0.45 (0.06%)	1.68 (0.04%)
Inorganic fertilizer	27.95 (1.77%)	15.70 (1.83%)	3.28 (0.43%)	15.14 (0.40%)
Control	31.56 (1.99%)	35.59 (4.16%)	8.19 (1.08%)	29.40 (0.77%)

with landfill leachate it accumulates minimum metal concentration in flowers. Furthermore, when the heavy metal accumulation in flowers is observed under treatment with fertilizers and under the control conditions, the cumulative accumulation of flowers is 62.07 and 104.74 mg/kg, respectively, which is much higher than 5.14 mg/kg of cumulative accumulation in treatment with leachate.

16.5.3 Economic evaluation

Because of the in-situ character of plant-dependent techniques in phytoremediation and the lower cost of sludge disposal generated as a result of chemical treatment, phytoremediation methods are considered more economical (Seo *et al.*, 2009). However, Włóka *et al.* (2019) pointed out that the procedural cost of phytoremediation-based technologies is high because of factors such as types and degree of pollution, local environmental, urban, environmental conditions, and availability of infrastructure. Furthermore, each phytoremediation method is designed case by case, where the procedure includes preliminary evolution of contaminant, terrain type, and local environmental conditions.

Thus, to estimate the cost of the phytoremediation procedure in the current study it is assumed that the cost was incurred on manpower for the plantation of marigolds, and the cost of seeds, transplantation, and fertilizers. Furthermore, for estimating the total income generated the cost of selling marigolds is estimated at the market price in New Delhi, India which is Rs. 42/kg. The expenditure incurred and the income generated per hectare of landfill is enumerated in Table 16.5.

Table 16.5 Estimation of total expenditure and total income generated with phytoremediation with marigold.

Total Cost Incurred in Phytoremediation				
Types of Treatment	**Manpower (at Rs. 500/day) for 180 Days**	**Cost of Seeds + Transplantation (Rs. 3,000 for 10 kg)**	**Cost of Fertilizers**	**Total Cost Incurred**
Leachate	900,000	30,000 for seeds + 30,000 for transplantation	Nil	960,000
Fertilizer	900,000	30,000 for seeds + 30,000 for transplantation	20,000	980,000
Natural	900,000	30,000 for seeds + 30,000 for transplantation	Nil	960,000
Total income generated with phytoremediation				
Types of Treatment	Income through flower selling (44,400 plants/ ha)			
Leachate	6,433,560			
Fertilizer	6,041,952			
Natural	5,855,472			

For evaluating the economic sustainability, the refund is calculated using equation (16.1). Based on the calculation it was identified that the refund for marigolds grown in leachate, with fertilizer, and under the control conditions are 5.702, 5.165, and 5.09, respectively. Thus, it can be concluded that marigold farming is highly profitable in all three cases. When fertilizers are added the growth potential and number of flowers increases when compared with the control conditions but the net refund increases by 0.1. However, when the marigold plants are treated with landfill leachate, plants are provided with trace metals in addition to nitrogen, phosphorous, and potassium leading to better flower production, which increases the refund by 0.7 compared to the control conditions and by 0.5 compared with fertilizers.

16.6 CONCLUSION

When the symbiotic relationship between landfill leachate toward supporting marigold growth, and the phytoremediation potential of marigold from leachate-contaminated soil was evaluated, and the applicability of marigold for phytoremediation when tested based on economic refund, the following observations were made:

(i) The order of growth in marigolds (stem, root, and leaves) follows control > leachate > inorganic fertilizers; thus, the presence of heavy metals in leachate-contaminated soil hinders the growth of marigolds. However, the chlorophyll content follows the order leachate > inorganic fertilizers > control; thus, plants are more productive when treatments are applied to them.
(ii) The total number of flowers harvested during the growth period when treated with leachate, inorganic fertilizers, and control conditions are 345, 324, and 314 respectively. Thus, the number of flowers harvested increases by 10% when leachate is applied to the marigold plant.
(iii) When total metal removal is considered, maximum remediation is observed in the order leachate > control > inorganic fertilizer, this can be attributed to the root length which is the least in the case of inorganic fertilizers. In the presence of leachate, marigold removes 50% of the total heavy metals present in the soil, with fertilizer remediation 34.3%, whereas under natural conditions the removal is only 19.5%.
(iv) Flowers are the most economical part of marigold plants, and the accumulation of heavy metals in the flowers is a cause of health concern. The heavy metal accumulation in the flowers follows the order control > inorganic fertilizers > leachate. Thus, treating with landfill leachate makes marigold flowers safer for use.
(v) The economic potential of the three treatments is evaluated and the refund values are 5.702, 5.165, and 5.09 for marigolds grown in leachate, with fertilizers, and under control conditions, respectively.

The study suggests that marigold acts as an effective phytoremediator for landfill leachate and has a high economic refund when treated with landfill leachate. Thus, the study will be a value-add for policymakers in ensuring sustainable development by minimizing the production of chemical, toxics and hazardous waste.

ACKNOWLEDGMENT

This study was supported through a project titled 'Leachate Transport Modelling for Gazipur landfill site and suggesting Ameliorative measures' (Grant Ref: SP-28/2019-20/PDS-19) under National Hydrology Project.

Conflict of Interest

The authors declare that the research was conducted in the absence of any commercial or financial relationships that could be construed as a potential conflict of interest.

Authors' Contribution

Anjali Bhagwat: conceptualization, data curation, formal analysis, investigation, methodology, visualization, writing – original draft, writing – review and editing. Chandra Shekhar Prasad Ojha: supervision and writing – review and editing.

Data Availability

The datasets used and/or analyzed during the current study are available from the corresponding author on reasonable request.

REFERENCES

Abbas Z., Arooj F., Ali S., Zaheer I. E., Rizwan M. and Riaz M. A. (2019). Phytoremediation of landfill leachate waste contaminants through floating bed technique using water hyacinth and water lettuce. *International Journal of Phytoremediation*, **21**(13), 1356–1367, https://doi.org/10.1080/15226514.2019.1633259

Abrile M. G., Fiasconaro M. L., Gervasio S., Antolín M. C. and Lovato M. E. (2021). Evaluation of the sensitivity and production of flowers in *Tagetes erecta* L. exposed to high doses of sodium from irrigation with landfill leachates. *International Journal of Recycling Organic Waste in Agriculture*, **10**(1), 63–75.

Andresen E., Peiter E. and Küpper H. (2018). Trace metal metabolism in plants. *Journal of Experimental Botany*, **69**(5), 909–954, https://doi.org/10.1093/jxb/erx465

Awan B., Sabeen M., Shaheen S., Mahmood Q., Ebadi A. and Toughani M. (2020). Phytoremediation of zinc contaminated water by marigold (*Tagetes Minuta* L.), *Central Asian Journal of Environmental Science and Technology Innovation*, **1**(3), 150–158, https://doi.org/10.22034/CAJESTI.2020.03.04

Barasarathi J., Auwal H., Pariatamby A., Hamid, F. S. and Uche E. C. (2022). Phytoremediation of leachate contaminated soil: a biotechnical option for the bioreduction of heavy metals induced pollution in tropical landfill. *Environmental Science and Pollution Research*, **29**, 22069–22081, https://doi.org/10.1007/s11356-021-17389-3

Bhagwat A., Kumar R., Ojha C. S. P., Sharma M. K., Pant A., Sharma B. and Tyagi J. V. (2021). Assessing efficiency and economic viability in treating leachates emanating from the municipal landfill site at Gazipur, India. Environmental Science and Pollution Research, **30**, 71813–71825 (2023), https://doi.org/10.1007/s11356-021-16724-y

Bhagwat A., Ojha C. S. P., Pant A. and Kumar R. (2023). Interaction among heavy metals in landfill leachate and their effect on the phytoremediation process of Indian Marigold. *Journal of Hazardous, Toxic, and Radioactive Waste,* **27**(1), 04022039, https://doi.org/10.1061/(ASCE)HZ.2153-5515.0000731

Choudhury M. R., Islam M. S., Ahmed Z. U. and Nayar F. (2016). Phytoremediation of heavy metal contaminated Buriganga riverbed sediment by Indian mustard and marigold plants. *Environmental Progress & Sustainable Energy*, **35**(1), 117–124, https://doi.org/10.1002/ep.12213

Ferroni L., Živčak M., Sytar O., Kovár M., Watanabe N., Pancaldi S., Baldisserotto C. and Brestič M. (2020). Chlorophyll-depleted wheat mutants are disturbed in photosynthetic electron flow regulation but can retain an acclimation ability to a fluctuating light regime. *Environmental and Experimental Botany*, **178**, 104156, https://doi.org/10.1016/j.envexpbot.2020.104156

Goswami S. and Das S. (2017). Screening of cadmium and copper phytoremediation ability of *Tagetes erecta*, using biochemical parameters and scanning electron microscopy–energy-dispersive X-ray microanalysis. *Environmental Toxicology and Chemistry*, **36**(9), 2533–2542, http://icar-ciari.res.in/ https://doi.org/10.1002/etc.3768

Kalousek P., Schreiber P., Vyhnánek T., Trojan V., Adamcová D. and Vaverková M. D. (2020a). Effect of landfill leachate on the growth parameters in two selected varieties of fiber hemp. *International Journal of Environmental Research*, **14**, 155–163.

Kalousek P., Schreiber P., Vyhnánek T., Trojan V., Adamcová D. and Vaverková M. D. (2020b). Effect of landfill leachate on the growth parameters in two selected varieties of fiber hemp. *International Journal of Environmental Research*, **14**(2), 155–163, https://doi.org/10.1007/s41742-020-00249-2

Khaleghi E., Arzani K., Moallemi N. and Barzegar M. (2012). Evaluation of chlorophyll content and chlorophyll fluorescence parameters and relationships between chlorophyll a, b and chlorophyll content index under water stress in *Olea europaea* cv. Dezful. *International Journal of Agricultural and Biosystems Engineering*, **6**(8), 636–639.

Madanan M. T., Shah I. K., Varghese G. K. and Kaushal R. K. (2021). Application of Aztec Marigold (*Tagetes erecta* L.) for phytoremediation of heavy metal polluted lateritic soil. *Environmental Chemistry and Ecotoxicology*, **3**, 17–22, http://doi.org/10.1016/j.enceco.2020.10.007

Romero C., Ramos P., Costa C. and Márquez M. C. (2013). Raw and digested municipal waste compost leachate as potential fertilizer: comparison with a commercial fertilizer. *Journal of Cleaner Production*, **59**, 73–78, https://doi.org/10.1016/j.jclepro.2013.06.044

Sanadi N. F. A., Fan Y. V., Lee C. T., Ibrahim N., Li C., Gao Y., Ong P. Y. and Klemeš J. J. (2019). Nutrient in leachate of biowaste compost and its availability for plants. *Chemical Engineering Transactions*, **76**, 1369–1374, https://doi.org/10.3303/CET1976229

Sang N., Han M., Li G. and Huang M. (2010). Landfill leachate affects metabolic responses of Zea mays L. seedlings. *Waste Management*, **30**(5), 856–862, https://doi.org/10.1016/j.wasman.2010.01.023

Seo J. S., Keum Y. S. and Li Q. X. (2009). Bacterial degradation of aromatic compounds. *International Journal of Environmental Research and Public Health*, **6**(1), 278–309.

Shaarani S. M., Mokhtar N. J., Arshad Z. I. M., Man R. C., Mudalip S. K. A. and Sulaiman S. Z. (2019). Co-composting landfill leachate with sugarcane bagasse for biofertilizer production. *AIP Conference Proceedings*, **2124**(1), 020032, https://doi.org/10.1063/1.5117092

Włóka D., Placek A., Smol M., Rorat A., Hutchison D. and Kacprzak M. (2019). The efficiency and economic aspects of phytoremediation technology using *Phalaris arundinacea* L. and *Brassica napus* L. combined with compost and nano SiO_2 fertilization for the removal of PAH's from soil. *Journal of Environmental Management*, **234**, 311–319, https://doi.org/10.1016/j.jenvman.2018.12.113

Zehra A., Sahito Z. A., Tong W., Tang L., Hamid Y., Khan M. B., Ali Z., Naqvi B. and Yang X. (2020). Assessment of sunflower germplasm for phytoremediation of lead-polluted soil and production of seed oil and seed meal for human and animal consumption. *Journal of Environmental Sciences*, **87**, 24–38, https://doi.org/10.1016/j.jes.2019.05.031

doi: 10.2166/9781789063318_0375

Chapter 17

Landfill leachate pollution index

Arun Kumar Thalla[1], Anakha Ambujan[1] and Vijay Kubsad[2]*

[1]Department of Civil Engineering, NITK Surathkal, Mangalore, Karnataka, India
[2]Washington State Department of Ecology, Spokane, Washington, USA
**Corresponding author: arunkumar@nitk.edu.in*

ABSTRACT

A major portion of wastes generated due to rapid industrialization and urbanization ends up in landfills. The leachate generated as a consequence of physical and chemical changes in landfills can pose danger to the water and soil in the proximity. Poorly designed, faulty monitoring of leachate-handling systems could lead to unfortunate contamination affecting human and livestock health as well as adversely affecting agriculture and soil. An urgent need to quantify the contamination caused due to faulty handling of hazardous waste landfill (HWL) leachate has been felt. The leachate pollution index (LPI) has been an essential tool for quantifying the leachate contamination potential. The existing index was developed based on criteria, in particular, the toxicity and availability of the pollutants in the leachate from municipal solid waste (MSW) landfill sites, and thus many pollutants which are present in HWL leachate in significant concentrations were screened out. However, with rapid industrialization, the quantities of hazardous waste generated and its dumping in HWLs and the toxicity associated or rendered with it is not dealt currently but there is an urgent need for the same. Thus, pollutants such as halogenated organic compounds, polycyclic aromatic hydrocarbons and metals such as selenium need to be considered in the formulation of LPI for hazardous waste landfills. In this chapter, a review of sources of hazardous wastes, their conventional handling methods, developing LPI for MSW landfill sites, and further how the same can be modified to develop LPI for HWLs are discussed. The procedure stated for LPI development in this study can serve as an essential tool for quantifying the leachate contamination potential, can be used as a comparison tool for different HWL leachates, furthermore, and for ranking landfill sites based on the contamination potential.

Keywords: Landfill leachate, hazardous waste, leachate pollution index.

17.1 INTRODUCTION

The rapid surge in population, and subsequent urbanization and industrialization in the last few decades have resulted in enormous quantities of solid waste, both municipal and hazardous waste (Gholami *et al.*, 2020). Hazardous waste is any waste that causes danger or is likely to cause a threat to health or the environment, whether alone or in contact with other waste or substances because of its physical, chemical, biological, reactive, toxic, flammable, explosive, or corrosive characteristics (Central Pollution Control Board, 2016). Industrial wastes are immensely hazardous in nature and have the potential to cause severe environmental problems.

On the basis of data reported by the Central Pollution Control Board (2021), in India, about a net quantity of 10.67 million MT has been generated during 2019–2020. As tabulated in the report by Basel Convention (2018), the average generation of hazardous waste from high-income countries during 2007–2015 has been in the range (min–max) of 134–148 kg/year per inhabitant, whereas for low-income countries, it is in the range of 12–29 kg/year per inhabitant. However, in developing countries, hazardous waste management is impaired by a lack of comprehensive legislation and unauthorized scrap yards dealing with e-waste (Mmereki *et al.*, 2016). Poor understanding and inappropriate disposal methods exercised during the handling and disposal of hazardous wastes are increasing significant health hazards and environmental pollution.

Generally, most of the wastes (municipal and hazardous) generated end up in landfills. The leachate generated from these landfills can be a potential threat to the water and soil in the proximity. In addition to this, a faulty design or a leachate monitoring facility could aggravate the risks affecting human and livestock health as well as adversely affecting agriculture and the soil. Therefore, there is a necessity to quantify the contamination caused by the faulty handling of hazardous waste landfill (HWL) leachate, and in this chapter, tools for quantifying the leachate contamination potential from municipal and HWLs are discussed.

17.1.1 Hazardous waste in the recent scenario

In the era of the COVID-19 pandemic, hazardous waste, mainly healthcare waste generation, is the primary focus. Healthcare waste is considered the second most hazardous waste globally (Das *et al.*, 2021). It includes hazardous and non-hazardous, such as sharps, human body parts, blood, chemical waste, pharmaceutical waste, and medical devices. During infectious disease outbreaks, the waste accumulation surges exponentially; therefore, as suggested in the literature (Ramteke & Lal, 2020), special care must be taken by respective management to avoid harmful impacts on the human beings and the environment. The types of waste released from healthcare facilities include:

- Non-hazardous waste: The waste generated from the waiting areas of these facilities; these wastes are stored in prescribed bags or containers, sealed, and moved or disposed of as per the regulations (WHO, 2020).

- Hazardous waste: Around 10–25% of the total healthcare waste is designated as hazardous waste (Chartier *et al.*, 2014). Healthcare hazardous waste can be classified as chemical wastes, pharmaceutical wastes, infectious wastes, and sharp (needle, syringe, sharp items, etc.) wastes.
- Chemical wastes: This type of waste accounts for ~3% of the waste produced from the total healthcare activities (Ilyas *et al.*, 2020). Wastes such as laboratory reagents, film developing reagents, expired/unused disinfectants, waste containing heavy metals (batteries, broken thermometers, blood pressure gauges, etc.) are considered chemical healthcare waste (Chartier *et al.*, 2014). There are several safer alternatives available to substitute some of these hazardous substances. However, there are many facilities in both developing and developed countries that still use these toxic chemicals, and have poor chemical waste management strategies (Das *et al.*, 2021).
    - Pharmaceutical waste: Used biological products for therapy and transdermal patches, and contaminated pharmaceuticals, including vaccines, can also be listed as pharmaceutical waste (Malsparo, 2020). As mentioned in Das *et al.* (2021), there has been a surge in the amount of pharmaceutical waste during the COVID-19 pandemic, which ultimately increased the waste from the pharmaceutical production industries.
    - Infectious waste: Infected personal protective equipment, that is, heavy-duty gloves, masks, goggles, and face shields, are also considered infectious waste. Waste generated from these materials has increased substantially during the COVID-19 pandemic (WHO, 2020). Waste that contains infective pathogens, which could be a potential source for the progression of the disease, is considered infectious healthcare waste. There is a huge challenge in managing these types of waste during the pandemic.
    - Sharps waste: This type includes used or unused syringes, infusion sets, scalpels, knives, blades, and broken glasses. Sharp wastes generated from healthcare should be treated with extreme care, and adequately managed during the pandemic (WHO, 2020).

Waste management during the COVID era has been a considerable challenge; infected healthcare solid wastes are segregated and packed by waste-handling employees in hospitals. Das *et al.* (2021) showed that the method of disposal of these healthcare wastes depends on the particular hospital and its waste management facilities. However, a commonly adopted approach is sterilizing by autoclave or irradiation before disposing of the segment in a licensed landfill. Increasing the recyclable fraction and reducing the disposal to landfills has been considered to aid sustainable waste management.

Disposing of the incineration ash and the waste into the unsecured landfill contaminates the water resource in the proximity (Ramteke & Lal, 2020). The pandemic has put a massive pressure on waste management systems. The typical disposal technique used for the disposal of hazardous waste is discussed in Section 17.1.2.

17.1.2 Disposal methods for hazardous waste

As suggested, disposal of hazardous waste is the final and the most crucial step for an effective hazardous waste management plan. Its effectiveness depends on the activities performed in the prior stages, that is, waste collection, prevention, minimization, storage, and so on. The choice of a particular disposal method would depend on many factors such as financial background, the technology available, and so on. However, in general, countries adopt more than one disposal option (Visvanathan, 1996). There are different disposal techniques available for the final stage of a waste management system. Some disposal techniques followed are incineration, immobilization, landfilling, off-shore, underground storage, and so on. The disposal of waste lies at the bottom of the hierarchy. The first stage would be to reduce the quantity of waste produced (reduction at source), the next would be to reuse the waste, and be recycled, and the final and least desirable option is to dispose of the waste (Central Pollution Control Board, 2016).

In developing countries, hazardous waste management systems lack a systematic approach to administer waste management programs: inability to effectively collect and manage wastes (Mmereki *et al.*, 2016). Reports have shown that some undefined portions of the hazardous waste are shipped legally or illegally from developed to developing countries (Visvanathan, 1996). However, countries such as India have mentioned the prerequisite for the type of waste exported or imported, which could be an initiative toward a regulated waste management system. The most apparent reason for the export of waste from developed countries to developing countries is the rising cost of disposing of hazardous waste in the home country (Orloff & Falk, 2003). A waste management system is mainly mismanaged in most developing countries; hence, a review of waste disposal techniques followed in context with developing countries is discussed in this section.

Incineration: The waste that cannot be recycled, reduced, or safely disposed into secured landfill sites is incinerated. This disposal method incorporates volume, and weight reduction, detoxification, and energy recovery (Visvanathan, 1996). For the current scenario, most countries are using this method, as incineration disinfects the hazardous healthcare waste produced. Cement kilns and other furnaces have been considered as alternative facilities for disposing of the enormous amount of waste produced (Das *et al.*, 2021). The gaseous products of this process are released into the atmosphere, and solid residue, which is toxic, will be landfilled. The by-products are hazardous; therefore, the disposal of the same needs to be controlled and regulated.

Immobilization: In this method, the toxic waste is mixed with materials that create a highly impermeable solid matrix; the mechanism involved can be either physical or chemical or physico- chemical processes. Material for creating such an impermeable matrix can be cement, epoxy, resins, polymeric sulfur, and so on. Some advantages of this method as noted by Visvanathan (1996) are, the material used for solidification is cheaply available, and is tolerant of chemical variations in sludge. The disadvantages would be that this method adds to the weight and makes it bulky. Also, low-strength cement waste mixtures are often vulnerable to leaching of pollutants.

Off-shore: This method can be used in either of these modes: ocean dumping or export. As mentioned earlier, the developed countries export a portion of the waste to developing countries, where the financial constraints are more relaxed compared to the hometown. Ocean dumping is a common practice in many countries. As the name suggests, in this method, the waste is discharged into the ocean. This method has been considered only as a short-term or medium-term option for waste disposal (Visvanathan, 1996).

Deep-well disposal: The solid waste is emplaced in a geological formation, which should be homogenous, dense, massive, and hydrologically and mechanically stable. The pollutants, together with infiltrating water, generate leachate, which could cause severe environmental impacts. Another form of disposal is the storage of these hazardous wastes in thick salt deposits. The waste before disposal needs to be pre-treated, stabilized, and stored (Visvanathan, 1996).

Landfilling: This disposal method has been the most commonly adopted method, as is evident in Orloff and Falk (2003); Maowei *et al.* (2021); and Visvanathan (1996). Countries have proposed standards for design, maintenance, and technical requirements for a safe landfill site operation. An environmentally and technically sound landfill must have a double-layer liner system with leak detection, a properly monitored leachate collection system, and proper drainage and treatment system. Landfilling is the most commonly adopted disposal technique; sufficient attention should be paid to design, operate, and monitor if public health and the environment need to be protected (Visvanathan, 1996). If the by-products released, such as the leachate, are handled recklessly, there are chances of extreme mishaps. The leachate released from hazardous waste disposal sites constitutes heavy metals, organics with less biodegradable fraction, inorganics, and conventional pollutants (Gautam & Kumar, 2021). Quantifying the contamination caused by faulty handling of leachate from these disposal sites is the primary focus of this study; hence, a review of the existing quantification techniques for quantifying the contamination from municipal solid waste (MSW) landfills has been discussed in the next section.

17.2 QUANTIFICATION OF POLLUTION USING EXISTING LEACHATE POLLUTION INDEX

The pollution index proposed by Kumar and Alappat (2003) has been established as an efficient tool for judging the contamination potential by the leachate released from MSW landfills. The leachate released from the landfill is influenced by the type of waste disposed of, the degree of composition, climate, age of the landfill, and many other factors (Katsiri *et al.*, 1999). The leachate from a landfill has the potential to pollute the soil and water in the proximity. Therefore, a need to quantify the contamination potential was felt, as the index could help understand the landfill site that would need immediate attention. The methodology for the development of the index was carried out in four phases. The first phase required expertise from researchers, academicians, authority officials, engineers, and so on.

In Kumar and Alappat (2003), ~80 panelists were selected for carrying out the first two phases. The first phase was to establish the pollutants for inclusion in the index. The steps involved in the first phase are listed below:

- A list of 50 pollutants was given to the panelists, and they were requested to designate each pollutant as 'Do not include' or 'Undecided' or 'Include.'
- The pollutants designated 'Include' were asked to be rated again on a scale of '1'–'5,' where '1' was used for pollutants with relatively low significance, and '5' was used for pollutants with relatively high relevance.
- The panel suggested few inclusions; however, these pollutants didn't make into the final list.
- Eighteen pollutants were finalized after the first stage of the procedure. Some of the pollutants with high significance were stroke off from the list. The studies suggested that those pollutants were present in trace levels in MSW landfill leachate.

The second phase of the study was to develop the sub-index curve; the panelists were again asked to plot the curve on a graph sheet: the sub-index score in the range of '0–100' were marked on the ordinate, and the concentration of each pollutant was marked on the abscissa. The responses received were averaged out, and the final graphs were proposed. The third phase was the derivation of weights, and these were obtained by dividing the significance rating obtained for the selected pollutant by the arithmetic sum of the significance rating of the selected pollutants. These result in weights which add up to a sum of '1.'

The final or fourth phase of the development is aggregating the weights, and the scores obtained for each pollutant with respect to their concentration. The weighted linear sum aggregation method, as demonstrated in Ott (1978), was used. After following all four phases, an index with the potential to quantify the contamination was obtained. As mentioned earlier, the LPI can be a potential tool for monitoring the leachate trends over the lifetime of a landfill site, and so forth. The general LPI, however, fails in communicating the details about the strength of pollutant groups. Therefore, a sub-index was formulated.

17.2.1 Formulation of the sub-index

The proposed index was divided into three sub-classifications. The splitting of the index made the modified index more informative, and this modified index provides a better notion of the strength of the pollutant group. The procedure involved has been briefly explained below:

- The selected pollutants were categorized into three groups: 'Organic,' 'Inorganics,' and 'Heavy metals.'
- Around 4 out of the final 18 pollutants were grouped under organic, 5 under inorganics, and the rest were grouped as heavy metals.
- The weights for each group were calculated such that the sum of the weights came out to be '1.'

The sum obtained by adding the weights obtained in the earlier study were used to aggregate the LPI obtained for each group. A case study to show the application of the modified sub-index has been carried out by Kumar and Alappat (2005a). A need for such an index for estimating the contamination from leachate released from HWL has been felt. The need and the background study for the index developed in the study are discussed in the next section.

17.3 HAZARDOUS WASTE LANDFILL LEACHATE POLLUTION INDEX: NEED AND BACKGROUND STUDY

As mentioned earlier, in developing countries, the management and disposal are impaired due to the lack of comprehensive regulations. Hazardous waste management includes possession, transportation, handling, storage, and finally disposal. The hazardous wastes are treated using improper techniques and unsafe methods; these are treated in unlicensed facilities using faulty maintained landfill sites (Mmereki *et al.*, 2016). In some cases, hazardous wastes are exported to developing countries by the developed countries (Thanh *et al.*, 2010), leading to severe life-threatening environmental impacts.

Literature suggests that the leachate from HWLs is more dangerous than that released from MSW landfills. The reason is attributed to the type of waste disposed of and the composition of the leachate. The wastes disposed into MSW landfills have more biodegradable fraction than the wastes disposed into HWLs. Many literature studies have carried out the characterization and quantification of contamination for municipal waste leachate, but almost negligible efforts have been made to quantify contamination of leachate produced from HWLs (Gautam & Kumar, 2021).

Many developed countries have flawed management systems, and the current scenario has put pressure on waste disposal and management systems. In India, as per the Bio-Medical Waste (Management and Handling) Rules (2016), medical wastes are either sterilized or incinerated, and disposed of into HWLs. In many other countries, waste disposal has been of concern in the current scenario; therefore, the waste expected to landfills might increase, leading to uncontrolled release of infected leachate and other harmful impacts. Therefore, a need to establish a technique to quantify contamination caused by leachate released from a faulty managed landfill has been felt. Thus, the study focuses on developing a general hazardous waste landfill leachate pollution index (HWLLPI) and a modified sub-HWLLPI.

The existing leachate contamination tool concepts have been used as a base to develop the HWLLPI. However, the reasons why the existing tool would not be a sensible choice to quantify HWL leachate, and the background of the development of the index are explained in detail in Sections 17.3.1 and 17.3.2.

17.3.1 Existing LPI, not a reasonable choice for quantifying pollution caused by HWL leachate

The existing LPI has been a potential tool for quantifying the contamination caused due to mismanaged leachate (Kumar & Alappat 2005a). This index helps in comparing leachate contamination potential without actually comparing the individual concentration of pollutants. This index also provides an insight on which landfill site needs immediate attention, as the remedial and preventive measures cannot be undertaken for all the existing landfill sites due to financial constraints (Kumar & Alappat, 2003). This technique has also been useful in monitoring the leachate trends at a given landfill over a period of time. This technique can be used for ranking landfill sites (Kumar & Alappat, 2005b). This technique, as mentioned earlier, has been a potential tool when the leachate is released from an MSW landfill site.

In the development of the existing tool, ~50 pollutants were selected in the initial stage, which was then reduced to 18 pollutants after the first phase, that is, screening of final pollutants for inclusion. During this stage, the panel of experts suggested a list of 14 new pollutants apart from the initial 50. These included manganese, sodium, polychlorinated biphenyls (PCB), adsorbable organic halides, and so on, but these pollutants didn't make into the final list as they received low significance. Pollutants such as toluene also received low significance (Kumar & Alappat, 2003). Many other pollutants with a high significance value, such as cadmium and selenium, were excluded from the final list. Cadmium had a significance value of ~3.759 and selenium had a significance value of ~3.311, whereas toluene had a significance value of 2.744. These pollutants were excluded from the final list, as these pollutants were found in trace amounts in the leachate from MSW landfill sites. Pollutants such as total iron had a significance value of ~2.830, still it made into the list as the presence of a large concentration of iron in leachate from MSW landfills has been noted.

However, the leachate from HWL sites has shown a significant concentration for cadmium, selenium, halogenated organics, toluene, and so on. The concentrations noted as per studies carried out by Environmental Protection Agency (1982); Ghassemi *et al.* (1984); Pavelka *et al.* (1993); and Yasuhara *et al.* (1999) in the leachate from HWL were very high. Therefore, using an index that ignores the mentioned pollutants and several other pollutants of higher significance for HWL leachate would not be a sensible choice. Pollutants such as chlorobenzene, chloroform, tetrachloroethene, and dichloromethane were also present in significant concentrations. Therefore, these were grouped as 'halogenated organic compounds,' and were added to the list.

The existing LPI was established using the traditional Delphi technique, which didn't incorporate the vagueness of the judgments made; however, the index developed in this study uses two fuzzy-integrated tools for screening and assigning weights. As mentioned by Gautam and Kumar (2021), leachate from HWLs are several times dangerous than the leachate from MSW. Hence, using an index meant specifically for MSW landfill leachate and for HWL leachate would most probably present a false sense of security as the pollutants that are not covered in the index will not be considered for the calculations.

The index developed in this study had included conventional parameters such as pH, total dissolved solid (TDS), total kjeldahl nitrogen (TKN), and so on. However, these were excluded from the list as there were more threatening pollutants that outperform these in terms of pollution. Thus, using the existing LPI for quantifying the contamination from a faulty handling of leachate from HWLs would not be a sensible choice. The pollutants in the index and the background on how the index was developed are discussed in the following section.

17.3.2 Background for the formulation of the HWLLPI

As the need for the development of the HWLLPI was felt, the pollutants for the index were finalized; 40 pollutants were selected for the initial screening. As mentioned by Gautam and Kumar (2021), there were no recent studies carried out showing the trend of pollutants in the leachate from HWLs. Therefore, pollutants were selected from a list of literature (Environmental Protection Agency, 1982; Ghassemi *et al.*, 1984; Pavelka *et al.*, 1993; Yasuhara *et al.*, 1999). The literature in recent years (Gautam *et al.*, 2019) has focused on removing only a few pollutants, and didn't provide any information on the characteristics of the pollutants in the leachate samples. Therefore, selection of the 40 pollutants for the initial phase was based on the literature as mentioned earlier. In the report from Environmental Protection Agency (1982), an insight into the maximum, minimum, and average concentrations of all the pollutants from ~30 hazardous landfill sites was obtained. Pavelka *et al.* (1993) provided leachate data from ~18 commercial HWL sites. The leachate samples in this study were analyzed for ~62 volatile compounds, 107 semi-volatile compounds, 16 metals, 28 pesticides, herbicides, and insecticides, and 17 other chemicals. Yasuhara *et al.* (1999) provided a notion on the organic compounds, and certain inorganic elements in ~11 landfill sites in and around Japan. Details of characteristics of ~30 leachates from ~11 disposals were discussed by Ghassemi *et al.* (1984).

Another addition to the selection of the pollutants was adopted from Central Pollution Control Board (2010). Here, discharge limits of five pollutants, namely adsorbable organic halogens (AOXs), polycyclic aromatic hydrocarbons, benzene, toluene, and xylene were mentioned that needs to be taken into consideration when leachate from HWLs is to be discharged. These reports provided an insight into the pollutants to be selected for the initial phase of the index development. The concepts of decision-making tools for assigning weights and screening were conveniently incorporated in the study. These methods include the imprecisions in the traditional tools (Liu *et al.*, 2020). The methodology of the development is explained in Section 17.4.

The details of leachate characteristics from Gautam and Kumar (2021) were referred to for understanding the application of the index developed. The index developed in this study was used to calculate the contamination index for all the three leachate samples and obtain a more informative notion on the contamination; the modified sub-LPI for the same was also calculated. In developing and developed countries, the index developed could be used to understand which hazardous landfill site needs immediate attention.

17.4 HWLLPI: DEVELOPMENT OF THE INDEX

A technique to quantify the contamination potential of leachate from MSW landfills has been developed; however, using the index for studying leachate with entirely different characteristics would not be a sensible choice that led to this study. As the nature of waste disposed of is quite different, the process inside the landfill is different. Thus, the nature of leachate released is also considerably different (Gautam & Kumar, 2021). A technique to quantify the extent of contamination caused by a faulty handling of HWL leachate would be an important tool for developing countries where the landfills are recklessly managed, and as well as for the developed countries. The developed index for HWLs would be an essential tool for quantifying contamination, comparing leachate samples, and ranking based on the contamination potential.

For the establishment of the index, a step-by-step methodology was followed. The decision-making tools used to carry out the first two steps were more recent and advanced techniques. The development of this index included (1) methods to finalize the pollutants for the index, (2) a method to assign weights to the pollutants to understand the significance of each pollutant over the other, (3) plotting a sub-index curve for marking the scores with respect to the concentration, and (4) finally aggregating the weights and the sub-index scores. The decision-making tools and the aggregation method are discussed in Sections 17.4.1–17.4.4.

The development of the index includes four phases:

- Finalizing the pollutant for inclusion in the index
- Deriving weightage of the selected pollutant
- Plotting the sub-index curves
- Aggregation of weights and the scores to develop LPI

The first and the foremost step to the development of the index is selecting experts with expertise in the field of research. The experts may include academicians, scientists, regulatory officials, engineers, and so on. The next step would be to use a suitable tool to convert the judgments given by the selected experts into understandable numerical terms. The first phase of the development procedure starts with screening the pollutants for inclusion in the index.

17.4.1 Screening the pollutants for inclusion in the index

A decision-making tool was used for the screening of the most contaminating pollutants from a list of pollutants mentioned in the literature by Pavelka *et al.* (1993); Environmental Protection Agency (1982); Yasuhara *et al.* (1999); and Ghassemi *et al.* (1984). A fuzzy-integrated Delphi technique was used for screening pollutants. The Delphi technique was applied as a tool and method for finalizing criteria (here, pollutants) or for forecasting, using a series of questionnaires for data collection from a panel of experts in the field of research. On the basis of the framework proposed by Habibi *et al.* (2015), after selecting the panel of experts, the questionnaire was circulated for judgments.

A traditional Delphi technique didn't consider the vagueness in the judgments and was time-consuming. To overcome this shortcoming, the concept of integrating the conventional Delphi method and fuzzy set theory was proposed. Habibi *et al.* (2015) suggested that the fuzzy Delphi technique is more consistent with the human linguistic and vague description, making it a better decision-making tool. For carrying out the fuzzy Delphi technique, the following algorithm by Habibi *et al.* (2015) needs to be followed:

- Identifying an appropriate spectrum for selecting the expressions for judgments

- Aggregating the judgments given by all the experts for each pollutant
- Defuzzification
- Fixing a threshold value, and screening.

After the selection of the level of scale, the judgments were converted to fuzzy numbers. Various fuzzy sets are available, namely triangular fuzzy number, trapezoidal fuzzy number, intuitionistic fuzzy set, and so on (Liu *et al.*, 2020). The triangular fuzzy set was selected for the study, as suggested by Liu *et al.* (2020).

The next step was to consolidate all the fuzzy numbers assigned by each expert for each pollutant. There are several methods for carrying out the same, namely mean method, max–min method, consensus degree method, and so on. A brief review of methods available is presented in Section 17.4.2.

17.4.1.1 Methods for aggregating the judgments given by experts for each pollutant

(a) *Mean methods:* The geometric mean and arithmetic mean methods have been extensively used by researchers. Yang *et al.* (2008) and Chen and Yang (2011) have adopted the geometric mean method for aggregating fuzzy numbers. Ayhan and Kilic (2015) and Viswanadham and Samvedi (2013) have used the arithmetic mean method for the same.
As explained by Liu *et al.* (2020), the arithmetic mean method emphasizes the average. The study also suggested that there should be no extreme values in the judgment due to its sensitivity. The geometric mean method also emphasizes the average, but this method is less affected by extreme value.

(b) *Max–min method:* This method, unlike the mean method, includes the best and the worst judgments. A fuzzy triangular number, F, is written as $F =$ (lower, middle, upper), then the upper and lower bounds of the aggregated triangular number are the maximum and minimum values of the judgments. The middle value is calculated either by the geometric or arithmetic mean method. Hsu *et al.* (2010) used arithmetic mean for the middle value, and Cheng *et al.* (2009) and Wu and Fang (2011) have used geometric mean value. As summarized by Liu *et al.* (2020), this method includes the best and the worst, and thus, introduces further fuzziness into the scenario.

Here, the max–min method with geometric mean was selected for consolidating the judgments. After obtaining the aggregated fuzzy number, the next process was to convert the obtained fuzzy number to a more understandable value. For defuzzification, again, many methods are available; centroid methods and the extension of these methods are to name some. A brief review of these methods is provided below.

17.4.1.2 Methods for defuzzification

Centroid methods: A centroid method, also termed the center of area or center of gravity method, has been the most prevalent method for defuzzification.

A centroid method generally used by most of the researchers is the method in which the arithmetic mean of the lower, upper, and middle values of the fuzzy number is calculated. However, Awasthi *et al.* (2018) and Kar (2014, 2015) have used equations (17.1)–(17.3). An extension of the centroid method would be to find the maximum value among the three mentioned equations. Another extension of the centroid method used by Tzeng and Junn-Yuanteng (1993) is given in equation (17.4):

(i) Centroid methods/center of area method:

$$DF_j = \frac{X_j + Y_j + Z_j}{3} \tag{17.1}$$

$$DF_j = \frac{X_j + 2Y_j + Z_j}{4} \tag{17.2}$$

$$DF_j = \frac{X_j + 4Y_j + Z_j}{6}, \quad j = 1,2,\ldots,m; \tag{17.3}$$

(ii) Modified center of area method:

$$DF_j = \left[\frac{(Z_j - X_j) + (Y_j - Z_j)}{3}\right] + X_j \tag{17.4}$$

After selecting an appropriate method for defuzzification, the next step was to fix a threshold value and select the criteria (here, pollutants). The threshold value set depends on the researcher (Habibi *et al.*, 2015). The criteria (here, pollutants) greater than or equal to the threshold values are confirmed; and values less than the threshold are removed from the study. The threshold value set for the current study was set 7. After finalizing the pollutants, the next phase of the index's development was to assign weights and find the significance of each pollutant.

17.4.2 Deriving weightage for the selected pollutant

In many professional scenarios, experts need to prioritize one alternative from a set of alternatives by setting weights and understanding the significance; for example, selecting a supplier or a technology (Liu *et al.*, 2020). Here, in the index development, a decision-making tool was needed for assigning weights to the finalized set of pollutants. There were many methods available for structuring these types of problems. Analytic hierarchy process, technique for order of preference by similarity to ideal solution, and data envelopment analysis are some available methods. The current study focuses on using the fuzzy-integrated analytic hierarchy process for understanding the significance of each pollutant.

Traditional analytic hierarchy process (AHP) has been extensively studied and is used as a decision-making tool for complex scenarios, where people work together to make decisions when human perceptions and judgments have a long-term repercussion. It outperforms by the ease of use, structuring problems systematically, and calculating weights (Liu *et al.*, 2020). Fuzzy sets were integrated with traditional techniques for handling imprecision. The procedure for structuring the problem systematically follows the below steps; these steps have been adopted from Liu *et al.* (2020):

- First, establishing a comparison matrix
- Aggregating the multiple judgments
- Establishing fuzzy weights
- Defuzzification of the fuzzy weights

A comparison matrix is a matrix where elements are filled as per the judgments received. A sample comparison matrix is given in Table 17.1. The judgments from the panel of experts on the significance of each pollutant over the other pollutants were aggregated first, and then filled into the comparison matrix. The aggregation methods available were the same as those available for the fuzzy Delphi technique. After filling the elements in the matrix, the next step would be to calculate the fuzzy weights. Different methods are available; a brief review of the methods is provided below:

- *Mean method*: There are two types of mean methods; per Liu *et al.* (2020), the geometric mean method is the widely used one to calculate fuzzy weights; the equation for the same is given in equations (17.5) and (17.6). Many researchers have used the arithmetic mean method for finding the fuzzy weights; the equation for which is given in equations (17.7) and (17.8). As summarized in Liu *et al.* (2020), the geometric mean method is better than the arithmetic method and is very simple to carry out. Both arithmetic mean and geometric mean methods are very simple to calculate as these only involve arithmetic addition, multiplication, or division.
- Arithmetic mean method:

$$Z_a = \frac{1}{m}(F_{a1} \oplus F_{a2} \oplus \cdots \oplus F_{am}); \quad \forall a \tag{17.5}$$

$$W_a = Z_a \varnothing (Z_1 \oplus Z_2 \oplus \cdots \oplus Z_m) \tag{17.6}$$

- Geometric Mean Method

$$Z_a = (F_{a1} \otimes F_{a2} \otimes \cdots \otimes F_{am})^{1/m}; \quad \forall a \tag{17.7}$$

$$W_a = Z_a \varnothing (Z_1 \oplus Z_2 \oplus \cdots \oplus Z_m) \tag{17.8}$$

Table 17.1 Sample three-criterion comparison matrix.

	Criterion A	Criterion B	Criterion C
Criterion A	(1,1,1)	F_{AB}	F_{AC}
Criterion B	$F_{BA} = 1/F_{AB}$	(1,1,1)	F_{BC}
Criterion C	$F_{CA} = 1/F_{AC}$	$F_{CB} = 1/F_{BC}$	(1,1,1)

Here, Z_a is the mean value for criterion '*a*' and W_a is the fuzzy weight for criterion '*a*.' The fuzzy operators used in equations (17.5)–(17.8) are described in equation (17.9). Assuming two triangular fuzzy numbers (TFNs) $A = (a, b, c)$, and $B = (m, n, o)$, then

$$\begin{aligned} &\oplus\text{: Addition of fuzzy numbers}: \ A \oplus B = (a+m,\ b+n,\ c+o); \\ &\otimes\text{: Multiplication of fuzzy numbers}: \ A \otimes B = (a \times m,\ b \times n,\ c \times o); \\ &\varnothing\text{: Division of fuzzy numbers}: \ A \varnothing B = (a/m,\ b/n,\ c/o) \end{aligned} \tag{17.9}$$

Many other are methods available: lambda-max method, eigenvector method, logarithmic least-squares method are to name some; these methods were complex as these involve multiple steps (Liu *et al.*, 2020). Therefore, the geometric mean method is selected for calculating the fuzzy weights of each pollutant. These fuzzy terms are not easy to compare; hence these need to be converted into intuitive terms. For converting these judgments to crisp values, a defuzzification process was performed. For defuzzification, the methods involved were the same as that mention for carrying out the fuzzy Delphi technique. The crisp values obtained were then normalized to obtain the final weights which were then used in further study.

17.4.3 Establishing sub-index curves for each pollutant

In this stage of the development of the index, sub-index curves for each pollutant were plotted. For calculating the sub-index scores, graphs were plotted with concentrations on the abscissa and the scores in the range of 0–100 on the ordinate. As reported by Kumar and Alappat (2003), many functions to relate sub-index score and the pollutant could be derived. Thus, no mathematical equations were proposed for plotting the graph. Instead, a set of rules were proposed for the development of the sub-index curves:

(1) The curve was plotted with 5 as the minimum score even if there is no contamination from the pollutant.
(2) From the literature (Gautam & Kumar, 2021; Ghassemi *et al.*, 1984; Han *et al.*, 2017; Kattel *et al.*, 2016; Pavelka *et al.*, 1993; USEPA, 1982; Yasuhara *et al.*, 1999), the range of concentration on the abscissa was fixed.
(3) A score of ~10 was assigned to the maximum value among the drinking water standard concentration specified by authorities World Health Organisation (2008) and USEPA (2018), and disposal standards by Central Pollution Control Board (2000).
(4) A few exceptions were made when the maximum value selected was way above the drinking water standards.
(5) The trend of the plots was set based on toxicity, weights assigned, maximum concentration, and average concentrations of each pollutant as per data from the literature (Gautam & Kumar 2021; Ghassemi *et al.*, 1984; Pavelka *et al.*, 1993; USEPA, 1982; Yasuhara *et al.*, 1999).

17.4.4 Aggregation of weights and the scores

The aggregation method has been a crucial part in the field of formulation of environmental indices. The type of aggregation function is selected based on the index application (Kumar & Alappat, 2004). As per Ott (1978), the aggregation functions usually consist either of the three forms, that is, additive form, multiplicative form, or the maximum or minimum operator form. There were many aggregation methods used in the formulation of environmental indices. The aggregation method used for the study was the same as that used for developing the existing index.

For this study, the significance of the pollutants has been considered; therefore, a method incorporating the weights was finalized. The weighted linear additive form was selected for the final aggregation. Kumar and Alappat (2004) concluded that the linear weighted sum aggregation function was the most suitable for the aggregation function for the development of the LPI.

17.4.5 Formulation of sub-HWLLPI

For making the general index more informative, the general index was split into different categories (Kumar & Alappat, 2005a). The decision-making tool used in the second phase of the development of HWLLPI paved a way to categorize the pollutants; the weights obtained were termed parent weights and local weights. These values helped in formulating the sub-HWLLPI. The product of the local weight and the parent weights provided the global weights, which were used for the formulation of the general HWLLPI.

17.5 CONCLUSION

On the basis of the studies carried out to quantify the contamination caused by using improper techniques and unsafe methods, the leachate released from landfill sites, has been evident that the existing LPI is a potential tool. It has been used as a comparison tool to establish which landfill might require immediate attention and rank the leachate based on the pollution. However, the existing index was developed considering many criteria of which availability of the pollutants in the leachate from MSW landfills was a vital criterion. Consequently, pollutants such as cadmium, selenium, and AOXs were excluded from the final list of pollutants for inclusion in the index. Literature studies have shown that these pollutants are present in significant concentrations in HWL leachate. This points to the fact that the existing LPI is an efficient tool for quantifying leachate contamination from MSW landfills.

Using the existing LPI for calculating the pollution potential of leachate from HWLs would not be a prudent choice. Studies also suggest that leachate from HWL sites is more toxic than that released from MSW landfills. Thus, a need for developing an index solely for leachate from HWLs was felt, hence this study. The following conclusions can be drawn:

- In this study, it is evident that based on the expertise, a single number can reflect the contamination from HWLs.
- The expertise provided by the panel of experts was converted to numerical terms using different methods. Unlike the existing LPI, the vagueness in

the judgments was incorporated as fuzzy-integrated tools were used for the conversion.

- The first questionnaire aimed to finalize the pollutants and the second focused on assigning weightage to the finalized pollutants.
- The fuzzy AHP tool helped in categorizing the pollutants, which further helped in formulating the modified sub-index.
- The sub-index curves plotted for each pollutant were based on specific rules.
- The aggregation method for the final calculation was the weighted linear sum aggregation method. As mentioned, the HWLLPI developed was also modified to obtain a more informative tool; the modified sub-HWLLPI helps evaluate the contamination potential of metals and organics separately.

REFERENCES

Awasthi A., Govindan K. and Gold S. (2018). Multi-tier sustainable global supplier selection using a fuzzy AHP-VIKOR based approach. *International Journal of Production Economics*, **195**, 106–117, https://doi.org/10.1016/j.ijpe.2017.10.013

Ayhan M. B. and Kilic H. S. (2015). A two-stage approach for supplier selection problem in multi-item/multi-supplier environment with quantity discounts. *Computers and Industrial Engineering*, **85**, 1–12, https://doi.org/10.1016/j.cie.2015.02.026

Basel Convention. (2018). *Global trends in Generation and transboundary movements of hazardous wastes and other wastes: Waste without frontiers II.* https://www.basel.int/Portals/4/Basel%20Convention/docs/pub/WasteWithoutFrontiersII.pdf

BMW_Rules. (2016). Ministry of Environment, Forest and Climate Change, Notification. The Gazette of India, Extraordinary, Part II, Section 3(i). http://www.iwma.in/BMW Rules, 2016.pdf

Central Pollution Control Board (CPCB). (2016). Hazardous Waste (Management, Handling and Transboundary Movement) Rules, 2016. Government of India, New Delhi.

CPCB. (2000). Environmental Standards for Ambient Air, Automobiles, Fuels, Industries, and Noise. Government of India, New Delhi, pp. 55–56.

CPCB. (2010). Protocol for Performance Evaluation, and Monitoring of the Common Hazardous Waste Treatment Storage, and Disposal Facilities Including Common Hazardous Waste Incinerators. Government of India, New Delhi.

CPCB. (2021). National Inventory on Generation, and Management of Hazardous, and Other Wastes. Government of India, New Delhi.

Chartier Y., Emmanuel J., Pieper U., Rushbrook P., Stringer R., Townend W., Wilburn S. and Zghondi R. (2014). Safe Management of Wastes From Health-Care Activities. World Health Organisation, Geneva, Switzerland.

Chen Z. and Yang W. (2011). An MAGDM based on constrained FAHP, and FTOPSIS, and its application to supplier selection. *Mathematical and Computer Modelling*, **54**(11–12), 2802–2815, https://doi.org/10.1016/j.mcm.2011.06.068

Cheng J. H., Lee C. M. and Tang C. H. (2009). An application of fuzzy Delphi, and fuzzy AHP on evaluating wafer supplier in semiconductor industry. *WSEAS Transactions on Information Science and Applications*, **6**(5), 756–767.

Das A. K., Islam M. N., Billah M. M. and Sarker A. (2021). COVID-19 pandemic and healthcare solid waste management strategy – a mini-review. *Science of The Total Environment*, **778**, 146220, https://doi.org/10.1016/J.SCITOTENV.2021.146220

Environmental Protection Agency. (2018). 2018 edition of the drinking water standards and health advisories tables.

Gautam P. and Kumar S. (2021). Characterisation of hazardous waste landfill leachate and its reliance on landfill age and seasonal variation: a statistical approach. *Journal of Environmental Chemical Engineering*, **9**(4), 105496, https://doi.org/10.1016/J.JECE.2021.105496

Gautam P., Kumar S. and Lokhandwala S. (2019). Advanced oxidation processes for treatment of leachate from hazardous waste landfill: a critical review. *Journal of Cleaner Production*, **237**, 117639, https://doi.org/10.1016/j.jclepro.2019.117639

Ghassemi M., Quinlivan S. and Bachmaier J. (1984). Characteristics of leachates from hazardous waste landfills. *Journal of Environmental Science and Health. Part A: Environmental Science and Engineering*, **19**(5), 579–620, https://doi.org/10.1080/10934528409375180

Gholami M., Torkashvand J., Rezaei Kalantari R., Godini K., Jonidi Jafari A. and Farzadkia M. (2020). Study of littered wastes in different urban land-uses: an 6 environmental status assessment. *Journal of Environmental Health Science and Engineering*, **18**(2), 915–924, https://doi.org/10.1007/s40201-020-00515-7

Habibi A., Jahantigh F. F. and Sarafrazi A. (2015). Fuzzy Delphi technique for forecasting, and screening items. *Asian Journal of Research in Business Economics and Management*, **5**(2), 130, https://doi.org/10.5958/2249 7307.2015.00036.5

Han Y. J., Kim P. R., Lee G. S., Lee J. I., Noh S., Yu S. M., Park K. S., Seok K. S., Kim H. and Kim Y. H. (2017). Mercury concentrations in environmental media at a hazardous solid waste landfill site and mercury emissions from the site. *Environmental Earth Sciences*, **76**, 1–14, https://doi.org/10.1007/s12665-017-6700-z

Hsu Y. L., Lee C. H. and Kreng V. B. (2010). The application of fuzzy Delphi method, and fuzzy AHP in lubricant regenerative technology selection. *Expert Systems with Applications*, **37**(1), 419–425, https://doi.org/10.1016/j.eswa.2009.05.068

Ilyas S., Srivastava R. R. and Kim H. (2020). Disinfection technology and strategies for COVID-19 hospital and bio-medical waste management. *Science of The Total Environment*, **749**, 141652, https://doi.org/10.1016/J.SCITOTENV.2020.141652

Kar A. K. (2014). Revisiting the supplier selection problem: an integrated approach for group decision support. *Expert Systems with Applications*, **41**(6), 2762–2771, https://doi.org/10.1016/j.eswa.2013.10.009

Kar A. K. (2015). A hybrid group decision support system for supplier selection using analytic hierarchy process, fuzzy set theory, and neural network. *Journal of Computational Science*, **6**, 23–33, https://doi.org/10.1016/j.jocs.2014.11.002

Katsiri K., Bosdogianni A. and Christoulas D. (1999). Prediction of leachate quality from sanitary landfills. *Journal of Environmental Engineering*, **125**, 831–838.

Kattel E., Kivi A., Klein K., Tenno T., Dulova N. and Trapido M. (2016). Hazardous waste landfill leachate treatment by combined chemical and biological techniques. *Desalination and Water Treatment*, **57**(28), 13236–13245, https://doi.org/10.1080/19443994.2015.1057539

Kumar D. and Alappat B. J. (2003). A technique to quantify landfill leachate pollution. Proceedings of the 9th International Waste Management Landfill Symposium, October. p. 400.

Kumar D. and Alappat B. J. (2004). Selection of the appropriate aggregation function for calculating leachate pollution index. *Practice Periodical of Hazardous, Toxic, and Radioactive Waste Management*, **8**(4), 253–264, https://doi.org/10.1061/(ASCE)1090-025X(2004)8:4(253)

Kumar D. and Alappat B. J. (2005a). Analysis of leachate pollution index, and formulation of sub-leachate pollution indices. *Waste Management & Research*, **23**(3), 230–239, https://doi.org/10.1177/0734242X05054875

Kumar D. and Alappat B. J. (2005b). Evaluating leachate contamination potential of landfill sites using leachate pollution index. *Clean Technologies and Environmental Policy*, **7**(3), 190–197, https://doi.org/10.1007/s10098-004-0269-4

Liu Y., Eckert C. M. and Earl C. (2020). A review of fuzzy AHP methods for decision-making with subjective judgments. *Expert Systems with Applications*, **161**, 113738, https://doi.org/10.1016/j.eswa.2020.113738

Malsparo. (2020). *Pharmaceutical Waste Management, Treatment, and Disposal.* https://www.malsparo.com/pharm.htm

Maowei W., Jiahao L., Qifen L. and Yongwen Y. (2021). Current status of hazardous waste disposal technology. *IOP Conference Series: Earth and Environmental Science*, **760**(1), 012034, https://doi.org/10.1088/1755-1315/760/1/012034

Mmereki D., Baldwin A. N., Hong L. and Li B. (2016). The management of hazardous waste in developing countries. In book: Management of Hazardous Wastes, 39–51.

Orloff K. and Falk H. (2003). An international perspective on hazardous waste practices. *International Journal of Hygiene and Environmental Health*, **206**(4–5), 291–302, https://doi.org/10.1078/1438-4639-00225

Ott W. R. (1978). Environmental Indices: Theory, and Practice, Ann Arbor Science Publishers, Ann Arbor, MI.

Pavelka C., Loehr R. C. and Haikola B. (1993). Hazardous waste landfill leachate characteristics. *Waste Management*, **13**(8), 573–580, https://doi.org/10.1016/0956-053X(93)90017-Q

Ramteke S. and Lal B. (2020). Case studies in chemical, and environmental engineering novel coronavirus disease 2019 (COVID-19) pandemic: considerations for the biomedical waste sector in India. *Chemical and Environmental Engineering*, **2**, 100029.

Shuckrow A. J., Pajak A. P. and Touhill C. J. (1982). Management of hazardous waste leachate. Environmental Protection Agency, Municipal Environmental Research Laboratory, Office of Research and Development, U.S.

Thanh N. P., Matsui Y. and Fujiwara T. (2010). Household solid waste generation and characteristic in a Mekong Delta city, Vietnam. *Journal of Environmental Management*, **91**(11), 2307–2321, https://doi.org/10.1016/j.jenvman.2010.06.016

Tzeng G. H. and Junn-Yuanteng J. Y. (1993). Transportation investment project selection with fuzzy multiobjectives. *Transportation Planning and Technology*, **17**(2), 91–112, https://doi.org/10.1080/03081069308717504

Visvanathan C. (1996). Hazardous waste disposal. *Resources, Conservation, and Recycling*, **16**(1–4), 201–212, https://doi.org/10.1016/0921-3449(95)00057-7

Viswanadham N. and Samvedi A. (2013). Supplier selection based on supply chain ecosystem, performance, and risk criteria. *International Journal of Production Research*, **51**(21), 6484–6498, https://doi.org/10.1080/00207543.2013.825056

WHO. (2008). WHO guidelines for drinking-water quality. **35**(5), 307–312.

World Health Organisation (WHO). (2020). Water, sanitation, hygiene, and waste management for SARS-CoV-2, the virus that causes COVID-19. pp. 1–11.

Wu C. H. and Fang W. C. (2011). Combining the fuzzy analytic hierarchy process, and the fuzzy Delphi method for developing critical competences of electronic commerce professional managers. *Quality and Quantity*, **45**(4), 751–768, https://doi.org/10.1007/s11135-010-9425-6

Yang J. L., Chiu H. N., Tzeng G. H. and Yeh R. H. (2008). Vendor selection by integrated fuzzy MCDM techniques with independent and interdependent relationships. *Information Sciences*, **178**(21), 4166–4183, https://doi.org/10.1016/j.ins.2008.06.003

Yasuhara A., Shiraishi H., Nishikawa M., Yamamoto T., Nakasugi O., Okumura T., Kenmotsu K., Fukui H., Nagase M. and Kawagoshi Y. (1999). Organic components in leachates from hazardous waste disposal sites. *Waste Management & Research*, **17**(3), 186–197, https://doi.org/10.1177/0734242X9901700304

doi: 10.2166/9781789063318_0393

Chapter 18

Prevalence of antibiotics and antibiotic resistance genes in landfill leachate

Neelam Gunjyal[1], Chandra Shekhar Prasad Ojha[1] and Vinay Kumar Tygai[2]*

[1]*Department of Civil Engineering, Indian Institute of Technology Roorkee, Roorkee 247667, India*
[2]*Environmental Hydrology Division, National Institute of Hydrology, Roorkee 247667, India*
**Corresponding author: ngunjyal@ce.iitr.ac.in*

ABSTRACT

Landfills are an essential source of antibiotics and antibiotic resistance genes (ARGs) and can potentially risk humans and the environment. Landfills receive heavy metals, organic matter, microbial consortia, antibiotics, antibiotic residue, and biomedical waste, which enhance the prevalence of antibiotics, ARGs, and antibiotic-resistant bacteria (ARBs). The critical problem is the transfer of antibiotics and ARGs in groundwater, surface water, food web, and the surrounding environment through landfill leachate. In this chapter, we have discussed several studies on antibiotics and ARGs in landfill leachates, such as a review of the source of antibiotics and ARGs in landfills, a discussion of the correlation between antibiotics, antibiotic residue, and heavy metals on ARGs, a summary about the impact of various physicochemical parameters, environmental, and social factors which influences the antibiotics and ARGs, landfill age, and effect of antibiotics and ARGs on groundwater and surface water. Finally, this chapter fills the current gaps not examined until now. Especially in developing countries such as India, it is essential to research antibiotics and ARGs in landfill leachate.

Keywords: Landfill leachate, antibiotic resistance genes (ARGs), antibiotics, landfill age, heavy metals, antibiotic resistance.

18.1 INTRODUCTION

Landfill leachate is a crucial source and reservoir of antibiotics and antibiotic resistance genes (ARGs), which poses a threat to human health and the environment. Since the introduction of penicillin in 1929, antibiotics have been beneficial for human health and animals. It is the life protector in the veterinary field and it improves animal growth rate (Sarmah *et al.*, 2006;

Zhou *et al.*, 2013a). There are various types of antibiotics, such as macrolides, sulfonamides, lincosamides, glycopeptides, and quinolones, which are widely used in animal husbandry (Sarmah *et al.*, 2006), aquaculture, and to cure the infection in medical specialty (Bai *et al.*, 2022; Zhou *et al.*, 2013b). There are various advantages of antibiotics in human health: antibiotics work very fast, and most of the antibiotics are active within a few minutes; they can quickly kill toxic bacteria, and they can reduce the growth of harmful bacteria in our gut; they prevent the infection while undergoing surgeries; and most of the antibiotics are taken orally, and they are easy to take at any time. But when the concentration of antibiotics attains a certain level, it could be a very potential risk to animals, humans, and bacterial community and it is called as overuse of antibiotics. The overdose and misuse of antibiotics cause diarrhea, stomach upset, and feeling restless in humans. The resistance of the bacteria (conversion of good bacteria into bad in our gut) is the leading cause of the overuse of antibiotics, and it is called antibiotic resistance (AR).

Worldwide, China is the largest producer and consumer of antibiotics. They consume almost 1.6 lakh ton annually; out of which 50,000 ton antibiotics contain inactive drugs and metabolites discharged into the surrounding environment (Zhang *et al.*, 2015; Zheng *et al.*, 2012). India is the third largest producer and consumer of antibiotics worldwide after United States. Hence, antibiotics and AR are the global threat for humans and the environment. The overuse of antibiotics inhibited the growth of good bacteria in a human gut. Moreover, the growth of the microorganisms in soil, sediment, and water is inhibited by the environmental antibiotics and it also affects the microbial activity and structure (Su *et al.*, 2017). The accumulation of antibiotics in the environment affects humans and animals via food chain and provides unhealthy gut such as allergies and food poisoning. Some drugs of antibiotics literally are carcinogenic and mutagenic, which seriously interfere with various physical functions of human beings (Christian *et al.*, 2003; Watanabe *et al.*, 2010).

The antibiotics and antibiotic residue are also very harmful for the environment. Generally, antibiotic residues are released into the environment via urine, feces (Lienert *et al.*, 2007; Sarmah *et al.*, 2006), and wastewater discharge, in addition to ARGs and human commensal microbes (Sommer *et al.*, 2009). Antibiotics and ARGs are potentially risk to the human beings and natural environment in the 21st century because they are spread by horizontal gene transfer (HGT) across bacteria. According to the World Health Organization (WHO), AR has become a major challenge to global health (WHO, 2019). Antibiotics and ARGs have been detected in various areas such as sewage treatment plants, influents, effluents of sewage treatment plants (Gothwal & Shashidhar, 2015; Jechalke *et al.*, 2014), rivers, soil, sediments (Garcia-Armisen *et al.*, 2011; Luo *et al.*, 2010), municipal solid waste (MSW), landfills, pharmaceutical industries, pig farms, broiler feedlots (Wu *et al.*, 2010; Zhu *et al.*, 2013), and so on. Moreover, MSW landfills have become an important hotspot reservoir for antibiotics and ARGs (Song *et al.*, 2016). Landfills are the important hotspots for the generation of ARGs, antibiotics, and spread of antibiotic-resistant bacteria (ARBs). The amount of the concentration of the antibiotics is certainly higher in landfill leachate as compared to the landfill refuse.

In landfills, emerging contaminants, antibiotics, and ARGs generally raised concern. Moreover, many findings have been reported on antibiotics and ARGs in landfills, but still reviews on this topic are limited. The aim of this chapter is to study origin of antibiotics and ARGs in landfills and abundance and characterization of antibiotics, ARGs, and bacterial community in landfill leachate, identifying the numerous quality factors of ARGs in landfills, study the landfill age and its impact on antibiotics and ARGs, and study impact of landfill leachate on groundwater and surface water in terms of antibiotics and ARGs in landfills. Finally, future research discussions are mentioned.

18.2 SOURCE OF ANTIBIOTICS AND ARGS IN LANDFILLS

18.2.1 Mentioning the distribution of antibiotics in landfills worldwide

MSW is directly disposed into landfill sites without any segregation and separation, and it is one of the major origins of generation of antibiotics and ARGs in the ecological environment (Anand *et al.*, 2021). The origin and distribution of antibiotics in landfills are given in Table 18.1. Excessive consumption of antibiotics is rising in the global level. On the observation of antibiotic consumption from 2000 to 2015 in 76 countries and the project has been expanded through 2030, it is shown that the consumption of antibiotics is increasing rapidly in both high-income countries (HICs) such as United States, Italy, and France, and low-middle-income countries (LMICs) such as India, China, and Pakistan (Klein *et al.*, 2018). Although, HICs also consume antibiotics, in LMICs the consumption of antibiotics was predicted to be at the highest level. In 2015, India has become the highest antibiotic consumption in the world, China has become second highest antibiotic consumption in the world, and United States has become the third highest antibiotic consumption in the world (Jabbar *et al.*, 2019). The origin and transferred mechanism of antibiotic dissemination in the environment is shown in Figure 18.1.

There is also a probability that the abundance of ARGs in the environment is due to the following causes: (1) overuse and abuse of antibiotics, (2) some antibiotics does not have a good metabolic competence to absorb entirely, (3) the residue of the antibiotics was absorbed by urine, feces, wastewater discharge, and fertilizers (Chen *et al.*, 2017), (4) household waste directly enters into landfills which contaminate the soil, groundwater by leaching or runoff discharge (Chen *et al.*, 2017), and surface water, and (5) waste and unused antibiotics, and emerging contaminants (Zhang *et al.*, 2022).

18.2.2 Abundance of ARGs and dissemination of ARGs in landfills

Till date, researchers have focused on antibiotics and ARGs present in wastewater effluents (Pei *et al.*, 2006), sewage discharge, soil, and sediment (Pei *et al.*, 2006;

Table 18.1 Distribution and concentration of antibiotics in landfill leachate.

Sl. No.	Antibiotics	Source of Antibiotics	Site of Sampling	Concentration of Antibiotics	Unit	References
1	Macrolides, sulfonamides, tetracyclines	Landfill leachate	Yuhang, Panan, and Anji in Zhejiang province, China	Avg. of (SDZ, SMZ, SMT, SM1, SM2, TC, OTC, CTC, ETM, RTM) (0–1-year-old) = 4,875.96 ± 771.20, (3–5-year-old) = 1,487.99 ± 158.68, (5–10-year-old) = 1,133.68 ± 154.83	ng/L	Liu *et al.* (2022)
2	Fluoroquinolones (FQs) and β-lactams (BLs)	Landfill leachate	6 provinces of China	FQs: OFL = nd–21,033, NOR = nd–23,267, ENR = 17.95–4,027, PEF = nd–25.37	ng/L	You *et al.* (2018)
3	Sulfonamide (SA), quinolone (FQ), tetracycline (TC), macrolide (ML), and chloramphenicol (CP)	Landfill leachate	Shanghai, China	MLs = 3,561, FQs = 975, SAs = 402, CPs = 141, and TCs = 83	ng/L	Wu *et al.* (2015)
4	Sulfamethoxazole (SMX), tetracycline (TC), and oxytetracycline (OTC)	Landfill refuse	Xian, China	OTC = 100.9, TC = 63.8, SMX = 47.9	ng/L	Song *et al.* (2016)

5	Lincosamide: clindamycin (CLI) and lincomycin (LIN). Macrolide: azithromycin (AZ[illegible]), clarithromycin (CLAR), erythromycin (ERY), and tylosin (TYL) Sulfonamide: sulfamethoxazole (SMX) and sulfamethazine (SMZ)	Landfill leachate	Singapore	SMZ = 62–438	ng/L	Yi *et al.* (2017)
6	Sulfonamides, fluoroquinolone, and macrolides,	Landfill refuse	Zhejiang, China	FQs = 1–1,400, MLs = 0.5–130, SAs = 1–16	ng/g	Yao *et al.* (2020)
7	Sulfamethazine, sulfamethoxazole, and erythromycin	Landfill leachate	United States	ETM = 204, SDX = 183, SMX = 861	ng/L	Masoner *et al.* (2016)

(i) Sulfonamides (SAs): sulfamethoxazole (SMX), sulfamethazine (SMZ), sulfadiazine (SDZ), sulfadimethoxine (SDX), sulfadimethoxine (SMT), chlortetracycline (CTC), roxithromycin (RTM); (ii) tetracyclines (TCs): tetracycline (TC), oxytetracycline (OTC); (iii) quinolones (FQs): norfloxacin (NOR), enrofloxacin (ENR), pefloxacin (PEF), ofloxacin (OFL) and (iv) macrolides (MLs): erythromycin (ETM).

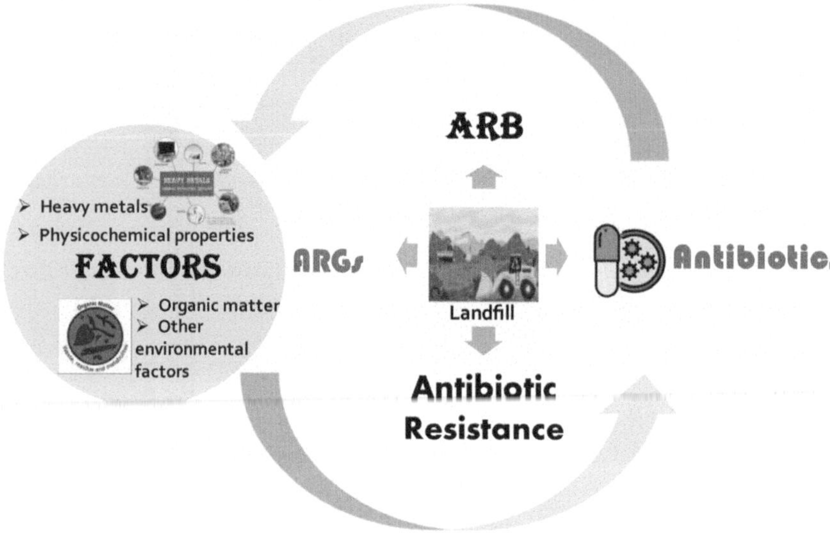

Figure 18.1 Origin of antibiotics and dissemination of antibiotics in the environment (modified from Pan & Chu (2017) and Zhang *et al.* (2022).

Zuccato *et al.*, 2010), but none have studied the presence of antibiotics and ARGs in landfills. Some researchers found that the concentration of antibiotics in landfills is higher than the effluent of wastewater, soil, and sediment (Clarke *et al.*, 2015; Sui *et al.*, 2017). However, they found the concentration of antibiotics in leachate is higher than that of landfills. Therefore, the order of antibiotic concentration is leachate > landfill > effluent > wastewater discharge > sediment > soil. A quantitative measurement of antibiotic concentration was carried out by (i) liquid chromatography-tandem mass spectrometry (LC-MS/MS), (ii) high- performance LC-MS/MS (HPLC-MS/MS), and ultra-performance LC-MS/MS. Currently, LC-MS/MS is used widely in the global level, because of its great performance and sensitivity (Zhang *et al.*, 2022).

The dissemination of ARBs and ARGs in landfill leachate is the main cause of excessive use of antibiotics and its residue (Song *et al.*, 2016). The mobile leachate, which is being generated throughout the landfilling process, is the main cause of the dissemination and distribution of ARBs and ARGs (Wu *et al.*, 2015). Although there are numerous studies on ARGs in water, soil, and sediment, there is a scarcity of studies on landfill leachate (Chen *et al.*, 2017). We need to evaluate the ARGs in landfill leachate, and there are various methods for that such as culture isolation, polymerase chain reaction (PCR), quantitative polymerase chain reaction (qPCR), DNA microarray, and metagenomics analysis. Many researchers evaluated the ARGs in landfill leachate by different methods, but there are some limitations for PCR and qPCR, due to the lack of primers and low-throughput sequencing. To overcome the limitations of PCR and qPCR, metagenomic sequencing is a powerful tool to determine the ARGs and ARBs in landfill leachate. Currently, increasing contamination due to ARGs is a global risk; hence it is necessary to focus on the primers so that we could measure ARGs in various environments. The abundance and distribution of ARGs in landfill leachate are given in Table 18.2.

18.3 CORRELATION BETWEEN ANTIBIOTICS, ANTIBIOTIC RESIDUE, AND HEAVY METALS ON ARGs

The presence of antibiotics, antibiotic residue, and heavy metals in the environment poses problem in the global world (Välitalo *et al.*, 2017). Commonly, antibiotics are used as feed preservatives for agricultural purposes, but the residue of antibiotics affects human beings and animals which are dependent on the farms. The unabsorbed part of antibiotic residue is also the main factor which affects the AR. Most of the researchers found a correlation between antibiotics and ARGs in landfill leachate and surface soil (Sun *et al.*, 2016). Sometimes many antibiotics are positively correlated with their corresponding ARGs (p value <0.05 and $r^2 = 0.74–0.89$) in landfill leachate and the same outcome is reported in the wastewater treatment process (Gao *et al.*, 2012). Simultaneously, researchers found that the antibiotics are weakly correlated with ARGs in landfill leachate (Yao *et al.*, 2020) and same is also reported in the sewage treatment process (Gao *et al.*, 2012). It indicates that antibiotics might not be the primary factor which affects AR. Hence, we need to focus more on the other factors that affect resistance, such as heavy metals, physicochemical properties, and organic matter of leachate.

The pollution of heavy metals is also associated with ARGs. The pollution of heavy metals is also shown in the landfill leachate. Most of the heavy metals such as Zn, Cu, and As are used as a feed additive for human beings and for veterinary and agricultural purpose. The unabsorbed residues of antibiotics and heavy metals are accrued by the natural environment, and it provides a prolonged period of co-selection pressure of the microbes for AR (Liu *et al.*, 2022; Lu *et al.*, 2020). Most of the calculated ARGs significantly correlate with the heavy metals (for instance, $p < 0.05$) such as Cd and Cr (Wu *et al.*, 2015). It means, most of the antibiotics, heavy metals, and ARGs present in the MSW landfill leachate, which is the major source of the proliferation of antibiotics and ARGs in the environment.

18.4 IMPACT OF SEVERAL VITAL FACTORS ON ANTIBIOTIC CONCENTRATION AND ARG PROFILES IN LANDFILLS

There are various important factors to increase the antibiotic concentration and ARG profiles in landfills: physicochemical factors, heavy metals, organic and inorganic matter, and other environmental and social factors. The correlation of the various factors with antibiotics, ARGs, and ARBs in detail is shown in Figure 18.2.

Table 18.2 Abundances and distribution of ARGs in landfill leachate.

Sl. No.	ARGs	Source of ARGs	Site of Sampling	Abundance of ARGs	Unit	References
1	SRGs: *sul*1, *sul*2	Landfill leachate	Shanghai, China	*sul*R = 6.06 ± 0.6	$\log_{10}$(copies/mL)	Wu *et al.* (2019)
2	*mex*F and *tet*Q	Landfill leachate	Yuhang, Panan, and Anji in Zhejiang province, China	*mex*F = 11.92 ± 0.22, *tet*Q = 9.57 ± 1.32	$\log_{10}$(gene copies/L)	Liu *et al.* (2022)
3	BRGs (bla_{OXA10}, *Amp*C, bla_{CTX-M}, bla_{OXY}, bla_{SHV})	Landfill leachate	6 provinces in China	*qnr*A = 0.04, *qnr*B = −4.95, *qnr*D = −5.31), *qnr*S = −6.85 bla_{OXA10} = −3.41), *amp*C = −6.13, $bla_{CTX-M=}$ −4.44, $bla_{SHV=}$ −6.23, $bla_{OXY=}$ −5.72	ARGs/16S	You *et al.* (2018)
4	SRGs (*sul*1, *sul*2) TRGs (*tet*Q, *tet*W) MRGs (*erm*B, *mef*A)	Landfill leachate	Shanghai, China	*sul*2 = −1.37, *erm*B = −1.77, *sul*1 = −2.58, *tet*M = −2.72, and *tet*Q = −3.05	$\log_{10}$(copies/16S-rDNA)	Wu *et al.* (2015)
5	SRGs (*sul*1), TRGs (*tet*O)	Landfill refuse	Xian, China	*sul*1 = −3.06), *tet*O = −4.37	$\log_{10}$(ARGs/16S)	Song *et al.* (2016)
6	SRGs (*sul*1, *sul*2, *dfr*A), TRGs (*tet*O), BRGs (bla_{NMD1}, bla_{KPC}, bla_{CTX}), MRGs (*erm*B), AmRGs (*aac*6), QRGs (*qnr*A)	Landfill leachate	Singapore	*sul*1 = 1.995×10^5, *sul*2 = 5.269×10^4, *aac*6 = 180, *tet*O = 37.5, *dfr*A = 34.4, *qnr*A = 26.5, *erm*B = 3.6	$\log_{10}$(gene copies/mL)	Yi *et al.* (2017)
7	Fluoroquinolone ARGs: *qnr*D, *qnr*S, *mex*F Macrolide ARGs: (*mef*A, *erm*A, and *erm*B) Sulfonamide ARGs: *sul*1 and *sul*2	Landfill refuse	Zhejiang, China	*sul*1 = 2.19×10^{-3} to 1.04×10^{-1} *sul*2 = 4.61×10^{-3} to 2.88×10^{-1} *qnr*D, *qnr*S, and *mex*F = 8.19×10^{-6} to 1.27×10^{-1} *erm*A, *erm*B, and *mef*A = 4.94×10^{-6} to 2.16×10^{-1}	copies/16S rRNA	Yao *et al.* (2020)
8	16S rRNA, *int*I, and *cint*I1	Landfill refuse	Xiamen, China	ARGs: 2.5×10^9 to 1.27×10^{11}	copies/cell	Chen *et al.* (2017)

(i) Sulfonamide resistance genes (SRGs); (ii) beta-lactam resistance genes (BRGs); (iii) tetracycline-resistant genes (TRGs); (iv) MLS (macrolide-lincosamide-streptogramin) resistance genes (MRGs); (v) aminoglycoside resistance genes (AmRGs); and (vi) quinolone resistance genes (QRGs).

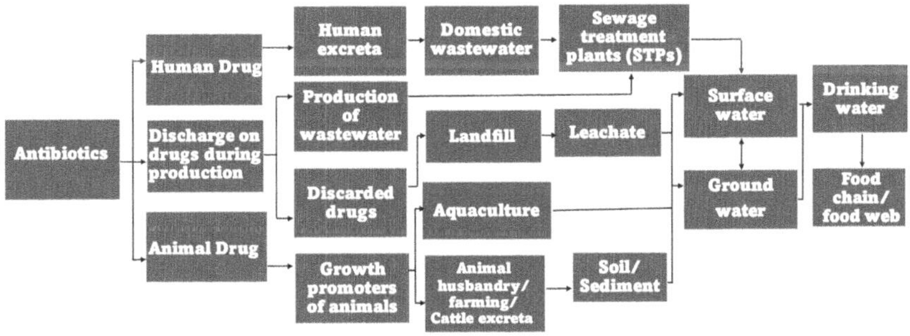

Figure 18.2 Relationship between heavy metals, physicochemical factors, and other environmental factors with antibiotics, ARGs and ARBs.

18.4.1 Physicochemical factors

There are various physicochemical parameters, such as dissolved oxygen (DO), pH, conductivity, chemical oxygen demand (COD), biological oxygen demand (BOD), total nitrogen (TN), and total phosphorus (TP). In landfill leachate, high BOD, COD, TN, and TP indicate the high organic matter, and it is one of the major causes to generate bacterial resistance. Some of the researchers found a positive correlation between pH and ARGs and a negative correlation of conductivity, COD, BOD, TN, and TP with ARGs (Zhao *et al.*, 2018). The *sul*R and *int*I genes were also correlated positively with DO and pH in landfill leachate (Koczura *et al.*, 2016). Most of the studies found that COD, TN, and TP correlated positively ($p < 0.05$) with the intended genes (Sun *et al.*, 2016), and some studies show that the BOD and COD were significantly negatively ($p < 0.05$) correlated with ARGs (Yu *et al.*, 2016). Hence, these studies revealed that the physicochemical parameters contribute to ARGs in landfills.

18.4.2 Heavy metals

Heavy matter contributes pollution to the environment, and it is the major co-selection agent to increase the pollution and ARGs in the environment (Li *et al.*, 2017). Most heavy metals such as Pb, Ni, As, Zn, Cu, Cd, and Cr are normally identified in the landfill leachate and these heavy metals are generally related to the environmental ARGs (Chen & Zhang 2013; Jia *et al.*, 2013). Out of these seven heavy metals, Cu, Cr, Ni, and Zn have a great relationship with human activities (Ji *et al.*, 2012) because Cu, Zn, and As are used in agriculture as a feed additives which are directly correlated with humans and animals (Lu *et al.*, 2020; Zhu *et al.*, 2013). Sometimes, heavy metals might interact with the antibiotics, and some heavy metals such as Cd and Cr dominate the antibiotics and it might overrule the effect of antibiotics on ARGs. It is observed that the correlation of antibiotics and ARGs vanished when the Cd and Cr heavy metals are dominated. It indicates that the ARGs are almost dependent on the heavy metals in the landfill leachate, and it increases the dissemination of ARGs in the MSW landfill leachate (Alonso *et al.*, 2001; Czekalski *et al.*, 2014).

18.4.3 Organic pollutants

Organic pollutants are also the main factors to proliferate and enrich ARGs in landfill leachates (Lv *et al.*, 2014; Rysz *et al.*, 2013; Zhou *et al.*, 2014). It is observed that the humic acids were significantly positively correlated with ARG concentrations (Yu *et al.*, 2016). With the presence of organic matter, certain microbial communities are present in landfills and organic matter also boosts the HGT and it stimulates the ARG diffusion in the aquatic environment (Jiao *et al.*, 2017). Overall, it signifies the positive significant correlation found between the organic matter and antibiotics and ARGs.

18.4.4 Environmental and other social factors

The antibiotics and ARG profiles in leachates were prompted by various parameters, and the contamination of antibiotics and ARGs might be lessened and monitored by altering the environmental aspects (Liu *et al.*, 2022). Some of the earlier studies showed that the other social and environmental factors such as seasonal variation, age of the landfill (Zhao *et al.*, 2018), human emigration activities, use of harmful chemical and pesticides, globalization and populace expansion affect the antibiotics and ARGs in landfill leachate (You *et al.*, 2018). Moreover, we need further research to understand the mechanisms of other social and environmental factors to increase the AR in the environment.

18.5 EFFECT OF LEACHATE QUALITY PARAMETERS AND LANDFILL AGE ON ANTIBIOTIC CONCENTRATION AND ARGS

There are various factors to determine the distribution characteristics of antibiotics and ARGs in the landfill leachate, such as heavy metals, seasonal variation, leachate quality parameters, volume of waste (Wu *et al.*, 2017), population of the vicinity, landfill age, and other prompting factors, but landfill age is also one of the main factors (Liu *et al.*, 2022; Yi *et al.*, 2017). The landfill leachate has a huge and different quality parameter due to the seasonal variation (summer, monsoon, and winter).

With the increasing age of the landfill, the nitrogen and carbon content decreases and on the contrary, the DO and pH value increases, indicating the degradation of the MSW (Wu *et al.*, 2019). Moreover, the concentration of heavy metals such as Co, Pb, and Zn increases significantly ($p < 0.05$) with the landfill age (Yu *et al.*, 2016). The landfill age has a major impact on the leachate composition specifically on organics, nitrogen, and carbon (Kulikowska & Klimiuk 2008). The transformation and migration of the organic compounds within leachates is also dependent on the landfill age. The landfill age is also divided into three categories: old leachate (>20 years), middle leachate (~10 years), and young (<3 years). The landfill age is positively correlated with the abundance of ARGs and negatively correlated with the leachate quality parameters, such as NO_3-N, NH_4-N, TP, COD, BOD, conductivity, and total organic carbon. It indicates that the landfill age is inversely proportional to the landfill quality parameters (Shi *et al.*, 2021). Landfills of the same age show highly heterogenous nature, because they have distinct leachate quality (Wang *et al.*, 2015). Overall, various factors affect the profile of antibiotics and ARGs; hence, the pollution of antibiotics and ARGs might be reduced by the control and variation of the environmental factors.

18.6 IMPACT OF ANTIBIOTICS AND ARGs ON GROUNDWATER AND SURFACE WATER DUE TO THE LANDFILL LEACHATE

According to history, the groundwater is free from any contaminated pollutants due to the natural filtering capability of the subsurface environment, and it takes higher time for microbes or pollutants to travel into the groundwater. But currently, contamination, ARGs, and antibiotic residue contamination were found in the groundwater and surface water due to the improper management of wastewater disposal, MSW disposal, contamination of septic tank, landfill leachate, and for other reasons such as saturated soil (Jindal *et al.*, 2013). Hence, the strong impact of antibiotics and ARGs on groundwater and surface water due to the landfill leachate.

18.6.1 Occurrence of the antibiotics and ARG contamination in groundwater

There are several reasons for the occurrence of the groundwater contamination: (i) the first reason is due to the chemicals because they are more persistent, and it is extremely difficult to eradicate them from the groundwater due to the comparatively reduced oxidoreduction reactions and deficiency of photodegradation compared to the surface water. Therefore, the groundwater contamination has more risk as compared to the surface water. (ii) The second reason is that the groundwater sites in India are generally not lined because the water table depth of the groundwater depends on the percolation. Therefore, the unlined drains are the biggest threat for the contamination of groundwater, and it adds more risk to the environment (Kumar *et al.*, 2016). The groundwater and surface water near the MSW landfills are also the main hotspots of antibiotics and ARGs in the environment (Wang *et al.*, 2015). The impact of antibiotic residue and ARGs on groundwater and surface water is detrimental for human health and threat to the public health.

18.6.2 Assessment of the antibiotics and ARG contamination in groundwater and surface water

Currently, landfill is the easiest way to the disposal of MSW and it spreads antibiotics and ARGs into the environment (Chen *et al.*, 2017). There are various survey and monitoring methods to verify the contamination of the groundwater. The first step is to check the physicochemical parameters and the limits of WHO guidelines, but due to the landfill leachate, drains, and drain leachate, the groundwater parameter limits exceed the WHO permissible limits. After that, check the diversification of the ARGs, the prevalence of antibiotics and ARGs, and identification of the various genetic elements in groundwater with the help of high-throughput qPCR (Chen *et al.*, 2017). The contamination of the groundwater generally also depends on the landfill age and changes occurring constantly with the landfill age. The changes in the groundwater contamination not only depends on the hydrogeological situations, natural land-use characteristics, depth of groundwater, soil type, slope, and precipitation, but it also depends on the leaked leachate water which is directly influenced by the components of the MSW (Han *et al.*, 2016). Hence, the groundwater contamination depends on the change in the water quality of leachate which is generated in MSW landfills. In the preliminary landfill phase, the concentrations of contaminants in landfill leachate slowly become complicated and high level. After the concentration extends its highest value, it decreased and stabilizes to a certain constant value (Renou *et al.*, 2008).

18.7 CONCLUSION

According to this study and availability of the scientific literature which were discussed in this chapter, following outcomes can be noted:

- Landfills play an important role for the disposal and treatment of MSW, despite the fact that they are the main hotspots of dissemination of antibiotics and ARGs, they are also the leading cause for further transmission of AR.
- Most of the physicochemical parameters, antibiotics, heavy metals, ARGs, organic matter, and other social and environmental factors present in the MSW landfill leachate, which are the most important source of the proliferation of antibiotics and ARGs in the environment and furthermore, antibiotic residues can alter the composition, structure, and biological functions of microbial community, which induces AR in the landfill leachate.
- The landfill age is inversely proportional to the landfill quality parameters and the transformation and migration of the organic compounds within leachates is also dependent on the landfill age. Hence, it is also one of the main factors which affect the landfill leachate and induces AR in the vicinity of the landfill.
- The groundwater and surface water adjacent to the landfill are also the hotspots of antibiotics and ARGs in the environment and the contamination of the groundwater generally also depends on the landfill age and changes constantly occurring with the landfill age. Moreover, the impact of antibiotic residue and ARGs on groundwater and surface water is a threat to the public health.

There are some gaps which need to fulfill by the researchers: (a) researchers should pay attention to the different compositions of MSW and its effect on the antibiotics and ARGs individually, (b) most of the studies found in developed countries such as China, United States, Italy, and France, but we need to focus on the landfill leachate problems which are related to the antibiotics and ARGs in India also, and (c) to develop a risk assessment procedure for the AR in the environment and suggests an urgent need for surveillance of ARGs and development of practices to reduce the risk.

REFERENCES

Alonso A., Sánchez P. and Martínez J. L. (2001). Environmental selection of antibiotic resistance genes. *Environmental Microbiology*, **3**, 1–9, https://doi.org/10.1046/j.1462-2920.2001.00161.x

Anand U., Reddy B., Singh V. K., Singh A. K., Kesari K. K., Tripathi P., Kumar P., Tripathi V. and Simal-Gandara J. (2021). Potential environmental and human health risks caused by antibiotic-resistant bacteria (ARB), antibiotic resistance genes (ARGs) and emerging contaminants (ECs) from municipal solid waste (MSW) landfill. *Antibiotics*, **10**, 374, https://doi.org/10.3390/antibiotics10040374

Bai L., Tan Z., Gong H., Xu M., Li Z., Yue J., Liu L., Yang D. and Li R. (2022). Study on antibiotics, antibiotic resistance genes, bacterial community characteristics and their correlation in the landfill leachates. *Journal of Applied Microbiology*, **132**, 445–458, https://doi.org/10.1111/jam.15229

Chen H. and Zhang M. (2013). Occurrence and removal of antibiotic resistance genes in municipal wastewater and rural domestic sewage treatment systems in eastern China. *Environment International*, **55**, 9–14, https://doi.org/10.1016/j.envint.2013.01.019

Chen Q. L., Li H., Zhou X. Y., Zhao Y., Su J.-Q., Zhang X. and Huang F.-Y. (2017). An underappreciated hotspot of antibiotic resistance: the groundwater near the municipal solid waste landfill. *Science of the Total Environment*, **609**, 966–973, https://doi.org/10.1016/j.scitotenv.2017.07.164

Christian T., Schneider R. J., Färber H. A., Skutlarek D., Meyer M. T. and Goldbach H. E. (2003). Determination of antibiotic residues in manure, soil, and surface waters. *Acta Hydrochimica et Hydrobiologica*, **31**, 36–44, https://doi.org/10.1002/aheh.200390014

Clarke B. O., Anumol T., Barlaz M. and Snyder S. A. (2015). Investigating landfill leachate as a source of trace organic pollutants. *Chemosphere*, **127**, 269–275, https://doi.org/10.1016/j.chemosphere.2015.02.030

Czekalski N., Gascón Díez E. and Bürgmann H. (2014). Wastewater as a point source of antibiotic-resistance genes in the sediment of a freshwater lake. *The ISME Journal*, **8**, 1381–1390, https://doi.org/10.1038/ismej.2014.8

Gao P., Munir M. and Xagoraraki I. (2012). Correlation of tetracycline and sulfonamide antibiotics with corresponding resistance genes and resistant bacteria in a conventional municipal wastewater treatment plant. *Science of the Total Environment*, **421–422**, 173–183, https://doi.org/10.1016/j.scitotenv.2012.01.061

Garcia-Armisen T., Vercammen K., Passerat J., Triest D., Servais P. and Cornelis P. (2011). Antimicrobial resistance of heterotrophic bacteria in sewage-contaminated rivers. *Water Research*, **45**, 788–796, https://doi.org/10.1016/j.watres.2010.09.003

Gothwal R. and Shashidhar T. (2015). Antibiotic pollution in the environment: a review. *Clean – Soil, Air, Water*, **43**, 479–489, https://doi.org/10.1002/clen.201300989

Han Z., Ma H., Shi G., He L., Wei L. and Shi Q. (2016). A review of groundwater contamination near municipal solid waste landfill sites in China. *Science of the Total Environment*, **569–570**, 1255–1264, https://doi.org/10.1016/j.scitotenv.2016.06.201

Jabbar F. K., Grote K. and Tucker R. E. (2019). A novel approach for assessing watershed susceptibility using weighted overlay and analytical hierarchy process (AHP) methodology: a case study in Eagle Creek Watershed, USA. *Environmental Science and Pollution Research*, **26**, 31981–31997, https://doi.org/10.1007/s11356-019-06355-9

Jechalke S., Heuer H., Siemens J., Amelung W. and Smalla K. (2014). Fate and effects of veterinary antibiotics in soil. *Trends in Microbiology*, **22**, 536–545, https://doi.org/10.1016/j.tim.2014.05.005

Ji X., Shen Q., Liu F., Ma J., Xu G., Wang Y. and Wu M. (2012). Antibiotic resistance gene abundances associated with antibiotics and heavy metals in animal manures and agricultural soils adjacent to feedlots in Shanghai, China. *Journal of Hazardous Materials*, **235–236**, 178–185, https://doi.org/10.1016/j.jhazmat.2012.07.040

Jia M., Wang F., Bian Y., Jin X., Song Y., Kengara F. O., Xu R. and Jiang X. (2013). Effects of pH and metal ions on oxytetracycline sorption to maize-straw-derived biochar. *Bioresource Technology*, **136**, 87–93, https://doi.org/10.1016/j.biortech.2013.02.098

Jiao Y. N., Chen H., Gao R. X., Zhu Y.-G. and Rensing C. (2017). Organic compounds stimulate horizontal transfer of antibiotic resistance genes in mixed wastewater treatment systems. *Chemosphere*, **184**, 53–61, https://doi.org/10.1016/j.chemosphere.2017.05.149

Jindal T., Kumar A., Nirpen L., Ranjan A., Gulati K. and Thakur S. (2013). Leaching of overloaded and unlined drains through the ground water in Delhi. *Indian Journal of Environmental Protection*, **33**, 990–997.

Klein E. Y., Van Boeckel T. P., Martinez E. M., Pant S., Gandra S., Levin S. A., Goossens H. and Laxminarayan R. (2018). Global increase and geographic convergence in antibiotic consumption between 2000 and 2015. *Proceedings of the National Academy of Sciences of the USA*, **115**, E3463–E3470, https://doi.org/10.1073/pnas.1717295115

Koczura R., Mokracka J., Taraszewska A. and Łopacinska N. (2016) Abundance of class 1 integron-integrase and sulfonamide resistance genes in river water and sediment is affected by anthropogenic pressure and environmental factors. *Microbial Ecology* **72**, 909–916. https://doi.org/10.1007/s00248-016-0843-4

Kulikowska D. and Klimiuk E. (2008). The effect of landfill age on municipal leachate composition. *Bioresource Technology*, **99**, 5981–5985, https://doi.org/10.1016/j.biortech.2007.10.015

Kumar A., Ranjan A., Gulati K., Thakur S. and Jindal T. (2016). Assessment of chemical and microbial contamination in groundwater through leaching of sewage waste in Delhi, India. *Environmental Earth Science*, **75**, 1–11, https://doi.org/10.1007/s12665-015-5016-0

Li L. G., Xia Y. and Zhang T. (2017). Co-occurrence of antibiotic and metal resistance genes revealed in complete genome collection. *The ISME Journal*, **11**, 651–662, https://doi.org/10.1038/ismej.2016.155

Lienert J., Güdel K. and Escher B. I. (2007). Screening method for ecotoxicological hazard assessment of 42 pharmaceuticals considering human metabolism and excretory routes. *Environmental Science & Technology*, **41**, 4471–4478, https://doi.org/10.1021/es0627693

Liu H., Li H., Qiu L., Chen B., Wang H., Fang C., Long Y. and Hu L. (2022). The panorama of antibiotics and the related antibiotic resistance genes (ARGs) in landfill leachate. *Waste Management*, **144**, 19–28, https://doi.org/10.1016/j.wasman.2022.03.008

Lu L., Liu J., Li Z., Zou X., Guo J., Liu Z., Yang J. and Zhou Y. (2020). Antibiotic resistance gene abundances associated with heavy metals and antibiotics in the sediments of Changshou lake in the three gorges reservoir area, China. *Ecological Indicators*, **113**, 106275, https://doi.org/10.1016/j.ecolind.2020.106275

Luo Y., Mao D., Rysz M., Zhou Q., Zhang H., Xu L. and Alvarez P. J. J. (2010). Trends in antibiotic resistance genes occurrence in the Haihe river, China. *Environmental Science & Technology*, **44**, 7220–7225, https://doi.org/10.1021/es100233w

Lv L., Jiang T., Zhang S. and Yu X. (2014). Exposure to mutagenic disinfection byproducts leads to increase of antibiotic resistance in pseudomonas aeruginosa. *Environmental Science & Technology*, **48**, 8188–8195, https://doi.org/10.1021/es501646n

Masoner J. R., Kolpin D. W., Furlong E. T., Cozzarelli I. M. and Gray J. L. (2016). Landfill leachate as a mirror of today's disposable society: pharmaceuticals and other contaminants of emerging concern in final leachate from landfills in the conterminous United States. *Environmental Toxicology and Chemistry*, **35**, 906–918, https://doi.org/10.1002/etc.3219

Pan M. and Chu L. M. (2017). Fate of antibiotics in soil and their uptake by edible crops. *Science of the Total Environment*, **599–600**, 500–512, https://doi.org/10.1016/j.scitotenv.2017.04.214

Pei R., Kim S. C., Carlson K. H. and Pruden A. (2006). Effect of river landscape on the sediment concentrations of antibiotics and corresponding antibiotic resistance genes (ARG). *Water Research*, **40**, 2427–2435, https://doi.org/10.1016/j.watres.2006.04.017

Renou S., Givaudan J. G., Poulain S., Dirassouyan F. and Moulin P. (2008). Landfill leachate treatment: review and opportunity. *Journal of Hazardous Materials*, **150**, 468–493, https://doi.org/10.1016/j.jhazmat.2007.09.077

Rysz M., Mansfield W. R., Fortner J. D. and Alvarez P. J. J. (2013). Tetracycline resistance gene maintenance under varying bacterial growth rate, substrate and oxygen availability, and tetracycline concentration. *Environmental Science & Technology*, **47**, 6995–7001, https://doi.org/10.1021/es3035329

Sarmah A. K., Meyer M. T. and Boxall A. B. A. (2006). A global perspective on the use, sales, exposure pathways, occurrence, fate and effects of veterinary antibiotics (VAs) in the environment. *Chemosphere*, **65**, 725–759, https://doi.org/10.1016/j.chemosphere.2006.03.026

Shi J., Wu D., Su Y. and Xie B. (2021). Selective enrichment of antibiotic resistance genes and pathogens on polystyrene microplastics in landfill leachate. *Science of the Total Environment*, **765**, 142775, https://doi.org/10.1016/j.scitotenv.2020.142775

Sommer M. O. A., Dantas G. and Church G. M. (2009). Functional characterization of the antibiotic resistance reservoir in the human microflora. *Science (80-)*, **325**, 1128–1131, https://doi.org/10.1126/science.1176950

Song L., Li L., Yang S., Lan J., He H., McElmurry S. P. and Zhao Y. (2016). Sulfamethoxazole, tetracycline and oxytetracycline and related antibiotic resistance genes in a large-scale landfill, China. *Science of the Total Environment*, **551–552**, 9–15, https://doi.org/10.1016/j.scitotenv.2016.02.007

Su Y., Wang J., Huang Z. and Xie B. (2017). On-site removal of antibiotics and antibiotic resistance genes from leachate by aged refuse bioreactor: effects of microbial community and operational parameters. *Chemosphere*, **178**, 486–495, https://doi.org/10.1016/j.chemosphere.2017.03.063

Sui Q., Zhao W., Cao X., Lu S., Qiu Z., Gu X. and Yu G. (2017). Pharmaceuticals and personal care products in the leachates from a typical landfill reservoir of municipal solid waste in Shanghai, China: occurrence and removal by a full-scale membrane bioreactor. *Journal of Hazardous Materials*, **323**, 99–108, https://doi.org/10.1016/j.jhazmat.2016.03.047

Sun M., Ye M., Schwab A. P., Li X., Wan J., Wei Z., Wu J., Friman V.-P., Liu K., Tian D., Liu M., Li H., Hu F. and Jiang X. (2016). Human migration activities drive the fluctuation of ARGs: case study of landfills in Nanjing, Eastern China. *Journal of Hazardous Materials*, **315**, 93–101, https://doi.org/10.1016/j.jhazmat.2016.04.077

Välitalo P., Kruglova A., Mikola A. and Vahala R. (2017). Toxicological impacts of antibiotics on aquatic micro-organisms: a mini-review. *International Journal of Hygiene and Environmental Health*, **220**, 558–569, https://doi.org/10.1016/j.ijheh.2017.02.003

Wang Y., Tang W., Qiao J. and Song L. (2015). Occurrence and prevalence of antibiotic resistance in landfill leachate. *Environmental Science and Pollution Research*, **22**, 12525–12533, https://doi.org/10.1007/s11356-015-4514-7

Watanabe N., Bergamaschi B. A., Loftin K. A., Meyer M. T. and Harter T. (2010). Use and environmental occurrence of antibiotics in freestall dairy farms with manured forage fields. *Environmental Science and Technology*, **44**, 6591–6600, https://doi.org/10.1021/es100834s

WHO (2019). Ten threats to global health in 2019, https://www.who.int/news-room/spotlight/ten-threats-to-global-health-in-2019 (Accessed August 22, 2021)

Wu N., Qiao M., Zhang B., Cheng W.-D. and Zhu Y.-G. (2010). Abundance and diversity of tetracycline resistance genes in soils adjacent to representative swine feedlots in China. *Environmental Science & Technology*, **44**, 6933–6939, https://doi.org/10.1021/es1007802

Wu D., Huang Z., Yang K., Graham D. and Xie B. (2015). Relationships between antibiotics and antibiotic resistance gene levels in MSW solid waste leachates in Shanghai, China. *Environmental Science & Technology*, **49**, 4122–4128, https://doi.org/10.1021/es506081z

Wu D., Huang X. H., Sun J. Z., Graham D. W. and Xie B. (2017). Antibiotic resistance genes and associated microbial community conditions in aging landfill systems. *Environmental Science & Technology*, **51**, 12859–12867, https://doi.org/10.1021/acs.est.7b03797

Wu D., Wang B. H. and Xie B. (2019). Validated predictive modelling of sulfonamide and beta-lactam resistance genes in landfill leachates. *Journal of Environmental Management*, **241**, 123–130, https://doi.org/10.1016/j.jenvman.2019.04.026

Yao L., Li Y., Li Z., Shen D., Feng H., Zhou H. and Wang M. (2020). Prevalence of fluoroquinolone, macrolide and sulfonamide-related resistance genes in landfills from East China, mainly driven by MGEs. *Ecotoxicology and Environmental Safety*, **190**, 110131, https://doi.org/10.1016/j.ecoenv.2019.110131

Yi X., Tran N. H., Yin T., He Y. and Gin K. Y.-H. (2017). Removal of selected PPCPs, EDCs, and antibiotic resistance genes in landfill leachate by a full-scale constructed wetlands system. *Water Research*, **121**, 46–60, https://doi.org/10.1016/j.watres.2017.05.008

You X., Wu D., Wei H., Xie B. and Lu J. (2018). Fluoroquinolones and β-lactam antibiotics and antibiotic resistance genes in autumn leachates of seven major municipal solid waste landfills in China. *Environment International*, **113**, 162–169, https://doi.org/10.1016/j.envint.2018.02.002

Yu Z., He P., Shao L., Zhang H. and Lü F. (2016). Co-occurrence of mobile genetic elements and antibiotic resistance genes in municipal solid waste landfill leachates: a preliminary insight into the role of landfill age. *Water Research*, **106**, 583–592, https://doi.org/10.1016/j.watres.2016.10.042

Zhang Q. Q., Ying G. G., Pan C. G., Liu Y.-S. and Zhao J.-L. (2015). Comprehensive evaluation of antibiotics emission and fate in the river basins of China: source analysis, multimedia modeling, and linkage to bacterial resistance. *Environmental Science & Technology*, **49**, 6772–6782, https://doi.org/10.1021/acs.est.5b00729

Zhang R., Yang S., An Y., Wang Y., Lei Y. and Song L. (2022). Antibiotics and antibiotic resistance genes in landfills: a review. *Science of the Total Environment*, **806**, 150647, https://doi.org/10.1016/j.scitotenv.2021.150647

Zhao R., Feng J., Yin X., Liu J., Fu W., Berendonk T., Zhang T., Li X. and Li B. (2018). Antibiotic resistome in landfill leachate from different cities of China deciphered by metagenomic analysis. *Water Research*, **134**, 126–139, https://doi.org/10.1016/j.watres.2018.01.063

Zheng Q., Zhang R., Wang Y., Pan X., Tang J. and Zhang G. (2012). Occurrence and distribution of antibiotics in the Beibu Gulf, China: impacts of river discharge and aquaculture activities. *Marine Environmental Research*, **78**, 26–33, https://doi.org/10.1016/j.marenvres.2012.03.007

Zhou L. J., Ying G. G., Liu S., Zhang R.-Q., Lai H.-J., Chen Z.-F. and Pan C.-G. (2013a). Excretion masses and environmental occurrence of antibiotics in typical swine and dairy cattle farms in China. *Science of the Total Environment*, **444**, 183–195, https://doi.org/10.1016/j.scitotenv.2012.11.087

Zhou L. J., Ying G. G., Zhang R. Q., Liu S., Lai H.-J., Chen Z.-F., Yang B. and Zhao J.-L. (2013b). Use patterns, excretion masses and contamination profiles of antibiotics in a typical swine farm, South China. *Environmental Science: Processes & Impacts*, **15**, 802–813, https://doi.org/10.1039/c3em30682h

Zhou H., Meng A., Long Y., Li Q. and Zhang Y. (2014). An overview of characteristics of municipal solid waste fuel in China: physical, chemical composition and heating value. *Renewable and Sustainable Energy Reviews*, **36**, 107–122, https://doi.org/10.1016/j.rser.2014.04.024

Zhu Y. G., Johnson T. A., Su J. Q., Qiao M., Guo G.-X., Stedtfeld R. D., Hashsham S. A. and Tiedje J. M. (2013). Diverse and abundant antibiotic resistance genes in Chinese swine farms. *Proceedings of the National Academy of Sciences of the USA*, **110**, 3435–3440, https://doi.org/10.1073/pnas.1222743110

Zuccato E., Castiglioni S., Bagnati R., Melis M. and Fanelli R. (2010). Source, occurrence and fate of antibiotics in the Italian aquatic environment. *Journal of Hazardous Materials*, **179**, 1042–1048, https://doi.org/10.1016/j.jhazmat.2010.03.110

doi: 10.2166/9781789063318_0411

Chapter 19

Aerobic and anaerobic methods of landfill leachate treatment: limitations and advantages

Sandeep Singh[1], *Sandeep K. Malyan*[2], *Rajesh Singh*[1*], *Vinay Kumar Tyagi*[1] *and Sujata Kashyap*[3]

[1]*Environmental Hydrology Division, National Institute of Hydrology, Roorkee 247667, India*
[2]*Department of Environmental Studies, Dyal Singh Evening College, University of Delhi, New Delhi 110003, India*
[3]*Axa Parenterals Limited, Roorkee 247667, India*
**Corresponding author: rsingh.nih@gmail.com*

ABSTRACT

Landfilling is the most common and popular method to dispose of solid waste. The production of leachate from landfill sites is a major problem associated with the landfill technique of solid waste disposal. This chapter presents a general idea of the technical applicability of aerobic and anaerobic treatment of landfill leachate. This chapter reviews the performance, mechanism, application, limitations, and upgradation of various existing aerobic and anaerobic treatment techniques such as activated sludge (AS), sequencing batch reactors (SBRs), aerated lagoons, rotating biological contactors, upflow anaerobic sludge blanket (UASB) reactors, and constructed wetlands. The designing parameters such as hydraulic retention time (HRT) and organic loading rate and performance indicators are discussed. Among the aerobic and anaerobic treatment processes, AS, SBRs, and UASB are the most often used techniques. These techniques are effective in reducing chemical oxygen demand up to 90% and NH_3-N up to 80%. Generally, the selection of the effective treatment technique for landfill leachate depends on leachate characteristics, technical applicability, HRT, economics, and further environmental impacts.

Keywords: Landfill leachate, pollutants, aerobic, anaerobic, constructed wetlands

19.1 INTRODUCTION

In recent years, municipal solid waste (MSW) has emerged as one among the truly serious challenges to the world environment. Due to the mobilization of the population to urban areas, about an average of 1,000 m^3 MSW is generated in Asia annually (Hoornweg, 2000), and increased by ~3–4% annually per capita generation of solid waste from 0.5 to 2 kg per person per day (Mian *et al.*, 2017). Greenhouse gas emissions from MSW contribute to ~20–25% of total anthropogenic emissions (Adhikari *et al.*, 2006). As less specific measures are designed to prevent environmental contamination from solid waste, this MSW and its leachate exert harmful effects on the environment and human health. For sustainable waste management, a well-engineered landfill is required for the proper disposal of MSW with impervious liners, a leachate collection system, and a leachate treatment system to minimize soil and natural water pollution.

Biodegradable refuse with high moisture and organic contents such as food waste, garden waste, and so on are components of MSW. This type of waste is immediately decomposed by the physicochemical and biological processes when dumped into a landfill (Horner, 1994). Physical changes cause a change in volume and size, and biological changes involve decomposition by microbes. In total, 6% of landfill leachate is produced during the microbial decomposition of waste (Börjesson & Svensson, 1997), and it continues for over 50 years after the closure of the landfill (Kurniawan *et al.*, 2006, 2010). The landfill leachate characteristics vary with time based on various factors such as decomposition of solid waste, seasonal variation, hydrology of landfill site, pH, and moisture content (Kjeldsen *et al.*, 2002). In the monsoon season, the rainwater can seep through the landfill and contaminate groundwater with toxic pollutants that are present in leachate such as organic halides, heavy metals, xenobiotic compounds, and ammoniacal nitrogen (NH_3-N) which causes acute toxicity (Baun *et al.*, 2004; Bernard *et al.*, 1997; Silva *et al.*, 2004). The common characteristics of young and stabilized leachate are provided in Table 19.1 (Ehrig, 1988; Lo, 1996).

The landfill leachate is heavily polluted and therefore efficient control measures are required to remediate and control infiltration, else, it will lead to accelerated groundwater contamination with toxic pollutants from waste. As groundwater is a major source of public drinking water and irrigation, the risk of groundwater contamination becomes a foremost environmental concern these days (Tränkler *et al.*, 2005).

Table 19.1 Characteristics of young and stabilize leachate.

Parameters	Young Leachate (1–5 years)	Stabilized Leachate (5–10 years)
Composition (molecular weight)	Low molecular weight compounds (C_2–C_6 carboxylic acids)	High molecular weight compounds (humic and fulvic acids)
pH	<6.5	>7.5
NH_3-N (mg/L)	500–2,000	400–5,000
BOD (mg/L)	4,000–13,000	20–1,000
COD (mg/L)	6,000–60,000	5,000–20,000
TKN (mg/L)	0.1–2	NA
Heavy metals (mg/L)	>2	<2

The design and management of a landfill play a major role in leachate percolation. Good engineering practices adopted during landfill design result in reduced leachate production and seepage into the groundwater (Qasim & Chiang, 2017).

For the removal of organic pollutants from leachate, various physical and chemical treatments such as coagulation–flocculation (Amokrane *et al.*, 1997), chemical precipitation (Zhang *et al.*, 2009), ammonia stripping (Cheung *et al.*, 1997), adsorption (Halim *et al.*, 2010), and advanced oxidation processes such as ozonation and Fenton oxidation (Choi, 1998) have been used. However, these treatments have a limited success rate for the removal of targeted compounds from leachate, so the development of an alternative technology for leachate treatment is necessary.

Biological treatments such as sequencing batch reactors (SBRs) (Neczaj *et al.*, 2005), upflow anaerobic sludge blanket (UASB) reactors (Parawira *et al.*, 2006), natural treatment systems such as constructed wetlands (CWs), and leachate recirculation (Lee *et al.*, 2002) nowadays are widely employed to remove the chemical matter and gaseous emission from MSW due to their low cost

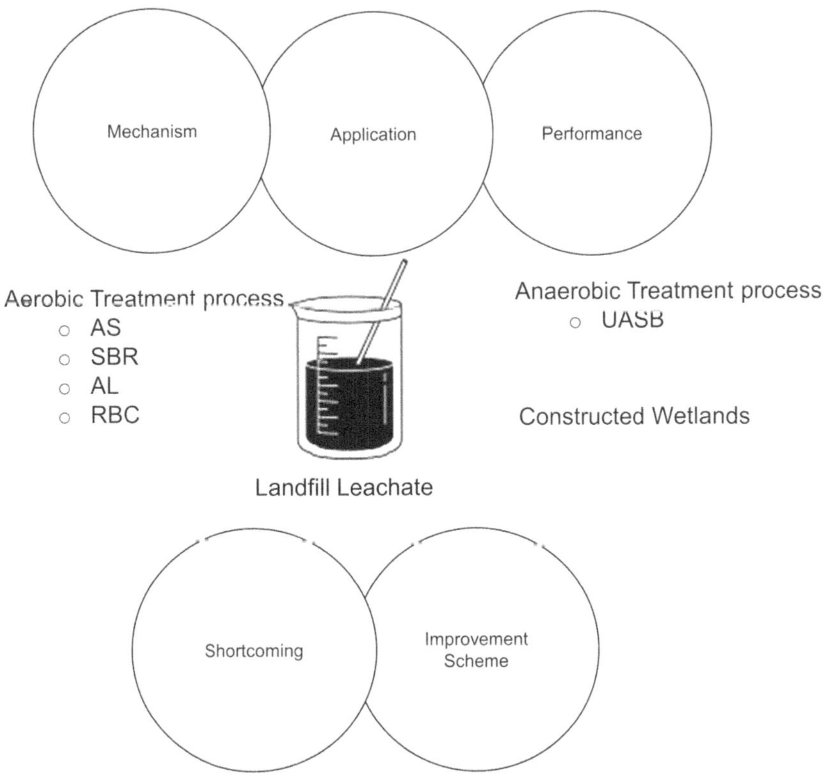

Figure 19.1 Aerobic and anaerobic landfill leachate treatment processes.

and applicability. Microbes break down the organic pollutants from leachate under either aerobic or anaerobic conditions (Devare & Bahadir, 1994). In the aerobic process, the target pollutants are converted into carbon dioxide, water, and biological products (biomass) in the presence of atmospheric oxygen and microorganisms. In anaerobic treatment, the major part of organic matter is converted into biogas (methane) and biological sludge in the absence of oxygen. The biological processes are very effective in the case of young leachate compared to stabilized leachate (Lema *et al.*, 1988). Few studies have been conducted to provide a comprehensive overview of the biological leachate treatment process and optimize the process to maximize the target compound removal.

This chapter presents an overview with a critical analysis of the technical application and treatment performance of biological processes covering their mechanism, application, performance, limitations, and improvement scheme (Figure 19.1).

19.2 AEROBIC TREATMENT

Landfill leachate consists of high-organic and -inorganic pollutant loads and the biological treatment is economical, reliable, and easy to operate. Biological treatment approaches are generally differentiated into two types: aerobic and anaerobic. In aerobic treatment, microorganisms consume organic matter in the presence of oxygen and convert it into inorganic components and biomass. The most frequently used aerobic biological treatment technologies for landfill leachate treatment are activated sludge (AS), SBRs, aerated lagoons (ALs), rotating biological contactors (RBCs), and so on. AS processes are usually applied for the leachate generated from domestic wastes (Huang *et al.*, 2008; Lin *et al.*, 2000; Liu, 2003; Pan & Tseng, 2001; Qu *et al.*, 2019; Sachdeva *et al.*, 2000; Sillanpää & Oikari, 1996; Sorvari & Sillanpää, 1996; Stolarek & Ledakowicz, 2001; Virkutyte & Sillanpää, 2006). In AS processes, harmful pollutants are converted into harmless products such as carbon dioxide and water through different microorganisms. In this process, oxygen or air is blown into the wastewater and this aerated wastewater is biologically treated. AS has high concentrations of bacteria, protozoa, and fungi, which degrades the organic matter. Mojiri *et al.* (2021) quoted that the AS treatment processes are efficient in the complete degradation of organic matter. An AS process has some limitations such as longer aeration time for sludge settleability (Loukidou & Zouboulis, 2001), high-energy demand (Kjeldsen *et al.*, 2002), and microbial inhibition due to high ammonium-nitrogen concentration (Lema *et al.*, 1988). Few researchers have reported landfill leachate treatment by AS. Hoilijoki *et al.* (2000) in their lab-scale study investigated the application of AS on landfill leachate at different temperatures (5–10°C) and concluded that chemical oxygen demand (COD) 150–500 mg/L, biochemical oxygen demand (BOD) <7 mg/L, and NH_3-N < 13 mg/L is achievable in the leachate after treatment. They also concluded that the nitrification efficiency was enhanced with the accumulation of plastic carrier material (Hoilijoki *et al.*, 2000). Few other studies reported different percentage removal for COD, BOD, and NH_3-N under different operating conditions (Table 19.2).

Table 19.2 Removal efficiency of aerobic treatment processes under different operating conditions.

Technique Used	Leachate Volume (L)	HRT (day)	SRT (day)	COD Removal Efficiency (%)	NH_3-N Removal Efficiency (%)	Reference
AS	–	–	–	19.7	95	Zhang *et al.* (2019)
AS	5	20	–	98	99	Robinson and Maris (1985)
AS	5	20	–	41	90	Robinson and Maris (1985)
AS	5	1	32	53	98	Setiadi and Fairus (2003)
AS	2	–	3–4	95	–	Li and Zhao (1999)
AS	2	–	–	98	–	Li and Zhao (2001)
SBR + RO	15	–	–	97.4	>99	Anna Tałałaj *et al.* (2021)
SBR	–	0.5	–	68	–	Dollerer and Wilderer (1996)
SBR	–	1	–	38	99	Hosomi *et al.* (1989)
SBR	–	3.2	–	97	99	Zaloum and Abbott (1997)
SBR	5	1	–	97	99	Yalmaz and Öztürk (2001)
SBR	4	0.29	–	64	23	Uygur and Kargi (2004)
SBR	2.75	2	–	74	–	Timur and Özturk (1999)
SBR	8	20	–	90	–	Loukidou and Zouboulis (2001)
SBR	8	20	–	75	70	Zouboulis *et al.* (2001)
SBR	5	–	–	90	70	Neczaj *et al.* (2005)
SBR	0.5	12	–	83	–	Klimiuk and Kulikowska (2006)
SBR	6	0.25	–	–	100	Doyle *et al.* (2001)
SBR	5	–	–	62	–	Uygur and Kargi (2004)
SBR	18.8	1.25	–	74	–	Kargi and Pamukoglu (2004)
AL	–	–	–	99	99	Robinson and Maris (1985)
AL	–	–	–	–	77	Frascari *et al.* (2004)
AL	–	–	–	98	99.5	Robinson and Maris (1985)
AL	–	–	–	83	99.8	Robinson and Maris (1985)
RBC	33	–	–	–	95	Siegrist *et al.* (1998)
RBC	–	3	–	86	–	Robinson and Maris (1985)
RBC	–	–	–	85.65	99.92	Song *et al.* (2020)

For superior removal of NH_3-N, SBRs are used for landfill leachate treatment (Agamuthu, 1999; Diamadopoulos *et al.*, 1997). Some studies have reported up to 99% NH_3-N removal (Lo, 1996) and up to 75% COD removal (Dollerer & Wilderer, 1996; Kargi & Pamukoglu, 2004). The percentage removal of COD, BOD, and NH_3-N under different operating conditions using SBRs is present in Table 19.2. Biological nitrogen removal from landfill leachate can also be achieved through trickling filters (Aluko & Sridhar, 2013; Bove *et al.*, 2015; Renou *et al.*, 2008). Trickling filters are low-cost biofilters and can tolerate certain variations in landfill leachate composition and concentration (Bove *et al.*, 2015). The average depth of the trickling filter is ~2 m and it has ~50% of porosity (Bove *et al.*, 2015). Raw landfill leachate is fed from the top which passes through the bed and treated leachate is collected from the bottom (Bove *et al.*, 2015). Renou *et al.* (2008) concluded that the trickling filters can remove 90% of the ammoniacal nitrogen. Aluko and Sridhar (2013) also investigated the role of trickling filters and SBRs in leachate treatment and observed that the trickling filters removed 59.50% of ammonia, 76.69% of BOD, 71.96% of turbidity, and 73.17% of suspended solid. The advantage of using a trickling filter is that different pollutants (such as BOD, COD, ammonia, etc.) can be removed simultaneously, however, clogging of the filter is an issue (Nath & Debnath, 2022). Therefore, it is recommended to avoid the treatment of leachate having a high suspended load (Nath & Debnath, 2022).

For removing pathogens, organics, and inorganic matter from landfill leachate, ALs are generally preferred because of their low cost and high efficiency (Zaloum & Abbott, 1997). Mæhlum (1995) explored the combination of ALs with CWs for landfill leachate treatment and observed N, P, and Fe removal efficiency to be above 70%. ALs have a limitation in terms of their temperature dependency which affects microbial activity (Zaloum & Abbott, 1997).

RBCs have a series of closely spaced circular plastic disks containing microbial films of diameter ~3.5 cm. When landfill leachate flows between these closely mounted disks, the biomass developed on the surface of the disk absorbs and disintegrates the organic toxins. This RBC technique is generally used for the treatment of leachate having low strength and it has several benefits such as low-energy consumption, less space requirement, low sludge generation, and high pollutant removal (Qasim & Chiang, 2017). However, RBCs are not appropriate for high-strength leachate and temperature-dependence processes. High-capital cost limits the RBCs applicability for high-flow and high-strength wastewater and more research is needed to reduce the associated capital cost to increase the acceptability of this technique.

Although aerobic leachate treatment techniques exhibit promising results, certain drawbacks are also associated with them such as high-operating costs, odor issues, high-sludge generation, and so on. The advantages and disadvantages of the aerobic treatment processes are summarized in Table 19.3.

Table 19.3 Advantages and limitations of the aerobic and anaerobic leachate treatment processes.

Sl. No.	Type of Treatment	Target Removal	COD Removal (%)	NH_3-N Removal (%)	Advantage	Disadvantage	Reference
Aerobic							
1	SBR	NH_3-N and organic material	90–97	99–100	Clarifier not required, sludge generation: used as fertilizer	Odor, high-energy consumption	Klimiuk and Kulikowska (2005)
2	AL	NH_3-N and organic material	83–98	99	Low cost	Long retention time, algal growth, odor, high-energy consumption	Frascari *et al.* (2004)
3	Nitrification	Nitrogen compound	30.35	90	Low cost, high nitrification rate	High-energy consumption, limited applicability	Kabdasli *et al.* (2000)
4	AS	NH_3-N and organic material	95–98	90–99	Sludge generation: used as fertilizer	High cost, high retention time	Dollerer and Wilderer (1996)
Anaerobic							
1	UASB	NH_3-N and organic material	91–98	50–80	Methane production, sludge settling	Long treatment time, ammonia toxicity	Gülşen and Turan (2004)

19.3 ANAEROBIC TREATMENT

In anaerobic treatment, microbes grow in the absence of oxygen and convert organics into CO_2, CH_4, and other metabolite products (Price *et al.*, 2003). Some of the widely used anaerobic treatment processes for landfill leachate are digesters, UASB reactors, anaerobic filters, hybrid-bed filters, fluidized-bed reactors, and anaerobic–pulsed-bed filters integrated with aerobic sequencing batch reactors. It is a suitable and effective technique for waste valorization from young leachate (Fang & Chui, 1993). In some laboratory-scale studies for performance evaluation, 50–90% COD removal was obtained at different temperatures (Bull *et al.*, 1983; Sung *et al.*, 1997). The most broadly applied anaerobic technique for landfill leachate treatment is UASB (Lin *et al.*, 2000), which is used for the removal of organic matter and ammoniacal nitrogen from landfill leachate, and the generation of methane as a valuable product. The UASB process has high treatment efficiency and short hydraulic retention time (HRT) resulting in low capital and operating costs (Lin *et al.*, 2000). At intermediate organic loading rates (6–19.7 g/L/day COD), up to 92% COD removal efficiency was obtained by Kennedy and Lentz (2000). Few studies conducted at low temperatures concluded lower pollutant removal efficiency (García *et al.*, 1996; Kettunen & Rintala, 1998). The percentage removal of COD and NH_3-N under different operating conditions by using UASB is presented in Table 19.4. The high removal efficiency results indicate a high rate of treatment at a low cost with valuable end products (García *et al.*, 1996). Occasionally, the UASB reactors underperform due to free ammonia toxicity (Gülşen & Turan, 2004), which can be minimized by maintaining the pH at $\sim$7. During the anaerobic treatment of landfill leachate with the UASB technique, organic material is converted to carbon dioxide and methane, and organic nitrogen is converted to ammonia which causes ammonia toxicity resulting in inhibition and sudden failure of the treatment process.

19.4 CONSTRUCTED WETLANDS

CWs are artificially biologically engineered structures to represent natural wetlands for the degradation of the desired contaminants present in polluted water (Malyan *et al.*, 2021a; Singh *et al.*, 2023). CWs use the synergistic effect of different types of bacteria attached to the plant roots and substrates for the degradation of pollutants present in the leachate (Muthukumaran, 2022). CWs are natural treatment systems for on-site removal of contaminants, which use emergent plants such as *Phragmites australis*, *Typha latifolia*, *Typha angustifolia*, *Cyperus papyrus*, *Cyperus alternifolius*, *Spider lily*, *Heliconia*, and so on. HRT, water flow regimes, and microbial consortia are the main factors affecting the performance of CWs (Malyan *et al.*, 2021a). Due to its ecosystem characteristics, it is a widely used technology for the treatment of complex pollutants such as landfill leachate. Due to its natural aerobic and anaerobic interface microbial biomass sustainability, CWs degrade the desired toxic pollutants present in the landfill leachate. The vegetation root zone in CWs allows the natural reduction of many contaminants by microbial degradation processes. CW systems are widely used in the United States, Canada, Slovenia, and France (Urbanc-Berčič,

Table 19.4 UASB efficiency under different operation conditions.

Technique Used	Leachate Volume (L)	HRT (day)	SRT (day)	COD Removal Efficiency (%)	NH_3-N Removal Efficiency (%)	Reference
HUASB	–	–	–	–	89.6	Kannah *et al.* (2022)
APBF + aeSBR	–	8	–	97	–	Tyagi *et al.* (2019)
UASB	0.8	2.9	–	98	–	Parawira *et al.* (2006)
UASB	–	2	–	80	–	Calli *et al.* (2006)
UASB	9	9	–	87	–	Marañón *et al.* (2006)
UASB	–	9	–	87	–	Marañón *et al.* (2006)
UASB	2.5	4	–	98	–	Ağdağ and Sponza (2005)
UASB	2.8	6.6	–	90	–	Fang *et al.* (2005)
UASB	13	1	–	90	–	Gülşen and Turan (2004)
UASB	41	3.9	–	96	–	Shin *et al.* (2001)
UASB	7.8	2	–	90	–	Inanc *et al.* (2000)
UASB	10	3	–	94	–	Timur *et al.* (2000)
UASB	1.4	1	–	91	–	Kennedy and Lentz (2000)
UASB	10	0.5	–	85	–	Timur and Özturk (1999)
UASB	13	1.5	–	68	–	Griffin *et al.* (1999)
UASB	–	3	–	75	–	Kettunen and Rintala (1998)
UASB	0.7	6	–	80	–	Sung *et al.* (1997)
UASB	2	–	–	98	50	Diamadopoulos *et al.* (1997)
UASB	–	0.5	–	70	–	Berrueta *et al.* (1996)
UASB	0.3	0.4	–	75	80	Kettunen *et al.* (1996a, 1996b)
UASB	–	0.1	–	91	–	García *et al.* (1996)
UASB	0.2	2.1	–	95	–	Kettunen *et al.* (1996a, 1996b)
UASB	–	12	–	85	–	Robinson and Maris (1985)

APBF, anaerobic–pulsed-bed filter; aeSBR, aerobic sequencing batch reactor; HUASB, hybrid upflow anaerobic sludge blanket.

1996). Bulc (2006) investigated the feasibility of pilot-scale hybrid (vertical plus horizontal flow) CWs (311 m^2) with cattails and reeds for the treatment of landfill leachate in Slovenia for 7 years and observed a significant reduction in the pollution load (BOD_5 – 59%; COD – 50%; ammoniacal nitrogen – 51%, total phosphorous – 53%; and Fe – 84%) indicating suitability of CWs for landfill leachate treatment. High-organic load in raw landfill leachate hinders the use of membrane (ultrafiltration (UF), nanofiltration, and reverse osmosis (RO)) based treatment. Recently, Lam *et al.* (2023) investigated the feasibility of CWs in the pre-treatment of landfill leachate followed by scaling up UF and RO and concluded that CWs with absorbent materials in substrate media remove significant concentrations of pollutants leading to the applicability of membranes for the complete removal of pollutants from the landfill leachate. Low-cost absorbent materials such as biochar, zeolite, and so on (Malyan *et al.*, 2021b; Parihar *et al.*, 2022) can be added to the substrate media of CWs to enhance the removal potential of pollutants (Lam *et al.*, 2023). The removal of heavy metals such as nickel, zinc, lead, copper, chromium, and so on is also reported from landfill leachate through CWs (Fu *et al.*, 2022; Teng & Chen, 2022; Wdowczyk *et al.*, 2022). Biological pathway, sorption on the substrate media, filtration through the substrate, and uptake through plant root/shoot systems are major heavy metal removal mechanisms in CWs (Bakhshoodeh *et al.*, 2017; Malyan *et al.*, 2021a; Wdowczyk *et al.*, 2022). Landfill leachate is also contaminated with emerging pollutants such as personal care products, pesticides, microplastics, and so on (Kara *et al.*, 2022; Laiju *et al.*, 2022; Shen *et al.*, 2022; Yang *et al.*, 2022), and their removal through CWs is also possible. Yang *et al.* (2022) investigated the removal of hexachlorocyclohexane (HCH) and dichlorodiphenyltrichloroethane (DDT) from landfill leachate through multi-stage vertical subsurface CWs with three plant species (*P. australis*, *Canna indica*, and *Cyperus involucratus*) and observed 38.6% removal of total HCH and 82.7% of total DDT. The removal of the pesticides through CWs is affected mainly by the octanol-water partition constant (K_{ow}). Pesticides having higher K_{ow} get adsorbed on the biofilms and substrate of the CWs and get degraded under different microbial actions (Yang *et al.*, 2022). K_{ow} of DDT (5.87–6.91) is higher than K_{ow} of HCH (3.93–4.14), and therefore the removal of DDT is higher in CWs as compared to HCH (Yang *et al.*, 2022). Apart from microbial sorption and microbial degradation, plant uptake is another pesticide removal mechanism in the CWs (Malyan *et al.*, 2021a). The removal of pesticides due to the microbial processes in the substrate of CWs is higher as compared to the uptake by the plants (Yang *et al.*, 2022). DDT is adsorbed in the sediment/substrate where the microbial process is comparatively very high resulting in higher degradation, whereas on the contrary, HCH gets accumulated in the plant through the uptake mechanism resulting in a lower removal rate (Yang *et al.*, 2022). On the basis of above discussion, it can be concluded that CW is an efficient economical technology for landfill leachate treatment and requires low energy and operation cost. Sometimes, the use of CWs for leachate treatment results in high COD and low NH_3-N removal (Klomjek & Nitisoravut, 2005).

Although CW application for landfill leachate treatment has several merits, it still has some limitations. First, CWs require a higher area, and second, the plant species of CWs hyper-accumulate certain toxic non-biodegradable pollutants in their biomass, and the management of this polluted biomass is another issue. Human population is expanding with time and land availability is shrinking day by day resulting in high land costs in metro and megacities making the use of CW prohibitive. Recently, researchers are working toward more efficient CWs which can be constructed in smaller space (Chang *et al.*, 2016; Rampuria *et al.*, 2020). Rampuria *et al.* (2020) explored the feasibility of deep CWs for the removal of pollutants from wastewater and observed 75.71, 64.37, 66.29, 50.11, and 47.94% removal of BOD_5, COD, total kjeldahl nitrogen (TKN), total nitrogen (TN), and phosphate, respectively, with the requirement of smaller land. Further research is needed in the area of reducing the footprint and management of contaminated biomass to make CW technology more acceptable.

19.5 CONCLUSION

In the last few decades, extensive research has been undertaken worldwide on the amelioration of landfill leachate by biological treatment processes and it can be concluded that none of the individual biological treatment process is universally applicable for the removal of the targeted compounds from leachate. Both aerobic and anaerobic treatment processes have their advantages and disadvantages and can be complementary to each other in the treatment of landfill leachate.

Although many techniques can be applied for the treatment of landfill leachate, the suitability of the technique depends on the properties of leachate, cost-effectiveness, and our requirements. All of the above factors need to be taken into consideration while selecting appropriate biological technology for landfill leachate treatment.

REFERENCES

Adhikari B. K., Barrington S. and Martinez J. (2006). Predicted growth of world urban food waste and methane production. *Waste Management and Research*, **24**(5), 421–433, https://doi.org/10.1177/0734242X06067767

Agamuthu P. (1999). Characteristics of municipal solid waste. *Malaysian Journal of Science*, **18**(1), 99–108.

Ağdağ O. N. and Sponza D. T. (2005). Anaerobic/aerobic treatment of municipal landfill leachate in sequential two-stage up-flow anaerobic sludge blanket reactor (UASB)/completely stirred tank reactor (CSTR) systems. *Process Biochemistry*, **40**(2), 895–902, https://doi.org/10.1016/j.procbio.2004.02.021

Aluko O. O. and Sridhar M. K. (2013). Evaluation of leachate treatment by trickling filter and sequencing batch reactor processes in Ibadan, Nigeria. *Waste Management & Research*, **31**(7), 700–705, https://doi.org/10.1177/0734242X13485867

Amokrane A., Comel C. and Veron J. (1997). Landfill leachates pretreatment by coagulation–flocculation. *Water Research*, **31**(11), 2775–2782, https://doi.org/10.1016/S0043-1354(97)00147-4

Anna Tałałaj I., Bartkowska I. and Biedka P. (2021). Treatment of young and stabilized landfill leachate by integrated sequencing batch reactor (SBR) and reverse osmosis (RO) process. *Environmental Nanotechnology, Monitoring and Management*, **16**, 100502, https://doi.org/10.1016/j.enmm.2021.100502

Bakhshoodeh R., Alavi N., Majlesi M. and Paydary P. (2017). Compost leachate treatment by a pilot-scale subsurface horizontal flow constructed wetland. *Ecological Engineering*, **105**, 7–14, https://doi.org/10.1016/j.ecoleng.2017.04.058

Baun A., Ledin A., Reitzel L. A., Bjerg P. L. and Christensen T. H. (2004). Xenobiotic organic compounds in leachates from ten Danish MSW landfills – chemical analysis and toxicity tests. *Water Research*, **38**(18), 3845–3858, https://doi.org/10.1016/j.watres.2004.07.006

Bernard C., Colin J. R. and Anne L. D. D. (1997). Estimation of the hazard of landfills through toxicity testing of leachates. Comparison of physico-chemical characteristics of landfill leachates with their toxicity determined with a battery of tests. *Chemosphere*, **35**(11), 2783–2796, https://doi.org/10.1016/S0045-6535(97)00332-9

Berrueta J., Gutiérrez A. and Fueyo G. (1996). Anaerobic treatment of leachates in a pilot-scale UASB: strategy of start-up. *Journal of Chemical Technology and Biotechnology*, **67**(3), 302–314, https://doi.org/10.1002/(SICI)1097-4660(199611

Börjesson G. and Svensson B. H. (1997). Nitrous oxide emissions from landfill cover soils in Sweden. *Tellus, Series B: Chemical and Physical Meteorology*, **49**(4), 357–363, https://doi.org/10.3402/tellusb.v49i4.15974

Bove D., Merello S., Frumento D., Arni S. A., Aliakbarian B. and Converti A. (2015). A critical review of biological processes and technologies for landfill leachate treatment. *Chemical Engineering & Technology*, **38**(12), 2115–2126, https://doi.org/10.1002/ceat.201500257

Bulc T. G. (2006). Long term performance of a constructed wetland for landfill leachate treatment. *Ecological Engineering*, **26**(4), 365–374, https://doi.org/10.1016/j.ecoleng.2006.01.003

Bull P. S., Evans J. V., Wechsler R. M. and Cleland K. J. (1983). Biological technology of the treatment of leachate from sanitary landfills. *Water Research*, **17**(11), 1473–1481, https://doi.org/10.1016/0043-1354(83)90001-5

Calli B., Mertoglu B., Roest K. and Inanc B. (2006). Comparison of long-term performances and final microbial compositions of anaerobic reactors treating landfill leachate. *Bioresource Technology*, **97**(4), 641–647, https://doi.org/10.1016/j.biortech.2005.03.021

Chang T. J., Chang Y. S., Lee W. T. and Shih S. S. (2016). Flow uniformity and hydraulic efficiency improvement of deep-water constructed wetlands. *Ecological Engineering*, **92**, 28–36, https://doi.org/10.1016/j.ecoleng.2016.03.028

Cheung K. C., Chu L. M. and Wong M. H. (1997). Ammonia stripping as a pretreatment for landfill leachate. *Water, Air, and Soil Pollution*, **94**(1–2), 209–221, https://doi.org/10.1007/bf02407103

Choi H. J. (1998). Evaluation of Fenton's Process for the Treatment of Landfill Leachate. Ph.D. Thesis in Civil and Enviromental Engineering, University of Delaware, United States (ProQuest Dissertations Publishing, 1998. UMI Number: 9906834).

de Carluccio M., Fiorentino A. and Rizzo L. (2020). Multi-barrier treatment of mature landfill leachate: effect of Fenton oxidation and air stripping on activated sludge process and cost analysis. *Journal of Environmental Chemical Engineering*, **8**(5), 104444, https://doi.org/10.1016/j.jece.2020.104444

Devare M. and Bahadir M. (1994). Biological monitoring of landfill leachate using plants and luminescent bacteria. *Chemosphere*, **28**(2), 261–271, https://doi.org/10.1016/0045-6535(94)90123-6

Diamadopoulos E., Samaras P., Dabou X. and Sakellaropoulos G. P. (1997). Combined treatment of landfill leachate and domestic sewage in a sequencing batch reactor. *Water Science and Technology*, **36**(2–3), 61–68, https://doi.org/10.1016/S0273-1223(97)00370-3

Dollerer J. and Wilderer P. A. (1996). Biological treatment of leachates from hazardous waste landfills using SBBR technology. *Water Science and Technology*, **34**(7–8 pt 4), 437–444, https://doi.org/10.1016/S0273-1223(96)00776-7

Doyle J., Watts S., Solley D. and Keller J. (2001). Exceptionally high-rate nitrification in sequencing batch reactors treating high ammonia landfill leachate. *Water Science and Technology*, **43**(3), 315–322, https://doi.org/10.2166/wst.2001.0152

Ehrig H. J. (1988). Treatment of sanitary landfill leachate. *Wasser und Boden*, **40**(10), 570–574.

Elmaadawy K., Liu B., Hassan G. K., Wang X., Wang Q., Hu J., Hou H., Yang J. and Wu X. (2022). Microalgae-assisted fixed-film activated sludge MFC for landfill leachate treatment and energy recovery. *Process Safety and Environmental Protection*, **160**, 221–231, https://doi.org/10.1016/j.psep.2022.02.021

Fang H. H. P. and Chui H. K. (1993). Maximum COD loading capacity in UASB reactors at 37°C. *Journal of Environmental Engineering*, **119**(1), 103–119, https://doi.org/10.1061/(asce)0733-9372(1993)119:1(103)

Fang H. H. P., Lau I. W. C. and Wang P. (2005). Anaerobic treatment of Hong Kong leachate followed by chemical oxidation. *Water Science and Technology*, **52**(10–11), 41–49, https://doi.org/10.2166/wst.2005.0677

Frascari D., Bronzini F., Giordano G., Tedioli G. and Nocentini M. (2004). Long-term characterization, lagoon treatment and migration potential of landfill leachate: a case study in an active Italian landfill. *Chemosphere*, **54**(3), 335–343, https://doi.org/10.1016/j.chemosphere.2003.08.013

Fu G., Zhou S., Zhao Y., Li Z., Xu Y. and Guo Z. (2022). Removal efficiency of heavy metals such as lead and cadmium by different substrates in constructed wetlands. *Processes*, **10**(12), 2502, https://doi.org/10.3390/pr10122502

García H., Rico J. L. and García P. A. (1996). Comparison of anaerobic treatment of leachates from an urban-solid-waste landfill at ambient temperature and at 35°C. *Bioresource Technology*, **58**(3), 273–277, https://doi.org/10.1016/S0960-8524(96)00114-9

Griffin P., Jennings P. and Bowman E. (1999). Advanced nitrogen removal by rotating biological contactors, recycle and constructed wetlands. *Water Science and Technology*, **40**(4–5), 383–390, https://doi.org/10.1016/S0273-1223(99)00521-1

Gülşen H. and Turan M. (2004). Anaerobic treatability of sanitary landfill leachate in a fluidized bed reactor. *Turkish Journal of Engineering and Environmental Sciences*, **28**(5), 297–305, https://dergipark.org.tr/en/pub/tbtkengineering/article/145161

Halim A. A., Aziz H. A., Johari M. A. M., Ariffin K. S. and Adlan M. N. (2010). Ammoniacal nitrogen and COD removal from semi-aerobic landfill leachate using a composite adsorbent: fixed bed column adsorption performance. *Journal of Hazardous Materials*, **175**(1–3), 960–964, https://doi.org/10.1016/j.jhazmat.2009.10.103

Hoilijoki T. H., Kettunen R. H. and Rintala J. A. (2000). Nitrification of anaerobically pretreated municipal landfill leachate at low temperature. *Water Research*, **34**(5), 1435–1446, https://doi.org/10.1016/S0043-1354(99)00278-X

Hoornweg D. (2000). 'What a waste': solid waste management in Asia. *Industry and Environment*, **23**(1–2), 65–70. https://agris.fao.org/agris-search/search.do?recordID=US2012400855

Horner P. C. (1994). Landfilling of waste: leachate. *Geotextiles and Geomembranes*, **13**(10), 704, https://doi.org/10.1016/0266-1144(94)90070-1

Hosomi M., Matsusige K., Inamori Y., Sudo R., Yamada K. and Yoshino Z. (1989). Sequencing batch reactor activated sludge processes for the treatment of municipal landfill leachate: removal of nitrogen and refractory organic compounds. *Water Science and Technology*, **21**(12 pt 5), 1651–1654, https://doi.org/10.2166/wst.1989.0135

Huang X., Sillanpää M., Duo B. and Gjessing E. T. (2008). Water quality in the Tibetan plateau: metal contents of four selected rivers. *Environmental Pollution*, **156**(2), 270–277, https://doi.org/10.1016/j.envpol.2008.02.014

Inanc B., Calli B. and Saatci A. (2000). Characterization and anaerobic treatment of the sanitary landfill leachate in Istanbul. *Water Science and Technology*, **41**(3), 223–230, https://doi.org/10.2166/wst.2000.0075

Islam M., Xu Q. and Yuan Q. (2020). Advanced biological sequential treatment of mature landfill leachate using aerobic activated sludge SBR and fungal bioreactor. *Journal of Environmental Health Science and Engineering*, **18**(1), 285–295, https://doi.org/10.1007/s40201-020-00466-z

Kabdasli I., Tunay O., Öztürk I., Yılmaz S. and Arikan O. (2000). Ammonia removal from young landfill leachate by magnesium ammonium phosphate precipitation and air stripping. *Water Science and Technology*, **41**(1), 237–240, https://doi.org/10.2166/wst.2000.0034

Kannah R. Y., Rohini K. B., Gunasekaran M., Gokulakrishnan K., Kumar G. and Banu J. R. (2022). Prediction of effective substrate concentration and its impact on biogas production using artificial neural networks in hybrid upflow anaerobic sludge blanket reactor for treating landfill leachate. *Fuel*, **313**, 122697, https://doi.org/10.1016/j.fuel.2021.122697

Kara N., Sari Erkan H. and Onkal Engin G. (2022). Characterization and removal of microplastics in landfill leachate treatment plants in Istanbul, Turkey. Analytical Letters, 1–14.

Kargi F. and Pamukoglu M. Y. (2004). Adsorbent supplemented biological treatment of pre-treated landfill leachate by fed-batch operation. *Bioresource Technology*, **94**(3), 285–291, https://doi.org/10.1016/j.biortech.2004.01.003

Kennedy K. J. and Lentz E. M. (2000). Treatment of landfill leachate using sequencing batch and continuous flow upflow anaerobic sludge blanket (UASB) reactors. *Water Research*, **34**(14), 3640–3656, https://doi.org/10.1016/S0043-1354(00)00114-7

Kettunen R. H. and Rintala J. A. (1998). Performance of an on-site UASB reactor treating leachate at low temperature. *Water Research*, **32**(3), 537–546, https://doi.org/10.1016/S0043-1354(97)00319-9

Kettunen R. H., Hoilijoki T. H. and Rintala J. A. (1996a). Anaerobic and sequential anaerobic–aerobic treatments of municipal landfill leachate at low temperatures. *Bioresource Technology*, **58**(1), 31–40, https://doi.org/10.1016/S0960-8524(96)00102-2

Kettunen R. H., Pulkkinen E. M. and Rintala J. A. (1996b). Biological treatment at low temperatures of sulfur-rich phenols containing oil shale ash leachate. *Water Research*, **30**(6), 1395–1402, https://doi.org/10.1016/0043-1354(96)00016-4

Kjeldsen P., Barlaz M. A., Rooker A. P., Baun A., Ledin A. and Christensen T. H. (2002). Present and long-term composition of MSW landfill leachate: a review. *Critical Reviews in Environmental Science and Technology*, **32**(4), 297–336, https://doi.org/10.1080/10643380290813462

Klimiuk E. and Kulikowska D. (2005). The influence of operational conditions in sequencing batch reactors on removal of nitrogen and organics from municipal landfill leachate. *Waste Management and Research*, **23**(5), 429–438, https://doi.org/10.1177/0734242X05058579

Klimiuk E. and Kulikowska D. (2006). Organics removal from landfill leachate and activated sludge production in SBR reactors. *Waste Management*, **26**(10), 1140–1147, https://doi.org/10.1016/j.wasman.2005.09.011

Klomjek P. and Nitisoravut S. (2005). Constructed treatment wetland: a study of eight plant species under saline conditions. *Chemosphere*, **58**(5), 585–593, https://doi.org/10.1016/j.chemosphere.2004.08.073

Kurniawan T. A., Lo W. h. and Chan G. Y. S. (2006). Radicals-catalyzed oxidation reactions for degradation of recalcitrant compounds from landfill leachate. *Chemical Engineering Journal*, **125**(1), 35–57, https://doi.org/10.1016/j.cej.2006.07.006

Kurniawan T. A., Lo W., Chan G. and Sillanpää M. E. T. (2010). Biological processes for treatment of landfill leachate. *Journal of Environmental Monitoring*, **12**(11), 2032–2047, https://doi.org/10.1039/c0em00076k

Laiju A. R., Gandhimathi R. and Nidheesh P. V. (2022). Removal of pharmaceutical and personal care products in landfill leachate treatment process. *Current Opinion in Environmental Science & Health*, **31**, 100434.

Lam T., Yang X., Ergas S. J. and Arias M. E. (2023). Feasibility of landfill leachate reuse through adsorbent-enhanced constructed wetlands and ultrafiltration–reverse osmosis. *Desalination*, **545**, 116163, https://doi.org/10.1016/j.desal.2022.116163

Lee C. M., Lin X. R., Lan C. Y., Lo S. C. L. and Chan G. Y. S. (2002). Evaluation of leachate recirculation on nitrous oxide production in the Likang landfill, China. *Journal of Environmental Quality*, **31**(5), 1502–1508, https://doi.org/10.2134/jeq2002.1502

Lema J. M., Mendez R. and Blazquez R. (1988). Characteristics of landfill leachates and alternatives for their treatment: a review. *Water, Air, and Soil Pollution*, **40**(3–4), 223–250, https://doi.org/10.1007/BF00163730

Li X. Z. and Zhao Q. L. (1999). Inhibition of microbial activity of activated sludge by ammonia in leachate. *Environment International*, **25**(8), 961–968, https://doi.org/10.1016/S0160-4120(99)00068-9

Li X. Z. and Zhao Q. L. (2001). Efficiency of biological treatment affected by high strength of ammonium-nitrogen in leachate and chemical precipitation of ammonium-nitrogen as pretreatment. *Chemosphere*, **44**(1), 37–43, https://doi.org/10.1016/S0045-6535(00)00382-9

Lin C. Y., Chang F. Y. and Chang C. H. (2000). Co-digestion of leachate with septage using a UASB reactor. *Bioresource Technology*, **73**(2), 175–178, https://doi.org/10.1016/S0960-8524(99)00166-2

Liu Y. (2003). Chemically reduced excess sludge production in the activated sludge process. *Chemosphere*, **50**(1), 1–7, https://doi.org/10.1016/S0045-6535(02)00551-9

Lo I. M. C. (1996). Characteristics and treatment of leachates from domestic landfills. *Environment International*, **22**(4), 433–442, https://doi.org/10.1016/0160-4120(96)00031-1

Loukidou M. X. and Zouboulis A. I. (2001). Comparison of two biological treatment processes using attached-growth biomass for sanitary landfill leachate treatment. *Environmental Pollution*, **111**(2), 273–281, https://doi.org/10.1016/S0269-7491(00)00069-5

Mæhlum T. (1995). Treatment of landfill leachate in on-site lagoons and constructed wetlands. *Water Science and Technology*, **32**(3), 129–135, https://doi.org/10.1016/0273-1223(95)00613-3

Malyan S. K., Yadav S., Sonkar V., Goyal V. C., Singh O. and Singh R. (2021a). Mechanistic understanding of the pollutant removal and transformation processes in the constructed wetland system. *Water Environment Research*, **93**(10), 1882–1909, https://doi.org/10.1002/wer.1599

Malyan S. K., Kumar S. S., Fagodiya R. K., Ghosh P., Kumar A., Singh R. and Singh L. (2021b). Biochar for environmental sustainability in the energy–water–agroecosystem nexus. *Renewable and Sustainable Energy Reviews*, **149**, 111379, https://doi.org/10.1016/j.rser.2021.111379

Marañón E., Castrillón L., Fernández Y. and Fernández E. (2006). Anaerobic treatment of sludge from a nitrification–denitrification landfill leachate plant. *Waste Management*, **26**(8), 869–874, https://doi.org/10.1016/j.wasman.2005.08.002

Mian M. M., Zeng X., Nasry A. a. N. B. and Al-Hamadani S. M. Z. F. (2017). Municipal solid waste management in China: a comparative analysis. *Journal of Material Cycles and Waste Management*, **19**(3), 1127–1135, https://doi.org/10.1007/s10163-016-0509-9

Mojiri A., Zhou J. L., Ratnaweera H., Ohashi A., Ozaki N., Kindaichi T. and Asakura H. (2021). Treatment of landfill leachate with different techniques: an overview. *Water Reuse*, **11**(1), 66–96, https://doi.org/10.2166/wrd.2020.079

Muthukumaran M. (2022). Advances in bioremediation of nonaqueous phase liquid pollution in soil and water. In: Biological Approaches to Controlling Pollutants, S. Kumar and M. Z. Hashmi, Woodhead Publishing (an imprint of Elsevier), Sawston, United Kingdom, pp. 191–231, https://doi.org/10.1016/B978-0-12-824316-9.00006-9

Nath A. and Debnath A. (2022). A short review on landfill leachate treatment technologies. *Materials Today: Proceedings*, **67**, 1290–1297, https://doi.org/10.1016/j.matpr.2022.09.109

Neczaj E., Okoniewska E. and Kacprzak M. (2005). Treatment of landfill leachate by sequencing batch reactor. *Desalination*, **185**(1–3), 357–362, https://doi.org/10.1016/j.desal.2005.04.044

Pan S. C. and Tseng D. H. (2001). Sewage sludge ash characteristics and its potential applications. *Water Science and Technology*, **44**(10), 261–267, https://doi.org/10.2166/wst.2001.0636

Parawira W., Murto M., Zvauya R. and Mattiasson B. (2006). Comparative performance of a UASB reactor and an anaerobic packed-bed reactor when treating potato waste leachate. *Renewable Energy*, **31**(6), 893–903, https://doi.org/10.1016/j.renene.2005.05.013

Parihar P., Chand N. and Suthar S. (2022). Treatment of high nutrient-loaded wastewater in a constructed floating wetland with different configurations: role of lantana biochar addition. *Sustainability*, **14**(23), 16049, https://doi.org/10.3390/su142316049

Price G. A., Barlaz M. A. and Hater G. R. (2003). Nitrogen management in bioreactor landfills. *Waste Management*, **23**(7), 675–688, https://doi.org/10.1016/S0956-053X(03)00104-1

Qasim S. R. and Chiang W. (2017). Sanitary landfill leachate: generation, control and treatment. CRC press, Florida, United States, 1–7, https://doi.org/10.1201/9780203740217

Qu B., Zhang Y., Kang S. and Sillanpää M. (2019). Water quality in the Tibetan plateau: major ions and trace elements in rivers of the 'water tower of Asia'. *Science of the Total Environment*, **649**, 571–581, https://doi.org/10.1016/j.scitotenv.2018.08.316

Rampuria A., Gupta A. B. and Brighu U. (2020). Nitrogen transformation processes and mass balance in deep constructed wetlands treating sewage, exploring the anammox contribution. *Bioresource Technology*, **314**, 123737, https://doi.org/10.1016/j.biortech.2020.123737

Renou S., Givaudan J. G., Poulain S., Dirassouyan F. and Moulin P. (2008). Landfill leachate treatment: review and opportunity. *Journal of Hazardous Materials*, **150**(3), 468–493, https://doi.org/10.1016/j.jhazmat.2007.09.077

Robinson H. D. and Maris P. J. (1985). The treatment of leachates from domestic waste in landfill sites. *Journal of the Water Pollution Control Federation*, **57**(1), 30–38, https://doi.org/10.1016/0043-1354(83)90010-6

Sachdeva V., Tyagi R. D. and Valéro J. R. (2000). Production of biopesticides as a novel method of wastewater sludge utilization/disposal. *Water Science and Technology*, **42**(9), 211–216, https://doi.org/10.2166/wst.2000.0209

Setiadi T. and Fairus S. (2003). Hazardous waste landfill leachate treatment using an activated sludge–membrane system. *Water Science and Technology*, **48**(8), 111–117, https://doi.org/10.2166/wst.2003.0459

Shen M., Xiong W., Song B., Zhou C., Almatrafi E., Zeng G. and Zhang Y. (2022). Microplastics in landfill and leachate: occurrence, environmental behavior and removal strategies. *Chemosphere*, **305**, 135325, https://doi.org/10.1016/j.chemosphere.2022.135325

Shin H. S., Han S. K., Song Y. C. and Lee C. Y. (2001). Performance of UASB reactor treating leachate from acidogenic fermenter in the two-phase anaerobic digestion of food waste. *Water Research*, **35**(14), 3441–3447, https://doi.org/10.1016/S0043-1354(01)00041-0

Siegrist H., Reithaar S. and Lais P. (1998). Nitrogen loss in a nitrifying rotating contactor treating ammonium rich leachate without organic carbon. *Water Science and Technology*, **37**(4–5), 589–591, https://doi.org/10.1016/S0273-1223(98)00164-4

Sillanpää M. and Oikari A. (1996). Assessing the impact of complexation by EDTA and DTPA on heavy metal toxicity using microtox bioassay. *Chemosphere*, **32**(8), 1485–1497, https://doi.org/10.1016/0045-6535(96)00057-4

Silva A. C., Dezotti M. and Sant'Anna G. L. (2004). Treatment and detoxification of a sanitary landfill leachate. *Chemosphere*, **55**(2), 207–214, https://doi.org/10.1016/j.chemosphere.2003.10.013

Singh S., Maithani C., Malyan S. K., Soti A., Kulshreshtha N. M., Singh R., Brighu U., Gupta A. B., Kumar J., Yadav S., Singh O. and Goyal V. C. (2023). Comparative performance and 16S amplicon sequencing analysis of deep and shallow cells of a full scale HFCW having sequentially decreasing depths reveals vast enhancement potential. *Bioresource Technology Reports*, **22**, 101404, https://doi.org/10.1016/j.biteb.2023.101404

Song J., Zhang W., Gao J., Hu X., Zhang C., He Q., Yang F., Wang H., Wang X. and Zhan X. (2020). A pilot-scale study on the treatment of landfill leachate by a composite biological system under low dissolved oxygen conditions: performance and microbial community. *Bioresource Technology*, **296**, 122344, https://doi.org/10.1016/J.BIORTECH.2019.122344

Sorvari J. and Sillanpää M. (1996). Influence of metal complex formation on heavy metal and free EDTA and DTPA acute toxicity determined by daphnia magna. *Chemosphere*, **33**(6), 1119–1127, https://doi.org/10.1016/0045-6535(96)00251-2

Stolarek P. and Ledakowicz S. (2001). Thermal processing of sewage sludge by drying, pyrolysis, gasification and combustion. *Water Science and Technology*, **44**(10), 333–339, https://doi.org/10.2166/wst.2001.0655

Sung M. S., Chang D. and Lee H. Y. (1997). Performance improvement of an unstable anaerobic leachate treatment system in an industrial waste landfill. *Water Science and Technology*, **36**(12), 333–340, https://doi.org/10.1016/S0273-1223(97)00743-9

Teng C. and Chen W. (2022). Technologies for the treatment of emerging contaminants in landfill leachate. *Current Opinion in Environmental Science & Health*, **31**, 100409.

Timur H. and Özturk I. (1999). Anaerobic sequencing batch reactor treatment of landfill leachate. *Water Research*, **33**(15), 3225–3230, https://doi.org/10.1016/S0043-1354(99)00048-2

Timur H., Ozturk I., Altinbas M., Arikan O. and Tuyluoglu B. S. (2000). Anaerobic treatability of leachate: a comparative evaluation for three different reactor systems. *Water Science and Technology*, **42**(1–2), 287–292, https://doi.org/10.2166/wst.2000.0327

Tränkler J., Visvanathan C., Kuruparan P. and Tubtimthai O. (2005). Influence of tropical seasonal variations on landfill leachate characteristics – results from lysimeter studies. *Waste Management*, **25**(10), 1013–1020, https://doi.org/10.1016/j.wasman.2005.05.004

Tyagi V. K., Liu J., Poh L. S. and Ng W. J. (2019). Anaerobic–aerobic system for beverage effluent treatment: performance evaluation and microbial community dynamics. *Bioresource Technology Reports*, **7**, 100309, https://doi.org/10.1016/j.biteb.2019.100309

Urbanc-Berčič O. (1996). Constructed wetlands for the treatment of landfill leachates: the Slovenian experience. *Wetlands Ecology and Management*, **4**(3), 189–197, https://doi.org/10.1007/bf01879237

Uygur A. and Kargi F. (2004). Biological nutrient removal from pre-treated landfill leachate in a sequencing batch reactor. *Journal of Environmental Management*, **71**(1), 9–14, https://doi.org/10.1016/j.jenvman.2004.01.002

Virkutyte J. and Sillanpää M. (2006). Chemical evaluation of potable water in Eastern Qinghai Province, China: human health aspects. *Environment International*, **32**(1), 80–86, https://doi.org/10.1016/j.envint.2005.05.022

Wdowczyk A., Szymańska-Pulikowska A. and Gałka B. (2022). Removal of selected pollutants from landfill leachate in constructed wetlands with different filling. *Bioresource Technology*, **353**, 127136, https://doi.org/10.1016/j.biortech.2022.127136

Wiszniowski J., Robert D., Surmacz-Gorska J., Miksch K. and Weber J. V. (2006). Landfill leachate treatment methods: a review. *Environmental Chemistry Letters*, **4**, 51–61, https://doi.org/10.1007/s10311-005-0016-z

Yalmaz G. and Öztürk I. (2001). Biological ammonia removal from anaerobically pre-treated landfill leachate in sequencing batch reactors (SBR). *Water Science and Technology*, **43**(3), 307–314, https://doi.org/10.2166/wst.2001.0151

Yang C., Wang B., Wang H., He Z., Pi Y., Zhou J., Liang T., Chen M., He T. and Fu T. (2022). Removal of organochlorine pesticides and metagenomic analysis by multi-stage constructed wetland treating landfill leachate. *Chemosphere*, **301**, 134761, https://doi.org/10.1016/j.chemosphere.2022.134761

Zaloum R. and Abbott M. (1997). Anaerobic pretreatment improves single sequencing batch reactor treatment of landfill leachates. *Water Science and Technology*, **35**(1), 207–214, https://doi.org/10.1016/S0273-1223(96)00898-0

Zhang T., Ding L. and Ren H. (2009). Pretreatment of ammonium removal from landfill leachate by chemical precipitation. *Journal of Hazardous Materials*, **166**(2–3), 911–915, https://doi.org/10.1016/j.jhazmat.2008.11.101

Zhang F., Peng Y., Wang Z. and Jiang H. (2019). High-efficient nitrogen removal from mature landfill leachate and waste activated sludge (WAS) reduction via partial nitrification and integrated fermentation–denitritation process (PNIFD). *Water Research*, **160**, 394–404, https://doi.org/10.1016/j.watres.2019.05.032

Zouboulis A. I., Loukidou M. X. and Christodoulou K. (2001). Enzymatic treatment of sanitary landfill leachate. *Chemosphere*, **44**(5), 1103–1108, https://doi.org/10.1016/S0045-6535(00)00343-X

doi: 10.2166/9781789063318_0429

Chapter 20

Landfill leachate management

Mandeep Singh[1], Muntjeer Ali[1], Nehaun Zargar[2], Vinay Kumar Tyagi[3], Absar Ahmad Kazmi[1] and C. S. P. Ojha[1]*

[1]*Department of Civil Engineering, Indian Institute of Technology, Roorkee 247667, Uttarakhand, India*
[2]*Thapar Institute of Engineering and Technology, Patiala 147004, India*
[3]*Environmental Hydrology Division, National Institute of Hydrology, Roorkee 247667, Uttrakhand, India*
**Corresponding author: vinayiitrp@gmail.com; vtyagi.nihr@gov.in*

ABSTRACT

Landfill leachate management is a critical aspect of waste management, as it involves the collection and treatment of liquids that form when rainwater and other liquids seep through solid waste in landfills which can pose a threat to the environment and public health if not properly managed. Landfill leachate management involves the use of various technologies to treat the liquid that seeps through waste in landfills. Some common technologies include physical treatment such as screening, sedimentation, and filtration; chemical treatment which involves adding chemicals to stabilize and neutralize the leachate; biological treatment which uses microorganisms to break down organic matter; evaporation/incineration which reduces the volume of leachate; and membrane processes such as reverse osmosis and ultrafiltration which remove impurities. The most effective technology or combination of technologies depend on the characteristics of the leachate and local regulations. In some cases, a combination of technologies may be required to adequately treat the leachate and achieve the desired level of treatment. This chapter provides an overview of the sources and composition of landfill leachate, its potential impacts, and the current state of treatment methods. The chapter highlights the importance of proper leachate management and the need for continued research and development of innovative treatment technologies to improve the environmental performance of landfills.

Keywords: Landfill leachate, treatment, waste management, membrane.

20.1 INTRODUCTION

Landfill leachate management refers to techniques and practices used to control and treat the liquid that is generated when solid waste decomposes in a landfill.

Leachate is a mixture of water, organic and inorganic compounds, heavy metals, pathogens, and other pollutants that can pose a serious threat to the environment and human health if not properly managed. Effective management of landfill leachate involves minimizing its generation and potential for migration into the environment and properly treating and disposing of it to prevent soil and groundwater pollution. This requires a comprehensive understanding of the composition and characteristics of leachate, as well as appropriate technologies and treatment methods to remove the pollutants. Landfill leachate management is an important aspect of solid waste management and is critical for ensuring the protection of the environment and public health.

Landfills for urban solid waste are frequently utilized for non-hazardous sludge as well as domestic, commercial, and industrial solid waste. Despite its potential for environmental harm, sanitary landfilling is nevertheless used in waste management programs (Mojiri *et al.*, 2017). Sanitary landfilling often has cheaper operating expenses compared to alternative processes, such as incineration (Gotvajn & Pavko, 2015). After being dumped, waste may undergo several biological and physicochemical changes, generating leachate, a highly contaminated effluent. Such waste can contaminate neighboring groundwater, surface water, and soil (Zamri *et al.*, 2017).

According to Chávez *et al.* (2019), landfill leachate frequently contains large amounts of organic pollutants, heavy metals, hazardous materials, ammonia, and inorganic materials as well as refractory compounds such as humic substances. It also frequently has a high chemical (COD) and biological oxygen demand (BOD) (Eggen *et al.*, 2010). Depending on how quickly they deteriorate, the temperature, the hydrology, and how old they are, leachate from landfills can have a variety of characteristics. Inadequate landfill leachate management is frequently linked to ecological damage and health problems. In open dumping and sanitary landfills, reducing dangers to the environment and public health is a critical problem (Xaypanya *et al.*, 2018). The most effective primary strategies for treating landfill leachate include biological treatments as well as chemical and physical methods. A detailed analysis of landfill leachate's features, effects, and remediation options is required. This chapter fulfills the purpose of offering such a critical evaluation.

20.2 LANDFILL LEACHATE CHARACTERIZATION

The liquid that drains from solid waste in a landfill is called landfill leachate, and it includes a variety of contaminants, including pathogens, heavy metals, and organic and inorganic chemicals. To establish the composition of landfill leachate and its possible effects on the environment and public health, its physical, chemical, and biological characteristics must be analyzed. Effective treatment and management solutions for the leachate are designed using this information. The presence of heavy metals and other harmful contaminants, as well as pH, conductivity, total organic carbon, nitrogen, and phosphorous levels, may all be assessed as important factors during characterization. For landfill leachate to be effectively managed and treated to minimize soil and groundwater pollution, it must first be accurately characterized.

When water permeates waste in a landfill, it generates leachate, which transmits some types of contaminants (Mojiri *et al.*, 2017). Four major categories of contaminants, including organic contaminants and substrates, inorganic compounds, heavy metals, total dissolved solids (TDS), and color, may be used to categorize the pollutants found in municipal landfill leachate. Landfill leachate may be categorized into three major classes (Table 20.1) based on its age: young, intermediate, and old. Leachate from 'young' landfills (i.e., the acid phase) is characterized by low pH values, large concentrations of volatile acids, and simply decomposed organic waste. Leachate methane production and pH are high in mature landfills (i.e., the methanogenic phase), and the predominant organic components are humic and fulvic fractions. The pH of young landfill leachate is usually <6.5, whereas the pH of aged landfill leachate is found to be >7.5. Additionally, it was noted that leachates with high volatile fatty acid (VFA) concentrations have low pH. The pH of stabilized leachate is in the range of 7.5–9. The pH of the leachate increases to an alkaline condition as a result of the methane-producing bacteria consuming more partially ionized free VFAs during the anaerobic breakdown of the wastes (Gautam & Kumar, 2021; Hussein *et al.*, 2019). However, owing to the characteristics of waste are dependent on nations, there is a little variance in some other research. The properties of landfill leachate throughout the world are provided in Table 20.2. According to Table 20.2, the cities with the highest concentrations of landfill leachate were Riyadh, Saudi Arabia, with Fe 167.6 mg/L, and China, with COD 28,000 mg/L.

20.2.1 Color and TDS

Color in leachate can be caused by various factors such as organic matter, heavy metals, and dyes present in the waste. The color can indicate the presence of specific pollutants and help identify the potential health and environmental hazards associated with the leachate. However, it is important to note that the

Table 20.1 Leachate characteristics and treatability based on the landfill age.

Sl. No.	Parameters	Unit	Age (Years)		
			Young (0–5)	Intermediate (5–10)	Old (>10)
1	pH	–	<6.5	6.5–7.5	>7.5
2	COD	mg/L	>10,000	5,000–10,000	<5,000
3	BOD_5/COD		0.5–1.0	0.1–0.5	>0.1
4	NH_4-N	mg/L	<400	–	>400
5	HM	Low to high	Medium to low	Low	Low
6	VFA/HFA	%	VFA: 80%	VFA: 5–30%, +HFA	HFA: 80%
7	Biodegradability	Low to high	High	Low	Medium

HM, heavy metals; VFA, volatile fatty acids; HFA, humic and fulvic acids.

Table 20.2 Leachate characteristics across the globe.

Sl. No.	Leachate Sources	COD (mg/L)	BOD (mg/L)	BOD/ COD	Ammoniacal Nitrogen (mg/L)	Heavy Metals (mg/L)					Site Location	Reference
						Fe	Mn	Zn	Cd	Ni		
1	Concentrated leachate	28,000 3.50	950	0.04	3.50	30.00	4.03	17.80	NR	3.70	MSW incineration plants, China	Ren *et al.* (2018)
2	Semi-aerobic	935	83	0.09	483	7.9	NR	0.6	NR	NR	Pulau Burung, Malaysia	Kamaruddin *et al.* (2015)
3	–	6,140	558	0.09	1,856	NR	NR	NR	0.01	NR	Heimifeng, Changsha, China	Hu *et al.* (2016)
4	Covered landfill	24,040	15,021	0.59	2,281	10.37	NR	0.96	NR	0.95	Istanbul Kömürcüoda landfill, Turkey	Akgul *et al.* (2013)
5	–	2,350	NR	NR	310	NR	NR	0.05	0.02	0.54	Sivas, Turkey	Atmaca (2009)
6	Sanitation landfill	2,305	105	0.04	1,240	NR	NR	NR	NR	NR	Beijing, China	
7	Semi-aerobic	1,343	96	0.07	NR	3.41	0.17	2.3	NR	0.17	Matuail landfill, Bangladesh	Jahan *et al.* (2016)
8	–	10,400	1,500	0.14	NR	11.16	NR	3.00	0.03	1.33	Mavallipura landfill, India	
9	–	17,003	NR	NR	NR	167.61	10.83	0.18	NR	0.50	Riyadh city, Saudi Arabia	Al-Wabel *et al.* (2011)

10	Semi-sanitary	3,380	760	0.22	1,150	NR	NR	1.35–1.60	0.13–0.3	NR	Nonthaburi landfill, Thailand	Xaypanya *et al.* (2018)
11	Concentrated landfill leachate	1,281 7,700	NR 1,300	– 0.16	14.2 1,780	NR 10.03	0.692 NR	– 1.06	– NR	0.233 NR	Jiangsu province, China Xiangtan, China	Cui *et al.* (2018) Hu *et al.* (2011)
12	–	3,308–3,540	823–1,274	0.24–0.35	1,036–1,197	NR	NR	NR	NR	NR	Nam Binh Duong, Vietnam	Le Luu (2020)
13	–	781	1,16	0.14	212	21	NR	NR	NR	NR	Jones county municipal landfill, Iowa, USA	Nivala *et al.* (2007)
14	Sanitation landfill	4,737	NR	NR	1,897	NR	NR	NR	NR	NR	Virginia, USA	Iskander *et al.* (2017)
15	NR	765	70	0.09	342	2.6	NR	0.07	NR	NR	Saint-Rosaire's city, Québec, Canada	Oumar *et al.* (2016)
16	Old and active landfills	1,380	NR	NR	665.2	NR	NR	NR	0.004	NR	Akuševec landfill, Zagreb, Croatia	Dolar *et al.* (2016)

(*Continued*)

Table 20.2 Leachate characteristics across the globe. (*Continued*)

Sl. No.	Leachate Sources	COD (mg/L)	BOD (mg/L)	BOD/COD	Ammoniacal Nitrogen (mg/L)	Heavy Metals (mg/L)					Site Location	Reference
						Fe	Mn	Zn	Cd	Ni		
17	Operated for 2 years (very young) non-hazardous waste, non-fermentable waste	260	47	0.18	187	NR	NR	NR	NR	NR	France	Ricordel and Djelal (2014)
18	–	3,847	388	0.11	3,158.98	21.50	NR	NR	1.7	NR	Ouled Fayet landfill site, Algeria	Boumechhour *et al.* (2013)
19	Sanitation landfill	4,425–4,860	433–588	0.09–0.12	NR	NR	NR	NR	NR	NR	Sao Carlos, Brasil	Ferraz *et al.* (2014)
20	–	1,013	NR	NR	398.02	6.84	0.42	NR	6.26	NR	Guaratinguetá, Brasil	Peixoto *et al.* (2018)

color alone may not be a reliable indicator of leachate quality and further analysis such as chemical testing is needed to determine the exact composition and potential hazards. Water may change color from yellow to dark brown due to the breakdown of certain organic components, such as humic acid. The production of color and turbidity by chemicals and particles is emphasized by Gotvajn and Pavko (2015).

TDS show how various cations and anions, including calcium, chloride, magnesium, sodium, potassium, and bicarbonates, have an integrated impact on water and waste. Additionally, TDS may prevent or reduce the biological decomposition of dissolved organic carbon and can be generated from modest quantities of dissolved organic matter. High electrical conductivity and TDS may be used to identify dissolved organic and inorganic compounds in samples, according to Hussein *et al.* (2019). Inorganic salts and dissolved organics constitute the majority of TDS. The level of mineralization is determined by the quantity of TDS, and a higher TDS concentration may alter the physical and chemical properties of the receiving water. Because it alters the ionic makeup of water, an increase in salinity brought on by an increase in TDS concentration also increases toxicity. Similar to ammonia, TDS is spread across a wide range in tropical monsoon climate, tropical rainforest climate, and tropical savanna climate (Lindamulla *et al.*, 2022).

20.2.2 Organic and inorganic pollutants, and heavy metals

Organic and inorganic pollutants, as well as heavy metals, are commonly found in landfill leachate and can pose a significant threat to human health and the environment. Organic pollutants include a wide range of chemicals such as pharmaceuticals, personal care products, pesticides, and polychlorinated biphenyls. Inorganic pollutants include heavy metals such as lead, cadmium, and arsenic, which are toxic and can persist in the environment for many years. Heavy metals can also leach into groundwater and contaminate drinking water sources. Landfill leachate can also contain high levels of nitrogen and phosphorous, leading to eutrophication and the growth of harmful algae in aquatic ecosystems. It is essential to properly manage and treat landfill leachate to minimize the risk of these pollutants entering the environment. Leachate's organic composition changes depending on the types of waste it contains, how old the landfill is, and the weather (Mojiri *et al.*, 2016a). Many organic compounds can be found in landfill leachate and urban solid waste (Scandelai *et al.*, 2019). Dissolved organic matter, which accounts for 80% of all organic components in landfill leachate, is often made up of VFAs and refractory humic chemicals (Jiang *et al.*, 2019). It is possible that traditional biological processes cannot effectively breakdown such refractory organic materials. BOD_5 and COD are two indicators of dissolved organics (Samadder *et al.*, 2017). In each zone, that is, tropical monsoon climate, tropical rainforest climate and tropical savanna climate zones, the young leachates had greater BOD/COD ratios compared to the older leachates. The BOD/COD ratio is a measure of the amounts of biodegradable organics in leachate. Biodegradable waste decomposes quickly, hence BOD concentrations decrease over time at a faster rate compared to COD. The BOD/COD ratio is therefore used to estimate the age of landfills. Leachate from modern waste disposal facilities has a BOD/COD ratio of 0.5–1.0, whereas leachate from older facilities has a BOD/COD ratio of <0.1 (Tałałaj *et al.*, 2019).

Additionally, landfill leachate may include persistent organic contaminants. According to Scandelai *et al.* (2019), leachate frequently contains a variety of medium and low polarity organic compounds, including amines, alcohols, carboxylic acids, aldehydes, benzothiazolone, ketones, phenols, chlorinated benzenes, phosphates, nitrogen compounds, pesticides, and aromatic and polyaromatic hydrocarbons. Pharmaceuticals, personal care items, surfactants, plasticizers, fire retardants, insecticides, and nanomaterials are among the contaminants of rising concern that may be discovered in many municipal landfills and need to be managed (Ramakrishnan *et al.*, 2015). Anions and cations are found in inorganic macro components such as sulfates, chloride, iron, ammonia, aluminum, and zinc (Agbozu *et al.*, 2015). Landfill leachate typically contains vast quantities of chemicals, of which 80–95% are inorganic and ~52% are organic. Chloride (Cl), nitrites and nitrates, cyanide (CN), sulfides (S), and sulfates are among the inorganic ions (SO_4^{2-}). Ammonia and ferrous are additional components of inorganic cations. Heavy metals are among the most hazardous pollutants in landfill leachate. The separation of non-hazardous trash from hazardous waste before dumping into a landfill is unusual in the majority of poor nations (Edokpayi *et al.*, 2018); hence, a number of heavy metals with significant concentrations have been recorded in landfill leachates (Chuangcham *et al.*, 2008). Although removing heavy metals may be difficult, we have chosen to concentrate more on removing metals from landfill leachate in our research. According to Dan *et al.* (2017), the most common heavy metals found in landfill leachate include chromium, manganese, cadmium, lead, iron, nickel, and zinc. Younger (acetogenic) leachate frequently contains more metals compared to older leachate (Dan *et al.*, 2017).

Leachate contains a large amount of chlorine (Cl), which is also very movable, inert, and never biodegrades. As a result, it may be used as a leachate plume tracer element as well as a potent indication of pollution. The presence of significant amounts of soluble salts from likely anthropogenic sources, such as kitchen waste from homes, restaurants, and hotels, might result in high levels of chlorides in leachate (Abunama *et al.*, 2021; Hussein *et al.*, 2019).

Emerging contaminants are substances that have recently been proven to be present in large quantities in the environment and have been recognized as potentially hazardous to the environment or to human health, but there is insufficient information to evaluate their risk. In landfill leachate, emerging pollutants include: pharmaceuticals and personal care products, per- and polyfluoroalkyl substances, microplastics, nanoparticles, endocrine-disrupting compounds, pesticides and herbicides, bisphenol A, fire retardants, genetically modified organisms and so on (Ramakrishnan *et al.*, 2015).

20.3 METHODS FOR LANDFILL LEACHATE TREATMENT

A sufficient understanding of the features of landfill leachate is necessary to understand the variable performance observed in treating landfill leachate utilizing biological, physical, or physicochemical techniques. In Figure 20.1, typical landfill leachate treatment techniques are shown. Leachate treatment techniques grouped under the categories of biological, physical, and chemical, as well as integrated treatment techniques. In addition to these, leachate-channeling techniques, such as reusing leachate and treating it along with household sewage, are also used to reduce the pollutant load in landfill leachate (Lindamulla *et al.*, 2022).

The biological degradation of contaminants is a result of metabolic processes of microorganisms. Due to their low cost, biological methods are frequently used to remove nutrients, such as ammonia, and organic compounds; nevertheless, these techniques may not be able to successfully remove heavy metals and non-biodegradable organics. Techniques for biological purification can be categorized as aerobic or anaerobic depending on whether the biological processing medium requires oxygen. Among aerobic biological technologies that may be used to treat landfill leachate, it is possible to identify conventional-activated sludge processes, sequencing batch reactors (SBRs), rotating biological contactors, and moving-bed biofilm reactors. Among them, the most popular

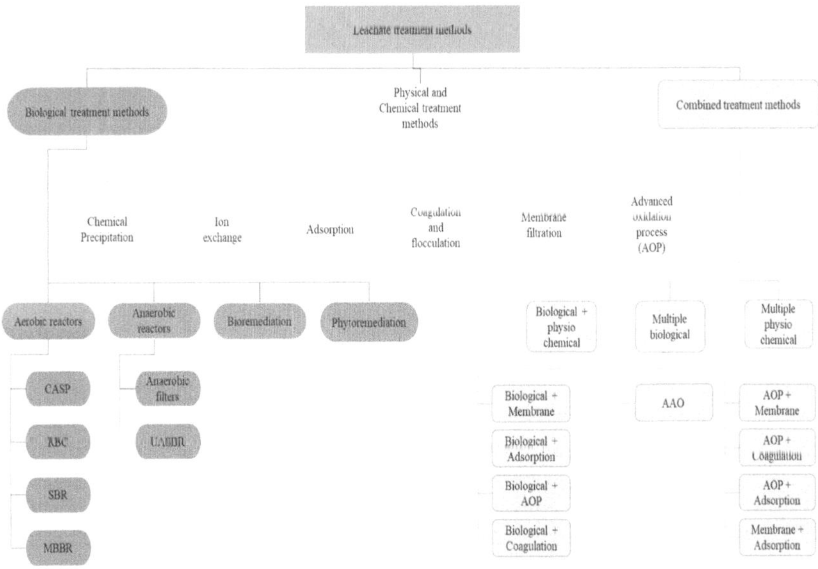

Figure 20.1 Typical leachate treatment techniques (AOP, advanced oxidation process; CASP, conventional activated sludge process; RBC, rotating biological contactor; SBR, sequencing batch reactor; MBBR, moving-bed biofilm reactor; AAO, anaerobic anoxic aerobic process).

anaerobic techniques are anaerobic filters and an up-flow anaerobic sludge blanket (UASB) reactor.

Among the physical–chemical techniques under investigation for landfill leachate treatment include chemical precipitation, advanced oxidation processes (AOPs), coagulation–flocculation membrane filtration, ion exchange, adsorption, and electrochemical treatment.

It is frequently preferable to use a biological treatment method for leachates with high BOD/COD ratios. On the contrary, poor biodegradability of stabilized landfills shows that physical and chemical treatments, including membrane separation techniques, are superior to biological ones. Young leachate requires biological treatments, but older leachate may be treated using physicochemical techniques. Recently, it has been demonstrated that combining biological processes with physicochemical treatments can be very effective. Other integrated physicochemical methods and combined physicochemical/biological techniques are shown to be less efficient than activated sludge/reverse osmosis and combined coagulation–flocculation/nanofiltration, respectively. The landfill leachate methods used in tropical climates are provided in Table 20.3. In the context of leachate treatment in tropical nations, the application of chemical approaches, in particular the use of AOPs, can be noted.

20.3.1 Co-treatment with wastewater

Leachate is diluted with domestic wastewater to make it more biologically treatable and to balance its nutrient profile. To treat landfill leachate and urban wastewater simultaneously, used an SBR. In total, 93%, 83%, 70%, and 83% of ammonia, nitrite, COD, and turbidity were eliminated using a hydraulic retention time of 6 days and a landfill leachate concentration of 20% v/v. According to Wang *et al.* (2018), the most extensively investigated approach for leachate co-treatment with municipal wastewater in a laboratory setting is the activated sludge process employing SBRs. The ease of system setup and the adaptability of SBR control and operation for research objectives are likely to be responsible for this.

Using three different volumetric mixing ratios, El-Gohary *et al.* (2016) investigated the efficacy of anaerobic, aerobic, and anaerobic–aerobic co-treatment of young landfill leachate with municipal wastewater. In all reactors, rising organic matter removal efficiency is noted as the mixing ratio increases, but the anaerobic–aerobic combination performed the best. At higher leachate mixing ratios, they also noted enhanced ammonium concentrations and nitrite deposition in aerobic and anaerobic–aerobic reactors. Mojiri *et al.* (2017) highlighted that it is challenging to compare landfill leachate treatments with those used for household wastewater because of the high COD and BOD/COD ratios. Therefore, to treat leachate, a hybrid system should be used. It is projected that these procedures will be utilized more often in wastewater treatment plants (WWTPs) as a consequence of advancements in process engineering, materials science, microbiological monitoring instruments, sensor technology, data processing, control, and automation for the co-treatment of leachate and municipal wastewater. Currently there are not many studies that are focused as to how these technologies are used (Dereli *et al.*, 2021).

Table 20.3 Leachate treatment technologies.

Treatment Technology	Conditions Used	Removal Efficiency	Reference
Bioremediation	Using garbage enzyme		Madera-Parra (2016)
Constructed wetland	Using *Colocasia esculenta*, *Gynerium sagittatum*, and *Heliconia psittacorum*	COD: 67%	Madera-Parra (2016)
Anaerobic reactor	Anaerobic sequencing batch biofilm reactor	COD: 71%	Contrera *et al.* (2018)
Aerobic reactor	Aerobic sequencing batch reactor (ASBR)	COD: 43% Ammonia: 96%	Lim *et al.* (2016)
AOP	Fenton process	COD: 79%	Maslahati Roudi *et al.* (2018)
AOP	Ozone/catalyst ($ZrCl_4$)	COD: 88% Ammonia: 79%	Abu Amr *et al.* (2017)
AOP	Vermiculite/ozonation	COD: 17%	Braga *et al.* (2020)
Adsorption	Silica nanoparticle	COD: 77%	Pavithra and Shanthakumar (2017)
Coagulation/flocculation	Tannin-based natural coagulant	COD: 55% Ammonia: 91%	Banch *et al.* (2019)
Coagulation/flocculation	Polyaluminum chloride and *Dimocarpus longan* seeds as flocculant	COD: 62%	Aziz *et al.* (2018)
Coagulation/flocculation	Red earth as coagulant	COD: 67% Ammonia: 43%	
AOP/adsorption (ion exchange)	Supercritical water oxidation (ScWO)/zeolite	COD: 74%	Scandelai *et al.* (2020)
AOP/coagulation	Electro oxidation and coagulation	COD: 90%+ Ammonia: 90%–	

(Continued)

Table 20.3 Leachate treatment technologies. (*Continued*)

Treatment Technology	Conditions Used	Removal Efficiency	Reference
Coagulation/membrane filtration	Integration of ultrafiltration membrane process with chemical coagulation	Turbidity: 73%	Nazia *et al.* (2021)
AOP/coagulation flocculation	UV-based sulfate radical oxidation process/coagulation–flocculation	COD: 91% Ammonia: 5%	
AOP/adsorption	MAC/ozonation	COD: 74%	
AOP/adsorption	Adsorption/photo-Fenton ozone	Ammonia: 94.5% COD: 95.1% Color: 95% HA (ABS_{254}): 97.9%	
AOP/coagulation/adsorption	Electrocoagulation/fiber filtration	COD: 94% As: 87% Fe: 96% P: 86%	Li *et al.*, (2014)
AOP	Electro-catalytic ozonation	COD: 3,381 mg/L BOD: 1,521 mg/L	
AOP/adsorption (ion-exchange)	ScWO/zeolite	Ammonia: 98% Nitrite: 100% Nitrate: 98% Color: 98% Turbidity: 98% COD: 74%	Scandelai *et al.* (2020)

AOP/biological method	Kefir grains/Ag-doped TiO_2 photocatalytic process	COD: 83.3% Ammonia: 70% Cd: 100% Ni: 94% Zn: 62.5% Mn: 53.1% Cu: 47.5%	
AOP/coagulation	Coagulation/photo-Fenton	COD: 68% Color: 97% HA (ABS_{254}): 83%	
AOP	Fenton process	COD: 97.8%	
Advance oxidation process/coagulation–flocculation	Coagulation–flocculation/microelectrolysis–Fenton processes	COD: 90.2% HA: 93.7%	
AOP/biological/adsorption	Electro-ozonation/adsorbent augmented SBR	COD: 88.2% Color: 96.1% Ni: 73.4%	Mojiri *et al.* (2017)
AOP/coagulation	EO/coagulation	Color, turbidity, and ammonia: >90%	de Oliveira *et al.* (2019)
AOP /adsorption	$UV_{solar}/O_3/H_2O_2/S_2O_8^{2-}$/zeolite	COD: 36% Ammonia: 99%	
AOP/coagulation–flocculation	UV-based sulfate radical oxidation process/coagulation–flocculation	COD: 91%	
AOP	Ozone/catalyst ($ZrCl_4$)	Color: 100% COD: 88% Ammonia: 79%	Abu Amr *et al.* (2017)

(Continued)

Table 20.3 Leachate treatment technologies. (*Continued*)

Treatment Technology	Conditions Used	Removal Efficiency	Reference
AOP	Vermiculite/ozonation	COD: 16.5% Color: 40.5%	Braga *et al.* (2020)
AOP/adsorption	MAC/ozonation	COD: 72% HA: 91%	
Adsorption methods	Activated carbon (oat hulls)	COD and color: 100%	
Adsorption methods	Activated carbon (coffee wastes)	COD: 51% Ammonia: 32.8% Chlorine: 66% Bromine: 81% Copper: 91.7%	Chávez *et al.* (2019)
Adsorption methods	Zero-valent iron nanofibers/reduced ultra-large graphene oxide (ZVINFs/rULGO)	COD: 93.6% Ammonia: 84.8%	
Adsorption methods	Silica nanoparticle	COD: 77.3% Color: 82.5%	Pavithra and Shanthakumar (2017)
Adsorption methods	Zeolite feldspar mineral composite adsorbent	COD: 49% Ammonia: 45%	
Adsorption methods	Amino acid-modified bentonite	COD: 65.5–92.1%	
Adsorption methods	MS@GG	Pb: 99.2%	
Coagulation/flocculation	Tannin-based natural coagulant	COD: 53.5% Ammonia: 91.3% TSS: 60.2% Fe: 89.7% Zn: 94.6% Cu: 94.1% Cr: 89.9% Cd: 17.2% Pb: 93.7% As: 86.4%	Banch *et al.* (2019)

Coagulation/ flocculation	Polyaluminum chloride and *D. longan* seeds as flocculants	COD: 66.9% Color: 98.8% SS: 99.5%	Aziz *et al.* (2018)
Coagulation/ flocculation	Red earth as coagulant	COD: 66.9% Ammonia: 43.3% Turbidity: 96.2%	Zainol *et al.* (2018)
Coagulation/ flocculation	Ferric chloride as coagulant and a cationic flocculant AN 934-SH polyelectrolytes as flocculant	COD: 45%	
Membrane	Using membrane processes of NF and RO	COD: 94.6% Ammonia: 88.9%	
Bioremediation with the fungi	Using *Aspergillus flavus*	COD: 15.5–48.5% BOD: 45.4–81.6% Ammonia: 50–98.8%	
Bioremediation	Using *B. panacihumi* strain ZB1	COD: 40% Ammonia: 50% Mn: 40% Cu: 60% Se: 52%	
Bioremediation with microalgae	Using *Chlorella* sp.	COD: 60% Ammonia: 90%	
Bioremediation with microalgae	Using *Chlamydomonas* sp. SW15aRL	Ammonia: 83%	Paskuliakova *et al.* (2018)
Bioremediation/ enzyme	Using garbage enzyme	Leachate pollution index: 74.7%	

(*Continued*)

Table 20.3 Leachate treatment technologies. (*Continued*)

Treatment Technology	Conditions Used	Removal Efficiency	Reference
Phytoremediation/ wetland	Using *C. esculenta*, *G. sagittatum* and *H. psittacorum*	COD: 67% Cd: 80% Pb: 40% Hg: 50%	Madera-Parra (2016)
Phytoremediation	Using *Imperata cylindric*	COD: 75% Pb: 56.3% Cd: 16.2% Zn: 6.5%	
Phytoremediation/ wetland	Using *Typha latifolia* and *Canna indica*	COD: 81% Ammonia: 60%	
Wetland/ co-treatment	Using *Typha domingensis*	COD: 86.7% Ammonia: 99.2% Color: 90.3% Ni: 86% Cd: 87.1%	
Bioreactor/ membrane	Membrane bioreactor + activated sludge Membrane bioreactor + indigenous leachate bacteria	COD: 93% Fe: 71% Zn: 78%	Azzouz *et al.* (2018)
Bioreactor/ membrane	Membrane bioreactor	COD: 63% TOC: 35% Ammonia: 98% Phosphorous: 52%	
Bioreactor/ membrane	Membrane bioreactor	Ammonia: >98% TN: >90%	
Bioreactor/air stripping	Air stripping, and aerobic and anaerobic biological processes	COD: 80% Ammonia: 78%	

Bioreactor/ coagulation	SBR and coagulation	Color: 85.8% COD: 84.8% Ammonia: 94.2% TSS: 91.8%	
Bioreactor	Anaerobic sequencing batch biofilm reactor	COD: >70%	Contrera *et al* (2018)
Bioreactor	Aerobic sequencing batch reactor (ASBR)	COD: 30% Ammonia: 65%	Lim *et al*. (2016)
Integrated bioreactor	Partial-denitrification and anammox	TN: 95%	
Integrated bioreactor	Partial nitrification, simultaneous anammox and denitrification	TN: 98.7%	
Bioreactor/ membrane	DM bioreactor	Ammonia: 98% TN: 90%	
Bioreactor/ membrane	Activated sludge process/RO	COD: 99% Ammonia: 99%	Tałałaj *et al*. (2019)

Humic Acid (HA); Total Nitrogen (TN); Total Organic Carbon (TOC).

20.3.2 Coagulation–flocculation

Adding a coagulant causes colloidal particles to become less stable during the coagulation process. The unstable particles typically flocculate into large, bulky floccules after coagulation to enhance the particle size, which facilitates easier settlement (Mojiri *et al.*, 2021). Rui *et al.* (2012) proposed that $FeCl_3$ is a better coagulant than $Al_2(SO_4)_3$. The maximal COD removal of 56% was achieved with 0.8 g/L of $FeCl_3$ at an initial concentration of 5,690 mg/L and at pH 4.8, as contrary to 39% with 0.4 g/L of aluminum sulfate. Assou *et al.* (2016) found that hydrated iron hydroxides precipitate more readily than the corresponding aluminum flocs at pH values >9, which leads to more effective removal of pollutants than that at lower pH levels. $FeCl_3$ was also found to be more effective than $Al_2(SO_4)_3$ for the removal of COD. Li *et al.* (2010) investigated the use of coagulation, flocculation, and adsorption processes in combination to stabilize landfill leachate. With only one coagulation process, very high levels of suspended solids (SS) (93%) and turbidity (99%) were removed; however, COD and toxicity were only moderately affected, that is 53–70% and 74%, respectively. The COD and toxicity reduction may be increased to 86% and 78%, respectively, by combining coagulation/flocculation with the adsorption process. The greatest benefit of this treatment, according to Lippi *et al.* (2018), is its effectiveness in removing organic materials, suspended particles, and humic acids. The cost of chemicals and the handling of generated sludge are downsides, though.

20.3.3 Adsorption

Adsorption is a process by which a concentrated version of an adsorbate, usually as a gas or liquid, adheres to the surface or pores of a solid adsorbent. The effectiveness of the adsorbent influences whether the adsorption process is successful or unsuccessful. In comparison to other treatment techniques, the adsorption process has several benefits, including high removal efficiency, low cost, ease of application, minimal sensitivity to pH and temperature, and environmental friendliness. There are two types of adsorption processes: physical and chemical. If the attraction forces between the adsorbent and the adsorbate are van der Waals forces, then physical adsorption occurs. Assuming that the force of attraction between the adsorbate and the adsorbent is about equal to the strength of chemical bonds, then chemical adsorption takes place (Reshadi *et al.*, 2020). Talebi *et al.* (2020) investigated a variety of influencing elements, including agitation speed, activated carbon dosage, pH, duration, and temperature to enhance the VFA adsorption process from fermented leachate. They also suggested a future study that would examine the financial advantages of recovering acetic and butyric acid from landfill leachate, which would call for careful consideration and analysis.

Modified clinoptilolites, according to Hedayati *et al.* (2021), were superior adsorbents for the removal of polyaromatic hydrocarbon (PAH) from landfill leachate. They also suggested that additional research is required to characterize these adsorbents in a systematic manner using various surfactants. Ishak *et al.* (2017) examined that when the coagulation–flocculation combined with the Fenton reaction was carried out under somewhat acidic conditions, the totol organic carbon (TOC) removal was more favorable. They also suggested that the Fenton reaction may be a feasible and useful technique for treating stabilized landfill leachate because it has been used in the full-scale treatment of several wastewaters. On the basis of few research studies, it appears that magnetic adsorbents have more environmental effects than originally thought because of the possibilities of regeneration and reuse. According to Reshadi *et al.* (2020), much more investigation is necessary before reliable conclusions can be drawn.

20.3.4 Membrane

Selective filtering of influents across a range of pore sizes is a characteristic of membrane separation. Low total energy requirements, simplicity, and high efficiency are benefits of employing membranes. The major membrane techniques used in landfill leachate treatment are reverse osmosis, microfiltration, dynamic membranes, nanofiltration, and ultrafiltration (Mojiri *et al.*, 2021). Chen *et al.* (2021) investigated and thoroughly assessed the application characteristics of membrane separation technology in the leachate treatment of three widely used membrane separation processes, nanofiltration, reverse osmosis, and nanofiltration + reverse osmosis. Baker *et al.* (2015) reported that reverse osmosis technology was employed at Seneca Meadows landfill to treat leachate because it has the advantages of low initial investment cost, the capacity to treat huge quantities of leachate while staying within the legal limitations for effluents, and the removal of more than 95% of contaminants. Visvanathan *et al.* (2007) studied the leachate treatment using aerobic thermophilic membrane bioreactors. The removal efficiency of BOD increased from 0.39 to 0.65 fractions, whereas for COD it increased from 62% to 79%. For ammonia concentration, there was a decrease in removal efficiency from 75% to 60%. Membrane bioreactor (MBR) and reverse osmosis systems have been effectively used to remove organic, inorganic, and heavy metal components from leachate (Robinson, 2005). One disadvantage of the membrane separation technique is the generation of secondary pollutants. Membrane-concentrated leachate needs to be taken into consideration (Chen *et al.*, 2021).

20.3.5 Advanced oxidation processes

AOPs use oxidation reactions to convert organic pollutants into carbon dioxide, water, and inorganic ions. There are two phases in AOPs. The oxidation process takes place in the first phase. Strong oxidants or hydroxyl radicals are produced at this stage. These oxidants react with the organic pollutants present in the effluent at the second stage (Pawar & Gawande, 2015). AOPs can be divided into five categories based on the source of hydroxyl radicals: (1) ozone-based processes, (2) H_2O_2-based processes, (3) sonochemical oxidation, (4) electrochemical oxidation, and (5) heterogeneous photocatalysis. Han *et al.* (2020) reviewed most recent developments in leachate treatment and disposal strategies. They also concluded that excellent degradation of refractory contaminants in leachate is achieved when AOPs and graphitic carbon nitride are combined. Rocha *et al.* (2011) examined various advanced oxidation treatment methods (heterogeneous and homogeneous photocatalytic processes) for leachates employing natural sun radiation as an ultraviolet (UV) photon source. They concluded that at alkaline pHs, the heterogeneous photocatalytic reactions have low efficacy for mineralizing the leachate, necessitating longer reaction times. Also, compared to the heterogeneous photocatalytic, the photo-Fenton reaction has a substantially higher degradation rate, that is, 20 times higher. Wu *et al.* (2004) reported that modern ozone-based techniques may remove 90% of the color of leachate. Even though a fairly large dosage of ozone is used, the elimination of TOC may still be constrained. This finding suggests that ozone-based AOPs should be a great substitute before adopting biological treatment methods. Pisharody *et al.* (2022) concluded that a possible approach to entirely eliminate the micropollutants or transform them into non-toxic compounds is to use AOPs, which are based on both hydroxyl and sulfate radicals. Guo *et al.* (2020) studied three UV-catalyzed AOPs. Results showed that the treatment efficiencies of all the three procedures were greater than those of single-treatment processes. For the treatment of landfill leachate, a number of physical/chemical treatment techniques have been integrated to increase removal efficiency and decrease energy consumption. Some of the identified systems are discussed below.

20.3.5.1 AOPs combined with membranes

Overall separation performance may be improved by integrating membrane filtration with AOPs to effectively reduce membrane-fouling issues. Kamal *et al.* (2022) summarized that between 1.5% and 99.5% of COD may be removed biologically, between 9% and 96% using AOPs, and between 13% and 99.9% using membrane separation. Performance may be enhanced when bio-treatment and membrane filtration are used together; but there are disadvantages, including greater capital and operating costs.

20.3.5.2 AOPs combined with coagulation

By changing the molecular structure of residue organics, this integrated approach can reduce the percentage of organic contaminants and improve the biodegradability of wastewater. Because of this, the coagulation–AOP process may efficiently extract organic matter from refractory wastewater while still being cost-effective. Chen *et al.* (2019) studied a coagulation–ozonation technique which was developed to handle the concentrate that remained after landfill leachate underwent membrane treatment. Investigations were carried out on how coagulant type and initial pH affected treatment effectiveness. The coagulation method effectively eliminated numerous organics, according to the results. After that, ozone was used to continue treating the coagulation-resistant organic materials. This outstanding result demonstrated the viability of the combined coagulation–ozonation method for the elimination of refractory organic materials present in landfill leachate concentrate, hence facilitating further biological treatment.

20.3.5.3 AOPs combined with adsorption

AOPs and adsorption may be combined to more effectively remove pollutants, particularly metals from landfill leachate. Mojiri *et al.* (2021) studied an AOP combined with biochar adsorption, aiming to quantify the toxicity of leachate treatment in integrated systems. It was discovered that the removal efficiency of COD and TOC was 17% and 12% higher, respectively, when the Fenton process was carried out as system I at all reagent doses. Bello and Raman (2019) reported that AOPs can breakdown complex organic pollutants; however, total mineralization is typically impractical, and some intermediate contaminants are frequently produced. Therefore, these intermediates could be eliminated by combining AOPs and adsorption. Integrated H_2O_2–granular activated carbon can boost the biodegradable ratio by 116% while reducing COD by 97.3%.

20.3.5.4 Membrane filtration combined with adsorption

To remove suspended and colloidal materials from wastewater, adsorbents have been applied to membranes consecutively. This reduces the organic load and prevents membrane fouling. Zielińska *et al.* (2020) investigated how effectively a membrane method, both by itself and in combination with powdered activated carbon adsorption, could remove organic chemicals that are difficult to biodegrade from actually stabilized landfill leachate. When the membrane process was paired with addition of polyammonium chloride (PAC) for each membrane, the treatment was more effective. Accordingly, adsorption contributed the most to the treatment of leachate where COD and color removal efficiencies were found to be 73.1% and 94.4%, respectively. Leachate can be treated with integrated systems that are highly effective and need little energy.

Peyravi *et al.* (2016) concluded that powdered activated carbon increased the specific oxygen uptake rate of the activated sludge by serving as a reaction site for substrates and bacteria and by decreasing the harmful effects of COD on microorganism activities.

20.3.6 Bioreactors

In recent years, bioreactors have been used to treat wastewater because these processes are straightforward, reliable, and incredibly economical. However, their two main drawbacks are leachate toxicity for microbial communities and temperature (Mojiri *et al.*, 2021).

20.3.6.1 Aerobic bioreactors

A biological procedure that is used the most frequently is aerobic treatment. Continuous aeration and substantial established bacterial populations are characteristics of aerobic reactors. To remove organic materials from wastewater, the activated sludge process needs large concentrations of microorganisms, primarily bacteria, fungi, and protozoa. Biodegradable organic material can be effectively removed via the activated sludge technique by being entirely converted into carbon dioxide and water. For treating landfill leachate, SBRs are the most popular choice. An SBR has a number of time-oriented periodic steps, and it operates in batches, which might improve process effectiveness.

The requirement for large quantities of dissolved oxygen in biofilm reactors for denitrification is one of the key shortcomings of this method (Mojiri *et al.*, 2021).

20.3.6.2 *Anaerobic bioreactors*

The biological breakdown of organic or inorganic waste without the presence of oxygen molecules is a component of anaerobic treatment. Concentrated organic wastewater can be effectively treated biotechnologically using anaerobic methods. Due to their low sludge and biogas output, these techniques are both energy-efficient and eco-friendly. With the high COD and high BOD/COD ratio of landfill leachates, anaerobic approaches often exhibit higher landfill leachate treatment performance compared to aerobic treatment techniques.

Mojiri *et al.* (2021) reported that the extended retention time, sensitivity to temperature fluctuations, and low elimination effectiveness of this method are its main limitations. For the purification of landfill leachate, an anaerobic-activated sludge process may need an UASB and expanded granular sludge blanket reactors. Wastewater passes through a sludge bed with strong microbiological activity in a UASB reactor.

20.3.7 Bioremediation

Bioremediation is a process in which environmental contaminants are removed biologically. Microbial remediation (aerobic or anaerobic reactions are used to break down organic contaminants in leachate), phytoremediation (through plant root absorption and adsorption, environmental contaminants can be transferred, digested, metabolized, and stored), and combined remediation are the three basic types of bioremediation. It primarily relies on the regular metabolic processes and movement patterns of plants, animals, or microbes to collect, degrade, transform, or adsorb environmental contaminants. The microorganisms and accessible electron acceptors will determine whether this process is aerobic or anaerobic. The organism itself performs the roles of both a bio-accumulator and a bio-purifier during this purification process. Bioremediation is a cost-effective and eco-friendly technique (Tan *et al.*, 2022). Morris *et al.* (2018) studied that leachate's common components, such as ammonia, phosphate, and nitrate, can all be utilized by microorganisms that have been isolated from the leachate.

The removal percentage of ammonia in that study was 90% (high), whereas removal percentages for phosphate and nitrate were 67% and 63%, respectively (lower). Firmicutes, Actinobacteria, Proteobacteria, *Brevibacillus panacihumi* strain ZB1, and *Pseudomonas putida* are some of the prominent bacteria that have been used to treat landfill leachate (Morris *et al.*, 2018). *Chlorella vulgaris*, *Scenedesmus obliquus*, *Nannochloropsis salina*, *Desmodesmus* spp., *Nannochloropsis* sp., *Scenedesmus* sp., and *Stigeoclonium* sp. are some of the microalgae used as bioremediation agents (Martínez-Ruiz *et al.*, 2022). Martínez-Ruiz *et al.* (2022) concluded that leachates can be treated by removing various nutrients and contaminants using bioremediation with microalgae cultivation. This method requires minimal financial outlay, requires little maintenance, and is easy to use. By cultivating microalgae and producing biomass, the primary goal of using them is to remove contaminants from the environment, such as heavy metals. Tan *et al.* (2022) reported that bioremediation is in line with the 'green' and 'ecological' development trends and is significant in many ways for sustaining sustainable and healthy development. The pure green purification of landfill leachate, therefore, has significant potential for bioremediation technology.

20.3.7.1 Phytoremediation

Biological interactions between bacteria, wetland plant roots, soil, air, sun, and water are all included under phytoremediation technique. This form of treatment is a plant-based planned technique for cleaning wastewater as it flows through the area of artificially constructed wetlands to deal with a variety of pollutants, such as trace elements, organic compounds, and radioactive contaminants in water and soil. There are various methods for phytoremediation such as phytoextraction, phytodegradation, phytometabolism, phytovolatilization, rhizometabolism, and phytostabilization (Nissim *et al.*, 2021). The phytoremediation approach meets tertiary treatment criteria with no operational expenses, little maintenance, improving the landscape, no odor issues, and creating a natural habitat.

Nissim *et al.* (2021) studied that high landfill leachate loads appear to be more advantageous for poplar than for willow. Willow was less efficient than poplar in decreasing the concentration of BOD_5, COD, and As. Song *et al.* (2018) suggested that *Typha angustifolia* should be used to treat landfill leachate. They also suggested that while planning for remediation wetland water depth, soil texture, water quality, and water velocity need to be considered. The genera *Scenedesmus*, *Chlamydomonas*, and *Chlorella*, as well as cyanobacteria and other species, have all been used to remediate landfill leachate (Paskuliakova *et al.*, 2018).

20.3.7.2 Constructed wetlands

Engineered systems termed 'constructed wetlands' are used to emulate natural wetlands (Figure 20.2). These engineered systems purify wastewater by minimizing leachate pollutant concentration while simulating natural wetland processes (Bakhshoodeh *et al.*, 2020). TDS, organic matter, total suspended solids (TSS), SO_4^{2-}, heavy metals, phosphorous, nitrogen, and non-specific toxic constituents are the leachate characteristics of concern (Tchobanoglous & Kreith, 2002). The free-water surface wetland and the subsurface flow wetland are the two main categories of artificial wetlands based on the presence or absence of the water on the surface. Additionally, multiple hybrid constructed wetlands can be created by merging various types of constructed artificial wetlands. Constructed wetlands are also classified on the basis of the flow direction as vertically constructed wetlands and horizontally constructed wetlands (Prathap *et al.*, 2014). To prevent damage to the natural wetlands and other aquatic resources, constructed wetlands are typically constructed on uplands and outside of floodplains or floodways (Prathap *et al.*, 2014). Emergent wetland plants such as *Schoenoplectus*, *Typha*, *Phragmites*, and *Cyperus* are commonly planted on permeable substrata such as gravel in a wetland system (Mojiri *et al.*, 2021; Schiopu and Gavrilescu, 2010).

Bakhshoodeh *et al.* (2020) reported that removal efficiency for BOD_5 for installed wetlands for landfill leachate is in the range of 60–80%, with vertical flow constructed wetlands and free-water surface constructed wetlands having the highest removal. Also, hybrid-constructed wetlands were most effective at removing ammonia-N, showing a removal efficiency of 60–75%. However, the overall removal rates for TSS and total phosphorous for all constructed wetlands were 50–65% and 55–80%, respectively, with horizontal flow constructed

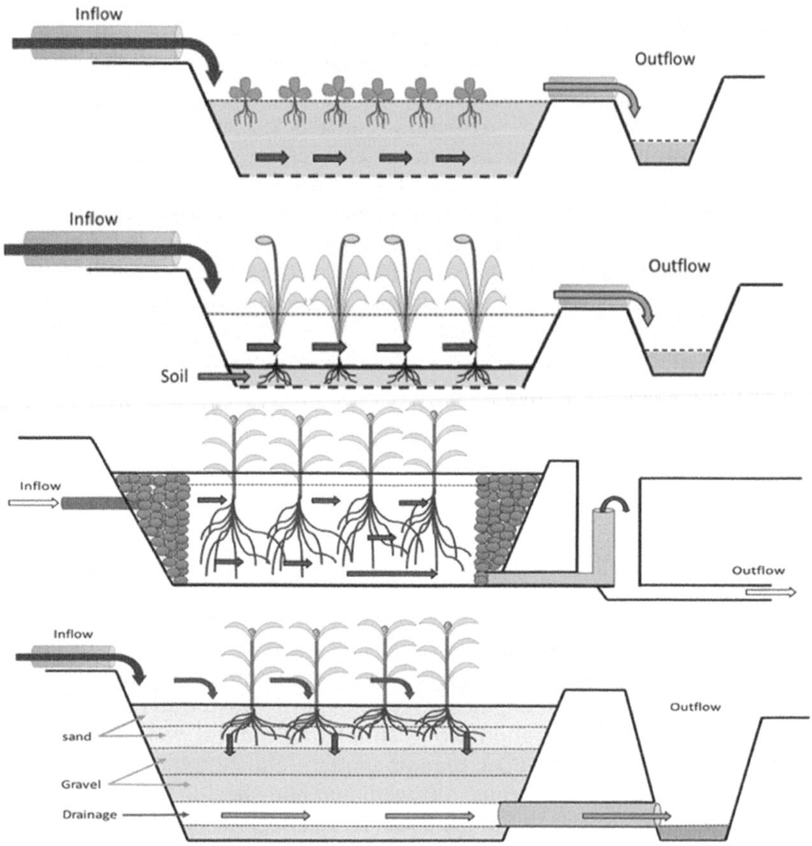

Figure 20.2 Various types of constructed wetlands for wastewater treatment.

wetlands and hybrid constructed wetlands being the most effective. Not many research studies exist that document heavy metal removal by utilizing constructed wetlands, so it is difficult to draw any clear conclusions. Vertical flow constructed wetland appears to have been most effective at removing various kinds of heavy metals. The removal of heavy metals has been reported to be between 15% and 95%, with *Phragmites* sp. plants proved to be the most effective species. Metal removal through constructed wetland systems may be impacted by several mechanisms, including the adsorption of soil or substrates together with particulates and soluble organics, the precipitation of insoluble salts, and the uptake of aquatic plants and microbes (Mojiri *et al.*, 2021). Careful consideration and case-by-case investigation of design specifications are required when treating landfill leachate with constructed wetlands. The design parameters for a specific system must be determined from experimental data that consider both the soil conditions and the leachate characteristics (Tchobanoglous & Kreith, 2002).

Tchobanoglous & Kreith (2002) studied that TDS, TN, BOD, and COD concentrations were reduced by 15%, 40%, 65%, and 60%, respectively, using their trial unit. By employing constructed wetlands, operating and maintenance costs can be reduced by 80%. The constructed wetland saves 57% over the traditional design for a similar influent discharge.

20.3.8 Combined physical–chemical treatment methods with biological treatment methods

20.3.8.1 Adsorption combined with biological treatment method

The amount of an adsorbent used in wastewater treatment processes is reduced as a result of the adoption of the adsorption technique and the biological method (Mojiri *et al.*, 2021). Yi *et al.* (2018) reported that adsorbents with a high ammonia nitrogen removal capacity and increased adsorbent dosages would improve removal efficiency. The leachate samples underwent combined biological and adsorption treatment. Nearly 40% and 40–50% of COD and ammonia nitrogen, respectively, are removed by pure biological treatment from leachate samples. For leachate management, combined adsorption and biological treatment have been used due to cost and efficiency considerations. More than 70% of the ammonia nitrogen from leachate wastewater was removed by combined treatment. Lim *et al.* (2016) studied that in 7 days of treatment, advanced SBR could eliminate 30% and 65% of COD and ammoniacal nitrogen, respectively. After that, the ammoniacal nitrogen and COD contents of leachate were polished using an efficient adsorbent called zeolite in a secondary treatment phase. The obtained results are encouraging, with the elimination of ammoniacal nitrogen and COD being increased by 96% and 43%, respectively, after the leachate was further strengthened by zeolite by adsorption. The removal of heavy metals including chromium, vanadium, and aluminum was another benefit of this combined biological–physical treatment approach.

20.3.8.2 AOPs combined with biological treatment methods

As a pre-treatment, incorporation of AOP procedures produces biological post-treatment intermediate pollutants that are easily biodegradable. As a result, it has a good effect on the treatment of wastewater such as landfill leachate (Mojiri *et al.*, 2021). To reduce the acute cytogenotoxicity of municipal landfills, Klauck *et al.* (2017) investigated the effectiveness of various advanced oxidation techniques and biological systems. An effluent with physicochemical characteristics compliant with Brazilian laws and fit for disposal in water bodies was produced by the electrochemical oxidation combined with ozonation treatment. Although none of the treatments was able to completely reduce the acute toxicity, electrochemical oxidation combined with ozonation and photo-electrochemical oxidation were able to remove the genotoxicity. For the treatment of landfill leachate, the advanced oxidation method using heterogeneous photocatalysis was combined with seeded bioreactors with various inoculum types. According to the results of heterogeneous photocatalysis, 50–84% of the initial COD at pH 5 was reduced. However, without additional treatment, this treated leachate cannot be repurposed or released into the environment.

20.3.8.3 *Membranes combined with biological treatment methods*

It is highly challenging to treat landfill leachate, especially aged landfill leachate, with conventional biological treatment methods. A membrane bioreactor is shown to be a promising substitute in this regard. In comparison to traditional biological systems, excellent removal rates for BOD, COD, and ammonia of ~90%, 75%, and 90% or more are possible, respectively. It has been proven that MBRs are efficient at removing micropollutants (Ahmed & Lan, 2012). Leachate from sanitary landfills is treated using a technique that incorporates an UASB, an SBR, continuous micron filtering, and reverse osmosis with average removal efficiencies for COD, TN, and BOD_5 of 99.8%, 99.5%, and 99.8%, respectively; the integrated technique has produced successful operational outcomes.

20.3.8.4 *Coagulation combined with biological treatment methods*

When using biological treatment techniques, flocculation and coagulation can be used as pre- and post-treatment methods (Mojiri *et al.*, 2021). Wang *et al.* (2009) concluded that the landfill leachate treated with a combination method that incorporates coagulation, Fenton oxidation, and biological aerated filtration following SBR treatment, produced outstanding COD removal efficiency and color removal outcomes, that is, 75 mg/L and <10°, respectively. Abood *et al.* (2014) studied that in comparison to other combinations used for leachate treatment around the world, the combination of agitation–polyfloroalkyl substances (PFS) coagulation–SBR–filtration displayed remarkable treatment results in the overall elimination of COD and NH_3-N, that is, 97.4% and 99.2%, respectively.

20.3.9 Nitrification and denitrification

First, under aerobic conditions, autotrophic bacteria oxidize ammonium and nitrite to produce nitrate termed nitrification; next, under anoxic conditions, heterotrophic microorganisms decrease nitrate to produce nitrogen gas termed denitrification. The two-step autotrophic reactions that constitute the nitrification process include the conversion of ammonium to nitrite and the subsequent oxidation of nitrite to nitrate in the presence of oxygen. Ammonia-oxidizing bacteria such as *Nitrosomonas*, *Nitrosococcus*, and *Nitrosospira* are involved in the oxidation of ammonia or ammonium to nitrite in the first step. *Nitrobacter*, *Nitrospira*, and *Nitrococcus* in particular, which are nitrite-oxidizing microbes, convert nitrite to nitrate in the usual sequence (Kabuba *et al.*, 2022).

Ilies and Mavinic (2001) concluded that when ambient temperatures decrease to 148°C, the nitrification process seemed to be unaffected (at least not significantly); nevertheless, as the temperatures decrease to 108°C, the nitrification percentage decreased by 10–30. However, during the overall process final ammonia reduction was still just ~50%. Carley and Mavinic (1991) studied that after complete nitrification for the four leachate treatment systems, the overall ammonia-N removal efficiency was >98%. Systematically examined the effects of two aeration modes: (1) reducing dissolved oxygen and (2) increasing dissolved oxygen for nitrogen removal and N_2O emission in the partial nitrification and denitrification processes. An increase in the dissolved oxygen reduced total nitrogen by 78%.

20.3.9.1 Anammox (anaerobic ammonium oxidation)

The anammox process uses nitrite as the electron acceptor to directly oxidize ammonium to nitrogen in the absence of oxygen using carbon dioxide as a source for growth (Mao *et al.*, 2017). COD and heavy metals can affect anammox activities, therefore, they are usually combined with other treatment processes.

Xu *et al.* (2010) concluded that to treat the leachate from urban landfills, an SBR system may incorporate the partial-nitrification process, the anammox process, and the heterotrophic denitrification process. Isaka *et al.* (2008) utilized a gel carrier containing anammox bacteria to achieve stable nitrogen removal capability at low temperatures. Six potential genera, including *Candidatus jettenia, Candidatus anammoxoglobus, Candidatus brocadia, Candidatus scalindua, Candidatus anammoximicrobium,* and *Candidatus kuenenia,* constitute the monophyletic group of anammox bacteria (Mojiri *et al.*, 2021).

20.4 CONCLUSIONS

As leachate demonstrates significant potential for environmental effects on groundwater and surface water pollution and constitutes a possible health concern for both surrounding ecosystems and human populations, understanding its composition is essential for anticipating the long-term effects of landfills. Physical, chemical, and biological processes are the main leachate treatment methods used in landfills. But in order to achieve stringent quality requirements for the direct discharge of leachate into surface water, integrated methods of treatment must be developed. Integrated treatment approaches including co-treatment with wastewater are particularly advised due to the high pollutant concentrations in landfill leachate and its limited biodegradability. One of the most active areas of study is the oxidation of the leachate, along with membrane technologies, adsorption, coagulation, and artificial wetlands. The studies on oxidation-based processes such as Fenton's reagent and catalyzed application of persulfate, as well as coagulation-based techniques such as electrocoagulation and peroxicoagulation, are leading future research topics. Additionally, research on membrane fouling and initiatives to improve membrane functions will be among the subjects of attention in the years to come.

REFERENCES

Abood A. R., Bao J., Du J., Zheng D. and Luo Y. (2014). Non-biodegradable landfill leachate treatment by combined process of agitation, coagulation, SBR and filtration. *Waste Management*, **34**(2), 439–447, https://doi.org/10.1016/j.wasman.2013.10.025

Abu Amr S. S., Zakaria S. N. F. and Aziz H. A. (2017). Performance of combined ozone and zirconium tetrachloride in stabilized landfill leachate treatment. *Journal of Material Cycles and Waste Management*, **19**, 1384–1390, https://doi.org/10.1007/s10163-016-0524-x

Abunama T., Moodley T., Abualqumboz M., Kumari S. and Bux F. (2021). Variability of leachate quality and polluting potentials in light of leachate pollution index (LPI) – a global perspective. *Chemosphere*, **282**, 131119, https://doi.org/10.1016/j.chemosphere.2021.131119

Agbozu I. E., Oghama O. E. and Odhikori J. O. (2015). Physico-chemical characterization and pollution index determination of leachates from Warri waste dumpsite, southern Nigeria. *Journal of Applied Sciences and Environmental Management*, **19**(3), 361, https://doi.org/10.4314/jasem.v19i3.4

Ahmed F. N. and Lan C. Q. (2012). Treatment of landfill leachate using membrane bioreactors: a review. *Desalination*, **287**, 41–54, https://doi.org/10.1016/j.desal.2011.12.012

Akgul D., Aktan C. K., Yapsakli K. and Mertoglu B. (2013). Treatment of landfill leachate using UASB–MBR–SHARON–anammox configuration. *Biodegradation*, **24**, 399–412, https://doi.org/10.1007/s10532-012-9597-y

Al-Wabel M. I., Al Yehya W. S., Al-Farraj A. S. and El-Maghraby S. E. (2011). Characteristics of landfill leachates and bio-solids of municipal solid waste (MSW) in Riyadh city, Saudi Arabia. *Journal of the Saudi Society of Agricultural Sciences*, **10**(2), 65–70, https://doi.org/10.1016/j.jssas.2011.03.009

Assou M., El Fels L., El Asli A., Fakidi H., Souabi S. and Hafidi M. (2016). Landfill leachate treatment by a coagulation–flocculation process: effect of the introduction order of the reagents. *Desalination and Water Treatment*, **57**(46), 21817–21826, https://doi.org/10.1080/19443994.2015.1127779

Atmaca E. (2009). Treatment of landfill leachate by using electro-Fenton method. *Journal of Hazardous Materials*, **163**(1), 109–114, https://doi.org/10.1016/j.jhazmat.2008.06.067

Aziz H. A., Rahim N. A., Ramli S. F., Alazaiza M. Y., Omar F. M. and Hung Y. T. (2018). Potential use of *Dimocarpus longan* seeds as a flocculant in landfill leachate treatment. *Water*, **10**(11), 1672, https://doi.org/10.3390/w10111672

Baker P., Pannuci D. and Weis R. (2015). Seneca Meadows landfill. Leachate pre-treatment using reverse osmosis.

Bakhshoodeh R., Alavi N., Oldham C., Santos R. M., Babaei A. A., Vymazal J. and Paydary P. (2020). Constructed wetlands for landfill leachate treatment: a review. *Ecological Engineering*, **146**, 105725, https://doi.org/10.1016/j.ecoleng.2020.105725

Banch T. J., Hanafiah M. M., Alkarkhi A. F. and Abu Amr S. S. (2019). Factorial design and optimization of landfill leachate treatment using tannin-based natural coagulant. *Polymers*, **11**(8), 1349, https://doi.org/10.3390/polym11081349

Bello M. M. and Raman A. A. A. (2019). Synergy of adsorption and advanced oxidation processes in recalcitrant wastewater treatment. *Environmental Chemistry Letters*, **17**, 1125–1142, https://doi.org/10.1007/s10311-018-00842-0

Boumechhour F., Rabah K., Lamine C. and Said B. M. (2013). Treatment of landfill leachate using Fenton process and coagulation/flocculation. *Water and Environment Journal*, **27**(1), 114–119, https://doi.org/10.1111/j.1747-6593.2012.00332.x

Braga W. L. M., de Melo D. H. A., de Morais D., Samanamud G. R. L., França A. B., Quintão C. M. F., Loures C. C. A., de Urzedo A. P. F. M., Naves L. L. R., de Freitas Gomes J. H. and Naves F. L. (2020). Optimization of the treatment of sanitary landfill by the ozonization catalysed by modified nanovermiculite in a rotating packed bed. *Journal of Cleaner Production*, **249**, 119395, https://doi.org/10.1016/j.jclepro.2019.119395

Carley B. N. and Mavinic D. S. (1991). The effects of external carbon loading on nitrification and denitrification of a high-ammonia landfill leachate. *Research Journal of the Water Pollution Control Federation*, 51–59.

Chávez R. P., Pizarro E. C. C. and Galiano Y. L. (2019). Tratamento de lixiviado de aterro sanitário com carvão ativado obtido de resíduos de café. *Engenharia Sanitaria e Ambiental*, **24**, 833–842, https://doi.org/10.1590/s1413-41522019178655

Chen W., Gu Z., Wen P. and Li Q. (2019). Degradation of refractory organic contaminants in membrane concentrates from landfill leachate by a combined coagulation–ozonation process. *Chemosphere*, **217**, 411–422, https://doi.org/10.1016/j.chemosphere.2018.11.002

Chen W., Gu Z., Ran G. and Li Q. (2021). Application of membrane separation technology in the treatment of leachate in China: a review. *Waste Management*, **121**, 127–140, https://doi.org/10.1016/j.wasman.2020.12.002

Chuangcham U., Wirojanagud W., Charusiri P., Milne-Home W. and Lertsirivorakul R. (2008). Assessment of heavy metals from landfill leachate contaminated to soil: a case study of Kham Bon landfill, Khon Kaen province, NE Thailand. *Journal of Applied Sciences*, **8**(8), 1383–1394, https://doi.org/10.3923/jas.2008.1383.1394

Contrera R. C., Culi M. J. L., Morita D. M., Rodrigues J. A. D., Zaiat M. and Schalch V. (2018). Biomass growth and its mobility in an AnSBBR treating landfill leachate. *Waste Management*, **82**, 37–50, https://doi.org/10.1016/j.wasman.2018.10.006

Cui Y. H., Xue W. J., Yang S. Q., Tu J. L., Guo X. L. and Liu Z. Q. (2018). Electrochemical/peroxydisulfate/Fe^{3+} treatment of landfill leachate nanofiltration concentrate after ultrafiltration. *Chemical Engineering Journal*, **353**, 208–217, https://doi.org/10.1016/j.cej.2018.07.101

Dan A., Fujii D., Soda S., Machimura T. and Ike M. (2017). Removal of phenol, bisphenol A, and 4-*tert*-butylphenol from synthetic landfill leachate by vertical flow constructed wetlands. *Science of the Total Environment*, **578**, 566–576, https://doi.org/10.1016/j.scitotenv.2016.10.232

de Oliveira M. S., da Silva L. F., Barbosa A. D., Romualdo L. L., Sadoyama G. and Andrade L. S. (2019). Landfill leachate treatment by combining coagulation and advanced electrochemical oxidation techniques. *ChemElectroChem*, **6**(5), 1427–1433, https://doi.org/10.1002/celc.201801677

Dereli R. K., Clifford E. and Casey E. (2021). Co-treatment of leachate in municipal wastewater treatment plants: critical issues and emerging technologies. *Critical Reviews in Environmental Science and Technology*, **51**(11), 1079–1128, https://doi.org/10.1080/10643389.2020.1745014

Dolar D., Krešimir K. and Strmecky T. (2016). Hybrid processes for treatment of landfill leachate: coagulation/UF/NF–RO and adsorption/UF/NF–RO. *Separation and Purification Technology*, **168**, 39–46, https://doi.org/10.1016/j.seppur.2016.05.016

Edokpayi J. N., Durowoju O. S. and Odiyo J. O. (2018). Assessment of heavy metals in landfill leachate: a case study of Thohoyandou landfill, Limpopo province, South Africa. In: Heavy Metals, United Kingdom, vol. **16**, pp. 214–227.

Eggen T., Moeder M. and Arukwe A. (2010). Municipal landfill leachates: a significant source for new and emerging pollutants. *Science of the Total Environment*, **408**(21), 5147–5157, https://doi.org/10.1016/j.scitotenv.2010.07.049

El-Gohary F. A. and Kamel G. (2016). Characterization and biological treatment of pre-treated landfill leachate. *Ecological Engineering*, **94**, 268–274, https://doi.org/10.1016/j.ecoleng.2016.05.074

Ferraz F. D. M., Povinelli J., Pozzi E., Vieira E. M. and Trofino J. C. (2014). Co-treatment of landfill leachate and domestic wastewater using a submerged aerobic biofilter. *Journal of Environmental Management*, **141**, 9–15, https://doi.org/10.1016/j.jenvman.2014.03.022

Gautam P. and Kumar S. (2021). Characterisation of hazardous waste landfill leachate and its reliance on landfill age and seasonal variation: a statistical approach. *Journal of Environmental Chemical Engineering*, **9**(4), 105496, https://doi.org/10.1016/j.jece.2021.105496

Gotvajn A. Ž. and Pavko A. (2015). Perspectives on biological treatment of sanitary landfill leachate. *Wastewater Treatment Engineering*, **13**, 31–39.

Guo S., Wang Q., Luo C., Yao J., Qiu Z. and Li Q. (2020). Hydroxyl radical-based and sulfate radical-based photocatalytic advanced oxidation processes for treatment of refractory organic matter in semi-aerobic aged refuse biofilter effluent arising from treating landfill leachate. *Chemosphere*, **243**, 125390, https://doi.org/10.1016/j.chemosphere.2019.125390

Han M., Duan X., Cao G., Zhu S. and Ho S. H. (2020). Graphitic nitride-catalyzed advanced oxidation processes (AOPs) for landfill leachate treatment: a mini review. *Process Safety and Environmental Protection*, **139**, 230–240, https://doi.org/10.1016/j.psep.2020.04.046

Hedayati M. S., Abida O. and Li L. Y. (2021). Adsorption of polycyclic aromatic hydrocarbons by surfactant-modified clinoptilolites for landfill leachate treatment. *Waste Management*, **131**, 503–512, https://doi.org/10.1016/j.wasman.2021.06.033

Hu X., Wang X., Ban Y. and Ren B. (2011). A comparative study of UV–Fenton, UV–H_2O_2 and Fenton reaction treatment of landfill leachate. *Environmental Technology*, **32**(9), 945–951, https://doi.org/10.1080/09593330.2010.521953

Hu L., Zeng G., Chen G., Dong H., Liu Y., Wan J., Chen A., Guo Z., Yan M., Wu H. and Yu Z. (2016). Treatment of landfill leachate using immobilized *Phanerochaete chrysosporium* loaded with nitrogen-doped TiO_2 nanoparticles. *Journal of Hazardous Materials*, **301**, 106–118, https://doi.org/10.1016/j.jhazmat.2015.08.060

Hussein M., Yoneda K., Zaki Z. M. and Amir A. (2019). Leachate characterizations and pollution indices of active and closed unlined landfills in Malaysia. *Environmental Nanotechnology, Monitoring & Management*, **12**, 100232, https://doi.org/10.1016/j.enmm.2019.100232

Ilies P. and Mavinic D. S. (2001). The effect of decreased ambient temperature on the biological nitrification and denitrification of a high ammonia landfill leachate. *Water Research*, **35**(8), 2065–2072, https://doi.org/10.1016/S0043-1354(00)00477-2

Isaka K., Date Y., Kimura Y., Sumino T. and Tsuneda S. (2008). Nitrogen removal performance using anaerobic ammonium oxidation at low temperatures. *FEMS Microbiology Letters*, **282**(1), 32–38, https://doi.org/10.1111/j.1574-6968.2008.01095.x

Ishak A. R., Hamid F. S., Mohamad S. and Tay K. S. (2017). Removal of organic matter from stabilized landfill leachate using coagulation–flocculation–Fenton coupled with activated charcoal adsorption. *Waste Management & Research*, **35**(7), 739–746, https://doi.org/10.1177/0734242X17707572

Ishak A. R., Hamid F. S., Mohamad S. and Tay K. S. (2018). Stabilized landfill leachate treatment by coagulation–flocculation coupled with UV-based sulfate radical oxidation process. *Waste Management*, **76**, 575–581, https://doi.org/10.1016/j.wasman.2018.02.047

Iskander S. M., Novak J. T., Brazil B. and He Z. (2017). Simultaneous energy generation and UV quencher removal from landfill leachate using a microbial fuel cell. *Environmental Science and Pollution Research*, **24**, 26040–26048, https://doi.org/10.1007/s11356-017-0231-8

Jahan E., Nessa A., Hossain M. F. and Parveen Z. (2016). Characteristics of municipal landfill leachate and its impact on surrounding agricultural land. *Bangladesh Journal of Scientific Research*, **29**(1), 31–39, https://doi.org/10.3329/bjsr.v29i1.29755

Jiang Y., Li R., Yang Y., Yu M., Xi B., Li M., Xu Z., Gao S. and Yang C. (2019). Migration and evolution of dissolved organic matter in landfill leachate-contaminated groundwater plume. *Resources, Conservation and Recycling*, **151**, 104463, https://doi.org/10.1016/j.resconrec.2019.104463

Kabuba J., Lephallo J. and Rutto H. (2022). Comparison of various technologies used to eliminate nitrogen from wastewater: a review. *Journal of Water Process Engineering*, **48**, 102885, https://doi.org/10.1016/j.jwpe.2022.102885

Kamal A., Makhatova A., Yergali B., Baidullayeva A., Satayeva A., Kim J., Inglezakis V. J., Poulopoulos S. G. and Arkhangelsky E. (2022). Biological treatment, advanced oxidation and membrane separation for landfill leachate treatment: a review. *Sustainability*, **14**(21), 14427, https://doi.org/10.3390/su142114427

Kamaruddin M. A., Yusoff M. S., Aziz H. A. and Hung Y. T. (2015). Sustainable treatment of landfill leachate. *Applied Water Science*, **5**(2), 113–126, https://doi.org/10.1007/s13201-014-0177-7

Klauck C. R., Giacobbo A., de Oliveira E. D. L., da Silva L. B. and Rodrigues M. A. S. (2017). Evaluation of acute toxicity, cytotoxicity and genotoxicity of landfill leachate treated by biological lagoon and advanced oxidation processes. *Journal of Environmental Chemical Engineering*, **5**(6), 6188–6193, https://doi.org/10.1016/j.jece.2017.11.058

Le Luu T. (2020). Post treatment of ICEAS-biologically landfill leachate using electrochemical oxidation with Ti/BDD and Ti/RuO_2 anodes. *Environmental Technology & Innovation*, **20**, 101099, https://doi.org/10.1016/j.eti.2020.101099

Li W., Hua T., Zhou Q., Zhang S. and Li F. (2010). Treatment of stabilized landfill leachate by the combined process of coagulation/flocculation and powder activated carbon adsorption. *Desalination*, **264**(1–2), 56–62, https://doi.org/10.1016/j.desal.2010.07.004

Li W., Zheng P., Guo J., Ji J., Zhang M., Zhang Z., Zhan E. and Abbas G. (2014). Characteristics of self-alkalization in high-rate denitrifying automatic circulation (DAC) reactor fed with methanol and sodium acetate. *Bioresource Technology*, **154**, 44–50, https://doi.org/10.1016/j.biortech.2013.11.097

Lim C. K., Seow T. W., Neoh C. H., Md Nor M. H., Ibrahim Z., Ware I. and Mat Sarip S. H. (2016). Treatment of landfill leachate using ASBR combined with zeolite adsorption technology. *3 Biotech*, **6**, 1–6.

Lindamulla L., Nanayakkara N., Othman M., Jinadasa S., Herath G. and Jegatheesan V. (2022). Municipal solid waste landfill leachate characteristics and their treatment options in tropical countries. *Current Pollution Reports*, **8**(3), 273–287, https://doi.org/10.1007/s40726-022-00222-x

Lippi M., Ley M. B. R. G., Mendez G. P. and Junior R. A. F. C. (2018). State of art of landfill leachate treatment: literature review and critical evaluation. *Ciência e Natura*, **40**, 78, https://doi.org/10.5902/2179460X35239

Madera-Parra C. A. (2016). Treatment of landfill leachate by polyculture constructed wetlands planted with native plants. *Ingeniería y Competitividad*, **18**(2), 183, https://doi.org/10.25100/iyc.v18i2.2166

Mao N., Ren H., Geng J., Ding L. and Xu K. (2017). Engineering application of anaerobic ammonium oxidation process in wastewater treatment. *World Journal of Microbiology and Biotechnology*, **33**(8), 1–11.

Martínez-Ruiz M., Molina-Vázquez A., Santiesteban-Romero B., Reyes-Pardo H., Villaseñor-Zepeda K. R., Meléndez-Sánchez E. R., Araújo R. G., Sosa-Hernández J. E., Bilal M., Iqbal H. M. and Parra-Saldivar R. (2022). Micro-algae assisted green bioremediation of water pollutants rich leachate and source products recovery. *Environmental Pollution*, **306**, 119422, https://doi.org/10.1016/j.envpol.2022.119422

Maslahati Roudi A., Chelliapan S., Wan Mohtar W. H. M. and Kamyab H. (2018). Prediction and optimization of the Fenton process for the treatment of landfill leachate using an artificial neural network. *Water*, **10**(5), 595, https://doi.org/10.3390/w10050595

Mojiri A., Ziyang L., Hui W., Ahmad Z., Tajuddin R. M., Amr S. S. A., Kindaichi T., Aziz H. A. and Farraji H. (2017). Concentrated landfill leachate treatment with a combined system including electro-ozonation and composite adsorbent augmented sequencing batch reactor process. *Process Safety and Environmental Protection*, **111**, 253–262, https://doi.org/10.1016/j.psep.2017.07.013

Mojiri A., Zhou J. L., Ratnaweera H., Ohashi A., Ozaki N., Kindaichi T. and Asakura H. (2021). Treatment of landfill leachate with different techniques: an overview. *Journal of Water Reuse and Desalination*, **11**(1), 66–96, https://doi.org/10.2166/wrd.2020.079

Morris S., Garcia-Cabellos G., Enright D., Ryan D. and Enright A. M. (2018). Bioremediation of landfill leachate using isolated bacterial strains. *International Journal of Environmental Bioremediation & Biodegradation*, **6**(1), 26–35.

Nazia S., Sahu N., Jegatheesan V., Bhargava S. K. and Sridhar S. (2021). Integration of ultrafiltration membrane process with chemical coagulation for proficient treatment of old industrial landfill leachate. *Chemical Engineering Journal*, **412**, 128598, https://doi.org/10.1016/j.cej.2021.128598

Nissim W. G., Palm E., Pandolfi C., Mancuso S. and Azzarello E. (2021). Willow and poplar for the phyto-treatment of landfill leachate in Mediterranean climate. *Journal of Environmental Management*, **277**, 111454, https://doi.org/10.1016/j.jenvman.2020.111454

Nivala J., Hoos M. B., Cross C., Wallace S. and Parkin G. (2007). Treatment of landfill leachate using an aerated, horizontal subsurface-flow constructed wetland. *Science of the Total Environment*, **380**(1–3), 19–27, https://doi.org/10.1016/j.scitotenv.2006.12.030

Oumar D., Patrick D., Gerardo B., Rino D. and Ihsen B. S. (2016). Coupling biofiltration process and electrocoagulation using magnesium-based anode for the treatment of landfill leachate. *Journal of Environmental Management*, **181**, 477–483, https://doi.org/10.1016/j.jenvman.2016.06.067

Paskuliakova A., McGowan T., Tonry S. and Touzet N. (2018). Microalgal bioremediation of nitrogenous compounds in landfill leachate – the importance of micronutrient balance in the treatment of leachates of variable composition. *Algal Research*, **32**, 162–171, https://doi.org/10.1016/j.algal.2018.03.010

Pavithra S. and Shanthakumar S. (2017). Removal of COD, BOD and color from municipal solid waste leachate using silica and iron nanoparticles – a comparative study. *Global NEST Journal*, **19**, 122–130, https://doi.org/10.30955/gnj.002065

Pawar V. and Gawande S. (2015). An overview of the fenton process for industrial wastewater. *IOSR Journal of Mechanical and Civil Engineering*.

Peixoto A. L. D. C., Salazar R. F. D. S., Barboza J. C. D. S. and Izário Filho H. J. (2018). Characterization of controlled landfill leachate from the city of Guaratinguetá-SP, Brazil. *Revista Ambiente e Água*, **13**.

Peyravi M., Jahanshahi M., Alimoradi M. and Ganjian E. (2016). Old landfill leachate treatment through multistage process: membrane adsorption bioreactor and nanofitration. *Bioprocess and Biosystems Engineering*, **39**(12), 1803–1816, https://doi.org/10.1007/s00449-016-1655-0

Pisharody L., Gopinath A., Malhotra M., Nidheesh P. V. and Kumar M. S. (2022). Occurrence of organic micropollutants in municipal landfill leachate and its effective treatment by advanced oxidation processes. *Chemosphere*, **287**, 132216, https://doi.org/10.1016/j.chemosphere.2021.132216

Prathap M. G., Reymond M. D. J., Sudarsan M. J. and Deeptha V. T. (2014). Leachate treatment using constructed wetlands technique. *IOSR Journal of Environmental Science, Toxicology and Food Technology*, **8**, 07–09, https://doi.org/10.9790/2402-08540709

Ramakrishnan A., Blaney L., Kao J., Tyagi R. D., Zhang T. C. and Surampalli R. Y. (2015). Emerging contaminants in landfill leachate and their sustainable management. *Environmental Earth Sciences*, **73**, 1357–1368, https://doi.org/10.1007/s12665-014-3489-x

Ren X., Liu D., Chen W., Jiang G., Wu Z. and Song K. (2018). Investigation of the characteristics of concentrated leachate from six municipal solid waste incineration power plants in China. *RSC Advances*, **8**(24), 13159–13166, https://doi.org/10.1039/C7RA13259J

Reshadi M. A. M., Bazargan A. and McKay G. (2020). A review of the application of adsorbents for landfill leachate treatment: focus on magnetic adsorption. *Science of the Total Environment*, **731**, 138863, https://doi.org/10.1016/j.scitotenv.2020.138863

Ricordel C. and Djelal H. (2014). Treatment of landfill leachate with high proportion of refractory materials by electrocoagulation: system performances and sludge settling characteristics. *Journal of Environmental Chemical Engineering*, **2**(3), 1551–1557, https://doi.org/10.1016/j.jece.2014.06.014

Robinson A. H. (2005). Landfill leachate treatment. *Membrane Technology*, **6**, 6–12, https://doi.org/10.1016/S0958-2118(05)70435-3

Rocha E. M., Vilar V. J., Fonseca A., Saraiva I. and Boaventura R. A. (2011). Landfill leachate treatment by solar-driven AOPs. *Solar Energy*, **85**(1), 46–56, https://doi.org/10.1016/j.solener.2010.11.001

Rui L. M., Daud Z. and Latif A. A. A. (2012). Treatment of leachate by coagulation–flocculation using different coagulants and polymer: a review. *International Journal on Advanced Science, Engineering and Information Technology*, **2**(2), 1–11, https://doi.org/10.5121/ijait.2012.2201

Samadder S. R., Prabhakar R., Khan D., Kishan D. and Chauhan M. S. (2017). Analysis of the contaminants released from municipal solid waste landfill site: a case study. *Science of the Total Environment*, **580**, 593–601, https://doi.org/10.1016/j.scitotenv.2016.12.003

Scandelai A. P. J., Rigobello E. S., de Oliveira B. L. C. and Tavares C. R. G. (2019). Identification of organic compounds in landfill leachate treated by advanced oxidation processes. *Environmental Technology*, **40**, 730–741, https://doi.org/10.1080/09593330.2017.1405079

Scandelai A. P. J., Zotesso J. P., Jegatheesan V., Cardozo-Filho L. and Tavares C. R. G. (2020). Intensification of supercritical water oxidation (ScWO) process for landfill leachate treatment through ion exchange with zeolite. *Waste Management*, **101**, 259–267, https://doi.org/10.1016/j.wasman.2019.10.005

Schiopu A. M. and Gavrilescu M. (2010). Options for the treatment and management of municipal landfill leachate: common and specific issues. *Clean – Soil, Air, Water*, **38**(12), 1101–1110, https://doi.org/10.1002/clen.200900184

Song U., Waldman B., Park J. S., Lee K., Park S. J. and Lee E. J. (2018). Improving the remediation capacity of a landfill leachate channel by selecting suitable macrophytes. *Journal of Hydro-Environment Research*, **20**, 31–37, https://doi.org/10.1016/j.jher.2018.04.005

Tałałaj I. A., Biedka P. and Bartkowska I. (2019). Treatment of landfill leachates with biological pretreatments and reverse osmosis. *Environmental Chemistry Letters*, **17**, 1177–1193, https://doi.org/10.1007/s10311-019-00860-6

Talebi A., Razali Y. S., Ismail N., Rafatullah M. and Tajarudin H. A. (2020). Selective adsorption and recovery of volatile fatty acids from fermented landfill leachate by activated carbon process. *Science of the Total Environment*, **707**, 134533, https://doi.org/10.1016/j.scitotenv.2019.134533

Tan B., He L., Dai Z., Sun R., Jiang S., Lu Z., Liang Y., Ren L., Sun S., Zhang Y. and Li C. (2022). Review on recent progress of bioremediation strategies in landfill leachate – a green approach. *Journal of Water Process Engineering*, **50**, 103229, https://doi.org/10.1016/j.jwpe.2022.103229

Tchobanoglous G. and Kreith F. (2002). Handbook of Solid Waste Management. McGraw-Hill Education, New York.

Visvanathan C., Choudhary M. K., Montalbo M. T. and Jegatheesan V. (2007). Landfill leachate treatment using thermophilic membrane bioreactor. *Desalination*, **204**(1–3), 8–16, https://doi.org/10.1016/j.desal.2006.02.028

Wang X., Chen S., Gu X. and Wang K. (2009). Pilot study on the advanced treatment of landfill leachate using a combined coagulation, Fenton oxidation and biological aerated filter process. *Waste Management*, **29**(4), 1354–1358, https://doi.org/10.1016/j.wasman.2008.10.006

Wang K., Li L., Tan F. and Wu D. (2018). Treatment of landfill leachate using activated sludge technology: a review. *Archaea*, **2018**, 1–11.

Wu J. J., Wu C. C., Ma H. W. and Chang C. C. (2004). Treatment of landfill leachate by ozone-based advanced oxidation processes. *Chemosphere*, **54**(7), 997–1003, https://doi.org/10.1016/j.chemosphere.2003.10.006

Xaypanya P., Takemura J., Chiemchaisri C., Seingheng H. and Tanchuling M. A. N. (2018). Characterization of landfill leachates and sediments in major cities of Indochina peninsular countries – heavy metal partitioning in municipal solid waste leachate. *Environments*, **5**(6), 65, https://doi.org/10.3390/environments 5060065

Xu Z. Y., Zeng G. M., Yang Z. H., Xiao Y., Cao M., Sun H. S., Ji L. L. and Chen Y. (2010). Biological treatment of landfill leachate with the integration of partial nitrification, anaerobic ammonium oxidation and heterotrophic denitrification. *Bioresource Technology*, **101**(1), 79–86, https://doi.org/10.1016/j.biortech.2009.07.082

Yi E. X., Wee S. T., Lim C. K., Ibrahim Z. and Chan N. W. (2018). Combined adsorption and biological treatment for landfill leachate management. *Journal of Advanced Research in Fluid Mechanics and Thermal Sciences*, **50**(1), 26–31.

Zainol N. A., Pin L. B., Rashid N. A., Ghani A. A., Zailani S. N. and Rani A. L. A. (2018). Treatment of landfill leachate by coagulation–flocculation process using red earth as coagulant. *AIP Conference Proceedings*, **2030**(1), 020043.

Zamri M. F. M. A., Kamaruddin M. A., Yusoff M. S., Aziz H. A. and Foo K. Y. (2017). Semi-aerobic stabilized landfill leachate treatment by ion exchange resin: isotherm and kinetic study. *Applied Water Science*, **7**, 581–590, https://doi.org/10.1007/s13201-015-0266-2

Zielińska M., Kulikowska D. and Stańczak M. (2020). Adsorption–membrane process for treatment of stabilized municipal landfill leachate. *Waste Management*, **114**, 174–182, https://doi.org/10.1016/j.wasman.2020.07.011

Index

M

N

O

P

R

S

T

U

V

W

X

IWA Publishing's authorised EU representative for General Product Safety Regulations is Diane D'Arras, 15 rue Duret, 75116 Paris, France, e-mail: safety@iwap.co.uk.

Printed and bound by CPI Group (UK) Ltd, Croydon, CR0 4YY
12/05/2026
02108866-0007